Collected Experimental Papers of P. W. Bridgman

Volume IV

P. W. BRIDGMAN

Collected Experimental Papers

Volume IV

Papers 59–93

Harvard University Press
Cambridge, Massachusetts
1964

Distributed in Great Britain by Oxford University Press, London

Library of Congress Catalog Card Number 64–16060

Printed in the United States of America

CONTENTS

Volume IV

VARIOUS PHYSICAL PROPERTIES OF RUBIDIUM AND CAESIUM AND THE RESISTANCE OF POTASSIUM UNDER PRESSURE.

By P. W. Bridgman.

Received November 6, 1924. Presented October 8, 1924.

TABLE OF CONTENTS.

Introduction.

The alkali metals are the most compressible of the metals and it is therefore probable that they will show the most interesting and significant variations of physical properties under high pressure. I have previously published the results of various high pressure measurements on lithium, sodium, and potassium,[1] but have not examined the two heaviest and most compressible members of the series, rubidium and caesium. The reason for this omission has been two fold: the manipulations offer certain technical difficulties, and it is not easy to obtain a sufficiently large quantity of these two metals in a state of high purity.

In this paper are given new results on rubidium and caesium concerning those same properties which have been previously measured for the three lighter metals. These properties include in the first place a determination of the effect of pressure on melting up to 100° C., which involves the determination of the effect of pressure on the melting temperature and on the difference of volume between liquid and solid (from which the latent heat of melting can be calculated as a function of pressure and temperature up to 100°), secondly the electrical resistance of both solid and liquid phases as a function of pressure up to 12000 kg./cm.² at temperatures between 0° and 100°, and thirdly the compressibility to 15000 kg./cm.²

I have already published preliminary results on caesium which now appear to be incorrect.[2] I had apparently found a new modification of caesium at high pressures which has the abnormal property that its resistance increases with increasing pressure. It now appears that the apparent transition was due to an impurity, which separated out of solid solution at sufficiently high pressure, or else delayed the ordinary freezing; actually there is no discontinuity, but instead caesium has the unique property that its electrical resistance passes through a minimum with increasing pressure. This property, particularly because Cs is the most compressible of the metals, should be of considerable theoretical significance. Since this incorrect preliminary result I have made a large number of measurements on caesium in order to be quite sure of the result. In my earlier work I followed the universal previous practice of measuring the electrical properties of the metal enclosed in a glass capillary. But various experimental irregularities showed that the constraining effect of the glass might be of considerable importance (for although the stresses which these metals can support without yield are small, nevertheless the variations with stress of the various properties are unusually high), so that eventually I was driven to making measurements on bare wires of rubidium and caesium. In my previous work I had used bare wires of lithium and sodium, but the mechanical dificulties in the case of the much softer metal potassium had led me to use the conventional glass capillaries for it. With the experience now gained with these still softer metals, I have now returned to potassium, and have repeated the measurements of the pressure effects on the bare wire. I find that the pressure coefficient is not very much affected, but the conclusions which I had previously drawn as to the effect of pressure on the temperature coefficient of resistance must now be essentially modified. These new results on the pressure coefficient of resistance of potassium are also given in this paper.

Experimental Details.

Preparation of the Metals. The importance and the difficulty of obtaining these metals in a state of high purity is not usually realized, and many results have been published on material of insufficient purity. With care in the preparation it is not difficult to obtain material free from metallic impurities, but the removal of oxide is a matter of greater difficulty. The oxide dissolves in these two metals, and behaves in some respects like an ordinary metallic impurity, as in depressing the freezing point. The usual method of preparation is by heating the chloride in contact with metallic calcium. It is comparatively easy to get the chloride free from foreign metals, but it is just as important that the calcium be pure, and in particular it should not contain any of the other alkali metals, which are difficult to remove at any later stage. Prepared in this way, with moderate heating, there is little likelihood of the rubidium or caesium containing other metals, but it is almost certain to contain oxide. This must be removed by slow distillation at the lowest feasible temperature in very high vacuum. It is practically impossible to make a satisfactory distillation with a gas flame, but an electric oven should be used.

The caesium used in my preliminary experiments was obtained from several sources. I am indebted to Professor G. N. Lewis for a generous supply of the metal, to Professor G. P. Baxter for a large quantity of highly purified chloride, and to the Research Laboratory of the General Electric Company for extracting the metallic caesium from the salt with calcium. I also obtained several grams of the metal from the Foote Mineral Co. All of this preliminary material, however, did not give satisfactory results, the best method of manipulation not having been found. The final results were obtained with metallic caesium from Kahlbaum. As provided by them it was stated to be entirely free from any foreign metals, but was obviously not pure, as was evidenced by the long temperature range over which melting took place. The caesium was provided sealed into glass tubes under a heavy white mineral oil. From these tubes it was transferred under Nujol to a distilling arrangement of Pyrex glass, and the Nujol washed out with petroleum ether, leaving a little ether in the apparatus, so that the metal was at no time uncovered. The glass was then sealed and evacuated to as high a degree as possible with a diffusion pump, heating all parts of the apparatus. A preliminary distillation was made from the receiving chamber to the first bulb, in this way removing the coarse dirt. The glass container was then transferred to an electric oven, where the metal was distilled from one bulb to the next at a temperature so low that between one and two hours was required for a transfer. After each transfer the vacated bulb was sealed off from the remaining part of the apparatus. In all, five such distillations were made. At the farther end of the apparatus was the arrangement adapted to the special experiment in hand, and which will be described in detail later. This method of preparation was adopted for all the material used in the final measurements.

The best test of the purity of the material (the presence of oxide being difficult to establish by the ordinary methods of spectroscopic analysis) is the sharpness of the freezing point. The most convenient method is by measurements of the electrical resistance, but there is a danger here which must be especially mentioned, which has probably in the past given rise to illusory results. It is of course well known that the greater the purity of the metal the higher its temperature coefficient of electrical resistance, other things being equal, so that under ordinary conditions a high temperature coefficient is pretty good evidence of high purity. But near the melting point this is obviously no longer the case, because the resistance increases when the metal melts, and premature melting may be brought about by impurity, thus simulating an improperly high temperature coefficient. This source of error is important only for the low melting metals, and is of course especially important for rubidium and caesium. It is not sufficient, therefore, to measure the temperature coefficient of caesium between 0° and 20°, for example, and to infer from its high value the high purity, but the resistance must be measured all the way up to the melting point and into the liquid state, if the metal is in a glass capillary. The corners of the discontinuity on melting must not be rounded. In practice it is easy to distinguish a rounding of the corner due to premature melting from the upward curvature due to the normal accelerated increase of resistance with rising temperature, provided the purity is high, but it is more difficult or impossible if there is considerable impurity.

Measurements of the temperature dependence of resistance should preferably be made by the method of stationary temperatures, instead of by the method of continually varying temperatures which is so often used. This latter method of necessity introduces some rounding of the corners due to varying temperature lag when there are such thermal effects as found in melting, and can be made reliable only by varying the temperature so slowly as to become virtually a stationary temperature method. The stationary temperature method, on the other hand, suffers from the disadvantage of not giving the exact melting temperature, but only shuts it between limits, which may be made narrower by longer experiment or greater good fortune. In my experiments this disadvantage was avoided by a second method of studying the sharpness of melting, namely by varying the pressure at constant temperature, measuring the volume as a function of pressure. Since the pressure can be manipulated much more easily than the temperature, and furthermore since the melting pressure automatically establishes itself within a certain range, it is possible to find exactly the melting pressure at any temperature, or to find whether this pressure varies with the fractional part of the metal which is melted, and so to study the sharpness of freezing. By both these tests, the freezing of the rubidium and caesium was unusually sharp, and the melting point so found was unusually high and presumably is close to the value for the absolutely pure metal.

Rubidium was made in the same way as the caesium, except that I made no measurements with preliminary material, but the source of all my material was metallic rubidium sealed under oil into glass obtained from Kahlbaum. The distillation temperature is higher than for caesium, but not high enough to make necessary any change in the general method of manipulation. Because of its higher melting point it is more convenient to manipulate at room temperature.

Experimental Methods. No new methods were necessary in making the measurements, but those were employed which have already been described in full detail. The resistances to be measured were of the order of a small fraction of an ohm, so that the potentiometer method with four leads was used which has been previously employed in measuring the effect of pressure on small resistances.[3]

The melting curves were traced by the method of the discontinuity of volume already used in connection with many melting and transition curves.[4] The apparatus was smaller than used before, and was in one piece instead of two, which eliminated the necessity for some corrections, but was in principle the same as that used before. The temperature limit of these measurements, 100° C., was set by the particular form of apparatus used. It would be interesting to follow the melting curve to higher temperatures (I have been as high as 275° with bismuth), but this would have demanded the construction of special apparatus, and it did not seem to me that there were any questions likely to be answered by an extension of the range sufficient to justify the construction of special apparatus for only two substances. On one or two occasions I also obtained the melting pressure corresponding to a given temperature by measurements of the electrical resistance. These results agreed with the others.

The compressibility was measured by two methods. One was the method for measuring linear compressibility which I have previously applied to a number of metals.[5] There were difficulties because of the extreme softness of the metals, particularly with caesium. The second method is that for measuring the cubic compressibility by determining the position of the piston of the compressing apparatus, which I have applied to 12 liquids,[6] and which has later been applied at the Geophysical Laboratory to a number of solids.[7] The apparatus was on a smaller scale than that previously used, and was the same as that used recently in determining the compressibility of gases,[8] by a method in all essentials the same. The pressure range was also, as for gases, 15000 kg./cm.^{2}, instead of the more usual 12000. The measurements of the cubic compressibility and of the melting curve were made with the same apparatus and usually with the same filling of the apparatus. There were various difficulties in the compressibility measurements which make these the least satisfactory of this paper. It was a disappointment that these were not accurate enough to permit more than very rough statements about the behavior of the thermal expansion under high pressures.

Numerical Data.

The numerical data for the various effects, together with such detailed description as seems necessary, now follow.

Potassium.

Resistance. These measurements on potassium were made after the measurements on rubidium and caesium, and were suggested by the apprehension that the previous results found for potassium in a glass capillary might be in error because of a restraining action of the glass. The point of particular interest is the temperature coefficient of resistance at high pressures; I had found that beyond 6000 kg. the temperature coefficient of potassium decreases greatly.[9] This was a unique phenomenon, the temperature coefficient of all the other metals measured being approximately independent of the pressure. I connected the unusual decrease with the unusual compressibility of potassium, and expected that the same effect would be found in other metals at a pressure sufficiently high to produce in them a comparable volume compression. This supposed decrease of the temperature coefficient of potassium and the ideas suggested by it now turns out to be incorrect, the effect found being, as feared, due to the constraining effect of the capillary.

The potassium used in this investigation was from the same batch as that whose resistance, melting, and compressibility to high pressures has been previously measured. It has been kept since the former measurements sealed under Nujol in a small glass tube. For these measurements it was extruded under Nujol to a bare wire of about 1.5 mm. diameter. The four connections, two for current, and two for potential, were made by piercing through the potassium with very fine copper wires, looped into place. The potassium wire was then folded at the middle to the shape of a hairpin and placed in a small glass test tube filled with Nujol with the connections folded over the edge. Connections were now made to the three terminal plug in the regular way, so that everything could be screwed as one self-contained unit into the pressure apparatus. Before this was done the Nujol in the test tube was replaced with petroleum ether by repeatedly flushing out the tube with ether. This is necessary to avoid distortion effects at high pressures due to viscosity of the oil.

Measurements were made in the first place of the temperature coefficient of resistance at atmospheric pressure from 0° to 35° and back again. It is to be emphasized that no measurement, either of a pressure or a temperature effect, on these soft and chemically active metals is trustworthy unless a complete cycle is made, returning to the starting point, because of the great liability to permanent changes of resistance after a change of pressure or temperature. After the temperature cycle above, 0° to 35° to 0° again, there was a permanent increase of resistance of 5% of the total change produced by 35°. The mean of readings with increasing and decreasing temperature was taken as giving the true temperature coefficient. The relation between resistance and temperature is linear, and the average coefficient between 0° and 35° in terms of the resistance at 0° as unity was 0.00541. I previously found for potassium in glass between 0° and 51° a linear relation, and for the numerical value 0.00512, corrected for the volume expansion of the glass to give the temperature coefficient of specific resistance. The value 0.00541 found above for bare wire is the coefficient directly measured with terminals attached to the wire. The correction term to reduce from measured to specific resistance is the linear expansion. Assuming for this the value

0.00008, the best value for the temperature coefficient of specific resistance is 0.00549, about 7% higher than the value previously found on the same material. The difference is to be ascribed to the restraining action of the glass walls, which, as temperature is raised, exert a pressure which lowers the resistance. It is to be noticed that measurements in a glass capillary may have no internal evidence of being incorrect, but the measurements with increasing and decreasing temperature usually agree, with no sign of hysteresis. This is understandable over a moderate temperature range, but over a wider range it would be expected that the thermal stresses set up in the metal would produce flow, and so permanent alterations, with hysteresis.

The pressure measurements were made in the regular way with the regular apparatus up to 12000 kg. at 0°, after a preliminary seasoning application of pressure which produced a permanent alteration of resistance of 6% of the change due to pressure. The temperature coefficient at 5000 kg. and 11300 kg. was measured by raising temperature at each of these pressures from 0° to 30° and then lowering again to 0°. A readily determined correction must be applied for the small change of pressure accompanying the change of temperature. This procedure is much better than that previously adopted of determining the temperature coefficient from the results of complete pressure runs at different temperatures. The return to the initial resistance at 0° in the new measurements was complete within 0.3%, and considerable confidence may be felt in the temperature coefficients so found.

The results are shown in Table I, which gives the relative measured resistance as a function of pressure to 12000 kg. at 0°, and the temperature coefficient at 1, 5075, and 11300 kg.

TABLE I.

RESISTANCE OF POTASSIUM.

Pressure kg./cm.	Relative Resistance at 0° C.
0	1.000
1000	.844
2000	.729
3000	.638
4000	.564
5000	.502
6000	.451
7000	.408

TABLE I.—*Continued.*

Pressure kg./cm.	Relative Resistance at 0° C.
8000	.372
9000	.340
10000	.313
11000	.291
12000	.271

Temperature Coefficient of Resistance between 0° and 30° C.

At 1 kg./cm.°	0.00541
5075	.00459
11300	.00454

When compared with the previous results, these new pressure data show the effect of the constraint by the glass, as did also the temperature coefficient at atmospheric pressure. The pure pressure effect is much less affected than the temperature coefficient. At 12000 kg. the ratio of measured resistance to that at atmospheric pressure at 0° is 0.271, which corrects to 0.246 for the ratio of specific resistances. Previously for potassium in glass, the ratio 0.275 was found at 25° and 0.252 at 60°. At the higher temperature the value previously found agrees better with the new value than the ratio at the lower temperature. This is consistent with the explanation as due to the effect of the constraint, because at the higher temperature the internal viscosity is less and the internal stresses will be less. The difference between the old and the new results is of the order of 10% as far as the pressure effect goes. At lower pressures the agreement is closer, as would be expected because of the smaller viscosity of the metal at lower pressures.

The new result for the temperature coefficient of resistance at high pressures when compared with the previous results amounts however, to a result of a different order of magnitude. Previously I found at 12000 kg. a decrease of the temperature coefficient by a factor of 2.5 fold. The decrease now found is only from 54 to 45, and the very marked decrease formerly found beyond 6000 is now seen not to exist.

The final result of these new measurements on potassium is therefore to leave unchanged the generalization which I made after my first series of pressure measurements,[10], namely that the temperature coefficient is little affected by pressure. This is what would be expected in view of the fact that the temperature coefficient of all pure metals is nearly alike, because the same metal at two different pressures may be regarded as a special case of two different metals.

Rubidium.

Melting Data. Little need be added in the way of description of experimental detail to that already given, except the method of filling the apparatus. The chief difficulty is in determining the amount of rubidium, since it must be kept continually immersed in oil to avoid oxidation. Connected to the glass bulb in which the final distillation was made was a small test tube of about 6 mm. inside diameter. The rubidium was run into this, and then the top and bottom of the test tube broken off under Nujol. The metal was now extruded from the tube with a closely fitting plunger, so that a rod of rubidium 6 mm. in diameter was obtained. This was cut to an appropriate length, and the amount determined by weighing under Nujol. The density of the Nujol was independently determined and for the density of rubidium the value 1.532 of Richards and Brink[11] was assumed. From these data the amount of rubidium may be calculated. The rubidium was then placed inside a steel cup of test tube shape, and the total weight in air determined, from which the amount of oil included in the steel cup with the rubidium could be found (this last being necessary for the compressibility measurements to be made with the same set-up, but not for the melting determinations), and then it was mounted in the regular way in the high pressure cylinder with a known amount of kerosene. As already mentioned, the small apparatus used previously for the compressibility of gases was used for this, the amount of metallic rubidium and caesium available not being sufficient to allow the use of the larger apparatus, which would otherwise have been desirable because somewhat more accurate results may be found with it. The area of the piston of the small apparatus was 0.43 cm.^{2}

Nine points on the melting curve were found, ranging from 250 to 3600 kg., and from 44° to 96°. The melting was very sharp and affords gratifying evidence of the purity. Thus at the lowest pressure, 250 kg., there was no perceptible difference in melting pressure between 0.1 and 0.9 melted. The sensitiveness was such that a difference of 4 kg./cm.^{2} could have been detected, corresponding to 0.08°.

Doubtless the best value for the melting point at atmospheric pressure is to be obtained by an extrapolation of the results found at high pressures. The extrapolation is short, a matter of five degrees, and may be made with considerable certainty by graphical methods, because the curvature of the melting curve is not important. The melting temperature at atmospheric pressure found in this way is 38.7°. This is 0.25° higher than the result found from measurements of electrical resistance in glass capillaries (temperature at which the metal is half melted), and differs from the electrical result in the direction to be expected, since any internal stress not a hydrostatic pressure equally distributed between solid and liquid results in a lowering of the melting point. The melting point given above is somewhat lower than the highest previously recorded,[12] 39.00.°

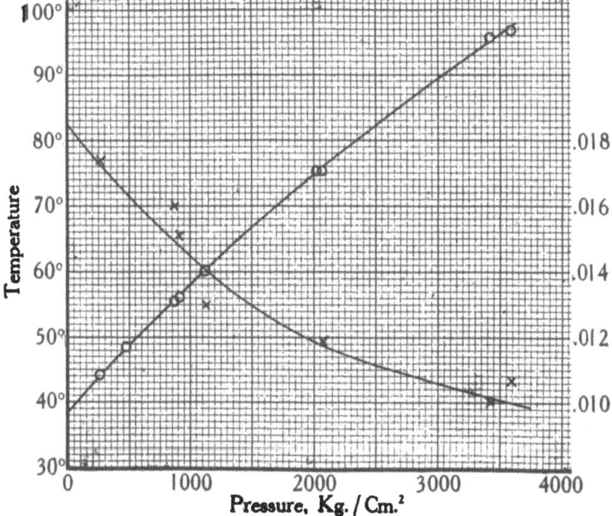

FIGURE 1. Melting data for Rubidium. The circles show the melting temperatures (scale on the left) as a function of pressure. The crosses show the observed changes of volume (scale on the right) in cm.³ per gm. as a function of pressure.

The changes of volume on melting were determined at the same time as the coordinates of the melting curve. The change of volume is small, so that the percentage accuracy is not high. However the main facts could be established that the change of volume becomes less at the higher pressures, and the direction of curvature is the normal one, namely convexity toward the pressure axis.

In Figure 1 are shown the experimentally determined coordinates of the melting curve, and in the same diagram the experimental changes of volume; the method of representation is the same as that previously used in my melting and transition determinations. In Table II are given the smoothed results obtained from these curves

TABLE II.

MELTING DATA FOR RUBIDIUM.

Pressure kg./cm.^{2}	Temperature Centigrade	dT/dp	ΔV cm.^{3}/gm.	Latent Heat kg. m/gm.
0	38.7°	.0209	.0185	2.76
500	48.7	192	163	2.73
1000	57.9	178	145	2.70
1500	66.5	166	130	2.66
2000	74.5	155	119	2.67
2500	82.0	146	112	2.72
3000	89.1	139	106	2.76
3500	95.9	134	101	2.78

for the coordinates of the melting curve, the change of volume, and computed from these the slope of the melting curve and the latent heat of melting. The latent heat varies along the melting curve only by small amounts, which is the behavior found for most substances.

The change of volume on fusion and the latent heat do not seem to have been previously determined, so that there are no values for comparison.

Resistance. The data to be covered by the resistance measurements are the effect of pressure at various temperatures on the resistance of the solid and liquid metal, and the change of resistance when the solid melts to the liquid at various points of the melting curve. For those measurements in which the solid alone is concerned the bare wire must be used, but for those which involve the liquid, a glass capillary was necessary.

The manipulation of the bare wire was essentially the same as that of potassium. It was extruded through a steel die under Nujol; the electrical connections were fine wires of silver, instead of copper, pierced through the rubidium wire. Measurements were made as follows. First the temperature coefficient of resistance at atmospheric pressure was determined. Then pressure measurements were made to 12000 kg. and back at 20°. This was the first application of pressure, and there was a comparatively large permanent increase of resistance. Then a pressure run was made to the maximum and back at 35°, which showed only a small permanent change, and then a pressure run at 0°, which showed the smallest set of all. Then measurements of the temperature coefficient under pressure were made at several high pressures by the cyclic method already described for potassium. Finally the bare wire was measured again for the temperature coefficient at atmospheric pressure, this time raising the temperature until the wire melted. A reading was obtained with the bare wire at 38.4°, but at 38.8° it melted and open circuited, indicating a melting temperature somewhere between these limits.

The results obtained with the bare wire were used in calculating the finally tabulated values of resistance at atmospheric pressure up to the melting point, and the resistance as a function of pressure to 12000 kg. at 0° and 35°. There is a small inconsistency in the finally tabulated results which I have not attempted to remove in that the temperature coefficients determined from the difference of the pressure data at 0° and 35° do not agree precisely with the directly determined coefficients by the cyclic method. The difference is partly perhaps due to the difference of temperature range, but there is doubtless also some outstanding experimental error. The results are best used as given; the pressure values in discussing the effect of pressure at 0° or 35°, and the directly measured temperature coefficients in discussing temperature effects at various pressures.

The measurements in the glass capillary followed the same general outline as previously. The capillary was small, perhaps 0.5 mm. inside diameter and 4 or 5 cm. long. It was filled while directly attached to the purifying and distilling apparatus, sealed off, and then the upper end broken under Nujol. Measurements were made in the first place of the resistance at atmospheric pressure as a function of temperature, beginning at 0° on the solid, and running to 95° on the liquid. The temperature coefficient in glass to 27° was 0.00477, somewhat less than the value found for the bare wire. The difference was in the same direction as was also found for potassium, but is not as great, corresponding perhaps to the greater mechanical softness of rubidium. At a temperature of 36.6° the resistance measurements indicate that 1% of the metal is melted; I did not attempt to determine the melting point more accurately by this method. The resistance of the liquid in glass is a linear function of temperature from the melting point up to 95°. The correction by which the measured resistances in the glass are reduced to specific resistances of the liquid are so small as to be almost negligible, amounting to an increase of only 1 in the last place at the highest temperature. From the measurements at atmospheric pressure the ratio of specific resistance of solid to liquid at atmospheric pressure was obtained, correcting for the very slight rounding of the corner by a graphical extrapolation.

With the capillary, pressure runs were made at 65° and 95° up to and just beyond the melting point, so that in addition to the pressure coefficient of the liquid we have the ratio of resistance of liquid to solid at the equilibrium pressure at these two temperatures. Finally, pressure runs were made on the frozen solid to 10000 kg. at 95°, and the temperature coefficient determined by a cyclic change of temperature between 22° and 95° at 5700 kg. The coefficient so found was 0.00427 in terms of the resistance at 0° (linear extrapolation), which is higher than the value found directly with the bare wire. At atmospheric pressure the bare wire has yielded higher coefficients; the difference may well be due to the difference of temperature range, 0° to 17° against 22° to 95°.

The final results are given in Table III. Here are given the relative resistances at different temperatures and pressures in terms of the resistance of the solid at 0° and atmospheric pressure as unity. Up to the melting point at atmospheric pressure the results tabulated are the measured results for the bare wire, with no correction for volume distortion. The tabulated value is to be increased by the linear thermal expansion in order to obtain the specific resistance. This correction will probably amount to 2% of the change due to change of temperature, but it is not known accurately enough to justify giving final values in terms of it. The values given for the resistance of the liquid have, on the other hand, been corrected for the changes of form of the glass envelope, so that these give the relative specific resistances starting from the melting point as the fiducial point. The corrections for the glass are small, and amount at the highest temperature at atmospheric pressure to only 1 in the last place, and at the highest pressure on the liquid (95° and 3500 kg.) to only 4 in the last place. At 95° the pressure measurements on the solid, having been made with the glass capillary, are of the specific resistance. The corrections by which the measured resistances on the bare wire may be reduced to specific resistances are large, because of the high compressibility of rubidium, and at 12000 kg. amount to a decrease of the tabulated resistance by 10%. The precise amount may be found from the values given later for the volume.

The ratio of resistance of liquid to solid is also given in Table III for 1, 1470, and 3425 kg. Since there is no change of volume of the container during melting at constant pressure and temperature, these values are the ratios of the specific resistances.

Cubic Compressibility. Two methods were used. The first was the method of determining linear compressibility already used for a number of metals. The apparatus and the methods of measurement

TABLE III.

RESISTANCE OF RUBIDIUM.

At atmospheric pressure		Behavior under Pressure				
Temp.	Rel. Resis.	Pressure kg./cm.²	Solid, bare wire 0° C.	Solid, bare wire 35°	Liquid, in glass 65°	Liquid, in glass 95°
0° C.	1.000 Solid	0	1.000	1.205	2.195	2.417
10	1.046	500			1.983	2.173
20	1.099	1000	.845	.982	1.823	1.984
30	1.165	1500			1.679	1.827
38.7	1.235	2000	.733	.840		1.695
38.7	1.990 Liquid	2500				1.578
40	2.000	3000	.648	.740		1.472
60	2.148	3425				1.381*
80	2.295					.879†
100	2.443	3500				.871
		4000	.583	.663		.820
		5000	.531	.602		.732
		6000	.490	.553		.663
		7000	.456	.514		.609
		8000	.428	.481		.565
		9000	.406	.455		.529
		10000	.387	.434		.499
		11000	.372	.418	* Liquid	
		12000	.360	.406	† Solid.	

Ratio, Resis. Liquid / Resis. Solid	0 kg.	1470 kg.	3425 kg
	1.612	1.608	1.571

Temperature Coefficient of Solid

Bare Wire, 0°–17°,	0.00481,	0 kg.	In Glass,
	.00385,	5700 kg.	22°–95°, 0.00427
	.00365,	11000 kg.	at 5730 kg.

were the same as previously used. The rubidium was pressed into the shape of a regular cylinder about 7 mm. long, and mounted in the apparatus for direct measurement without lever magnification.

Measurements were made to 12000 kg. at 0°. The difficulty with the method is a mechanical one due to the softness of the rubidium. There is danger of mechanical deformation during the assembling of the apparatus or during the application of pressure because of the viscosity of the transmitting medium. Because of the extreme compressibility of rubidium it is necessary to use specimens short compared with their length, and this may be responsible for a special sort of error due to deformation. The cylinder of rubidium is compressed between flat plattens of steel. Between the steel platten and the rubidium there is a thin film of oil. When pressure is applied the rubidium shrinks transversely more than the steel platten, so that there is relative slip tangentially to the platten. This slip is resisted by the viscosity of the oil. At high pressures, where the viscosity is great, this may result in an actual deformation of the rubidium, which becomes stretched transversely with respect to its natural figure, so that the accompanying decrease of length is too great, resulting in a too great measured compressibility. This is what was actually found, the compressibility determined by the linear method being greater above 8000 kg. than that determined by the other method.

The second method was by measuring the motion of the piston of the compression cylinder, as already explained. Two different fillings of the apparatus were used. The first employed about 4.5 gm. of rubidium, and was the same filling with which the melting curve data were obtained. This quantity of rubidium was so great, however, that the maximum pressure obtainable was 11000 kg. A second filling was therefore made, removing about 1 gm. of rubidium and replacing it by a steel core. This made it possible to reach 15000 kg. Runs with these two fillings were made at 50° and 95° especially to determine compressibility, in addition to the runs made with the first filling especially to determine the melting data. For some reason not definitely discovered the changes of volume obtained with the first filling were about 10% greater at the maximum than those obtained with the second. Since those obtained with the second filling agreed essentially with the results obtained by the other method, they were retained, and the first discarded. Attempts were made to obtain the thermal expansion as a function of pressure, but the measurements were not sufficiently accurate to give good values. The difficulties with the manganin measuring gauge are increased with the small apparatus and it was chiefly the uncertainty due to the shift of zero with changing temperature that was responsible for the lack of accuracy. Very likely this increased difficulty found with the small apparatus is not at all inherent in the size of the apparatus, but is due to the fact that the pressure range was 15000 and often 16000 kg. against the former 12000. This produces strains in the wire due to the viscosity of the transmitting medium which may result in a capricious shift of the zero. The greater pressure range is also responsible for a greatly increased elastic hysteresis in the deformation of the steel cylinder, which is not conducive to accuracy.

The final results for decrease of volume as a function of pressure to 15000 kg. at 50° C. are given in Table IV. Only three significant figures can be given, against the four which have been possible in most of the previous work. The volume at 0 pressure was obtained by extrapolation.

TABLE IV.

Volume of Rubidium under Pressure.

Pressure kg./cm.^{2}	Relative Volume at 50° C.
0	1.000
1000	.953
2000	.914
3000	.883
4000	.857
5000	.836
6000	.819
7000	.803
8000	.789
9000	.777
10000	.766
11000	.756
12000	.746
13000	.737
14000	.729
15000	.721

With regard to thermal expansion under pressure, one may make the following rough statement with some assurance; up to 7000 or 8000 kg. the thermal expansion is fairly constant, but beyond this it drops rather abruptly, falling to one-half or one-third its initial value at 15000.

The initial compressibility of rubidium to be deduced from the Table is 0.000052, greater than that of water. This compressibility is much higher than the only previously published value, 0.000040 by Richards.[13] The difference is doubtless due to impurity. Richards' rubidium contained so large an amount of potassium that it was liquid at room temperature, so that the compressibility measured was of the liquid alloy. From the densities, assuming no contraction on mixing, Richards estimated that one-third of the total was potassium, and the probable error seemed to him so high that in his summary he tabulated only one significant figure for the result.

Attempts were also made to measure the compressibility of the liquid, but the wandering of the zero prevented accuracy. From the difference of slope of the volume curves above and below the melting point it is possible to say that at 56° the liquid is about 0.000006 more compressible than the solid, and at 96° 0.0000045.

Caesium.

Melting Data. The general method of handling this substance was the same as for rubidium. In determining the quantity of caesium by weighing under Nujol, the density of caesium at 0° was assumed to be 1.88. This seems to be the most probable value to be deduced from the measurements of Richards [11] and others, but the accuracy is not as great as in the case of rubidium.

Four points on the melting curve were determined, ranging from 250 kg. and 35° to 3700 kg. and 95°. In addition, a point was determined from the resistance measurements at 96° which coincides within the sensitiveness of the readings with the value found by the method of changing volume. The melting was not quite as sharp as for rubidium, but the melting pressure at 35°, for example, varies by 40 kg. between 0.1 and 0.9 melted, corresponding to a temperature shift of 1.0°. Of course the temperature of beginning of freezing or end of melting is to be taken as the correct melting temperature. The melting point at atmospheric pressure, extrapolated from the high pressure readings as in the case of rubidium, was 29.7°. This is considerably higher than previous values, and is evidence of high purity.

The change of volume is of the same order of magnitude as for rubidium, and values even more satisfactory were obtained.

In Figure 2 are shown the observed coordinates of the melting curve and the changes of volume, plotted in the usual way. The smoothed results taken from these curves are shown in Table V,

TABLE V.

MELTING DATA FOR CAESIUM.

Pressure kg./cm.^{2}	Temperature Centigrade	dT/dp	ΔV cm.^{3}/gm.	Latent Heat kg m./gm.
0	29.7°	.0250	.0136	1.65
500	41.4	221	118	1.68
1000	51.9	199	105	1.72
1500	61.4	183	96	1.76
2000	70.2	168	88	1.80
2500	78.3	155	83	1.88
3000	85.7	141	78	1.99
3500	92.4	128	75	2.14
4000	98.5	117	72	2.29

together with the values computed from them for the slope of the curve and the latent heat of melting. This latter shows perhaps

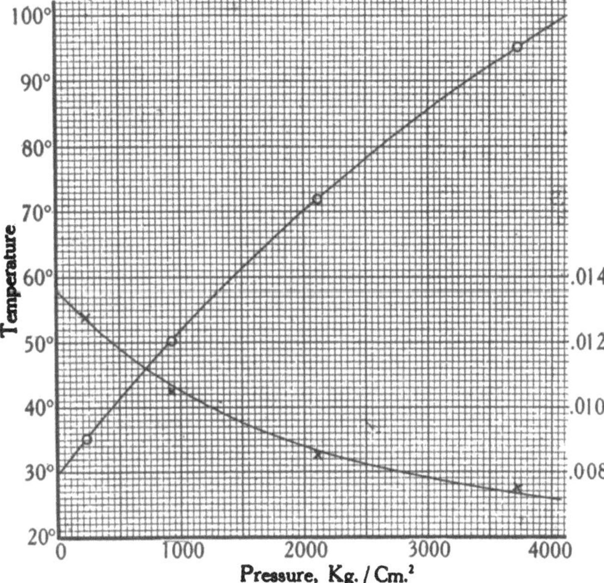

Figure 2. Melting data for Caesium. The circles show the melting temperatures (scale on the left) as a function of pressure. The crosses show the observed changes of volume (scale on the right) in cm.^{3} per gm. as a function of pressure.

more tendency than usual to change with pressure, becoming greater at the higher pressures.

Resistance. More time was spent in making the resistance measurements of caesium than all the other measurements of this paper. The measurements extended over three years, and were made on material from a variety of sources. The reasons for this have already been outlined and need not be reiterated. The final results to be given here were obtained with the purest material. There are no inconsistencies between the final and the preliminary results which are not explicable by impurities or improper manipulation.

The bare wire was obtained by extrusion out of the glass tube into which it was distilled after the final purification. One end of this tube was drawn out to the internal diameter desired for the wire (about 0.15 cm.), both ends of the tube were opened under Nujol, and the caesium was extruded by means of a plunger fitted with a closely fitting leather washer, so that the pressure which compelled the extrusion was transmitted to the caesium through an intermediate layer of oil. The metal is so soft that the extrusion was easily made holding the glass tube and the plunger in the hands. The extreme softness of course makes the various manipulations of assembly difficult, but ordinary care, without the use of special mechanical devices, is sufficient to surmount them. The leads were fine silver wires, as in the case of rubidium, and the subsequent manipulation was similar.

With this bare wire the temperature coefficient at atmospheric pressure was first determined from 0° to 10° and back to 0°. A pressure run was then made at 0°, locating carefully the position of the minimum of resistance, which was shown to be reversible by releasing the pressure a thousand kilograms or so after locating the minimum, and then proceeding to 12000 kg. with readings back to zero. Between 4000 and 6000 on the first increase of pressure there was a permanent increase of resistance of about 30%, due without question to distortion of the wire by the viscous transmitting medium. When corrected by this constant factor the initial results below the minimum of resistance agreed with the final results obtained with decreasing pressure. However, the pressure at which the minimum occurred was not the same with decreasing as with increasing pressure, but was about 400 kg. higher. I was never able to get entirely satisfactory values for the pressure of the minimum, which always varied somewhat. Doubtless the effect is due to internal strains set up by the transmitting medium. It is curious that more consistent values for the pressure of minimum resistance were found when the caesium was enclosed in a glass capillary than with the bare wire, although there were other fluctuations which were greater. The values listed as final in the following for the resistance of the solid under pressure are based essentially on the measurements obtained with decreasing pressure at 0° on the sample just described.

With this same bare wire the temperature coefficient of resistance was obtained at several pressures up to 11000 kg. by cyclic change of temperature from 0° to 10° to 0° again. The return to zero resistance was always exceedingly good, and these temperature coefficients may be accepted with considerable confidence.

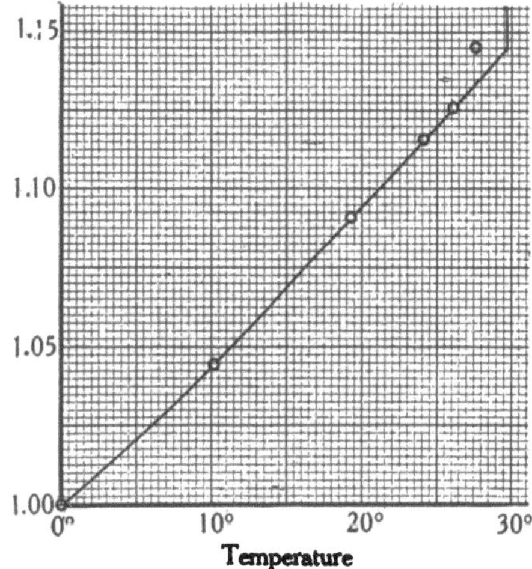

FIGURE 3. The resistance of solid caesium as a function of temperature at atmospheric pressure. Observe that up to 26° no premature melting is detectable.

The measurements on this bare wire were terminated by an attempt at a pressure run at higher temperatures, during which the wire was accidentally allowed to melt.

A number of readings were now obtained with a filling of a glass capillary, manipulated like the capillary of rubidium. First the temperature coefficient at atmospheric pressure was found below and through the melting point up to 95°. The values of the resistance up to the melting point shown in Figure 3 gave an idea of the sharpness of the melting. As before, the coefficient of the solid in glass was found to be less at atmospheric pressure than the coefficient of the bare wire; the bare wire over the common range, 10°, yielded a coefficient 10% higher. The results given in the final table for the resistance of the solid between 10° and the melting point are based on the values obtained with the glass capillary, corrected by a factor of 10%. The melting was sharp, as shown by the resistance measurements, at least judged by the usual standards. At 26.2° there was no premonition of melting, all the points lying on a smooth curve, but at 27.6° there was a jump in resistance corresponding to a premature melting of 1/60 part of the whole. Melting was completed somewhere between this temperature and 29.9°. These values are consistent with those found under pressure by the change of volume method. Above the melting point the change of resistance of the liquid in glass is linear with temperature up to 95° at least, with an error not more than 0.2%. The pressure coefficient of resistance of the liquid in glass was determined up to the melting point at 63°, and at 96° through the melting point and up to 11000 kg. The expected minimum of resistance of the solid shortly beyond the melting point was found. The temperature coefficient of the solid in glass was also found by the cyclic method between 20° and 95° at 9100 kg., and was considerably greater than found for the bare wire. This also agrees with the behavior of the temperature coefficient found for rubidium. Of course the value for the bare wire is to be preferred.

The ratio of resistance of liquid to solid at the normal melting point and at 96° was found to be 1.660 and 1.695 respectively. Again the variation of this ratio along the melting curve is slight compared with the variation of other properties.

The final results are shown in Table VI and Figure 4. The resistance of the solid at atmospheric pressure up to the melting point is first given. This is the measured resistance. To convert to specific resistance correction must be made for the linear thermal expansion, which is about 0.00011. Above the melting point the results tabulated are the specific resistance of the liquid, the directly observed values not differing appreciably from the corrected values. The pressure measurements on the solid at 0° are measured results for the bare wire. These results may be converted to specific resistances by correcting by the linear compressibility, to be given in the next section. At 12000 kg. this correction will produce a decrease of resistance of about 11%. The effect of the correction is to make the pressure of the minimum of specific resistance higher than the pressure of minimum of observed resistance.

In the table are also given the resistances of the liquid (specific) as a function of pressure up to the melting point at 63° and 96°. The

TABLE VI.
RESISTANCE OF CAESIUM.

At Atmospheric Pressure		Behavior under Pressure			
Temp.	Rel. Resis.	Pressure kg./cm.^{2}	Solid, Bare Wire at 0° C.	Liquid, in Glass 63.4°	Liquid, in Glass 95.8°
0° C.	1.0000	0	1.000	2.099	2.280
5	1.0247	500		1.934	2.094
10	1.0496	1000	.863	1.815	1.955
15	1.0748	1500		1.721	1.851
20	1.1002	2000	.779		1.770
25	1.1258	2500			1.721
29.7	1.1507 Solid	3000	.729		1.661
	1.910 Liquid	3500			1.625
30	1.912	4000	.709		
65	2.108	5000	.713		
100	2.304	6000	.728		
		7000	.756		
		8000	.794		
		9000	.833		
		10000	.879		
		11000	.931		
		12000	.994		

Ratio, $\frac{\text{Resis. Liquid}}{\text{Resis. Solid}}$	0 kg.	3780 kg.
	1.000	1.095

Temperature Coefficient of Solid.

Pressure kg./cm.^{2}	Bare Wire, 0°–10°	Solid in Glass
0	0.00496	Mean coefficient 0° to 96° at 9100 kg., extrapolated from results between 21° and 96° is 0.00463.
2000	434	
4000	388	
6000	366	
8000	367	
10000	386	
12000	418	

correction for changing from observed to specific resistance is 0.08% per 1000 kg.

The temperature coefficient of the solid bare wire between 0° and 10° is given as a function of pressure, and in Figure 5 are plotted the

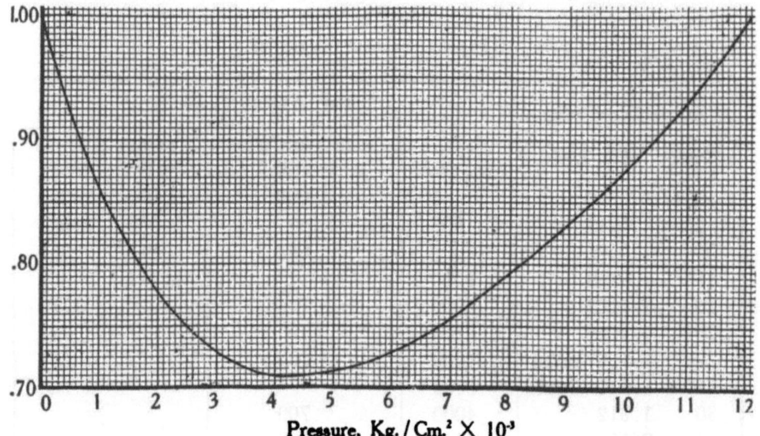

FIGURE 4. The relative resistance of solid caesium (bare wire) as a function of pressure at 0° C.

experimental values on which these are based. It will be seen that, like the resistance, the temperature coefficient passes through a minimum, but the pressure of the minimum is higher than the pressure of the minimum of resistance.

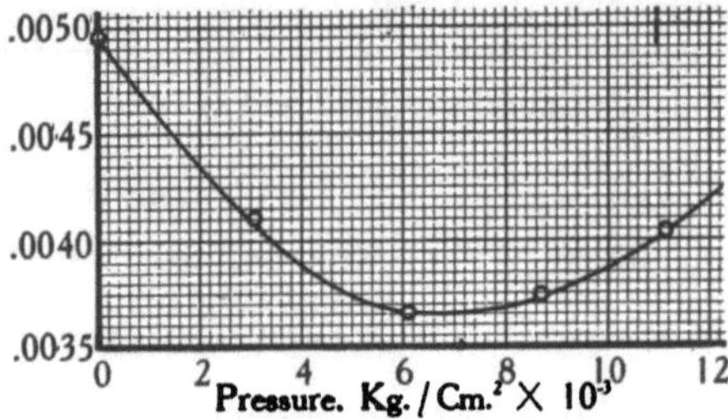

FIGURE 5. Mean temperature coefficient of resistance between 0° and 10° of bare caesium wire.

The ratio of resistance of liquid to solid is given at two points on the melting curve.

In Figure 6 is given the pressure of minimum resistance as a function of temperature. In this diagram are included some points obtained in some of the preliminary runs with caesium in glass. The agreement of the results so obtained with increasing and decreasing pressure seems to warrant their retention, although no other of the readings obtained with this material have been incorporated in the final results. It will be seen that these results show considerable spread; the probable explanation in terms of internal strains has already been given. In view of these irregularities it is probable that the values listed here for the resistance of the solid as a function of pressure have not a high degree of accuracy; there should be no such error in the resistances of the liquid.

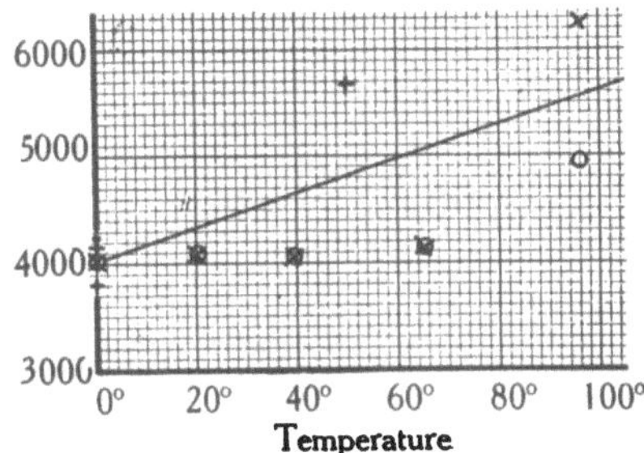

FIGURE 6. The pressure of the minimum resistance of caesium in kg/cm² plotted as ordinate against temperature. The open circles were obtained with decreasing pressure and the diagonal crosses with increasing pressure on caesium in glass. The rectangular crosses were obtained with increasing pressure and the dot with decreasing pressure on bare wire.

Compressibility. As in the case of rubidium, the compressibility was determined by the two methods of linear compressibility and cubic compressibility by piston displacement.

The difficulty found with the rubidium due to distortion in using the method of linear compressibility was very much accentuated in the case of caesium because of its greater softness, and the high pressure results had to be entirely discarded. Below 2000 kg., however, the results found by this method seem to be reliable, and they were used to supplement the results by the other method, which could not be extended below 1000 kg.

Measurements by the piston method were made with one filling of the apparatus, using 4.3 gm. of caesium, to a maximum pressure of nearly 16000 kg. at 50° and 95°, in addition to the low pressure runs made to obtain the melting data. The results finally given for the changes of volume in Table VII are based on the data obtained at 50°, these being somewhat more regular than the data at 95°. It was possible also to obtain readings on the compressibility of the liquid at 75°, which are also given in the Table. The liquid is not very much more compressible than the solid.

The values listed in the Table are changes of volume in cm.^{3} for 1.88 gm. of metal. The results given in this way were directly ob-

TABLE VII.

VOLUME DECREMENTS OF CAESIUM UNDER PRESSURE.

Pressure kg./cm.^{2}	Volume Decrement in cc. per 1.88 gm. 50° C., solid	75° C., liquid
0	.000	.000
500		.035
1000	.059	.063
1500		.086
2000	.102	.107
3000	.134	
4000	.161	
5000	.184	
6000	.204	
7000	.223	
8000	.240	
9000	.255	
10000	.269	
11000	.282	
12000	.294	
13000	.306	
14000	.317	
15000	.328	

tained from the experimental results without using any hypothetical values for the thermal expansion. 1.88 gm. is very approximately the amount of caesium that occupies 1 cm.^{3} at 0° C. at atmospheric pressure, so that the values listed in the Table are approximately fractional changes of volume.

The initial compressibility at 50° to be deduced from the Table is 0.0000070. Richards [13] gives for the average compressibility between 100 and 500 kg. 0.000060. He finds a more rapid decrease of compressibility with pressure than I do, giving for the average compressibility between 100 and 300 kg. 0.000067, and between 300 and

500 0.000053. A linear extrapolation would give 0.000081 for the initial compressibility. However, a detailed examination of Richards' data will show that his pressure coefficient of compressibility is in great doubt, and the agreement of our results is probably as close as could be expected.

Satisfactory numerical results could not be obtained for the variation of thermal expansion of the solid with pressure, but roughly, the behavior is of the same nature as that found for rubidium, the rapid decrease in expansion beyond 8000 or 9000 kg. being even more pronounced.

Discussion of Results.

We now have the data in hand for all the alkali metals, so that this discussion may well be concerned with a comparison of the properties of all of them.

The volume relations are the simplest, and first engage our attention. The volumes of all the alkali metals are shown in Figure 7 as a function of pressure. I have chosen somewhat unusual units for the volume, namely a volume proportional to the average space occupied by each extra-nuclear electron. This volume is obtained from the atomic volume (volume in cm.^{3} of a number of grams equal to the atomic weight) by dividing by the atomic number, or number of extra-nuclear electrons. This quantity I shall denote by the abbreviated expression "electronic" volume. The electronic volume seems somewhat better adapted to bring out certain relationships than the atomic volume. The values adopted for the electronic volumes at 0° C. at atmospheric pressure for Li, Na, K, Rb, and Cs are respectively 4.37, 2.15, 2.39, 1.507, and 1.284 cm.^{2}

The figure brings out an abnormality in the volume of potassium, which is between that of Na and Li, instead of between Na and Rb. This abnormality disappears, however, above 10000 kg., where the curves for Na and K cross. That the abnormality is to be ascribed to K and not to Na is made evident by some of the later curves.

The initial compressibilities at 0° of the 5 metals in order are $0.0_{5}87$, $0.0_{4}157$, $0.0_{4}355$, $0.0_{4}52$, and $0.0_{4}70$. By multiplying these by the electronic volumes we obtain numbers proportional to the actual loss under an external pressure of 1 kg./cm.^{2} of the average space occupied by 1 electron. The "electronic" compressibilities found in this way are $0.0_{4}38$, $0.0_{4}34$, $0.0_{4}85$, $0.0_{4}78$, and $0.0_{4}90$ respectively. It is interesting that the electronic compressibilities of the three heavier metals, in which the electrons are much more closely packed, is much larger than that of the two lighter metals, which have a much more open structure. This difference is the opposite of what one might expect. The difference is very much exaggerated if one calculates the atomic compressibility (ordinary compressibility

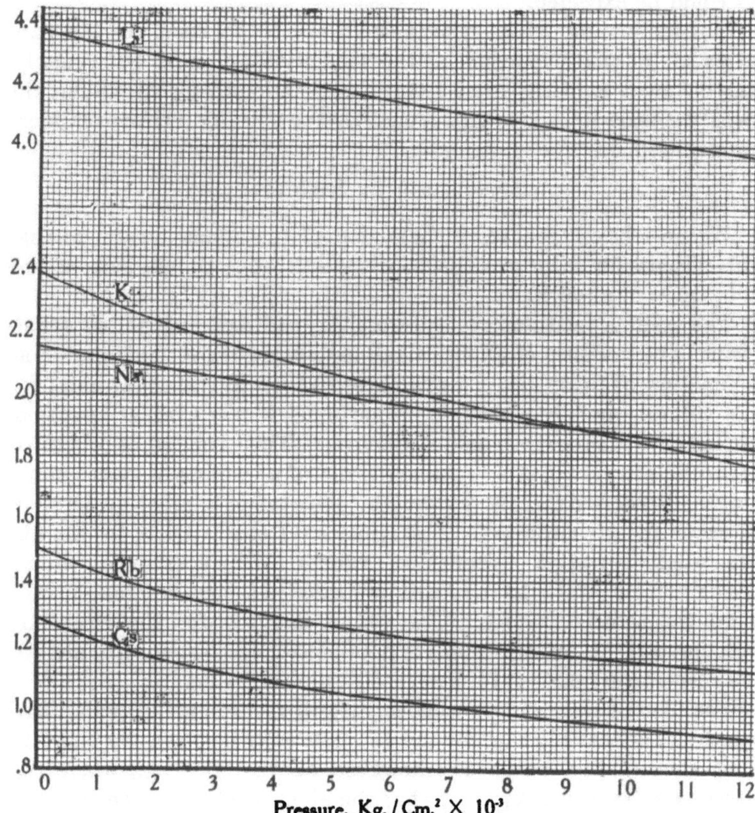

FIGURE 7. The electronic volumes (atomic volume divided by atomic number) in cm.³ at 0° C. of the five alkali metals as a function of pressure.

multiplied by atomic volume) in place of the electronic compressibility, the factor of variation through the series of five metals being 40 instead of 3.

Under high pressures, the anomaly disappears. The electronic compressibilities are respectively 0.0_430, 0.0_425, 0.0_450, 0.0_418, and

0.0₄19 at 12000 kg. Potassium occupies a highly anomalous position; except for this, the behavior under high pressure is as we would expect, the loss of volume per electron for a given increment of external pressure being less in the more complex and closely packed structure. The reversal of the effect at high pressures may be roughly described as a removal by the pressure of the "slack" which initially exists in the more complicated structures. Because the electronic compressibilities are much more nearly constant than the atomic compressibilities (both at 1 and 12000 kg.), it suggests itself that a more fruitful line of attack on the problem of the constitution of solids may be

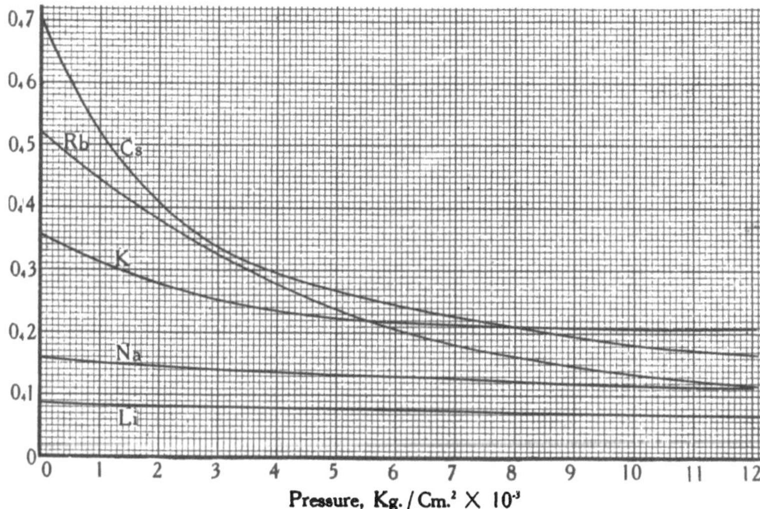

FIGURE 8. The instantaneous compressibilities $\frac{1}{v}\left(\frac{\partial v}{\partial p}\right)_t$ at 0° C. of the five alkali metals as a function of pressure.

found by neglecting the atomic structure and considering the solid merely as an aggregate of electrons (and nuclei). In fact, these volume relations show at high pressures a very rough similarity to the gas law, particularly those of Rb and Cs.

The instantaneous compressibility, $\frac{1}{v}\left(\frac{\partial v}{\partial p}\right)_t$, at 0° C. as a function of pressure is shown in Figure 8. This brings out strikingly again the anomalous position of K; it is evident from the figure that the anomaly is not to be ascribed to Na. Figure 8 has an important bearing on a suggestion of Professor Richards[14] with regard to compressibility at high pressures. His conception of internal pressure has led him to the view that a relatively incompressible metal at low pressures behaves like a more compressible metal at higher pressures. He applied this conception quantitatively to my data for Na and K, which were then available. An extrapolation equation for the volume of 1 gm. of K was found such that it fitted on smoothly at 18000 kg. with the volume of a certain weight of Na, starting from atmospheric pressure. The question now before us is whether this is a general relation between all the alkali metals. It is evident that if this were the case it should be possible to join smoothly together the curves of $\frac{1}{v}\left(\frac{\partial v}{\partial p}\right)_T$ for the different metals, merely by sliding them along the pressure axis, *without change of ordinate.* Now the figure makes it perfectly obvious that this is not the case, because the curves cross, whereas there would be no crossing if the supposed relation held. We conclude, therefore, that there are very important individual differences between the alkalis which become accentuated as the electronic structure becomes more tightly packed. The figure shows a crossing of the $\frac{1}{v}\left(\frac{\partial v}{\partial p}\right)_T$ curve of K by that of Rb and Cs, indicates that in another 1000 kg. the curve for Rb will cross that of Na, in another 5000 kg. Cs will probably cross, and it is not unlikely that Cs will cross Rb, and eventually both of these cross Li. This is all as one might expect; the surprising and inexplicable result is the persistent compressibility of K. This again lends color to the surmise that I have already expressed several times that K may have another polymorphic modification at high pressures.

The melting phenomena of the five metals next engage us. In Figure 9 are shown the melting curves. At atmospheric pressure the order of the melting points is the inverse of that of atomic weights (or atomic number) but again under pressure K is anomalous, and above 10000 kg. its melting point is higher than that of Na. Rb and Cs fall naturally in the sequence; the initial rise of melting point with pressure is greater than for the other metals, as one would expect. The fractional changes of volume on melting do not show the same regularity as some of the other properties, and I do not give a figure for them. At atmospheric pressure the order of the fractional changes is:

Metal	Fractional Change of Volume on Melting
Li	0.0060
Ka	.0231
Cs	.0256
Na	.0271
Rb	.0284

At higher pressures there is a complicated interchange of positions; the curve for Rb crosses that of Na and practically coincides with that

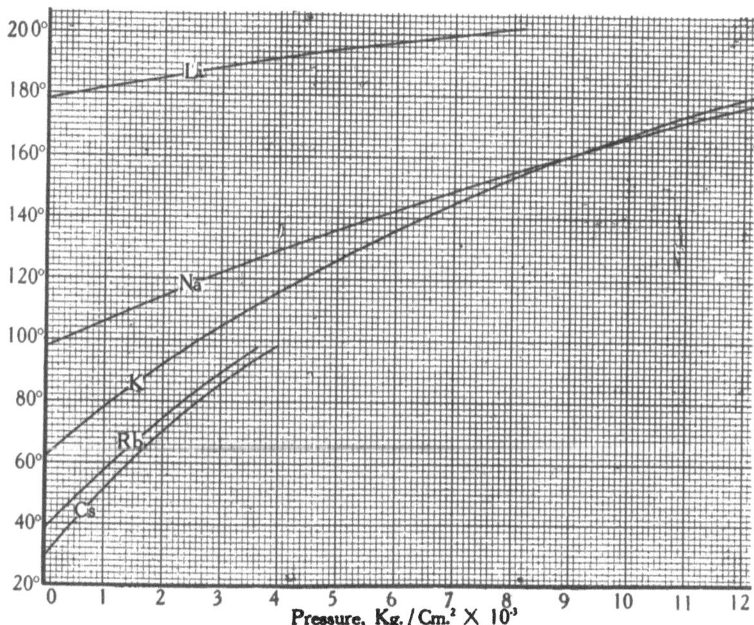

FIGURE 9. The melting temperatures as function of pressure of the five alkali metals.

of K from 2000 to 4000 kg., Cs drops below K but acts as though it might rise above it again, and the curve for K drops off at high pressures with abnormal rapidity, which is one of the pieces of evidence for a new polymorphic modification. With the exception of Li, the approximate equality of the fractional changes of volume on melting is interesting, and suggests a fundamental similarity.

The latent heats of melting are also interesting. At atmospheric pressure these are as shown.

Metal	Latent Heat of Melting, in kg. m. per gm. atom
Li	49.
K	215.
Cs	219.
Rb	236.
Na	271.

Again, except for Li, the approximate equality is striking. There is no connection here with the sequence of atomic weights, any more than there was for the change of volume. At high pressures, the behavior is complicated. The latent heat of Na is roughly constant, Cs rises rapidly, crossing Rb and Na, Rb is approximately constant but falls to a slight minimum at 2000 kg, while K again shows pronounced variation, rising to a maximum near 4000 and then falling rapidly to 18.9 at 12000 kg.

One may say in general that the melting data in the alkali group do not show as regular variations as do the compressibilities.

Finally we have to consider the electrical resistance under pressure. In Figure 10 are shown the relative resistances of the five metals at 0° C. as a function of pressure. In spite of their apparent complexity, there is an underlying regularity in the curves. There is a turning point for the electrical properties somewhere between K and Rb. This may be seen by plotting at any fixed pressure the decrement of resistance for the various metals as ordinate against atomic number (for example) as abscissa. The same sort of turning point is also shown on plotting the "conductivity per atom" at a definite reduced temperature against atomic number. This is described more in detail in my report to the fourth Solvay Conference on Metallic Conduction.

The diagram suggests the query whether all the alkali metals will ultimately, at pressure high enough, increase in resistance. This involves an extrapolation of the curves. Such an extrapolation is not easy; we may obtain a more satisfactory indication by extrapolating the instantaneous pressure coefficient $\frac{1}{R}\left(\frac{\partial R}{\partial p}\right)_T$. If the resistance has a minimum, the curve of the coefficient must cross the axis to become negative. It will be seen in Figure 11 that it is not improbable that Rb may cross the axis, and so have a minimum resistance, but it is much more uncertain in the case of K. The reversal of curvature shown by Cs on the negative side of the axis is, however, of great significance. If there should be a similar reversal for the other metals, they would all, of course, have minima.

It is an interesting question whether the minimum resistance of Cs is a function of the crystal structure of the solid, or whether it would be shown by the liquid also if pressure could be raised high

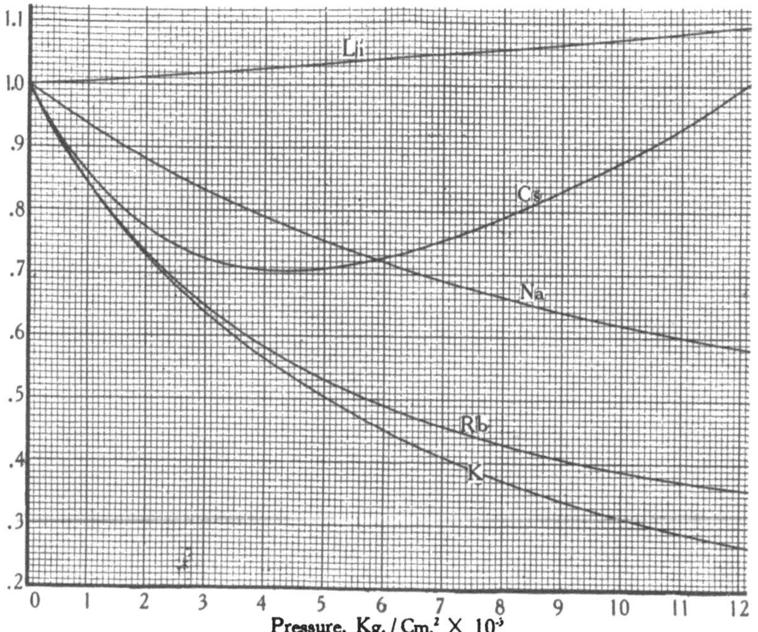

FIGURE 10. The relative resistances at 0° C. of the five alkali metals as a function of pressure.

enough without freezing. Figure 12 is an attempt to answer this question. Here are plotted the resistances under pressure of solid and liquid Cs at 95°, the resistance of the liquid being multiplied by 1/1.704 to remove the discontinuity at the melting point. The curves for solid and liquid are seen to join together continuously without change of slope at the melting point. It is therefore probable that the resistance of the liquid would also pass through a minimum if the pressure could be raised high enough without freezing, and we may draw the conclusion that the minimum does not involve the particular mechanism of the solid crystal.

The minimum resistance of Cs cannot but be of great significance for theories of conduction. It indicates that the conduction

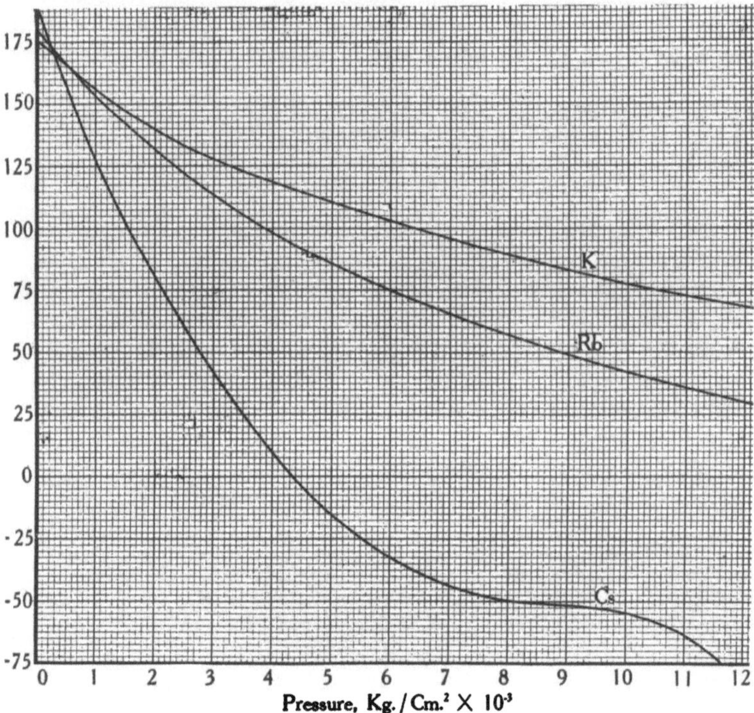

FIGURE 11. The instantaneous pressure coefficient of resistance $\frac{1}{R}\left(\frac{\partial R}{\partial p}\right)_t$ at 0° C. × 10^6 as a function of pressure for the three most compressible alkali metals.

mechanism is not simple, but that there are several factors involved, which may be affected by pressure in different ways. With regard to my own theory, which has been developed in several papers,[15] it merely drives me more strongly to a view which other phenomena have been more and more insistently suggesting. This is that the conduction electron, in its way through the metal is effectively confined to certain tracks through the atoms (probably in some cases between them) and that passage from atom to atom cannot be affected unless the tracks in adjacent atoms are in alignment. Under high pressures the atoms may in some cases, depending on the detailed structure of the atom, be so distorted or rotated that the tracks get out of alignment, and the resistance increases. Such a

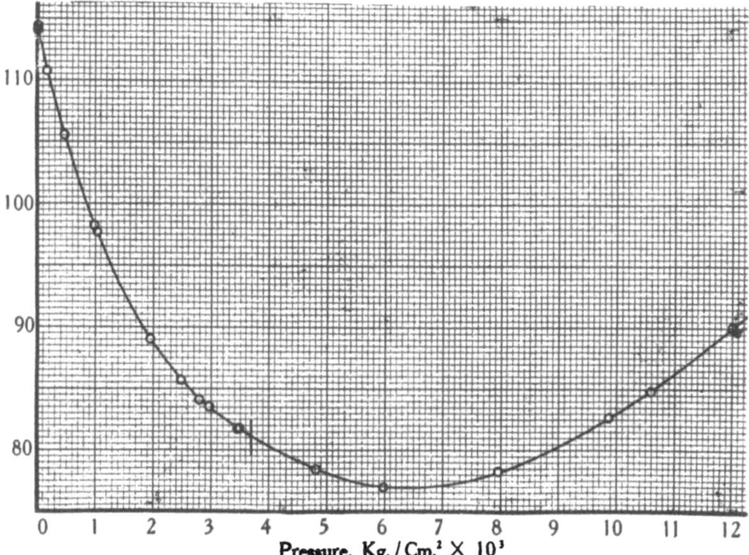

FIGURE 12. The resistance on an arbitrary scale as a function of pressure of liquid and solid caesium at 95°. The resistance of the solid has been multiplied by such a factor as to remove the discontinuity at the melting point, which takes place at the pressure shown by the vertical line. The smoothness of the curve indicates that the crystal arrangement plays a minor part in the resistance phenomena of this metal.

condition would be especially anticipated in caesium because of its high compressibility and complicated structure. The tracks themselves, it is natural to think, are the result of some sort of quantum phenomenon. It is possible that the track within the atom should itself be unfavorably modified by the distortion of the atom due to high pressure, but this would be expected to occur at considerably higher pressures than the peripheral effect, and may probably be disregarded in a first discussion.

Summary.

The following new data are presented for rubidium and caesium: the cubic compressibility to 15000 kg./cm.^{2}, the melting data (variation of melting temperature with pressure and the change of volume, permitting a calculation of the latent heats) to 100°, and the electrical resistance of both solid and liquid as a function of pressure between 0° and 100° to 12000 kg./cm.^{2}

In addition to the new data for rubidium and caesium, the electrical resistance of potassium under pressure has been redetermined; the new measurements were made with bare wire. The old measurements were made on potassium in a glass capillary, which exerted a constraining action. The new results do not show the great decrease of temperature coefficient at high pressures formerly found.

In the discussion, the properties of all the alkali metals are compared. The "electronic" volumes (atomic volume divided by the number of extra-nuclear electrons) is plotted as a function of pressure. Potassium is anomalous at low pressures, its volume being between that of lithium and sodium, but the anomaly is wiped out at high pressures. The compressibility of potassium is also anomalous, the rate of decrease at high pressures being much less than for the other metals. In general, the pressure volume relations of a compressible alkali at high pressures are *not* similar to those of a less compressible alkali at lower pressures. The melting of potassium is also anomalous; at high pressures its melting point becomes higher than that of sodium. Attention is directed to the fact that with the exception of lithium the fractional changes of volume on melting of all the alkali metals are nearly the same, as are also the latent heats per gm. atom. The small scale variations of these quantities with pressure are irregular.

The resistance of caesium is found to pass through a minimum with increasing pressure. Comparison of measurements on solid and liquid suggest that this minimum is probably not connected with the crystalline structure of the solid. Comparison with the resistance of the other alkalies under pressure suggests that at sufficiently high pressures the resistance of rubidium may also pass through a minimum, and if the reversal of curvature shown by caesium at high pressures should be a property of the other metals, all of them may have minima at sufficiently high pressures. The general significance of this for a theory of electronic conduction is discussed.

It is a pleasure to acknowledge the assistance received from Mr. Walter Koenig in making many of the readings of this paper.

The Jefferson Physical Laboratory,
Harvard University, Cambridge, Mass.

References.

[1] P. W. Bridgman, Phys. Rev. 3, 126–203, 1914; Proc. Amer. Acad. 56, 61–153, 1920; 57, 41–66, 1922; 58, 166–241, 1923.

[2] P. W. Bridgman, Proc. Amer. Acad. 58, 149–161, 1923.

[3] Second reference under 1.

[4] P. W. Bridgman, Proc. Amer. Acad. 47, 415, 1911; 51, 55–118, 1915.

[5] Fourth reference under 1.

[6] P. W. Bridgman, Proc. Amer. Acad. 49, 1–114, 1915.

[7] L. H. Adams, E. D. Williamson and J. Johnston, Jour. Amer. Chem. Soc. 41, 12–42, 1919; L. H. Adams and E. D. Williamson, Jour. Frank. Inst. 195, 475–529, 1923.

[8] P. W. Bridgman, Proc. Amer. Acad. 59, 171–211, 1924.

[9] Second reference under 1.

[10] P. W. Bridgman, Proc. Amer. Acad. 52, 64, 1917.

[11] T. W. Richards and F. N. Brink, Jour. Amer. Chem. Soc. 29, 117–127, 1907.

[12] E. Rengade, C. R. 156, 1897, 1913.

[13] T. W. Richards, Carnegie Inst. Wash. Pub. No. 76, 1907.

[14] T. W. Richards, Jour. Amer. Chem. Soc. 45, 422–437, 1923; 46, 1419–1436, 1924.

[15] P. W. Bridgman, Phys. Rev. 9, 269, 1917; 17, 161, 1921; 19, 114, 1922; Proc. Amer. Acad. 59, 119, 1923; Report of the Fourth Solvay Conference, Brussels, April, 1924, not yet published.

THE EFFECT OF TENSION ON THE TRANSVERSE AND LONGITUDINAL RESISTANCE OF METALS.

By P. W. Bridgman.

Received November 6, 1924. Presented October 8, 1924.

CONTENTS.

Introduction.

Although the effect of tension on the resistance of a metal when the current flows in the direction of the tension has been measured by a number of observers,[1] apparently the resistance to current flow at right angles to the direction of the tension has been measured only by Tomlinson,[2] and for only two metals. The matter is one of considerable interest for theories of metallic conduction, and deserves further attention. In the following I give the results of measurements of the transverse coefficient for several metals. I have also measured the longitudinal tension coefficient of the same samples; this is desirable because the longitudinal tension coefficient varies somewhat from sample to sample. In addition to the two tension coefficients, we have available the hydrostatic pressure coefficient[3] (not determined on the identical samples, but not nearly so variable with the specimen) so that the experimental basis for a discussion of the effect of changes of dimensions on resistance is now fairly well laid (at least for a few metals). It would be desirable if the effect of a pure shear could also be determined, but this is a matter of much greater experimental difficulty.

Method.

The experimental determination of the transverse coefficient is a matter of considerable difficulty. The method employed by Tomlinson[2] apparently leaves much to be desired. His specimen was in the form of a long narrow strip clamped by the long edges so that a tension could be applied crosswise of the strip. The current passed lengthwise of the strip, being led in and out by screw clamps, and the resistance was measured with a Wheatstone's bridge. The state of stress and strain in such a strip is evidently far from simple, because the lateral contraction is hindered, and there are complicated effects under the clamps, which may well constitute a considerable fraction of the whole. It would seem to be a desideratum of a good method that no part of the current which takes part in the measurements should be allowed to flow under clamps.

For the two metals (iron and zinc) which Tomlinson measured, he found that the transverse resistance decreases with tension, the opposite of the longitudinal effect.

In attempting to avoid the difficulties of Tomlinson's method, I adopted a method which demands two different sorts of measurement. The first is the geometrical mean of the longitudinal and transverse tension coefficients, while the second is the longitudinal coefficient alone. From these two data the transverse coefficient may be calculated.

Method of Measuring the Geometrical Mean of the Two Coefficients. In measuring the geometrical mean of the two coefficients, the metal in the form of a thin sheet (from 0.0075 to 0.015 cm. thick) was cut to a rectangle 5 by 10 cm., the two ends gripped by screw clamps, and tension applied by a simple lever arrangement lengthwise of the strip. By way of precaution the strip was insulated from the clamps by strips of mica, although special measurements showed that this precaution was not necessary. The electrical measurements were made on the central portion of the strip. Two sets of measurements were made, one transverse and the other longitudinal. To make either of these sets, a group of 4 contact points was employed, pressed against the strip by springs. These are shown in Figure 1. In making the transverse measurements current was led into the

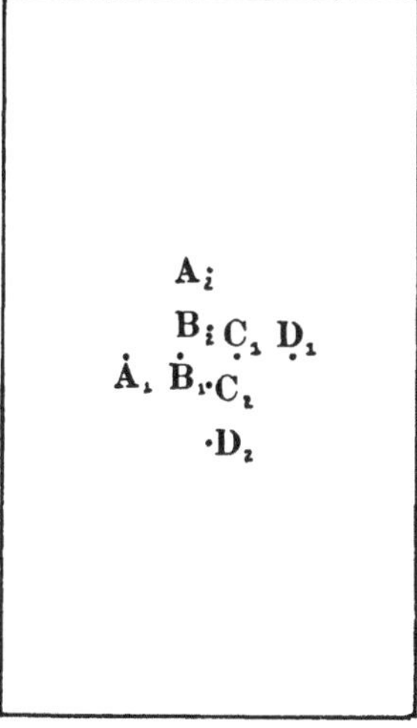

FIGURE 1. Shows the plate with the current and potential terminals. Either the horizontal or the vertical set was used. The current terminals are A and D and the potential terminals B and C. Tension is applied lengthwise of the plate.

plate at A_1 and out at D_1. The difference of potential between the two intermediate contact points B_1 and C_1 was determined on a potentiometer; in this way the specific resistance of the metal per unit of surface may be determined. An exactly similar set of measurements was made with the four vertically disposed contact points. Measurements made with and without tension give the effect of tension.

Without analysis, one might be inclined to think that the measurements with the transversely arranged points would give the transverse coefficient, and those with the longitudinal points the longitudinal coefficient, but detailed mathematical discussion, which will be given later, shows that this is not the case. Except for various corrections, both measurements should give exactly the same thing, namely the geometrical mean of the conductivity per unit surface longitudinally and transversely. At best, then, these two sets of measurements give merely a check on each other, which is not without value. As a matter of fact, however, the various corrections turn out to be so much larger with the longitudinal arrangement, that in making the final computations only those results were employed which were obtained with the transverse arrangement.

Some of the details of construction require comment. The contact points were large needles pressed against the metal strip with helical springs. The strip was backed with a piece of hard rubber to prevent its bending under the pressure of the contacts. The contact points, with the springs and the backing piece of rubber were constructed as one self-contained unit, which was independently supported, so as to float freely at the center of the metal sheet and follow, without introducing any stresses, the slight displacements due to applying and removing tension. A special holder was used in assembling the apparatus to avoid bending the thin metal sheet. The current and potential leads were flexible copper wire, about 0.015 cm. in diameter, soldered to the needles close to the points so as to minimize thermal e.m.f.'s. Because of the smallness of the effect it was necessary to use rather large currents, of the order of half an ampere. Since the resulting heating effects might be very serious, it was necessary to put the metal sheet with the contact points in a kerosene bath rapidly stirred. Measurements were attempted only at room temperature.

The spacing of the four points was maintained always the same by the use of a special jig; with this, holes were made in a piece of thin cardboard, by which the points were located, and then the cardboard was removed by tearing it apart along cuts previously made in it with a razor blade. Some such use of a jig was necessary, because the needles themselves had to be given some freedom of motion to follow the displacements due to tension. The points were equispaced at approximately 6 mm., making the distance between A_1 and D_1 1.8 cm., and similarly for the longitudinal points.

In terms of the potentiometer measurements the absolute value of the mean surface resistance may be found with and without tension. However, the chief interest of this work is not in values of absolute resistance, but in the percentage changes produced by tension, so that a somewhat simplified and more sensitive procedure in making most of the electrical measurements was adopted. The potential terminals B_1 and C_1 were connected to a senstive moving coil galvanometer (5 cm. deflection per micro-volt). Another e.m.f. was introduced into the same circuit and could be varied by a simple arrangement of variable shunts so that the e.m.f. due to B_1C_1 could be neutralized. The galvanometer zero was noted. Tension is now applied to the specimen; the resistance changes, the e.m.f. due to B_1C_1 changes, and the galvanometer balance is upset, giving a displacement, which is noted. The load is removed, and again the galvanometer zero noted. This is repeated a number of times, and the mean of the displacements taken. The change of resistance of the specimen corresponding to the galvanometer deflection was determined by finding the deflection caused by known changes of resistance in the balancing circuit. These resistances were so chosen that simple proportionality holds between the deflections and the changes of resistance.

This method would not be applicable unless equilibrium were reached very rapidly. Due to the thinness of the metal sheet and the rapid stirring, all temperature effects arising from the application of tension are dissipated in a time less than that required for the galvanometer to reach equilibrium, so that it is possible to make readings every half minute or less.

It will be seen that the method assumes that the current between the current leads A_1 and D_1 is unaffected by the tension. This was assured by the relative values of the various resistances. The current leads were fed by a storage battery of 12 volts through such a resistance as to cut the current to an ampere or less. The actual resistance between A_1 and D_1 is of the order of a few ten thousandths of an ohm, and the changes in resistance due to tension are of the order of a small fraction of a per cent of this, so that any changes in the feeding current may be entirely neglected.

In addition to these relative measurements of the effect of tension, absolute measurements with the potentiometer were made of the resistance under no load, from which the specific resistance of the metal was computed, and used as a check by comparing with the accepted values for the metal.

In applying tension to the specimen, considerable care is necessary to avoid any shock which might result in permanent set. This was accomplished by attaching the scale pan to the lever by a long helical spring. An arrangement of levers permitted the application and removal of the load by the observer at the galvanometer without changing his position.

Mathematical Theory of the Measurements of the Geometrical Mean Coefficient. The mathematical problem is one of flow in two dimensions in an anisotropic homogeneous substance, the anisotropy consisting of different conductivities in two perpendicular directions. Taking the x and y axes as the axes of principal conductivity (lengthwise and crosswise of our sheet under tension), we have

$$u_x = k_1^2 \frac{\partial \varphi}{\partial x}, \qquad u_y = k_2^2 \frac{\partial \varphi}{\partial y},$$

where u_x and u_y are the components of current, φ the potential, and k_1^2 and k_2^2 the principal conductivities. Since the condition is a steady one we also have the equation of continuity

$$\text{Div } u = \frac{\partial u_x}{\partial x} + \frac{\partial u_y}{\partial y} = 0,$$

which, on substituting from above, becomes

$$k_1^2 \frac{\partial^2 \varphi}{\partial x^2} + k_2^2 \frac{\partial^2 \varphi}{\partial y^2} = 0.$$

This may be reduced to the equation for an isotropic medium, the solution of which is known, by the change of variable

$$x = k_1 \xi, \qquad y = k_2 \eta$$

giving

$$\frac{\partial^2 \varphi}{\partial \xi^2} + \frac{\partial^2 \varphi}{\partial \eta^2} = 0.$$

The case in which we are interested is that of a source and a sink symmetrically situated on one of the median lines of a rectangle. We simplify our discussion at first by replacing the rectangle by an infinite plane. The solution may obviously be built up from two point singularities of the type

$$\varphi = -\log r,$$

where

$$r = \sqrt{(\xi - \xi_0)^2 + (\eta - \eta_0)^2}.$$

Consider the auxiliary ξ, η plane, with one point singularity at the origin ($\xi_0 = \eta_0 = 0$), the specific conductivity of the ξ, η plane being unity. Then the solution $\varphi = -\log r$ corresponds to a source at the origin of the auxiliary plane, current flowing uniformly from the source in radial lines to infinity, the total current flowing from the source being 2π. Changing back the variables now gives us a possible solution in the x, y plane, namely

$$\varphi = -\log \sqrt{\frac{x^2}{k_1^2} + \frac{y^2}{k_2^2}}.$$

We have to ask what is the corresponding current (I) flowing from the source in the x, y plane. The result may be obtained by integrating u_n over unit circle about the origin.

$$u_n = u_x \cos xn + u_y \cos yn$$

$$= u_x \frac{x}{r} + u_y \frac{y}{r}.$$

Now differentiating φ and substituting above,

$$u_x = \frac{x}{\frac{x^2}{k_1^2} + \frac{y^2}{k_2^2}}, \qquad u_y = \frac{y}{\frac{x^2}{k_1^2} + \frac{y^2}{k_2^2}},$$

$$u_n = \frac{r}{\frac{x^2}{k_1^2} + \frac{y^2}{k_2^2}}.$$

We have to form $I = \int u_n ds$, where ds is the element of arc of unit circle $(x^2 + y^2 = 1)$

$$ds^2 = dx^2 + dy^2, \qquad dy = -\frac{x}{y}\, dx,$$

$$ds = \frac{r}{y}\, dx,$$

$$I = 4\int_0^1 u_n \frac{r}{y}\, dx = 4\int_0^1 \frac{dx}{\sqrt{1-x^2}\left[\frac{1}{k_2^2} + x^2\left(\frac{1}{k_1^2} - \frac{1}{k_2^2}\right)\right]}.$$

This integral may be readily evaluated (See e.g. B. O. Pierce's Tables of Integrals, No. 158), giving

$$I = 2\pi k_1 k_2. \tag{1}$$

Now apply this to the special problem in hand, where we have a source and a sink of equal strength. The solution above applies, taking successively the origin at the source and sink, and using the minus sign for the sink. Our problem is to find, for an infinite sheet, the difference of potential between B_1 and C_1 due to a source at A_1

and a sink at D_1 (see figure 1). The distance A_1D_1 is denoted by 2β, and B_1C_1 by 2α. We have at once

Potential at B due to source at A is $-\log\frac{\beta-\alpha}{k_1}$.

Potential at B due to sink at D is $+\log\frac{\beta+\alpha}{k_1}$.

Total potential at B is $\log\frac{\beta+\alpha}{\beta-\alpha}$.

Potential at C due to source at A is $-\log\frac{\beta+\alpha}{k_1}$.

Potential at C due to sink at D is $+\log\frac{\beta-\alpha}{k_1}$.

Total potential at C is $\log\frac{\beta-\alpha}{\beta+\alpha}$.

Whence

$$\varphi_B - \varphi_C = 2\log\frac{\beta+\alpha}{\beta-\alpha}. \tag{2}$$

Now the effective resistance between B and C is defined (and measured) as $(\varphi_B - \varphi_C)/I$, which becomes

$$\frac{1}{\pi k_1 k_2}\log\frac{\beta+\alpha}{\beta-\alpha} \tag{3}$$

on substituting the values.

The essential feature of this solution is that k_1 and k_2 enter symmetrically, so that by this method only the geometrical mean of the conductivities in two mutually perpendicular directions can be determined.

In applying this mathematical analysis directly to the experimental arrangement two corrections are involved. In the first place, the actual three dimensional flow in the immediate neighborhood of the contact points by which current enters and leaves is replaced in the analysis by a current everywhere two dimensional. A precise mathematical discussion of the correction for this effect is complicated, and in view of the experimental irregularities there is probably little point in attempting it. The correction is probably beyond the limits of error, because of the thinness of the sheet compared with the distance between the current leads; this ratio was of the order of 1/150 for the metals used. The sign of the effect is to make the specific resistance calculated without correction too large. The effect of tension is to leave this (regarded as a per centage correction) unchanged for the transverse position of the points, but to decrease it slightly for the longitudinal position. The reason for this is that the ratio of the separation of the points to the thickness of the plate is unchanged by tension in the transverse position, but is changed in the longitudinal position.

The second correction is for the finite size of the plate, and is much more important. The rigorous solution for the finite plate may be obtained from the solution above for the infinite sheet by a simple application of the method of images. I owe this idea to a remark of Mr. B. O. Koopman. In applying the method, the plane is cut up into a set of rectangles similar to the original rectangle, with a source or sink in each rectangle in the same position as in the original rectangle. In the strip of rectangles laid off in the direction of the four points the signs alternate, the source being alternately to the right and the left of the sink, whereas in the strip running perpendicularly there is no alternation.

I have carried through an approximate evaluation of the correction for the transverse position of the points. The calculations are not too long with the actual dimensions of the metal sheet used in the experiment. The correction turns out to be approximately 12%, and in this particular case is contributed almost entirely by the phantom rectangles in the strip parallel to the direction of the points, the contributions of all the others nearly cancelling each other and being intrinsically smaller. As a matter of experiment, the mean of the specific resistance of all the metals measured, applying the formula for the infinite plane to the data obtained from the transverse position of the points, was 15% higher than the accepted values, sufficiently good agreement.

The calculation of the correction for the longitudinal position of the points is more difficult, because of slower convergence of the series, and I did not attempt it. It is evident from inspection, however, that the correction will be considerably larger than for the transverse position, and this agrees with the experimental fact that the uncorrected resistance from the longitudinal measurements was 25% high.

Besides determining the correction for the finite size of the sheet, we must also determine how the correction changes when tension is applied, thereby changing the dimensions of the plate, and also making the resistance anisotropic. Here the condition for the transverse position is very much simpler than for the longitudinal position. We have seen that the correction terms are contributed in the transverse position almost entirely by the phantom rectangles extending in the strip parallel to the four points. The terms in the potential arising from the images are logarithms of distances with plus and minus signs (sources and sinks), which may be written in terms of ratios of distances. Under tension, all the effective distances in the strip of phantom rectangles are changed uniformly because of the change of dimensions of the sample, and there is superposed another virtual change, also uniform, because of anisotropy. Hence the ratio of the distances (which enters the expression for the potential) are unchanged and hence the percentage correction on the absolute resistance is unchanged, so that the percentage change of resistance under tension may be calculated without correction from the formulas obtained for the infinite plane. For the longitudinal position, however, the contributions by the strips of phantoms running in both directions must be considered. The distortion is different in different directions, so that there is a correction. This is complicated to compute, and I have not attempted it. One may see by inspection, however, that the uncorrected change of resistance calculated from the longitudinal readings will be too large, and this agrees with the experimental facts.

In the actual calculation of the results, I have used only the data obtained with the transverse position of the points, demanding merely that the longitudinal readings should agree with what might be roughly expected.

Formula 3 yields the two dimensional specific resistance of the metal sheet. In calculating the ordinary resistance per cm. cube, the thickness of the sheet enters. The thickness is changed by tension, so that a correction has to be applied for it, the numerical magnitude of which is evidently σ/E, where σ is Poisson's ratio, and E Young's modulus. The correction is of such a sign that the actually measured proportional increase of resistance is greater than the proportional increase of specific resistance.

Method of Measuring the Longitudinal Coefficient. The measurements just described give the geometrical mean of the longitudinal and transverse coefficients. The straight longitudinal coefficient was measured in much the same way, using the same arrangement of levers, and kerosene bath, and electrical measuring apparatus. The specimen was a strip 1.25 cm. wide and 10 cm. long, cut lengthwise from the specimens the geometrical mean of which had just been measured. This strip was clamped at either end to screw clamps by means of which the tension was applied, and through which the current was led into and out of the specimen. The potential difference between two points 3 cm. apart in the center of the strip was taken off with needle points pressed against the strip. The dimensions of the strip ensure that at the center the lines of flow are straight and uniformly distributed, so that the change of resistance (measured in terms of galvanometer deflection as above) gives directly the longitudinal effect. The numerical magnitudes are much more favorable here than in the first case, so that it was possible to get the pure longitudinal coefficient with much greater accuracy than the geometrical mean.

In reducing the actually measured proportional changes of resistance to proportional changes of specific resistance, corrections must be applied for the increase of length due to tension and for the lateral contraction in both directions at right angles to the length. The magnitude of the correction is evidently $(1 + 2\sigma)/E$, and is in such a direction as to make the increase of specific resistance with tension less than the measured increase.

Experimental Details.

The experimental details were comparatively straightforward, but much care was necessary because the effects are so small. Greater sensitiveness is obtained by using a greater current, but an upper limit is soon reached because of the heating effects, which increase as the square of the current. The currents were chosen as near the upper limit as feasible. Much may be gained by properly directing the action of the stirrer by which the bath is kept at constant temperature. In applying load to the specimens much care must be taken to remain below the elastic limit, as the properties after even moderate overstrain may be greatly altered. This point is to be emphasized; I believe that in early work it has not been sufficiently regarded. The elastic limit of some of the softer metals in the thorougly annealed condition is surprisingly low. I found it impossible to obtain with this apparatus the effect on aluminum, although many attempts were made, because of the exceedingly low elastic limit of the annealed pure metal. The results obtained with gold and copper are also exceedingly uncertain from this cause. Silver gave less trouble. The harder metals, iron, nickel, palladium, platinum, gave entirely satisfactory readings, large enough not only to allow a determination of the average value of the coefficients, but also to allow an investigation of the linearity of the relation between stress and change of resistance. For theoretical use I believe it is safe to use the transverse coefficients of only the harder metals, but the longitudinal coefficients of all should be trustworthy.

Aluminum. This was of very unusual purity, containing not over 0.03% total impurity, and was obtained from the same source as the specimens for which I have previously measured the compressibility and the effect of tension on thermal conductivity.[4] It was annealed for several hours at 300°. It was so soft and the elastic limit so low that no measurements of the transverse effect with the four point apparatus could be obtained. A number of attempts were made to obtain the transverse effect, but in all cases loads had to be applied beyond the elastic limit to obtain anything measurable, and the attempt was abandoned. The exceeding of the elastic limit was shown not only by permanent change of dimensions, but also by failure of linearity between stress and change of resistance.

I was able, however, to get better measurements than hitherto of the pure longitudinal effect. This turns out to be very sensitive indeed to slight permanent stretch. The coefficient of a specimen slightly stretched beyond the elastic limit was only one third as great as the highest value obtained on a specimen whose elastic limit had not been exceeded. The maximum stress applied to the best specimen was only 130 kg./cm^2. Within the limits of error, which were somewhat large, the effect is proportional to the load. This is not the case, however, for aluminum which has been strained beyond the elastic limit, the effect departing markedly from linearity and becoming less at the higher stresses.

Two different specimens of aluminum which had apparently not been permanently stretched gave values of the pure longitudinal coefficient not very different from each other, and both were larger than the values which I have obtained before, and larger than values by other observers. The new values are doubtless to be preferred.

Gold. This was sheet 0.012 cm. thick, obtained from Baker and Co. and stated to be of the highest possible purity (guaranteed better than 99.9%). It was annealed by Baker, and also one piece was further annealed by me, but with no perceptible change in the coefficient. Measurements of the purely longitudinal effect were made on three different specimens with fairly concordant results. I cannot find that the pure longitudinal effect has been previously measured for this metal. The value obtained now is interesting because it is so high. The longitudinal effect is linear with tension with an error not more than 3 or 4% over the range of stress applied here, which was not more than 120 kg./cm². The transverse effect for gold was very much less than the longitudinal effect, but is so uncertain that I have not tried to assign any probable value to it.

Copper. Ordinary commercial sheet copper, annealed, was used. There were two different specimens, of thickness 0.012 and 0.20 cm. The measurements of the purely longitudinal effect on the two specimens agreed within 4%, and also agreed with a result which I have previously found for copper of unusually high purity, so that evidently the amount of impurity in commercial copper does not introduce any perceptible error here. Two measurements were also made of the geometrical mean of the two effects; these both agreed in giving a measured effect so small as to be zero within the limits of error. The check measurements with the four points in the longitudinal position behaved as they should for one of the samples, but for the other gave an effect of the wrong sign, for some reason which I have not been able to find. This is the only case of such a discrepancy in all this work, and throws considerable suspicion on the results obtained with the transverse position.

Silver. This was obtained from Baker and Co. of the highest purity, in the form of annealed sheets 0.012 cm. thick. Two samples were used. The behavior when stretched beyond the elastic limit is the same as was found for aluminum, namely the purely longitudinal effect becomes small; in this case about 65% of the effect when the limit is not exceeded. Measurements of the purely longitudinal effect on two different pieces gave coefficients agreeing within less than 1%. The effect is linear within 2% over a tension range of 250 kg./cm². The geometrical mean effect was measured on one specimen. It is linear with tension up to 300 kg./cm². within an experimental error of 10%.

Platinum. This was annealed sheet, 0.012 cm. thick, of the highest purity (99.9% or better) from Baker and Co. Only one specimen was used. The pure longitudinal effect agreed within about 3.5% with the value which I have previously found for platinum rod. It is linear to better than 1% to 400 kg./cm.², the highest tension used. The geometrical mean is linear within 4% up to 420 kg./cm.², and was large enough to give very comfortable measurements. The behavior for the longitudinal position of the points is as would be expected.

Palladium. This was similar in dimensions to platinum, was obtained from the same source, and was stated to be of the same purity (99.9% or better). There seems, however, for some reason to be considerable difficulty in getting pure palladium, so that I tried to get a check on the purity by measuring the temperature coefficient of resistance of a piece of palladium wire drawn by Baker from metal from the same lot. The coefficient was very low, only 0.0030, so that it would seem that the purity is probably not as high as estimated.

The purely longitudinal coefficient agreed within 4% with the value which I have previously found. The relation is linear with the stress to better than 1% up to 400 kg./cm.^{2}, the highest tension used. The geometrical mean coefficient was measured for three different loads to a maximum of 440 kg./cm.^{2} The irregularities were greater than usual, the maximum departure from the linear relation being 15%, but there is no reason to think that this is other than an effect of accidental errors, or that the effect is not actually linear.

Iron. I was not able to obtain pure iron in sheets thin enough for the purpose, and had to content myself with a commercial mild steel of low carbon content. The thickness was 0.010 cm. The dimensions of the original sheet were large enough to allow two samples to be cut, parallel and at right angles to the direction of rolling. This was not possible with the other metals.

The purely longitudinal effect of the two specimens agreed to three significant figures. The value is about 20% lower than I have previously found for iron of very high purity, and agrees essentially with the value found by Tomlinson for iron of no especial purity. The difference is evidently to be ascribed to the carbon content. It is interesting to note that in every case examined so far the effect of impurity is to lower the longitudinal tension coefficient of resistance (if there is any effect at all). Impurity also lowers the temperature coefficient and the pressure coefficient.

The geometrical mean effects for the two specimens did not agree, but was less by 25% for the specimen to which tension was applied in the direction of rolling.

Both effects for both specimens were linear with tension. Because of the largeness of the effect it was possible to make a rather careful examination of this point. The greatest departure of any reading, of which there were 12 in all, from linearity was 5%. The maximum load used was 250 kg./cm.^{2}

Nickel. Two specimens were used from two different sources. Both were what is known commercially as pure nickel, and were somewhat over 99% pure nickel. Both were in the form of sheet 0.012 cm. thick, and were annealed.

It is well known that the pure longitudinal effect is abnormal both with respect to sign, and departure from linearity, and hysteresis. The results obtained with any single specimen are a strong function of the past history. The disentanglement of all the complicated effects in nickel would constitute an elaborate study, and was much beyond the scope of this work. I contented myself here with merely determining whether the sign of the transverse effect would also be found to be abnormal.

The results found with the two specimens agreed in character, although the numerical agreement is not close. It is not worth while to try to reproduce here the complicated results found, but I will merely indicate the general nature.

Both purely longitudinal and geometrical mean effects show hysteresis, saturation (or more probably a maximum), and complicated dependence on the past history. The following numerical values were found for one of the samples. Under a load of 378 kg./cm.^{2} the measured purely longitudinal effect was a decrease of resistance of 0.00540. Correcting for distortion, this becomes a proportional decrease of specific resistance of 0.00570 for this load. Under the same load the geometrical mean of longitudinal and transverse effects was found to be -0.00189. Correcting for distortion this becomes -0.00200. This is less than half the pure longitudinal coefficient, so that it is evident that the transverse specific resistance increases under tension. Detailed calculation gives a proportional increase of transverse specific resistance by 0.00170 for the load of 378 kg./cm.^{2}

The chief conclusion to be drawn from these measurements on nickel is that the longitudinal and transverse coefficients are of opposite sign. Because of the abnormal character of nickel, I shall not attempt to discuss further the significance of the results.

NUMERICAL RESULTS.

In Table I are reproduced the numerical results found for all the metals except nickel. In column 1 are given the names of the metals, in column 2 the measured longitudinal effect, expressed as fractional change of resistance for a tension of 1 kg./cm.^{2}, in column 3 the meas-

TABLE I.

COLLECTED NUMERICAL RESULTS.

Metal	Longitudinal Fractional Change of Resistance, for 1 kg./cm.²		Geometrical Mean of Longitudinal and Transverse Fractional Change of Resistance for 1 kg./cm.²		Calculated Fractional Change of Transverse Resistance for 1 kg./cm.²
	Measured	Corrected for Distortion	Measured	Corrected for Distortion	
Al	$+5.8 \times 10^{-6}$ 6.9	$+4.0 \times 10^{-6}$			
Au	$+6.25 \times 10^{-6}$ 5.94 6.33	$+3.87 \times 10^{-6}$	Not greater than $+0.40 \times 10^{-6}$	$-.17 \times 10^{-6}$ (upper limit)	-4.5×10^{-6} (very uncertain)
Cu	$+3.01 \times 10^{-6}$ 3.15	$+1.75 \times 10^{-6}$	$-.1$ to $.0 \times 10^{-6}$	$-.3 \times 10^{-6}$	-2.4×10^{-6}(?)
Ag	$+5.06 \times 10^{-6}$ 5.03	$+2.86 \times 10^{-6}$	$+1.89 \times 10^{-6}$	$+1.41 \times 10^{-6}$	$-.04 \times 10^{-6}$
Pt	$+2.82 \times 10^{-6}$	$+1.78 \times 10^{-6}$	$+1.29 \times 10^{-6}$	$+1.06 \times 10^{-6}$	$+.34 \times 10^{-6}$
Pd	$+2.90 \times 10^{-6}$	$+1.37 \times 10^{-6}$	$+1.29 \times 10^{-6}$	$+.94 \times 10^{-6}$	$+.51 \times 10^{-6}$
Fe (1)	$+2.13 \times 10^{-6}$	$+1.42 \times 10^{-6}$	$+.99 \times 10^{-6}$	$+.86 \times 10^{-6}$	$+.30 \times 10^{-6}$
Fe (2)	$+2.13 \times 10^{-6}$		$+1.33 \times 10^{-6}$	$+1.20 \times 10^{-6}$	$+.78 \times 10^{-6}$

(1) Tension parallel to direction of rolling.

ured results of column 2 are corrected for the distortion by tension so as to give the change of specific resistance for a tension of 1 kg./cm.^{2} The correction by which column 3 is obtained from column 2 is $(1 + 2\sigma)/E$. Column 4 contains the measured change of resistance as given by the four point apparatus with the points in the transverse position. In column 5 the results of column 4 are corrected for the distortion produced by tension by subtracting the term σ/E, giving the geometrical mean of the effect of 1 kg./cm.^{2} on longitudinal and transverse resistance. Finally, in column 6 is given the effect of tension on specific resistance transverse to the direction of the tension. Column 6 is obtained by subtracting from column 5 half of column 3 and multiplying by 2. The essentially new results, to obtain which this investigation was made, are those contained in column 6; the results of column 3 are also necessary for any theoretical discussion.

The transverse coefficient may have either sign. The cases of negative sign are much less certain than the positive cases. The negative value for gold is very uncertain, that for silver is so small that within experimental error it might well be positive, and the value for copper is also very uncertain, as may be seen by examining the measured geometrical mean effects. It is perhaps significant that the negative coefficients are shown by the softer metals. The positive values for palladium, platinum and iron, are, however, much more certain experimentally; the effects were larger, larger stresses could be applied, and the corrections for distortions were smaller and more certain. The positive sign is the reverse of that found by Tomlinson in the two cases examined by him.

Theoretical Discussion.

In the first place, we are now in a position to answer a question of somewhat formal character raised in connection with the effect of tension on the resistance of the abnormal metals, namely whether it is possible to connect the changes of resistance directly with the changes of dimensions, changes of dimensions at right angles to the direction of current flow affecting the resistance differently from changes parallel to the flow. The coefficient of proportionality may be written k_l for parallel (longitudinal) changes, and k_t for perpendicular (transverse) changes. This involves in general writing for the change of resistance;

$$\frac{\Delta R}{R} = k_l \frac{\Delta \delta_l}{\delta_l} + k_t \left[\left(\frac{\Delta \delta_t}{\delta_t} \right)_1 + \left(\frac{\Delta \delta_t}{\delta_t} \right)_2 \right],$$

where $\Delta\delta_l/\delta_l$ denotes the strain in the direction of flow, and $\Delta\delta_t/\delta_t$ that at right angles to it, the latter appearing twice because there are two directions perpendicular to the flow. The equation as assumed involves two coefficients. The assumption may be checked by comparing the values of $\Delta R/R$ given by three independent kinds of experiment, which the data of this paper now for the first time place at our disposal, namely the changes of resistance under hydrostatic pressure (given by previous work), longitudinal changes under tension, and transverse changes under tension. For these three kinds of stress we have the following deformation:

Hydrostatic pressure

$$\frac{\Delta\delta_l}{\delta_l} = \frac{\Delta\delta_t}{\delta_t} = \frac{1}{3}\frac{1}{v}\left(\frac{\partial v}{\partial p}\right)_t p.$$

Tension, longitudinal effect

$$\frac{\Delta\delta_l}{\delta_l} = \frac{T}{E}, \quad \left(\frac{\Delta\delta_t}{\delta_t}\right)_1 = \left(\frac{\Delta\delta_t}{\delta_t}\right)_2 = -\frac{\sigma T}{E}. \qquad (T \text{ is tension.})$$

Tension, transverse effect

$$\frac{\Delta\delta_l}{\delta_l} = -\frac{\sigma T}{E}, \quad \left(\frac{\Delta\delta_t}{\delta_t}\right)_1 = \frac{T}{E}, \quad \left(\frac{\Delta\delta_t}{\delta_t}\right)_2 = -\frac{\sigma T}{E}.$$

Hence we have the following three equations for the two coefficients k_l and k_t.

Hydrostatic pressure

$$\frac{1}{p\frac{1}{3}\frac{1}{v}\left(\frac{\partial v}{\partial p}\right)_t} \cdot \frac{\Delta R}{R} = k_l + 2k_t.$$

Tension, longitudinal

$$\frac{E}{T} \cdot \frac{\Delta R}{R} = k_l - 2\sigma k_t.$$

Tension, transverse

$$\frac{E}{T} \cdot \frac{\Delta R}{R} = -\sigma k_l + k_t(1 - \sigma).$$

Putting p and T equal unity gives to $\Delta R/R$ the values of pressure or tension coefficient. The values used in checking the equations are given in Table II.

TABLE II.

Metal	Pressure Coefficient of Specific Resistance	Longitudinal Tension Coefficient of Specific Resistance	Transverse Tension Coefficient of Specific Resistance	Linear Compressibility $-\frac{1}{3}\frac{1}{v}\left(\frac{\partial v}{\partial p}\right)_t$	$\frac{1}{E}$	σ	k_l	k_t	$-\sigma k_l+(1-\sigma)k_t$ Obs.	$-\sigma k_l+(1-\sigma)k_t$ Calc.
Au	-3.33×10^{-6}	$+3.87\times10^{-6}$	-4.9×10^{-6}	$.193\times10^{-6}$	1.25×10^{-6}	.42	7.3	5.0	−3.92	−0.17
Ag	−3.83	+2.86	− .04	.329	1.26	.38	4.84	3.38	− .03	+ .26
Cu	−2.45	+1.75	−2.4	.240	.81	.34	4.20	3.00	−2.96	+ .55
Pd	−2.16	+1.37	+ .51	.173	.89	.39	4.62	3.94	+ .57	+ .60
Pt	−2.07	+1.78	+ .34	.120	.59	.39	7.0	5.1	+ .58	+ .38
Fe	−2.60	+1.42	+ .54	.196	.48	.28	5.2	4.05	+1.13	+1.45
Al	−4.87	+4.0		.448	1.42	.34				

The check is applied by comparing the observed value of $-\sigma k_l + (1 - \sigma)k_t$ with the calculated value. The check is probably as good as could be expected for Pt, Pd, and Fe, the metals for which the transverse coefficient was determined with the greatest accuracy, is not bad for Ag, considering that the calculated value is the difference of two numbers nearly equal, while for Au and Cu there is no agreement, but the experimental values were also exceedingly uncertain. I believe that these results give rather high probability to the legitimateness of the assumption made as to the connection between distortion and change of resistance, and that for the soft metals, for which the transverse effect is exceedingly difficult to measure, the transverse effect may probably be calculated from the hydrostatic pressure and longitudinal tension effects with greater accuracy than it can be measured.

The numerical values of k_l and k_t are of interest. In all cases they are positive (that is, increasing the distance between atoms increases resistance), and k_t is numerically less than k_l. It is perhaps at first unexpected that k_t is not zero. A reason is at once suggested by any free path theory of conduction, for the motion of those electrons which are in paths inclined to the e.m.f. is affected by transverse strain because the transverse strain has a component along the path.

This formal expression for the change of resistance in terms of strain allows considerable latitude in the underlying physical picture. For instance, if the free flight is from atom to atom, through the substance of the atom, as I have supposed in my theory for normal metals, the coefficient k_l would be expected to be the most important and positive, whereas if the path lies between the atoms, k_t would be most important and negative. The values found above correspond to the expectation for a normal metal.

It is one of the tasks of any complete theory to reproduce the experimental connection between these three effects, or to reproduce k_l and k_t. This I believe no theory is at present in a position to attempt, but we may get some indications of the relative magnitudes of the transverse and longitudinal effects. I have already given some discussion of this question in connection with the longitudinal tension coefficient. The conclusion there drawn was that the classical free electron theory, in which the electron is free to drift in the direction of an applied e.m.f. irrespective of the direction of the free path relative to the e.m.f., must be modified by picturing the electron as moving in something like a fixed groove, so that the velocity of drift imparted by the applied e.m.f. is directed along the original path, and is effectually produced only by the component of e.m.f. along the path.

This conclusion is much strengthened by the new evidence of this paper. There was, however, an error in the mathematical analysis of the former paper[5] which must be corrected. The problem is to integrate the effect of strain over all directions. The physical assumption made is that the change in length of any electronic free path is proportional to the geometrical elongation in that direction. In terms of my previous theory, this means a normal metal, in which there are no free paths between the atoms. It is also assumed that when a body is strained there is no redistribution of the relative number of electronic paths in different directions, nor any redistribution of velocity. This amounts to maintaining one of the fundamental assumptions of the classical theory, namely that after the free paths of the electrons have been terminated by collision, the new distribution of velocities is entirely at random. This means that the entire effect on resistance produced by strain is due to an alteration of length of free path.

Consider now the strain due to tension. In the direction of the tension there is an elongation

$$\frac{\delta l}{l}=\frac{T}{E},$$

and at right angles to the tension a contraction

$$\frac{\delta l}{l}=-\sigma\frac{T}{E}.$$

For any intermediate direction, making an angle θ with the tension, the elongation is

$$\frac{\delta l}{l}=\frac{T}{E}(\cos^2\theta-\sigma\sin^2\theta).$$

In the preceeding paper an incorrect expression, $\frac{T}{E}[1-(1+\sigma)\sin\theta]$ was used for this. We now assume that the free path, L, of an electron in any direction in the strained metal is related to the free path in the unstrained metal, L_0, by the equation

$$L=L_0\left(1+\alpha\frac{\delta l}{l}\right)=L_0\left[1+\alpha\frac{T}{E}(\cos^2\theta-\sigma\sin^2\theta)\right],$$

where α is an empirical constant, and may most easily be evaluated in terms of the pressure coefficient of resistance.

Consider now a metal to which an e.m.f. $\mathcal{E}$ is applied, and an electron moving in a free path making an angle φ with $\mathcal{E}$. The conventional free electron analysis gives for the velocity of drift in the direction of $\mathcal{E}$ the expression $\frac{1}{2}\frac{\mathcal{E}e}{m}\cdot\frac{L}{v}$, provided the force $\mathcal{E}$ produces full drift in its own direction. Here v is the normal electron velocity, supposed large compared with the velocity of drift. The result is obtained as follows. The time of free flight is L/v, the force acting is $\mathcal{E}e$, the acceleration is $\mathcal{E}e/m$, the velocity of drift at the end of free flight is $\frac{L}{v}\cdot\frac{\mathcal{E}e}{m}$, and hence the average velocity of drift is $\frac{1}{2}\frac{\mathcal{E}e}{m}\frac{L}{v}$. The total current is obtained by integrating, or

$$I = \int_0^{2\pi} d\psi \int_0^{\pi} n\frac{\mathcal{E}e}{m}\cdot\frac{L_0}{v}\sin\theta\left[1 + \alpha\frac{T}{E}(\cos^2\theta - \sigma\sin^2\theta)\right]d\theta,$$

where θ and ψ are ordinary polar coordinates, and n is the number of free electrons per unit volume per unit solid angle. The conductivity is obtained from this equation by dividing both sides by $\mathcal{E}$, or by setting $\mathcal{E}$ equal to unity.

But if the electron is not free to drift completely in the direction $\mathcal{E}$, but is constrained to maintain its original direction, then the component drift in the direction of $\mathcal{E}$ is $\frac{1}{2}\frac{\mathcal{E}e}{m}\cdot\frac{L}{v}\cdot\cos^2\varphi$, $\cos\varphi$ entering once because $\mathcal{E}$ must be resolved along the path, and again because the drift so produced must be resolved along $\mathcal{E}$. The conductivity is now

$$X = \int_0^{2\pi} d\psi \int_0^{\pi} n\frac{eL_0}{mv}\sin\theta\cos^2\varphi\left[1 + \alpha\frac{T}{E}(\cos^2\theta - \sigma\sin^2\theta)\right]d\theta.$$

Our problem is now to evaluate these expressions corresponding to different special cases. First consider the expression given by the hypothesis of unrestrained drift. It is at once evident that the conductivity is independent of the direction of $\mathcal{E}$, that is, the transverse change of resistance is the same as the longitudinal change. The reason of course, is that under the assumptions every electron makes a contribution to the total conductivity determined only by its time of free flight, and independent of the direction of $\mathcal{E}$. *This conclusion is directly contrary to the experimental evidence,* and demands that we abandon this conception of free flight. The same assumptions would also lead us to the conclusion that in a crystal the resistance must be the same in all directions, which again is directly contradicted by experiment.

For convenience of reference we record the result of carrying out the integration on the hypothesis of unrestricted drift. The result is

$$X_l = X_t \sim 1 + \frac{1}{3}\alpha\frac{T}{E}(1-2\sigma), \quad \text{unguided path.}$$

Turn now to the assumption of guided paths. We have to consider two cases, $\mathcal{E}$ along the tension, or at right angles. In the first case $\cos\varphi = \cos\theta$, and in the second $\cos\varphi = \sin\theta\cos\psi$. Hence we have respectively for the longitudinal and transverse conductivities

$$\left.\begin{aligned} X_l &\sim 1 + \frac{3}{5}\alpha\frac{T}{E}\left(1 - \frac{2\sigma}{3}\right) \\ X_t &\sim 1 + \frac{1}{5}\alpha\frac{T}{E}(1-4\sigma) \end{aligned}\right\} \text{guided path.}$$

For pure hydrostatic pressure, $L = L_0(1 + \alpha kp)$, where k is the linear compressibility. The two hypotheses give

$$Xp \sim \int_0^{2\pi} d\psi \int_0^{\pi} \sin\theta(1+\alpha kp)d\theta \sim 1 + \alpha kp, \quad \text{unguided path,}$$

$$Xp \sim \int_0^{2\pi} d\psi \int_0^{\pi} \sin\theta\cos^2\theta(1+\alpha kp)d\theta \sim 1 + \alpha kp, \quad \text{guided path.}$$

In applying this analysis, we have assumed n unaffected by the strain. But now it has been one of the assumptions of our theory that the number of free electrons *per atom* is constant. This means that in making experimental comparisons we must use the various coefficients " corrected for change of volume," as was done in the 1923 paper.

We may first apply the formulas given above to correct the previous results.[5] The ratio of pressure coefficient to longitudinal tension coefficient for the unguided path should be

$$\frac{3kE}{1-2\sigma}.$$

The sign of this does not agree with experiment, as already indicated.

On the guided path theory, the ratio is

$$\frac{5kE}{3\left(1-\frac{2\sigma}{3}\right)}.$$

Assuming for σ the mean value 1/3, this is $(15/7)kE$. The value previously found was $4.7kE$.

Using the new data, we now compare this expression with experiment. The results are given in Table III, columns 4 and 5. The agreement is not good. It is to be noticed, however, that the new value for the longitudinal tension coefficient of aluminum removes aluminum from a markedly exceptional position. In the last column of the table, is given the corrected transverse coefficient, X_t, for purposes of reference.

Returning now to the expressions for X_l and X_t on the guided path theory, we see that there is provision for a variation of sign. X_l is always positive for those normal metals whose resistance decreases under hydrostatic pressure. The reason is that $1-2\sigma/3$ is always positive, the maximum value of σ consistent with stability being 0.5. X_t on the other hand, is negative if σ is greater than 0.25. Now σ is greater than 0.25 for all the metals listed here, but nevertheless X_t is positive for the more certain ones. It is evident, therefore, that the picture of the free path given above does not correspond exactly to the facts, but it must be credited with making a step in the right direction in allowing either sign.

It is evident that a closer approach to experiment may be made by taking some sort of a mean between the guided and the unguided path formulas. The physical significance of this may be perhaps that both effects are present.

We turn now to another question connected with the tension coefficient, namely connection with crystal symmetry. The metals examined in this paper all belong to the cubic system crystallographically, and may therefore be expected to show a certain simplicity as compared with other metals. It should nevertheless be recognized that even with the cubic metals crystal structure has an effect, and that the quantities X_l and X_t may be expected to be different for samples cut in different directions from a single crystal. The reason of course is that the strain produced by tension is a function of the orientation. The values found above experimentally are mean

TABLE III.

COMPARISON OF EXPERIMENTAL AND CALCULATED RELATION BETWEEN PRESSURE AND LONGITUDINAL TENSION COEFFICIENTS OF RESISTANCE.

Metal	Pressure Coefficient of Specific Resistance Corrected by $\frac{1}{v}\left(\frac{\partial v}{\partial p}\right)_t$	Longitudinal Tension Coefficient of Specific Resistance corrected by $\frac{1-2\sigma}{E}, X_l$	Ratio, $\frac{\text{Corrected Press. Coeff.}}{\text{Corrected L. Ten. Coeff.}}$	$\frac{5}{3}\frac{kE}{1-\frac{2\sigma}{3}}$	Transverse Tension Coefficient of Specific Resistance, corrected by $\frac{1-2\sigma}{E}, X_t$
Al	-3.97×10^{-6}	$+3.55 \times 10^{-6}$	1.12	.68	——
Au	-2.75	$+3.65$	.75	.37	——
Ag	-2.83	$+2.56$	1.10	.58	$-.34 \times 10^{-6}$
Cu	-1.73	$+2.19$	.79	.64	$-.27$
Pd	-1.64	$+1.96$	.84	.44	$+.31$
Pt	-1.71	$+1.94$	.88	.46	$+.21$
Fe	-2.01	$+2.39$	.84	.84	$+.33$

values. The pressure coefficient of resistance, on the other hand, is independent of the orientation of the specimen. Experimentally we may have a manifestation of the effect in the different results obtained for iron parallel and perpendicular to the direction of rolling.

The general problem of determining the number of constants required to completely determine the resistance in all directions for any system of stress in a single crystal belonging to any crystallographic system does not seem to have been discussed.

SUMMARY

A new experimental method for measuring the change of resistance in a metal under tension when the direction of flow is at right angles to the tension (transverse coefficient) has been developed. The method gives immediately the geometrical mean of the transverse and longitudinal coefficients. An independent determination of the longitudinal coefficient permits a calculation of the transverse coefficient. Measurements are presented for 8 metals. Nickel is abnormal, as it also is with respect to the longitudinal coefficient. It is established that the signs of the two coefficients are opposite for Ni. The values found for the softer of the other metals, Al, Au, Ag, and Cu, are uncertain because of the smallness of the effect and the necessity for remaining below the elastic limit. It is probable that the transverse coefficients of some of these metals are negative. The results for Pd, Pt, and Fe are much more certain. The transverse coefficients of these are positive; this sign is the opposite of that found by the only previous observer, Tomlinson.

In the discussion it is shown that probably all changes of resistance due to deformation at constant temperature may be described in terms of two coefficients, one connecting the change of resistance with changes of dimensions at right angles to the current flow, and the other connecting with changes of dimension parallel to the flow. It is the task of theory to reproduce these two coefficients. It is pointed out that if some form of free path theory of conductivity is maintained it is not possible to suppose the electrons free to drift unrestrainedly in the direction of the applied force, for under such conditions the transverse and longitudinal coefficients are equal. It must be, therefore, that the electrons are constrained to a certain extent to move along guided paths in the metal. No theory is at present able to account satisfactorily for all the deformational effects. With regard to my own particular form of free path theory, a correction in a previous calculation of the longitudinal tension coefficient from the pressure coefficient is made by which the agreement between calculation and experiment becomes less good. It must be recognized, however, that this form of theory is so far in accord with the facts as it allows a transverse coefficient of either the same or opposite sign from that of the longitudinal coefficient.

Again it is a pleasure to acknowledge the assistance received from Mr. Walter Koenig.

The Jefferson Physical Laboratory,
Harvard University, Cambridge, Mass.

References.

[1] M. Cantone, Att. d. Lin. Rend. 6 (V) 175-182, 1897.
N. F. Smith, Phys. Rev. 28, 107-121, 1909.
H. Tomlinson, Trans. Roy. Soc. Lon., 174, 1-172, 1883.
W. E. William, Phil. Mag. 13, 635-643, 1907.
E. Zavattiero, Att. d. Lin. Rend. 29 (1), 48-54, 1920.
P. W. Bridgman, Proc. Amer. Acad., 57, 41-66, 1922; 59, 117-137 1923.
[2] Reference under 1.
[3] P. W. Bridgman, Proc. Amer. Acad. 52, 573-646, 1917; 58, 149-161, 1923.
[4] P. W. Bridgman, Proc. Amer. Acad. 58, 163-242, 1923; 59, 117-137, 1923.
[5] Second reference under 4, page 133.

THE VISCOSITY OF LIQUIDS UNDER PRESSURE

By P. W. Bridgman

Jefferson Physical Laboratory, Harvard University

Communicated September 10, 1925

A method has been developed by which the relative viscosity of liquids may be determined over a wide range of pressures at various temperatures. The method has been applied to 43 liquids in the pressure range between atmospheric and 12,000 kg./cm.^{2} and at 30° and 75°C. The results are summarized in tables 1 and 2. The results will be given in much greater detail elsewhere. In the detailed account will also be found certain data incidentally obtained, such as a number of new freezing points under pressure and the compressibility of glycerine to 12,000 kg.

In table 1 are the results for all the liquids except water. The viscosity increases so rapidly with pressure that it is convenient to tabulate log (viscosity) rather than viscosity itself. Opposite each substance in the table are two lines of data; the upper line gives data for 30°C. and the lower line for 75°C. The data tabulated are the logarithms of the ratios of the viscosity at the pressure and temperature in question to the viscosity at atmospheric pressure at 30°C. The frequently missing data in the table correspond to the fact that the substance is not liquid under the conditions in question, but is frozen by the high pressure. From the data tabulated the change of temperature coefficient of viscosity with pressure may be found. In table 2 are the data for water; the change of viscosity under pressure is comparatively small, so that it is convenient to tabulate directly relative viscosity rather than log viscosity.

Water is unique among the substances investigated in that, at low temperatures and pressures, its viscosity decreases with rising pressure instead of increasing. At low temperatures the viscosity passes through a pressure minimum and then increases. With increasing temperature the minimum flattens out, eventually disappears, and at temperatures above approximately 25° the viscosity increases with rising pressure from the beginning. This anomalous behavior of water has been already suspected from previous measurements of viscosity at low pressures. The anomaly is doubtless connected, as are many of the other anomalies of water with a high degree of association, which changes rapidly with pressure and temperature.

The viscosity of all the other 42 liquids increases uniformly with rising pressure. At low pressures the viscosity increases linearly with pressure, but beyond a pressure of the order of a thousand kilograms the rate of increase rapidly increases, so that a logarithmic curve is necessary to represent the entire course of the curve. The curve of log viscosity against pressure is at first concave toward the pressure axis, but above a few thou-

TABLE 1

LOG RELATIVE VISCOSITY AS A FUNCTION OF PRESSURE AT 30° AND 75°C.

SUBSTANCE	ATMOS.	$LOG_{10}\ \eta/\eta_0$ 2000 KG.	6000 KG.	12,000 KG.	SUBSTANCE	ATMOS.	$LOG_{10}\ \eta/\eta_0$ 2000 KG.	6000 KG.	12,000 KG.
Methyl	.000	.286	.616	.998	*i*-Pentane	.000	.559	1.175	1.947
Alcohol	9.769	.043	.334	.655		9.821	.408	.960	1.586
Ethyl	.000	.363	.829	1.390	*i*-Amyl	.000	.772		
Alcohol	9.657	.045	.473	.919	Decane	9.772	.463	1.334	
n-Propyl	.000	.494	1.131	1.915	Ethylene	.000			
Alcohol	9.598	.074	.610	1.223	Dibromide	9.756	.203		
n-Butyl	.000	.554	1.289	2.208	Ethyl	.000	.405	.837	1.323
Alcohol	9.548	.089	.690	1.396	Chloride	9.850	.285	.683	1.111
n-Amyl	.000	.607	1.448	2.495	Ethyl	.000	.387	.854	1.400
Alcohol	9.540	.105	.722	1.562	Bromide	9.806	.235	.653	1.123
i-Propyl	.000	.591	1.318	2.311	Ethyl	.000	.385	.888	1.549
Alcohol	9.505	.087	.701	1.424	Iodide	9.837	.227	.672	1.200
i-Butyl	.000	.696	1.655	2.898	Acetone	.000	.373	.804	
Alcohol	9.444	.075	.838	1.459		9.895	.245	.610	1.031
i-Amyl	.000	.686	1.636	2.952	Glycerine	.000	.497	1.346	
Alcohol	9.424	.065	.848	1.780		8.810	9.204	9.818	.628
n-Pentane	.000	.524	1.112	1.846	Ethyl	.000	.463	1.120	1.974
	9.811	.380	.908	1.493	Acetate	9.836	.253	.761	1.416
n-Hexane	.000	.561	1.224		*n*-Butyl	.000	.474	1.115	2.018
	9.803	.379	.961	1.646	Bromide	9.832	.273	.811	1.484
n-Octane	.000	.641	1.487		Cineole	.000			
	9.810	.390	1.080			9.654	.575		
Oleic Acid	.000				Diethyl-	.000	.761		
	9.419	.255			aniline	9.690	.259	1.250	
CCl_4	.000				Nitroben-	.000	.264[2]		
	9.760	.349			zene	Decomposes			
Chloroform	.000	.386	.884		Toluene	.000	.497	1.285	
	9.858	.251	.691			9.796	.267	.896	1.832
CS_2	.000	.307	.674	1.189	*o*-Xylene	.000	.577		
	9.875	.180	.527	.946		9.767	.292	1.087	
Ether	.000	.514	1.042	1.670	*m*-Xylene	.000	.529		
	9.878	.344	.806	1.311		9.799	.286	.983	
n-Amyl	.000	.708	1.685		*p*-Xylene	.000			
Ether	9.736	.364	1.125	2.007		9.797	.315		
Cyclohexane	.000				*p*-Cymene	.000	.626	1.859	
	9.723	.341[1]				9.800	.335	1.168	2.164[3]
Methyl	.000	.710	1.804		Eugenol	.000	1.081	3.007	
Cyclohexane	9.747	.434	1.335	2.582		9.429	.143	1.520	
Benzene	.000								
	9.765	.308							
Chloroben-	.000	.478	1.223						
zene	9.814	.245	.852						
Bromoben-	.000	.486							
zene	9.801	.228	.874						
Aniline	.000	.709							
	9.551	.102							

[1] At 1500 kg. [2] At 1000 kg. [3] At 10,000 kg.

sand kilograms the curve becomes nearly straight for most liquids, and for a number even reverses curvature slightly. The increase of viscosity under 12,000 varies from 10-fold for methyl alcohol to at least 10^7 fold (extrapolated) for eugenol. It thus appears that the effect of pressure on viscosity is greater than on any other physical property hitherto measured, and varies more with the nature of the substance. The temperature coefficient of viscosity increases with rising pressure; the increase is of the order of several fold under 12,000, and varies much with the liquid.

In general the results show a correlation between the increase of viscosity under pressure and the complication of the molecule. The increase is obviously less for such simple substances as methyl alcohol and CS_2, and is greatest for substances with such complicated molecules as Cineole. This effect is particularly well marked in the various series of homologous compounds, such as the alcohols.

TABLE 2

RELATIVE VISCOSITY OF WATER

PRESSURE KG./CM.^{2}	RELATIVE VISCOSITY 0°	10.3°	30°	75°
1	1.000	.779	.488	.222
500	.938	.755	.500	.230
1000	.921	.743	.514	.239
1500	.932	.745	.530	.247
2000	.957	.754	.550	.258
3000	1.024	.791	.599	.278
4000	1.111	.842	.658	.302
5000	1.218	.908	.720	.333
6000	1.347	.981	.786	.367
7000		1.064	.854	.404
8000		1.152	.923	.445
9000			.989	.494
10,000			1.058	
11,000			1.126	

Theoretically the viscosity of liquids is not well understood. It is, of course, well recognized that the mechanism of viscosity in a liquid is entirely different from that in a gas, as is shown by the fact that changes of pressure or temperature produce opposite changes in the viscosity of liquids and gases. Such theoretical discussions as there are, have placed special emphasis on the relation between viscosity and volume. Thus Phillips[1] has viscosity a function of volume only; that is, at 75° a liquid under a pressure sufficiently high to reduce its volume to the volume at atmospheric pressure at 30° would have the same viscosity as at 30° at atomspheric pressure. The data of this paper show that this relation is far from being satisfied. Brillouin[2] in his theory does not make viscosity a volume function only, but makes it vary under changes of temperature at constant volume very much less than under changes of temperature at constant pressure. He finds an expression for $(\partial\eta/\partial\tau)_v$ in terms of the compressibility and the velocity of sound in the substance. Tested against the data of this paper this expression fails by a factor which may amount to thousands of fold.

The very large pressure effects, the great variability of the pressure effect with the nature of the material, and the fact that the pressure effect is high in substances with complicated molecules suggests that there may be a very important element which is neglected in the usual explanations of the viscosity of liquids, namely an actual interlocking effect between the molecules. A complicated molecule may be thought to have geometrical projections which interfere with its free motion with respect to its neighbors. The restraint offered by such an interlocking will evidently be decreased by rising temperature, even at constant volume, and would be expected to be increased to a smaller or a very large extent (depending on the particular geometrical configuration) by decreases of volume under increasing pressure.

[1] H. P. Phillips, these PROCEEDINGS, **7**, 172–177 (1921).

[2] L. Brillouin, *Jour. Phys. Rad.* (VI), **3**, 326–340, 362–383 (1922).

THERMAL CONDUCTIVITY AND THERMO-ELECTROMOTIVE FORCE OF SINGLE METAL CRYSTALS

THERMAL CONDUCTIVITY AND THERMO-ELECTROMOTIVE FORCE OF SINGLE METAL CRYSTALS

By P. W. Bridgman

Jefferson Physical Laboratory, Harvard University

Communicated August 18, 1925

This note summarizes results which will be given in much greater detail elsewhere. The thermal conductivity at room temperature and the thermal e.m.f. in the range between room temperature and 100°C. is measured as a function of direction for single crystal rods of Bi, Zn, Cd and Sn. A few very incomplete results have also been obtained for Sb and Te, which will be found in the detailed paper.

Thermal Conductivity.—In figure 1 are plotted for Bi, Zn, Cd and Sn the reciprocals of the experimentally determined thermal conductivities against the specific electrical resistance in the same direction in which the thermal conductivity was measured. The experimental error is considerable, but nevertheless certain conclusions may be definitely drawn. In the first place there is the question whether the symmetry relations deduced by Voigt[1] for thermal conductivity are satisfied. Voigt's relations demand that the thermal conductivity have rotational symmetry about the axis for these four metals, and that the relation be such that the reciprocal of thermal conductivity be a linear function of electrical resistance. The heavy lines in figure 1 indicate what seem to me to be the best straight lines through the experimental points. It would seem that within the somewhat large experimental error the relation of Voigt is satisfied.

We have further to consider whether the Wiedemann-Franz relation between electrical resistance and thermal conductivity holds in detail for all directions in a crystal. In this case, the relation between electrical resistance and reciprocal thermal conductivity is not only linear, but the straight line must pass through the origin. In the figure the dotted line indicates what seems to me the best line through the points and the origin. It would seem that in the case of Cd and certainly in the case of Zn the experimental error is not large enough to allow a line through the origin, and that the conclusion may be drawn that the generalized Wiedemann-Franz law cannot hold for all directions in a crystal.

Thermo-electromotive Force.—The thermal e.m.f. between crystal rods of different orientations and copper was measured. For any single orientation the e.m.f. of the couple could be expressed in a two-power series in the temperature:

$$E_{\text{Cu—Crys.}} = at + \frac{b}{2}t^2.$$

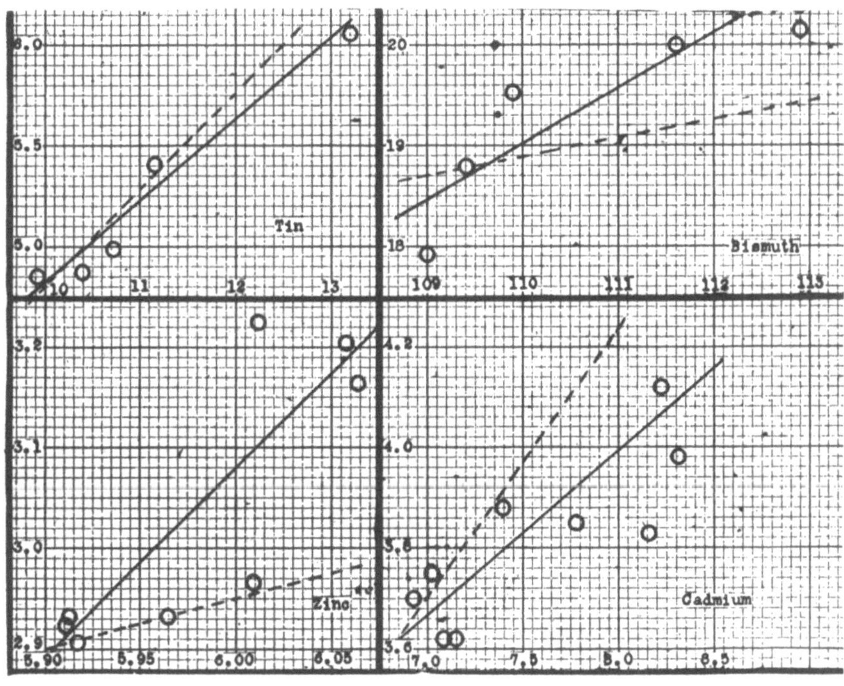

FIGURE 1

Reciprocal of the thermal conductivity (in gm. cal./cm. sec. °C.) as ordinates against specific electrical resistance in ohms per cm. cube × 10^6. The full lines show the best lines through the observed points, and the dotted lines the best lines through the origin. Voigt's symmetry relations demand that the points lie on the full lines, the generalized Wiedemann-Franz law demands that they lie on the dotted lines. The maximum possible range of resistances for directions ranging from perpendicular to parallel to the crystallographic axis is, respectively: Sn, 9.9 to 14.3; Bi, 109 to 138; Zn, 5.91 to 6.13; Cd, 6.80 to 8.30.

From this the Peltier heat when current passes from Cu to a crystal rod of the orientation in question is found by the conventional thermodynamic analysis to be

$$P_{\text{Cu—Crys.}} = (a + bt)\,\tau,$$

and the Thomson heat in the crystal when current is flowing in this particular direction is:

$$\sigma_{\text{Crys.}} - \sigma_{\text{Cu}} = b\tau.$$

Here t is ordinary Centigrade temperature and τ is absolute Centigrade temperature.

The experiments gave the coefficients a and b as a function of the direction in the crystal. The experimental accuracy is much greater here than for thermal conductivity, so that the experimental points need not

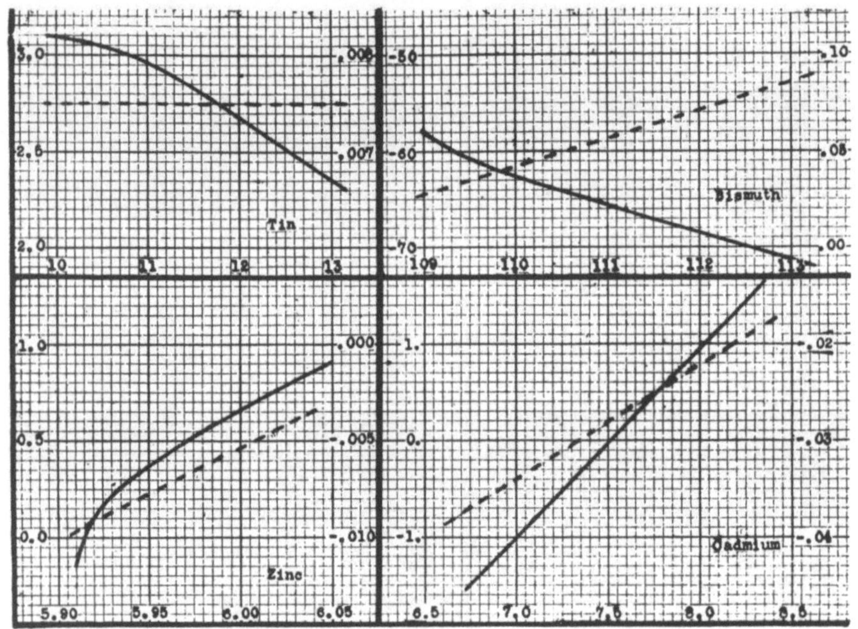

FIGURE 2

Plotted against specific electrical resistance $\times 10^6$ as abscissae are the coefficient a (Peltier heat against Cu at 0°C./273) shown as full lines with the scale at the left in microvolts, and the coefficient b (Thomson heat against copper /273) shown as the dotted line with the scale at the right in microvolts.

be given in detail. In figure 2 are plotted the coefficients a and b in microvolts as a function of the specific resistance in the direction in question. It appears from the diagram that the Thomson heat, which except for a constant is equal to b, is a linear function of the resistance, but that except in the case of Cd, a (or Peltier heat) is certainly not a linear function.

The sort of results to be expected theoretically appear to have been discussed only by Thomson[2] and Voigt.[3] Thomson's results are not given in detail, but only certain general conclusions are stated which Voigt states are in agreement with his own results. Voigt's analysis is given in complete detail. It appears on examination that Voigt's analysis applies only to the Thomson heat and not at all to the Peltier heat, which is by far the most important effect experimentally; it would appear by implication that Thomson did not consider this part of the phenomenon either. From Voigt's analysis we may conclude that the Thomson heat has the same sort of rotational symmetry as electrical resistance or reciprocal thermal conductivity. The linear relation between b and electrical resistance shown in figure 2 verifies this analysis.

We have now to consider the significance of the most important term, the Peltier heat absorbed reversibly when current passes across the surface of a crystal into an isotropic conductor. By means of very simple considerations to be given fully in the detailed paper, it may be shown that the heat absorbed at the surface of discontinuity does not depend at all on the orientation of the surface with respect to the crystal, but only on the direction of flow within the crystal before emergence. It is an immediate simple deduction that when the direction of current flow changes inside the crystal there is a reversible absorption or evolution of heat equal exactly to the difference of ordinary surface heat corresponding to the two directions of flow. This internal heat may be called the internal Peltier heat. I cannot find that it has been previously mentioned, although experimentally it may evidently be very large.

There is no simple connection between the sign of the internal Peltier heat and direction. In the case of Bi and Sn, heat is absorbed when the direction of current flow changes from parallel to the axis to perpendicular to the axis, whereas for Zn and Cd the sign is opposite.

There do not seem to be any such conditions of symmetry on the Peltier heat, either external or internal, as there are on the Thomson heat. Experimentally this conclusion is justified by the various shapes found for the curves of a against resistance. It does not even seem necessary that P have rotational symmetry. Experimentally the results have not sufficient accuracy to justify a final conclusion on this point, but the conclusion is at least justified that any departure from rotational symmetry is slight for these particular metals.

The mere existence of an internal Peltier heat would seem to have important bearings on our views of the nature of electrical conduction. I cannot see that any of our ordinary pictures of electrical conduction would lead us to expect a reversible absorption of heat on changing the direction of current flow. The dimensions of P are significant. We have here a definite absorption of heat *per electron* when the direction of flow is altered, independent of the magnitude of the current or the velocity of flow. This rules out any attempted explanation in terms of kinetic energy or momentum effects. It may be that we shall be driven to assume that the electron here functions as a dipole (magnetic), evidence for which has already been given by Compton, and that when moving in different directions in the non-isotropic force field of a crystal it assumes different orientations.

[1] W. Voigt, *Lehrbuch der Krystallphysik, Teubner, Leipzig,* 1910, p. 369.

[2] W. Thomson, *Edin. Proc.*, **3**, 255, 1854; *Edin. Trans.*, **21**, 153, 1857; *Phil. Mag.* (4) **11**, 379 and 433, 1856.

[3] W. Voigt, *l. c.*, p. 534.

LINEAR COMPRESSIBILITY OF FOURTEEN NATURAL CRYSTALS.

P. W. BRIDGMAN.

INTRODUCTION.

In this paper the methods already applied to a study of the compressibility of metals[1] are applied to a study of the compressibility of a number of natural crystals. The importance of a knowledge of the properties of single crystals requires no argument and is becoming increasingly recognized. Among the most important of these properties is the compressibility. Except for crystals belonging to the cubic system the compressibility is different in different directions, so that a complete characterization of the behavior of a non-cubic crystal under hydrostatic pressure demands a knowledge of the change of linear dimensions in several directions. This sort of information is not given by the ordinary methods of measuring compressibility. These give only the volume compressibility, that is, the average contribution to compressibility made by three mutually perpendicular directions, and are not able to analyze the total volume change into contributions made by the different directions. On the other hand, the method which I have applied to a study of the metals is precisely the method needed here, for it measures directly the change of length under hydrostatic pressure, the change of volume being found by a computation. Except for one erroneous attempt at a measurement on quartz, and a paper of my own on single metal crystals which is still in press, there are no previous measurements of the linear compressibility of crystals. Such knowledge as we have of the linear compressibility of crystals has been obtained indirectly from the elastic moduli determined by the conventional tension, bending, or torsion methods, which under practical conditions may lead to considerable error.

The range of the present measurements is that of my previous work on compressibility, that is, the pressure range is

12000 kg/cm², and measurements were made at 30° and 75°, thus giving the data for a calculation of the change with temperature of linear compressibility (or the change with pressure of the thermal expansion), and also for the change with pressure of the compressibility. It is evident that a wide pressure range is essential to obtain good values for the variation of the moduli.

THE METHOD AND THE MATERIAL.

In the following, seven cubic crystals are measured (each of these requiring a measurement in only one direction), three trigonal or hexagonal, requiring measurements in two directions (two only of these are complete, the other not being obtainable in sufficient size to give the second direction), one tetragonal (measured in one direction only), one orthorhombic, requiring three mutually perpendicular directions, and two monoclinic, of which one was measured in only one direction.

The same apparatus and the same method was used as in previous measurements of the compressibility of metals; these have already been described in sufficient detail. Only one feature of the manipulation need be reiterated here; before the two regular runs at 30° and 75° the crystal was subjected at 30° to a preliminary application of 2000 kg., releasing to zero, and then to 12000 kg. The metals had always shown minute irregularities in the zero which were smoothed out by this procedure; such a procedure seemed especially necessary with these natural crystals because of the possible effect of flaws. It was very gratifying that the irregularities of the zero proved no greater with these crystals than with the metals, with one or two exceptions to be noted later, giving good evidence of the sufficient homogeneity of the specimens. This was further checked by micrometer measurements of the length of the specimen before and after the pressure runs. There were no permanent changes of length within the error of reading, 0.0001 inch.

The measurement of linear compressibility makes more exacting demands on the size and perfection of the crystal than does the measurement of cubic compressibility. The material is measured in the form of little rods, of square section, sawn from the original crystal in the required direction. These rods should preferably be about 2.8 cm. long, in order to give the largest results obtainable with my particular apparatus. A number of specimens of this optimum length were obtained, but it is quite possible to use rods of smaller length. By the use of proper holders of iron, such as have been previously used and described in the work on metals, two shorter specimens may be piled together to give a compound specimen of the requisite length, or measurements may be made with less sensitiveness on a single shorter specimen. By the use of these same holders, specimens of anything less than the maximum diameter (6 mm.) accommodated by the apparatus may be used.

There are not many natural crystals which occur in sufficient size to give flawless pieces of the size required. The crystals used in this work were obtained from the collections of the Harvard University Museum. I am exceedingly indebted to Professor Palache for placing this material at my disposal, and for spending considerable time in going through the collections and selecting the best material. I am also indebted to Professor Palache for verifying or determining the location of the crystal axes, in order that the specimens might be cut with the proper orientation. Without this expert assistance I would not have had much confidence in my location of the axes of some of the more complicated crystals. In addition to the material described in this paper I have now on hand some fresh material from the University Museum, and I also have a number of specimens from the National Museum, which I hope to make the subject of a second paper.

The description of the individual specimens follows:

Cubic Crystals.

Fluorite, Westmoreland, N. H., transparent, of a light green color, single flawless specimen, 2.8 cm. long.

Magnetite, Mineville, N. Y., two pieces of lengths 0.9 and 1.8 cm.

Cobaltite, single piece, free from flaws, 1.7 cm. long.

Galena, Platteville, Wis., single flawless piece 1.8 cm. long.

Pyrite. It is easy to get apparently homogeneous pieces of this substance, but it was found on cutting them that there are very likely to be small cavities in the interior. Two specimens were used and two series of measurements made, the cavities visible to the eye in these specimens being small. The length of the two specimens were 2.4 and 2.8 cm. The material came from Leadville, Col.

Argentite, Freiberg, single piece, length 2.2 cm.
Sphalerite, Cananea, Mexico, single flawless piece, transparent, 2.8 cm. long.

Non-Cubic Crystals.

Tourmaline, trigonal. A single slender crystal was used cut perpendicular to the trigonal axis to a length of 2.8 cm. The compressibility was measured only parallel to the trigonal axis. The material was clear and flawless and of a light pink color.

Rutile, tetragonal. A single slender crystal was cut to a length of 2.4 cm. perpendicular to the length; the compressibility was measured only in the direction of the length of the natural crystal (tetragonal axis).

Quartz, trigonal. Two specimens are necessary, one parallel and the other in any direction perpendicular to the trigonal axis. The two rods were cut from two different crystals, each flawless, and of the full length, 2.8 cm. It was a surprise how much search was necessary to find a flawless piece large enough to give the transverse specimen.

Calcite, trigonal. Two specimens are necessary, as for quartz. The parallel and perpendicular specimens were obtained from different crystals, each clear and flawless. The specimen perpendicular to the axis was the full length, 2.8 cm., but it was not possible to find a crystal large enough to give the parallel specimen in one piece, and this was built of two, each 1.3 cm. long.

Celestite, orthorhombic, Put in Bay, Ohio. Three specimens are required, mutually perpendicular. A single large crystal was available, with numerous minute fissures, but there was a sufficient amount of unflawed material to give the requisite specimens. The specimen in the "a" direction (Dana's system of nomenclature) was single and 1.9 cm. long, that in the "b" direction was also single and 1.4 cm. long, but two pieces were necessary for the "c" direction, 0.9 and 0.8 cm. long.

Spodumene, variety Kunzite, monoclinic, San Diego County, Calif. Three specimens were used, along the three crystallographic axes. They were all cut from a single large crystal, which was not flawless, but gave a sufficient number of small flawless pieces. The crystal is columnar in structure, the "c" direction being that of the columns. The section perpendicular to the columns is that of a flattened hexagon. The "b" direction is that of the long axis of the hexagon. The "a" direction is perpendicular to "b," and makes an angle of 69.67° with the "c" direction. The "a" and the "b" specimens were both single pieces the full 2.7 cm. long, but the "c" specimen was built from two pieces 0.9 and 1.8 cm. long.

Crocoite, monoclinic, Tasmania. This was in the form of a single slender crystal, the length being parallel to the "c" direction. The specimen was cut to a length of 2.8 cm. by planes perpendicular to the natural length, and the linear compressibility in this direction only was measured.

NUMERICAL RESULTS.

The numerical results for most of these crystals can be given with sufficient detail in a table, since, with the single exception of quartz, the relation between change of length and pressure can be given by a two power series in the pressure. These results are collected in Table I. In column 1 is given the name and the crystal system of the material. In column 2, for those crystals which are not cubic, is given the orientation with respect to the crystal of the specimen measured. In columns 3 and 4 are given the two constants *a* and *b* of the two power series which gives the change of length as a function of pressure at 30°, and in columns 5 and 6 the same constants for 75°. The l_0 of the formula

$$\frac{\Delta l}{l_0} = a\,p - b\,p^2$$

is the initial length at atmospheric pressure and 30°. It is to be noticed that the fiducial length is the same at both 30° and 75°. This is different from the practice sometimes adopted, but to have used at 75° the length at atmospheric pressure and 75° as fiducial would have demanded a knowledge of the thermal expansion at atmospheric pressure, which we often do not possess. In any event the thermal expansion is small, and in most cases it makes no difference within experimental error which fiducial length is taken. In columns 7 and 8 are given the change of volume as a function of pressure at 30°, and in columns 9 and 10 at 75°, again two terms of the power development being sufficient. These volume changes are calculated from the changes of linear dimensions. The monoclinic system requires special discussion, to be given later in the text. In making the calculation it is not sufficient within experimental error to add the three changes of length, but the second degree terms in the changes of length affect

TABLE I.

Substance and System	Direction	Linear Compressibility, $\frac{\Delta l}{l_0} = ap - bp^2$ 30° a	30° b	75° a	75° b	Volume Compressibility, $\frac{\Delta V}{V_0} = ap - bp^2$ 30° a	30° b	75° a	75° b	Percentage Deviation of Single Reading from Smooth Curve 30°	75°
Fluorite, Cubic		4.019×10^{-7}	2.39×10^{-12}	4.126×10^{-7}	2.42×10^{-12}	12.06×10^{-7}	6.69×10^{-12}	12.38×10^{-7}	6.75×10^{-12}	.20 (3)	.05 (2)
Magnetite, Cubic		1.799	.70	1.792	.70	5.397	2.01	5.376	2.01	.49 (0)	.37 (1)
Cobaltite, Cubic		2.519	1.01	2.559	1.01	7.56	2.85	7.68	2.82	.20 (0)	.14 (0)
Galena, Cubic		6.122	2.48	6.311	2.78	18.37	6.33	18.93	7.14	.07 (2)	.29 (2)
Pyrite, 1 Cubic 2		2.233 2.253	.70 .70	2.241 2.236	.70 .70	6.696 6.759	1.95 1.95	6.723 6.708	1.95 1.95	.24 (0) .17 (0)	.16 (0) .19 (0)
Argentite, Cubic		Transition	above 9000	8.21	10.6			24.6	2.97		.21 (0)
Sphalerite, Cubic		4.27 Transition	.70 above 9000	4.19	.70	12.81	1.56	12.57	1.56	.46 (0)	.57 (0)
Tourmaline Hexagonal	‖ hex.* axis	4.663	2.53	4.611	2.39					.22 (1)	.30 (1)
Rutile, Tetragonal	‖ tetrag.* axis	1.038	.70	1.090	.70					.35 (0)	.56 (0)
Quartz, Trigonal	‖ trig. axis ⊥ trig. axis	See	Special	Table							
Calcite, Trigonal	‖ trig. axis ⊥ trig. axis	8.071 2.688	3.26 .70	8.157 2.770	3.51 .70	13.45	4.16	13.70	4.38	.30 (0) .27 (0)	.15 (0) .07 (0)
Celestite, ortho-rhombic	"a" "b" "c"	6.268 4.476 4.537	3.67 2.59 1.70	6.168 4.536 4.742	4.03 2.42 3.00	15.28	7.20	15.45	8.67	.18 (0) .27 (0) .55 (0)	.20 (0) .34 (0) .54 (0)
Spodumene, monoclinic	"a" "b" "c"	1.801 2.459 1.997	.70 .70 .70	1.826 2.587 1.938	.70 1.46 .70	6.26 ±	(See	text)		.69 (1) .20 (0) .10 (0)	1.48 (0) .14 (0) .11 (0)
Crocoite, monoclinic	"c"	4.978	3.78	5.102	4.22					.20 (0)	.16 (0)

*See the text for a calculation of the linear compressibility in the perpendicular direction.

the first degree terms in the change of volume. The precise relation is a matter of the most elementary algebra, and need not be written out in detail. Again, in calculating the volume compressibilities, the fiducial volume at both 30° and 75° has been taken as the atmospheric volume at 30°. In the last two columns, 11 and 12, are given figures from which an idea may be obtained of the accuracy of the measurements for each substance. The figures given are the average arithmetic deviations from a smooth curve of a single reading, expressed in percentages of the maximum effect at 12,000 kg.; in this maximum effect is included the additive correction due to the compressibility of the iron envelope. This differs from my usage in previous papers, in that formerly the additive correction due to the iron was not included in the maximum effect. But the new method of indicating the experimental irregularity gives immediately a much better idea of the probable accuracy of the final result, which is what we are mainly interested in, for obviously if the compressibility of the substance is very near to that of iron, as are a number of the crystals measured here, there may be comparatively large percentage irregularities in the *difference* of compressibility between iron and the substance, with only small percentage irregularities in the absolute compressibility. The figures given in parentheses in columns 11 and 12 are the number of readings that were discarded in taking the average. These discarded readings can usually be accounted for by some accident, such as failure of the temperature control of the bath, but it will be seen that in a very large majority of the runs it was not necessary to discard any readings at all. Each run normally consisted of fourteen readings, a series on the even thousands of kilograms with increasing pressure to 12,000 maximum, and with decreasing pressure on the odd thousands back to zero. A very rough, and probably highly unjustified application of the formulas for probable error suggests that the error in the change of length computed by the formulas at any pressure should be about one quarter of the error (deviation from a smooth curve) for a single observation.

Three of the substances measured require discussion in more detail than can be given in a table; these are: argentite, sphalerite, and quartz.

Argentite. At 30° there was a break in the curve for change of length against pressure above 9,000 kg. This was probably due to a polymorphic transition, but the precise nature of the break was not investigated in detail, because the method is not adapted to this sort of phenomenon; a polymorphic transition should be investigated by a method for measuring volume compressibility. The reality of some sort of internal change in the material was shown by the abnormally large shift of the zero after the initial application of 12,000 kg. and also by a permanent decrease of length after the run of 0.001 inch, ten times as much as the error of measurement, or the effect shown by other substances. The permanent change of length produced by the transition was responsible in some way for irregularity in the measurements below the transition. The specimen of argentite, being slender, was mounted in a sleeve of iron to maintain it in the proper direction, and after the change the specimen would no longer slip freely through the sleeve as it should. At 75° there was no transition, but the readings with increasing pressure, which were made immediately after the completion of the run with decreasing pressure at 30°, were irregular, being evidently affected by the constraint offered by the sleeve. The readings with decreasing pressure at 75° were, however, free from any such irregularity, and the constants listed in the table were obtained from this single run (8 observations in all). It will be seen that the percentage error of this run is no greater than for the normal substances. The irregularities below the transition point at 30° were not very serious, and one may certainly draw the conclusion that the compressibility at 30° is not different from that at 75° by an amount markedly different than for normal substances.

Sphalerite. Like argentite, this had a break in the curve at 30° above 9,000, which again is probably due to a polymorphic transition. The existence of this break was checked by repeating the run. At 75° the curve was perfectly continuous. The effect of the break is much less serious than it was with argentite, however, the permanent change of zero after the initial application of 12,000 being much less, and the permanent shortening of the specimen determined with the micrometer being five times less (still above experimental error). The readings below the transition at 30° were sufficiently smooth so that a good curve could be passed through them, and the constants given in the table under 30° are obtained from the curve below 9,000. At 75° there was no perceptible difference between the "up" and "down" points.

Quartz. Perfectly smooth readings were obtained with both specimens of this material; the necessity for detailed treatment arises from the fact that it was the only one of the crystals measured for which the relation between change of length and pressure could not be given within experimental error by an equation of the second degree. The results for quartz are therefore given in a special table, in which the change of length in the two directions is given at pressure intervals of 2,000 kg., and also the change of volume computed from the changes of length. The departure from a second

TABLE II.

Compressibility of Quartz.

Pressure kg/cm²	$\frac{\Delta l}{l_0}$, perpendicular to axis		$\frac{\Delta l}{l_0}$ parallel to axis		Volume change	
	30°	75°	30°	75°	30°	75°
0	0	0	0	0	0	0
2000	.00204	.00205	.00144	.00146	.00553	.00558
4000	397	396	281	284	1078	1081
6000	574	575	409	413	1565	1571
8000	742	747	528	532	2025	2040
10000	913	917	640	648	2486	2503
12000	1084	1092	748	758	2945	2970
verage rcentage viation m smooth curve	.09	.08	.24	.16	(No discards)	

degree curve of both perpendicular and parallel specimens is alike in that the maximum deviation from linearity occurs at a pressure less than the mean pressure of 6,000 kg. The departure from a second degree curve is much more marked in the specimen perpendicular to the axis; the deviation curve has its maximum at about 4,000 kg., and approaches zero at 12,000 kg. with a slight reversed curvature. The deviation from linearity is plotted in Figure 1; if the relations were of the second degree this curve should be a parabola, symmetrical about a maximum at 6,000. Within the region in which the curvature is reversed, approximately from 8,000 to 12,000, the linear compressibility of quartz in this direction *increases* with rising pressure. One is strongly reminded of the increase of compressibility of quartz glass shown over the entire pressure range, which has been described in a previous paper. It is of great interest that the volume compressibility of the quartz glass is less than that of the crystal, in spite of the fact that the volume of the glass is greater. This is doubtless connected with the abnormal increase of compressibility of quartz glass with rising pressure: at high pressures the behavior is reversed, and the glass becomes more compressible, as one would expect.

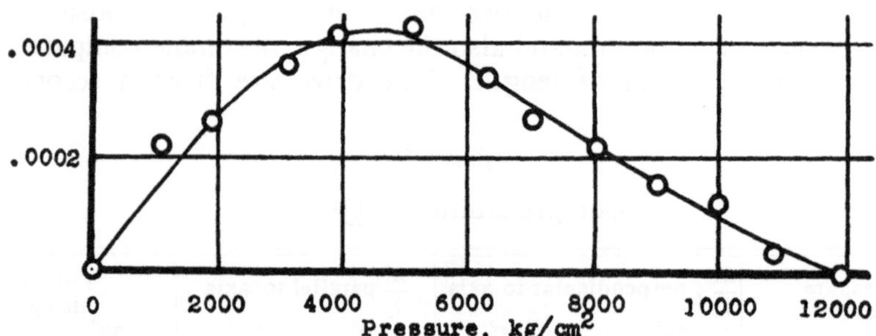

Figure 1.—The deviation from linearity, as a function of pressure, of $\Delta l/l_0$ for quartz at 30° perpendicular to the trigonal axis.

DISCUSSION OF RESULTS.

The first topic requiring discussion is a comparison of the numerical values found here with those of previous observers. By far the most extensive of the previous work has been done by Madelung and Fuchs,[2] who have determined the volume compressibility of a large number of minerals at 0°C over a pressure range of 150 kg. This range is too low to give the variation of compressibility with pressure, so that their compressibilities may be taken to be identical with the initial compressibility. The results found here may be compared with theirs by extrapolating linearly the constant a of the formulas of Table I to 0°, and decreasing their results by 1.9%, to change from dynes/cm^2 x 10^6, their unit of pressure, to kg/cm^2, the unit used in this work.

In addition to the measurements of Madelung and Fuchs there are measurements of the cubic compressibility of a few substances by the observers of the Geophysical Laboratory.[3] These are also shown in the Table. The cubic compressibility may also be found by calculation from the measurements of Voigt[4] on the elastic constants, the calculation having been made by Voigt himself. The values by these three observers are for approximately 20°.

The numerical results are summarized in Table III. Where several values are given for Madelung and Fuchs the meaning is that several different specimens were measured. It will be seen that in general the agreement of my results with those of Madelung and Fuchs is as good as could be expected. There are, however, several cases where the agreement is very poor.

TABLE III.

Comparison with other Values of Cubic Compressibility.

Substance	Initial Cubic Compressibility			
	This paper	Madelung and Fuchs	Geophysical laboratory	Voigt
Fluorite	1.19×10^{-6}	1.20×10^{-6} 1.24		1.16×10^{-6}
Magnetite	.541	.53 .56		
Galena	1.80	1.91 1.93		
Pyrite	.67	.70	.70	1.11
Argentite	2.46*	2.94		
Sphalerite	1.30	1.27 1.20		
Tourmaline		.82 .79		1.13
Rutile		.57		
Quartz	2.80	2.60 2.60 2.62 2.68	2.65	2.60
Calcite	1.33	1.30 1.32	1.36	1.51
Celestite	1.52	1.54 1.59 1.60		

*At 75°

The worst is for argentite, my compressibility at 75° being 2.46 against their 2.94 at 0°. This wide divergence is perhaps to be connected in some way with the transition which my specimen had experienced on the first application of pressure. The agreement for galena is not as good as it should be, my figure being 1.80 against 1.91 and 1.93 by them. But the most important case of disagreement is for quartz. My value, obtained by drawing a smooth curve through the high pressure points, is higher than theirs, 2.80 against an average of 2.63 for their four specimens. What is more, their value for quartz glass (not given in Table III) is higher than mine, being 2.64 against 2.56. Their quartz glass is slightly more compressible than the average of their quartz crystals, which is what would be expected in view of the difference of densities, whereas I find that the compressibility of the crystal with the smaller volume is abnormally greater than that of the glass with the larger volume. I do not believe that this abnormal behavior must necessarily be assumed to cast doubt on the validity of my results, because there is no doubt that the compressibility of quartz glass increases with increasing pressure, and there are many cases known in which the polymorphic modification of higher volume has the smaller compressibility. It may be mentioned that the difference of compressibility found by Madelung and Fuchs between crystal and glass is of a different order of magnitude from the difference of volumes; the compressibilities differ by 0.4% and the volumes by 20%.

The values of Madelung and Fuchs and of myself for the compressibility of quartz glass are both very much less than the value of Adams, Williamson, and Johnston,[3] which is 3.0×10^{-6}. The difference apparently is too large to be explained by error in the measurements, but must be due to a difference in the specimens. It would be expected that specimens with flaws would give the largest values.

With regard to the measurements of Voigt, the agreement of the calculated compressibility with the directly measured is very poor for pyrite and tourmaline. Without doubt the error arises from flaws in the specimens, which are much more harmful in the measurement of such constants as Young's modulus than in measuring compressibility.

Madelung and Fuchs give the compressibility of two crystals, tourmaline and rutile, for which I determined the compressibility in only one direction, and therefore was not able to calculate the volume compressibility. Assuming their results to be correct, which the general order of agreement for other substances seems to justify, we may combine their cubic compressibility with my linear compressibility in a single direction to get the linear compressibility in the perpendicular direction. In this way I compute for the linear compressibility of tourmaline perpendicular to the trigonal axis 1.7×10^{-7}, and for rutile perpendicular to the tetragonal axis 2.35×10^{-7}. The behavior of these two crystals are thus the reverse of each other, in that tourmaline is approximately 2.7 times more compressible along the trigonal axis than at right angles, whereas rutile is 2.3 times less compressible along the tetragonal axis than at right angles.

Next to the initial values of compressibility, the most important as well as the least affected by experimental error of the results obtained above are those for the change of compressibility with pressure. It is in the first place to be noticed that in every case the compressibility decreases with rising pressure (negative second degree term), so that these crystals in this respect behave like all other solids, liquids, and gases, with the exception of a few varieties of glass recently discussed. The magnitude of the decrease of compressibility with rising pressure is of the same order as that of the solids hitherto investigated. Of the crystals above, the minimum decrease of compressibility under 12,000 kg. is 4% by sphalerite, and the maximum is 18% in the "c" direction by crocoite. In general the percentage decrease of compressibility is greatest in those substances with the largest absolute compressibility, but the correlation is by no means close, as is instanced by the 16% decrease of compressibility of rutile parallel to the tetragonal axis, the absolute compressiblity of rutile in this direction being the least of any substance in the above list. Furthermore, in a single non-cubic crystal the percentage decrease of compressibility with pressure in different directions is by no means in the order of absolute compressibility, which means that not only do the axial ratios change under pressure, but that the relative order of the axial ratios may change under pressure. Such an inversion of axial ratios is far from taking place at any pressure within the range of these experiments, but the possibility of such an inversion has interesting theoretical implications. Consider, for instance, a tetragonal crystal at that particular pressure at which the axial ratios have become equal. At this pressure the crystal has the axial ratios and therefore the external shape of a cubic crystal, but it is not cubic, because the elastic constants, for example, do not have cubic symmetry. At this particular pressure, therefore, Neumann's law fails to hold. It is not inconceivable that the pressure of inversion might be atmospheric pressure itself, so that in such a case the crystal as observed under ordinary conditions would not satisfy Neumann's law. I do not know whether any such case has ever been observed; we would certainly expect it to be very rare.

The change with temperature of the compressibility may also be obtained from the above data, but with less certainty than the change with pressure. The normal behavior to be expected here is an increase of compressibility with rising temperature, and it is seen that this is usually the case, but by no means always. Sphalerite, tourmaline in the parallel direction, celestite in the "a" direction, and spodumene in the "c" direction show a decrease of compressibility with rising temperature which would certainly seem to be beyond experimental error. It is doubtful, however, whether the decrease found for magnetite and the second sample of pyrite can be claimed to be beyond the possibility of error.

Approximate Calculation of the Cubic Compressibility of Spodumene.

Spodumene belongs to the monoclinic system; this system differs from the others of higher symmetry in that the cubic compressibility cannot be found from the three linear compressibilities. The reason is that the linear compressibilities along three mutually perpendicular directions are necessary for the cubic compressibility, and the three *crystal* axes are not perpendicular in the monoclinic system. The three axes to which the elastic moduli are referred by Voigt are perpendicular. His Z and X axes are the "b" and "c" axes of the crystallographer. The "a" axis is in the plane perpendicular to "b," and in the case of spodumene makes an angle of 69.67° with "c." The linear compressibility along Z (or "b") is equal to the combination $s_{13} + s_{23} + s_{33}$ ($\equiv S_3$), and that along X (or "c") to the combination $s_{11} + s_{12} + s_{13}$ ($\equiv S_1$). The linear compressibility along "a" is given by

$$S_1 \cos^2 \alpha + S_2 \cos^2 \beta + S_6 \cos \alpha \cos \beta,$$

where α and β are the angles made by "a" with X and Y, S_2 is the combination $s_{12} + s_{22} + s_{23}$, and S_6 is $s_{16} + s_{26} + s_{36}$. Now the cubic compressibility is given by the combination $S_1 + S_2 + S_3$, and it cannot be found rigorously, because the three measured linear compressibilities involve the four S's in such a way that S_2 cannot be isolated from combination with S_6.

In the case of the particular substance, spodumene, of this paper, however, it is possible to make a fairly good approximation to the cubic compressibility because of the special relations for this substance. The linear compressibilities along the "a" and "c" axes are seen to be very closely alike. This suggests that in the plane perpendicular to "b" the deformation is probably nearly alike in all directions. We may take the average of the linear compressibilities in the "a" and "c" directions as the average for the plane perpendicular to "b," and obtain a rough value for the cubic compressibility by adding to twice this average compressibility the compressibility along "b" (which is nothing more than adding the "a," "b," and "c" compressibilities). This is the value given in the Table. The uncertainty is too great to justify attempting to get the temperature coefficient of the volume compressibility.

To get all the information possible from changes of length under hydrostatic pressure a fourth specimen should be cut for monoclinic crystals in the Y direction, thus permitting a complete determination of the combinations of moduli S_1, S_2, S_3, and S_6.

SUMMARY.

In this paper the linear compressibility of a number of natural crystals has been measured at 30° and 75° in the pressure range 0 to 12,000 kg./cm.^{2} The linear compressibility of the non-cubic crystals has been measured in several directions, so that the cubic compressibility may be calculated. For all substances except quartz, the relation between pressure and change of length may be represented by a second degree equation, the rate of change becoming less at the higher pressures. The compressibility also in general becomes less at the higher temperature, although there are several exceptions. For quartz, the relation is distinctly not of the second degree; in the direction perpendicular to the axis, the compressibility increases with increasing pressure above 8,000. Initially the compressibility of crystallized quartz is greater than that of quartz glass, which is not what would be expected from the volume relations, but at high pressures the relative magnitude of the compressibilities is reversed.

There seems to be no connection between the linear compressibility and the crystallographic axial ratios; the compressibility may be larger or smaller in the direction of the larger ratio. The change of dimensions under pressure may be such as to tend to make more nearly equal axial ratios initially dissimilar. This indicates the possibility of the failure of Neumann's law at some single specific pressure.

At 30° and above 9,000 kg. argentite and sphalerite experience some sort of change, probably a polymorphic transition.

I wish to acknowledge the assistance of Mr. T. E. Lane in making many of the readings.

The Jefferson Physical Laboratory,
Harvard University, Cambridge, Mass.

REFERENCES.

1. P. W. Bridgman, Amer. Acad. Arts and Sci., Proc. 58, 163-242, 1923.
2. E. Madelung und R. Fuchs, Ann. Phys. 65, 289-309, 1921.
3. L. H. Adams, E. D. Williamson and J. Johnston, Jour. Amer. Chem. Soc., 39, 41, 1919.
   L. H. Adams and E. D. Williamson, Jour. Frank. Inst. 195, 475-529, 1923.
4. W. Voigt, Krystal Physik, B. G. Teubner, Leipzig.

THE FIVE ALKALI METALS UNDER HIGH PRESSURE

THE FIVE ALKALI METALS UNDER HIGH PRESSURE

By P. W. Bridgman

Abstract

Effect of high pressure on the melting constants, electrical resistance and volume of the alkali metals.—(1) *New data for Rb and Cs*, presented in detail elsewhere, are given in three tables. Then, since similar data have previously been obtained for Li, Na and K, the results of a comparative study of all five alkali metals under pressure are given and discussed. (2) *Melting constants.* The general character of the melting phenomena is the same for the alkali metals as for other substances; the melting curve continues to rise indefinitely with increasing pressure, with neither maximum nor critical point. Above 10000 kg/cm² there is a reversal of the normal melting points of Na and K, and a moderate graphical extrapolation indicates other reversals also, so that at high pressures we may expect a complete reversal of the order of normal melting temperatures, Cs being the highest and Li the lowest. The fractional change of volume on melting is the same within 25 percent for Na, K, Rb, and Cs, as is also the latent heat of melting per gm atom; for Li the values are only about one-fourth as large. (3) *Electric resistance.* K shows the greatest relative decrease of resistance with pressure above 1400 kg/cm²; Rb occupies electrically an intermediate position between K and Cs. The effect of pressure on the resistance of Cs is unique in that there is a minimum (at 4000 kg/cm²). This minimum seems to have no connection with the crystal structure of the solid, but will probably be shown by the liquid also at higher pressures. At low pressures Cs has the maximum negative pressure coefficient of resistance and at high pressures the maximum positive coefficient of all pure metals measured. At high pressures the curve of pressure coefficient of resistance of Cs against pressure has a point of inflection which could not be predicted from the behavior at lower pressures; if other metals behave similarly above the present experimental range, it is possible that the resistance of all metals will eventually pass through a minimum. Such a change for Rb may be expected below 20000 kg/cm . This reversal of resistance may be an indication of the first beginning of a quantum break-down, suggested more strongly by the volume relations. The discontinuity of resistance at melting is approximately constant for all the alkalies. (4) *Volume.* Cs is the most compressible solid element yet measured directly. Instead of the atom the electron seems to be the significant unit of structure for volume since the volume per electron and the compressibility per electron do not vary greatly throughout the alkali series, whereas atomic volume and atomic compressibility vary greatly. At atmospheric pressure the electronic volume of K is abnormal, standing between that of Li and Na, but at high pressures there is a reversal. At high pressures the electronic compressibilities of all the alkalies are roughly equal (2.2 to 3.5×10^{-29} at 12000 kg/cm²), except for K, which is twice as compressible per electron (5.9×10^{-29}). The initial high relative compressibilities of Cs and Rb decrease rapidly with pressure, crossing that for K below 8000 kg/cm², while the high compressibility of K persists over a wide pressure range with comparatively small drop. All the evidence indicates an abnormally open electronic structure for K. (5) *Comparison with compressibility of electron gas.* Numeri-

cally the compressibility of all the alkalies is of the order of magnitude of that of a perfect gas under a high internal pressure, taking the electron as the gas unit. For K, the compressibility has passed a turning point within the experimental range, and appears to be ultimately headed for the perfect gas value. This may be the beginning of the quantum break-down, which is complete only at remotely high pressures. On the theoretical side, Schottky's theorem, in conjunction with the extrapolated behavior of compressibility, indicates the same break-down at high pressures.

Introduction

IN a paper now being published by the American Academy, data are given for several of the properties of rubidium and caesium under high pressure. Similar data have been previously published for Li, Na, and K,[1] so that the materials are now at hand for a systematic study of the effect of pressure on the properties of all the alkali metals. The data include the pressure-volume relations, and the effect of pressure on melting and electrical resistance. It is the purpose of this paper briefly to summarize the new data, and then to make a comparative study of the behavior of the five metals under pressure. Such a study may be expected to be of special significance because the effects of pressure on these metals are unusually large; Cs, for example, is by far the most compressible element solid under usual conditions, and is even more compressible over a wide range of pressure than the most compressible organic liquids, such as ether. From a study of these metals, if at all, we may expect light on the question of the ultimate behavior as pressure is increased indefinitely. Recent astronomical speculations deal with densities of 50000; such densities can exist only when the quantum structure of the atoms has entirely broken down and matter is resolved into a gas of electrons and nuclei. Some such break-down of the quantum orbits must of course take place at ordinary temperatures if the pressure is high enough; one of the questions that we shall consider here is whether the high pressure behavior of these metals suggests at all the beginning of such an atomic disintegration.

The New Data

Melting. The melting data for Rb and Cs are summarized in Table I. In broad features the behavior is the same as that found for all other substances under pressure; the curve of melting temperature against pressure is concave toward the pressure axis, the curve of the change in volume on melting, ΔV, against pressure is convex toward the pressure axis, and the latent heat is comparatively constant. The increase of the

[1] P. W. Bridgman, Phys. Rev. **3**, 154-157 (1914); **6**, 31 and 102 (1915); Amer. Acad. Proc. **56**, 67, 76, and 82 (1921).

latent heat of Cs under pressure is rather greater than usual. These new data on the most compressible metals give no reason to alter the conclusion previously drawn as to the character of the melting curve, namely that the melting curve probably rises indefinitely with continually decreasing curvature, but with neither maximum nor critical point.

TABLE I

Melting data for rubidium and caesium

Pressure (kg/cm²)	Rubidium Temp.	Rubidium ΔV (cm³/gm)	Rubidium Latent heat (kg · m/gm)	Caesium Temp.	Caesium ΔV (cm³/gm)	Caesium Latent heat (kg · m/gm)
1	38°.7	.0185	2.76	29°.7	.0136	1.65
500	48 .7	163	2.73	41 .4	118	1.68
1000	57 .9	145	2.70	51 .9	105	1.72
1500	66 .5	130	2.66	61 .4	96	1.76
2000	74 .5	119	2.67	70 .2	88	1.80
2500	82 .0	112	2.72	78 .3	83	1.88
3000	89 .1	106	2.76	85 .7	78	1.99
3500	95 .9	101	2.78	92 .4	75	2.14
4000				98 .5	72	2.29

A word should be said about the character of the melting. The melting points found for these metals are materially higher than those usually listed in the literature and the melting was very sharp, both evidence of high purity. It is more than usually difficult to obtain these metals pure; the impurity difficult to remove is not another metal, but oxide, which dissolves in the metal like a foreign metal and depresses the freezing point. Purification demands long slow distillation in high vacuum at the lowest possible temperature. The criterion of sharp freezing is one of the most delicate for high purity, and results should not be accepted for these metals unless the freezing is sharp. Thus the results recently published by Bidwell[2] on the thermo-electric properties of Rb and Cs, which he interprets as due to some sort of internal change in the metals, I believe are much more likely to be due to dissolved impurity. Bidwell's metals left much to be desired as to sharpness of freezing.

Electrical resistance. The data on electrical resistance are too numerous to summarize here, but the most important results, those for the solid metals, are summarized in Table II. It is to be emphasized that the resistance measurements on the solid were made on bare wires; internal stresses introduce too large errors if the metal is enclosed in a glass capillary as has usually been done. In addition to the data of Table II, the detailed paper gives the resistance of the liquid under pressure,

[2] C. C. Bidwell, Phys. Rev. 23, 357-376 (1924).

the temperature coefficients at atmospheric pressure, a study of the effect of pressure on the temperature coefficient of the solid, and the change with pressure of the discontinuity of resistance on melting.

By far the most important of the new results is the establishment of the minimum resistance of solid Cs. Initially, at atmospheric pressure, the resistance of Cs decreases more rapidly than does that of any other metal, but the rate of decrease rapidly diminishes, until near 4000 kg/cm^2 there is a flat minimum, and from here on the resistance increases smoothly. This is the only example of a minimum yet found.

TABLE II

Electrical resistance data for Rb and Cs

Pressure (kg/cm²)	Relative resistance of solid bare wire to values for 1 kg/cm², at 0°C		
	Rubidium		Caesium
	0°	35°	0°
1	1.000	1.205	1.000
1000	.845	.982	.863
2000	.733	.840	.779
3000	.648	.740	.729
4000	.583	.663	.709
5000	.531	.602	.713
6000	.490	.553	.728
7000	.456	.514	.756
8000	.428	.481	.794
9000	.406	.455	.833
10000	.387	.434	.879
11000	.372	.418	.931
12000	.360	.406	.994

In a previous paper I made a preliminary announcement that Cs has a second polymorphic modification at high pressures, and that the resistance of this modification increases with pressure. This was an error; the apparent discontinuity which I ascribed to a polymorphic modification turns out to be connected in some way with the separation of impurity and disappears entirely when the purity is sufficiently increased, so that we are left with the perfectly smooth passage of the resistance through a minimum.

The discontinuity of resistance between solid and liquid is approximately constant along the melting curve, decreasing slightly with rising pressure for Rb, and increasing slightly for Cs. The same approximate constancy of the ratio of resistance of solid to liquid has been found for other metals. The temperature coefficient of resistance of solid Rb decreases from 0.00481 at 1 kg/cm^2 to 0.00365 at 11000, but for solid Cs the temperature coefficient decreases from 0.00496 at 1 kg/cm^2 to 0.00366 at 6000, and then increases again to 0.00418 at 12000. This is the first example found of a temperature coefficient increasing with rising pressure; it is doubtless connected in some way with the minimum of resistance.

With the improved technique now available, results found previously for K were re-examined. Measurements on the temperature coefficient of K in glass had previously shown a very marked decrease beyond 6000 kg/cm^2,[3] a decrease shown by no other metal. The measurements were now repeated with bare wire, with the results that very little decrease was found beyond 6000, a result agreeing with that for all other metals. Doubtless the former results were due to the constraints exerted by the walls of the glass capillary.

Pressure-volume relations. The pressure-volume relations are summarized in Table III. The pressure range is here 15000 kg/cm^2 against the 12000 of my other work. The greater range was probably responsible for a somewhat lower accuracy, particularly at the upper end of the

TABLE III

Relative volumes for Rb and Cs
(Referred to values for 1 kg/cm^2)

Pressure (kg/cm^2)	Rubidium 50°C	Caesium 50°C
1	1.000	1.000
1000	.953	.941
2000	.914	.898
3000	.883	.866
4000	.857	.839
5000	.836	.816
6000	.819	.796
7000	.803	.777
8000	.789	.760
9000	.777	.745
10000	.766	.731
11000	.756	.718
12000	.746	.706
13000	.737	.694
14000	.729	.683
15000	.721	.672

range. In the detailed paper will also be found a few values for the volume of the liquid phase and also rough measurements of the thermal expansion. The data show that Cs is the most compressible of the solid elements directly measured (indirect inference, however, shows that at least solid hydrogen and helium are much more compressible,[4] but direct observations on these will not be feasible for a long time). At 12000 kg/cm^2 the volume of Cs has been reduced by very nearly the same amount as that of ether, but there is an essential difference in that the

[3] P. W. Bridgman, Amer. Acad. Proc. 56, 140 (1921).
[4] P. W. Bridgman, Amer. Acad. Proc. 59, 198 (1924).

compressibility of Cs is less affected by pressure than is that of ether. The initial compressibility of Cs is less than that of ether and its compressibility at 12000 greater, so that at higher pressures (20000 kg/cm², for example) the volume of Cs will be very materially less than that of ether.

Comparative Discussion

Melting data. In Fig. 1 are plotted in one diagram the melting points of the five alkali metals as a function of pressure. The most interesting

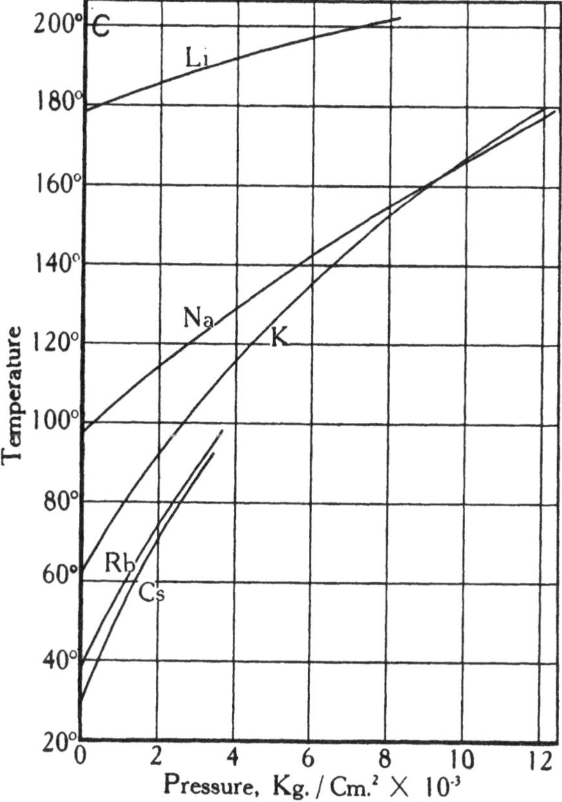

Fig. 1. The melting temperatures of the alkali metals as a function of pressure.

feature is the crossing of the curves for Na and K, so that above 10000 kg/cm² the melting point of K is higher than that of Na, reversing the normal order. The diagram suggests strongly that there are other crossings at higher pressures; it seems almost certain that K will cross Li, and highly probable that Na will also cross Li. It is more difficult to estimate the future course of the curves for Rb and Cs, because these are given over only a small range, but it is not at all improbable that these two curves will eventually cross all the others, so that at sufficiently high pressures the order of melting may be Cs, Rb, K, Na, and Li, the reverse of the usual order. This is not at all an unnatural state of affairs. At high pressures we may expect the atoms to become so packed in together that they find difficulty in melting; this packing effect will increase with the complexity of the atom, and the most complicated atom, Cs, will have the highest melting point.

The pressure at which this reversal of the ordinary melting points will be accomplished may be estimated by graphical extrapolation to be of the order of 25000 kg/cm². Such a pressure is comparatively low, much too low to produce those ultimate effects to be expected at extreme pressures when the quantum orbits break up. Under such pressures it will be shown that the volume relations of K differ from those of the other metals; it might be thought that the crossing of the melting curves is similarly an effect due to the unique behavior of K. This, however, I do not believe to be the case; the shape of the melting curves themselves, and also the fact that if melting temperature at atmospheric pressure is plotted against atomic number, K will be found to lie smoothly with the other metals, both suggest that K is not unusual with regard to melting. The peculiar volume relations of K to be discussed later do not seem strongly to affect the melting behavior.

The other melting data relating to change of volume and latent heat are also significant. It is evident that melting is an atomic phenomenon, and the five alkali metals should be compared on an atom for atom basis. Table IV shows the latent heat of melting at atmospheric pressure per gm atom. On passing from the solid to the liquid state approximately the same heat is absorbed per atom by all the alkalies (except Li). Under high pressure the latent heats vary in a complicated way, but not by large amounts, so that over the entire pressure range the generalization holds that the latent heat of melting per atom is approximately the same.

TABLE IV

Melting data at atmospheric pressure

Metal	Latent heat (kg · m/gm · atom)	Fractional change of volume
Li	49	.0060
K	215	.0231
Cs	219	.0256
Rb	236	.0284
Na	271	.0271

Table IV shows also that, except for Li, the fractional change of volume on melting is approximately the same, that is, on passing from the solid to the liquid state an atom of an alkali metal experiences a definite fractional increase of volume.

Lithium is unique in its melting phenomena. The reason doubtless is to be sought in the structure of the atom; the number of electrons is so small, only three, that the external symmetry of the atom must be much lower than that of the other four alkalies.

Electrical resistance. Unlike many of the other properties, the electrical resistance of the alkali metals does not vary smoothly in the same direction throughout the series. Thus the specific resistance at 0°C at atmospheric pressure decreases from Li to Na, and then increases again from K to Rb to Cs. The numerical values of the specific resistance are respectively 8.6×10^{-6}, 4.3, 6.1, 11.6, and 19. A somewhat more satisfactory basis for comparison is the "conductivity per atom," which is obtained by dividing the specific conductivity at a temperature one half the characteristic temperature by $N^{1/3}$, where N is the number of atoms per cc. This is discussed more in detail in my forthcoming Solvay Conference report. If "conductivity per atom" is plotted against atomic number, a comparatively smooth curve will be found with a very pronounced maximum between K and Rb. A similar reversal within the series is shown by the pressure effects. In Fig. 2 the relative resistance at 0° is plotted against pressure. It is evident that there is some sort of a turning point between K and Rb.

The resistance of most metals decreases with rising pressure, but there are seven metals whose resistance increases. Among some 50 metals studied, the series of the alkali metals includes the numerical extremes of maximum positive and maximum negative pressure coefficients of resistance.

It is of interest that the abnormal pressure effects of both Li and Cs do not seem to be associated with the crystal structure of the solid but are more intimately connected with the atoms themselves. This is shown by the fact that the resistance of Li increases under pressure in the liquid as well as in the solid state. A corresponding experiment could not be made on the resistance of liquid Cs because within the temperature range of these experiments the liquid freezes at a pressure below that of the minimum resistance. But at 97° the freezing pressure is very close to that of the minimum resistance. The resistance of solid and liquid at this temperature was plotted in one diagram as a function of pressure, multiplying the resistance of the solid by such a factor as to remove the discontinuity at the freezing point. In this way a curve was obtained passing perfectly smoothly through the melting point. The obvious inference is that the crystal structure of the solid has nothing to do with the minimum, and that at higher temperatures the resistance of the liquid as well will show a minimum.

A question of great importance for theories of resistance is whether the minimum resistance shown by Cs, the most compressible of the metals, will be shown by the other metals also at pressures high enough. This question can be answered only by some sort of extrapolation. Such an extrapolation is not easy from the curves of resistance given in

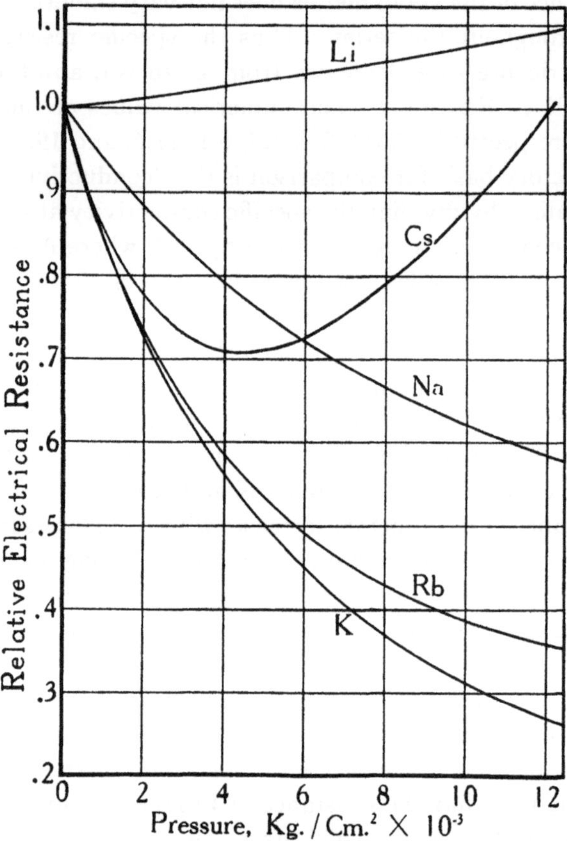

Fig. 2. The relative electrical resistances at 0°C of the alkali metals as a function of pressure.

Fig. 2. If, however, we plot $(1/w)(\partial w/\partial p)_\tau$ as a function of pressure, where w is resistance, we get something more suggestive as to the behavior at high pressure. For Cs, this quantity starts negative and crosses the axis at the pressure of the minimum, which is about 4400 kg/cm^2. The curvature against pressure is convex upward at pressures below 8000 kg/cm^2, but here there is a point of inflection, and above 8000, $(1/w)(\partial w/\partial p)_\tau$ increases at an accelerated rate with increasing pressure. If this inflection is characteristic of the behavior of other metals at still higher pressures, then the resistance of each of them must also ultimately pass through a minimum. The shape of the curve for Rb indicates that this may well occur at some pressure below 20000 kg/cm^2.

We next have to ask what is the theoretical significance of this minimum resistance. The theory of electrical resistance which I have developed[5] represents the conduction electrons as passing through the substance of the atoms and encountering resistance where they make the jump from atom to atom. A minimum resistance is to be thought of as due to the fact that as the pressure is increased a point is reached beyond which the electrons find it increasingly difficult either to make the jump from one atom to the next or to get through the atom. Such interference with the electron is not surprising in a complicated structure distorted at high pressure. The minimum resistance is perhaps in Cs to be regarded as the first shadowing forth of the ultimate disintegration of the quantum orbits under extreme pressure. It is significant that the effect is first found in the most compressible metal with a complicated atomic structure.

Volume relations. In discussing melting phenomena we have seen that the atom is the significant unit of structure, as is shown for example by the fact that the latent heat of melting per atom is nearly the same for the four heavy alkalies. This is as we would expect, since the atoms pass as a unit from the liquid to the solid phase. But now that we are to discuss volume relations, it will appear that the electron, the ultimate constituent of the atom, is the more significant unit of structure; this is suggested by the fact that the contributions made to the volume or to the compressibility per electron vary much less from one alkali to another than do the contributions per atom.

We first compare the volume per electron in the five metals as a function of pressure. The volume per electron is obtained from the volume per atom by dividing by the atomic number plus 1 (that is, we are obtaining the average volume occupied by one discrete piece of electricity, whether positive or negative, the nucleus counting as one). The volume per atom is the atomic weight multiplied by 1.66×10^{-24} (the weight of the atom of hydrogen) divided by the density. Thus we find, for example, the volume per electron of Li at atmospheric pressure is $6.94 \times 1.66 \times 10^{-24}/4 \times .53 = 5.43 \times 10^{-24}$ cm^3.

[5] P. W. Bridgman, Phys. Rev. **19**, 269-289 (1917); **17**, 161-194 (1921); **19**, 114-134 (1922); Report of the Solvay Conference, Brussels, 1924 (not yet published).

In Fig. 3 is plotted the volume per electron as a function of pressure. Li has the largest and Cs the smallest electronic volume, that is, the quantum orbits are closely packed in Cs but loosely packed in Li. The electronic volume of Li at atmospheric pressure is 2.6 fold greater than that of Cs. On the other hand, if we compared atomic volumes, we would find the volume of Li to be least, and that of Cs greatest, 5.3 times that of Li. The variability of volume throughout the series of the alkalies is thus twice as great when expressed in terms of the atom as when expressed in terms of the electron.

At atmospheric pressure the electronic volume of K is greater than that of Na, a reversal of order. At high pressures, however, the curves

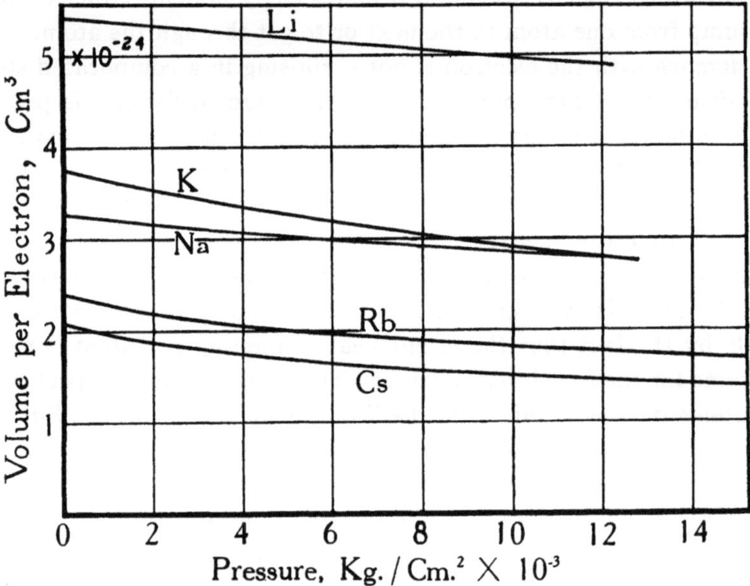

Fig. 3. The volume per electron in cm^3 of the alkali metals as a function of pressure.

for K and Na cross. This is the first of several pieces of evidence that the electronic structure of K is abnormally loose.

It is evident from Fig. 3 that the shapes of the curves of electronic volume against pressure are distinctly different for the different alkalies. This difference is brought out much more strikingly in Fig. 4 in which is plotted the instantaneous compressibility, $(1/v)(\partial v/\partial p)_\tau$, as a function of pressure. If the curves of volume against pressure were of the same shape, the curves of $(1/v)(\partial v/\partial p)_\tau$ against pressure would be the same for all the metals. This is evidently far from being the case. Not only are the absolute values of $(1/v)(\partial v/\partial p)_\tau$ quite different, but the shapes of the curves are very different. This has a bearing on speculations of Richards[6] as to the internal pressures of the alkalies. In endeavoring to find some rational basis for extrapolation, he made the assumption that the pressure-volume curve at low pressures of a comparatively incompressible alkali was of the same shape as the pressure-volume curve at high pressure of a more compressible alkali; that is, if all the ordinates of the pressure-volume curve of one metal were

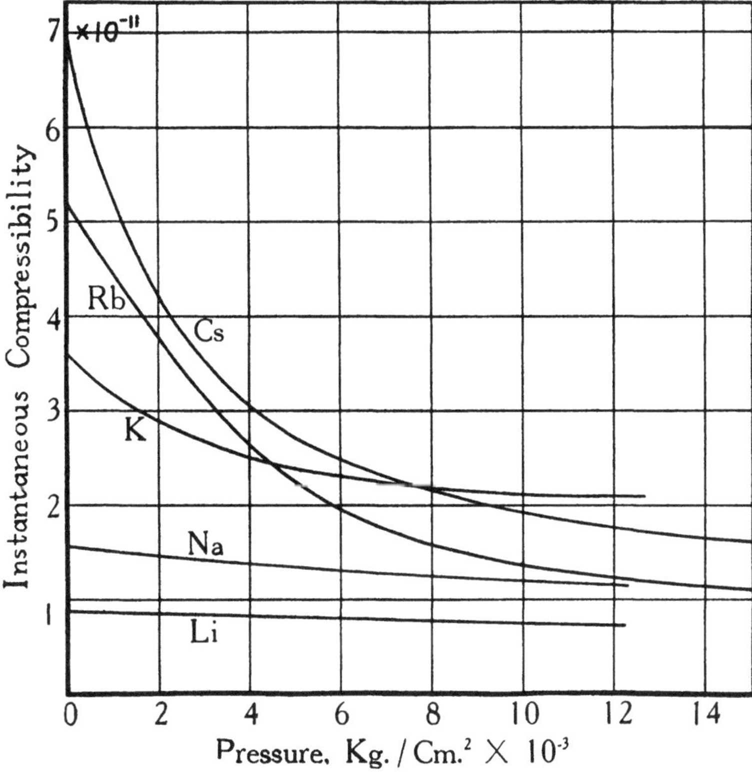

Fig. 4. The instantaneous compressibility $(1/v)(\partial v/\partial p)_T$ in Abs. C.G.S. units, of the alkali metals as a function of pressure.

multiplied by an appropriate constant and then the curve slid bodily along the pressure axis, the curve so found would coincide with the prolongation into a region as yet unreached experimentally of the pressure-volume curve of another alkali. In particular, he applied this method of extrapolation to Na and K. At the time that he did this, these data for Cs and Rb were not available. These new data show definitely that the assumed connection between the different metals

[6] T. W. Richards, Jour. Amer. Chem. Soc. **45**, 422-437 (1923).

does not exist, for if it did, the curves could not cross in the way they do. Richard's method of extrapolation can therefore be regarded as only roughly suggesting what to expect at high pressure; it is probable, however, that the approximation was sufficiently good for the purpose for which Richards used it.

The most striking feature of Fig. 4 is the crossing of the curves for Cs and Rb by that of K; this means that K retains its compressibility to an abnormal degree at high pressures. This again points to a looseness of internal structure. Another striking feature is the great progressive change in the character of the curves throughout the series. In 12000 kg/cm² the compressibility of Li drops to 0.83 of its initial value, whereas that of Cs drops to 0.23. The rapid drop of compressibility with increasing pressure is partly due, especially for the more compressible metals, to a taking out of "slack" from the structure by pressure. In the most complicated structure, that of Cs, the possibility of taking out slack is most rapidly exhausted, giving the most rapid drop of compressibility with increasing pressure.

The extent to which the "slack" in the structure is responsible for the initial behavior is further indicated by Table V which shows the compressibility per electron, that is, the actual loss in volume per electron under a pressure increase of 1 kg/cm², at 1 and 12000 kg/cm², for the 5 metals. Under 12000 the approximate constancy of the compressibility per electron for all the metals (except K) is striking, especially when compared with the initial variability. Li has the greatest compressibility per electron at high pressure, as we would expect because of

TABLE V

Volume compression per electron per kg/cm²

Metal	Electronic compression	
	Atms. pressure	12000 kg/cm²
Li	4.7×10^{-29}	3.5×10^{-29}
Na	5.1	3.2
K	13.5	5.9
Rb	12.5	2.2
Cs	14.5	2.5

its more open structure. However, the difference is not as great as might be expected. At 12000 the electrons are packed 3.3 times more loosely in Li than in Cs, but the compressibility is only 1.4 times greater. K is an outstanding exception, both in its initial compressibility (too great to lie in a smooth series with the others) and in its persistent compressibility at high pressures, about twice as high as the average of the others.

Finally we have to consider whether there is any indication within the pressure range reached experimentally of the beginning of the breakdown of the quantum structure which we may expect must be inevitable if the pressure can only be pushed high enough. An inquiry of this sort must, of course, be highly speculative. We take as our clue the recent speculations of astronomers that in the interiors of the stars matter may be disintegrated into a gas of electrons and nuclei.

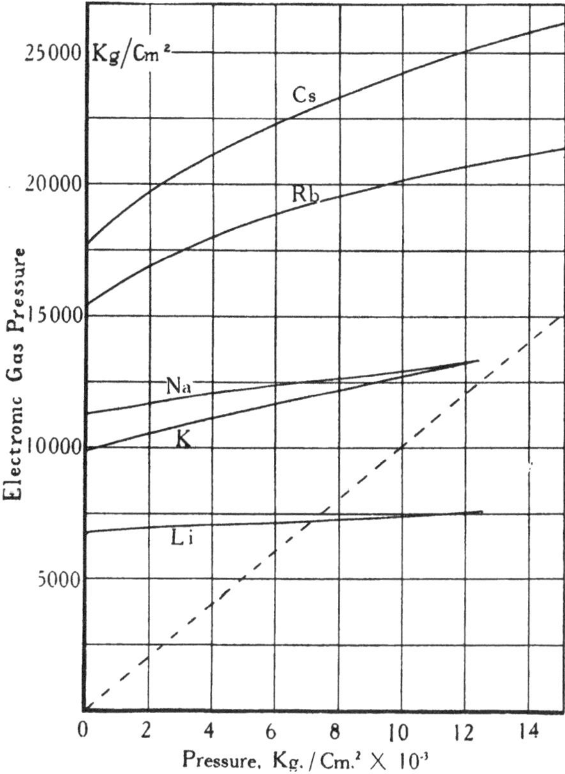

Fig. 5. The electronic gas pressures (that is, the pressures which would be exerted by the electrons and nuclei if they were in the condition of a perfect gas under the actual volume of the metal) of the alkali metals at 0°C as a function of external pressure.

It is in the first place to be noticed that the pressures which would be exerted by these alkali metals if electrons and nuclei constituted a perfect gas are not excessively high. The material for such a calculation is already at hand; we have merely to divide the reciprocal of volume per electron (number of electrons per cm³) by 2.7×10^{19}, the number of gas molecules per cm³ at 0°C at 1 atmosphere. In this way the curves of Fig. 5 were obtained. The pressures are of the order of tens of thousands of kilograms per cm², which is the order of experimentally obtainable pressures. It is therefore not extravagant to think that experimentally obtainable pressures may at least initiate the reversion to the perfect gas condition. The direction in which the quantum forces act is shown in the figure by the location of the gas pressure curve with respect to the 45° line. The curve for Li crosses this line at 7000; above 7000 the electrons in the perfect gas state would not be able to withstand the external pressure, so that if the quantum forces were removed the structure would collapse. Hence above 7000 the quantum forces distend the structure, whereas below 7000 they prevent it from flying apart. It is also evident from the figure that below 15000 kg the quantum forces in Na and K will change from an expansive to a compressive character. It seems probable that Rb and Cs will similarly change at considerably higher pressures.

When the structure breaks down completely into a perfect gas, the gas pressure curve will coincide with the 45° line, or at least will run parallel to it. The figure shows that any such state of affairs is remote, but there are two indications of the beginning of an approach to it. In the first place, it is evident that K comes nearest of the three lightest alkalies to satisfying this condition, because the slope of its gas pressure curve is closest to 45°. But if there is ever going to be coincidence or parallelism with the 45° line, it is evident that the gas pressure curve must be concave upwards, whereas at low pressures all the curves are concave downward, as is shown in the figure most unmistakably by the entire course of the curves of Rb and Cs. Now this necessary reversal of curvature is indeed shown, well beyond experimental error, both by K and Li, although the effect is not large enough to be well marked in the diagram, and probably is shown by Na also, although it is not so certain that experimental error may not enter here.

The effect is exhibited more clearly if we consider the compressibility instead of the volume. The question is whether the compressibility of the solid alkali metals has features in common with the compressibility of a perfect gas. The compressibility of a perfect gas is $(1/v)(\partial v/\partial p)_\tau = -1/p$. But when we attempt to apply this expression to our solid metals we encounter the difficulty of not knowing the internal pressure. We avoid this difficulty by eliminating p by dividing both sides of the equation by v_e, where v_e is the average volume occupied by one electron, using the gas law $pv_e = 1.35 \times 10^{-16}\tau$, and obtain:

$$\frac{1}{v_e}\left[\frac{1}{v}\left(\frac{\partial v}{\partial p}\right)_\tau\right] = \frac{-1}{1.35\times 10^{-16}\tau} = -2.4\times 10^{13} \text{ at } 300°\text{K}.$$

The material is at hand for a calculation of $(1/v_e)(1/v)(\partial v/\partial p)_\tau$, which we will call the "gas function," as a function of pressure. The results are shown in Fig. 6. It is significant that the gas function is of the order of magnitude of -2.4×10^{13}, which is the value it would have if the solid were to act like a perfect gas under some unknown but high pressure and in which the ***kinetic unit is the individual electron.*** If we assumed that the atom were the unit, the gas function would be of a different order of magnitude. The approach of the gas function to the value of a perfect gas is closest for Cs and most remote for Li. In speculating now as to the behavior at very high pressures, the shape of the curve for the gas function is significant. The curves for Cs and Rb are

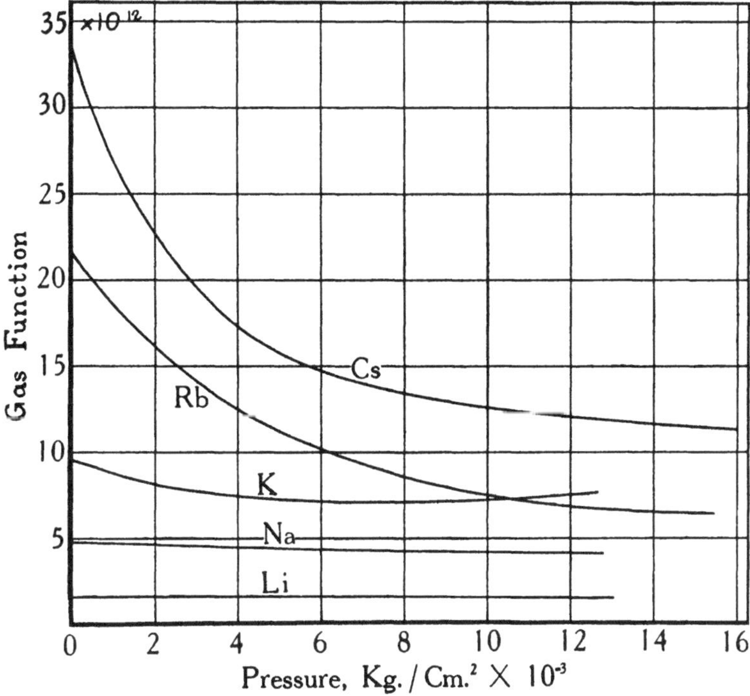

Fig. 6. The "gas function" $(1/v)(1/v)(\partial v/\partial p)_T$ of the alkali metals as a function of external pressure. If the metals acted like a perfect gas of electrons and nuclei, the gas function would have the constant value 24×10^{12}.

still receding from the perfect gas value at the highest pressures reached experimentally, whereas for K a reversal has been reached at 7000 or 8000 kg/cm^2, and above this pressure the gas function of K has started to approach more closely to the value for a perfect gas. It is tempting to see in this the beginning of the quantum breakdown.

Apart from this experimental evidence, we may obtain some theoretical light on the question of the ultimate behavior under exceedingly high pressures from a theorem due to Schottky.[7] He finds for any system in which there are forces between the elements given by the ordinary inverse square laws of electromagnetism and in which the elements move under these forces as if they had mechanical mass, whether or not the motion is in addition subjected to quantum forces, the following equations:

$$\left(\frac{\partial \bar{L}}{\partial p}\right)_\tau = 3v + \tau\left(\frac{\partial v}{\partial \tau}\right)_p + 4p\left(\frac{\partial v}{\partial p}\right)$$

$$\left(\frac{\partial \bar{E}}{\partial p}\right)_\tau = -3v - 2\tau\left(\frac{\partial v}{\partial \tau}\right)_p - 5p\left(\frac{\partial v}{\partial p}\right)_\tau ,$$

where $\bar{L}$ is the average internal kinetic energy of motion of the elements, and $\bar{E}$ is the average electromagnetic potential energy. The total internal energy $\bar{U}$ of the system is of course given by $\bar{U} = E + \bar{L}$. The equations in the form above were not given by Schottky, but may be deduced from equations given by him by using the ordinary thermodynamic relation

$$\left(\frac{\partial \bar{U}}{\partial p}\right)_\tau = -\tau\left(\frac{\partial v}{\partial \tau}\right)_p - p\left(\frac{\partial v}{\partial p}\right)_\tau$$

At 0°K we have rigorously

$$\left(\frac{\partial \bar{L}}{\partial p}\right)_\tau = 3v + 4p\left(\frac{\partial v}{\partial p}\right)_\tau$$

$$\left(\frac{\partial \bar{E}}{\partial p}\right)_\tau = -3v - 5p\left(\frac{\partial v}{\partial p}\right)_\tau$$

These relations are also approximately true at other temperatures, for on substituting numerical values for ordinary solid metals it appears that the term $\tau(\partial v/\partial \tau)_p$ is unimportant. Now in the equations as last given the term v is positive and $p(\partial v/\partial p)_\tau$ is negative. At zero pressure only the term v contributes, and $\bar{L}$ increases with increasing pressure and $\bar{E}$ diminishes. But with increasing pressure v diminishes and $p(\partial v/\partial p)_\tau$ increases. Will the second term eventually become larger than the first? We attempt to answer this by plotting $-(p/v)(\partial v/\partial p)_\tau$ against pressure for the five alkali metals. Fig. 7 shows the result. If $-(p/v)(\partial v/\partial p)_\tau$ becomes equal to 0.6, the direction of variation of $\bar{E}$ changes, and if it reaches 0.75, the direction of variation of $\bar{L}$ changes.

[7] W. Schottky, Phys. Zeits. **21**, 232 (1920).

Again K is the most suggestive metal. For it, $(p/v)(\partial v/\partial p)_\tau$ is nearly linear against pressure; as a matter of fact at low pressures the curve is slightly concave toward the pressure axis, but at high pressures it becomes slightly convex. $(p/v)(\partial v/\partial p)_\tau$ therefore rises at an accelerated rate as pressure increases, and the critical value 0.6 will be reached at a pressure near 30000 kg. The curves for the other metals are also flattening out at a rapid rate, so that is it probable that for all the alkalies the critical value will be eventually exceeded.

What now is the significance of the change of L and E? Initially the state of affairs is plain. In general an increase of L means an increase of rapidity of motion of the electrons in their orbits, which means an

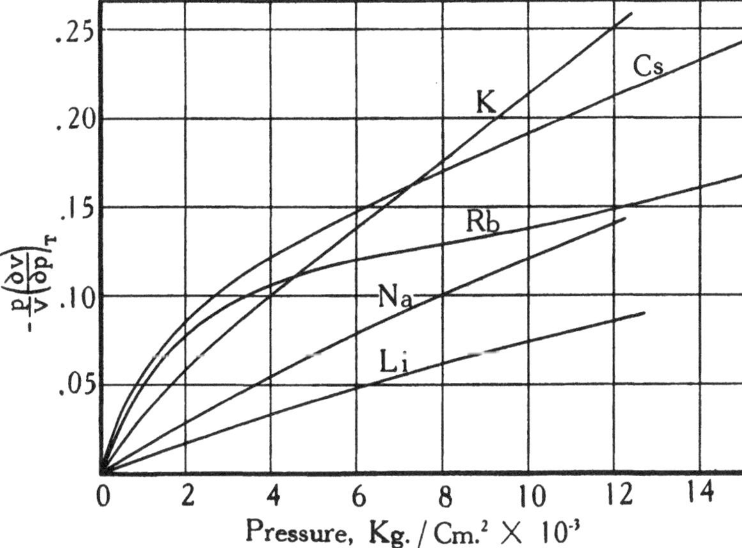

Fig. 7. The function $-(p/v)(\partial v/\partial p)_\tau$ of the alkali metals as a function of external pressure.

approach on the average to the attracting nuclei. The same approach to the nuclei which increases L brings the electrons into stronger parts of the fields of the nuclei and thus at the same time provides for the decrease of E. This approach to the nuclei may be thought of as signifying a compression of the atoms under pressure, as Schottky suggested, or it may be thought of as follows. As we proceed through the series of the chemical elements from simpler to more complex, it is well recognized that the electron orbits penetrate more and more deeply toward the nucleus, and at the same time become more eccentric. The same sort of thing may very naturally be produced also when we push the electron orbits into closer juxtaposition by increasing pressure, so that the initial increase of L and decrease of E with increasing pressure is accounted for by a deeper penetration of the orbits toward the nuclei. But now at higher pressures, how are we to account for a decrease of L and an increase of E, if according to our argument this means orbits described at greater distances from the nuclei? The difficulty is overcome if the electron orbits break down, allowing an approach to a more uniform distribution in space of electrons and nuclei. The electrons now do not penetrate so closely to the attracting nuclei, so that on the average L is smaller and at the same time E increases, both in spite of a continued decrease of total volume. But this sort of break-down means an approach to the condition of a gas. It thus appears that the dilemma to which Schottky's theorem, in conjunction with the experimental behavior of the compressibility of the alkali metals, forces us, is resolved by a break-down of the quantum orbits at sufficiently high pressures, and an approach to the condition of a gas.

Jefferson Physical Laboratory,
Harvard University,
October 4, 1925.

[Values of compression for Na and K are incorrect owing to an error in changing from $\Delta l/l$ to $\Delta V/V$.]

THE EFFECT OF PRESSURE ON THE VISCOSITY OF FORTY-THREE PURE LIQUIDS.

By P. W. Bridgman.

Presented October 14, 1925. Received November 16, 1925.

TABLE OF CONTENTS.

Introduction.

Among the non-thermodynamic properties of a liquid, the viscosity is one of those which might be expected to be of most significance, and is one which has been most studied experimentally. There are various lines of evidence which suggest that the viscosity of a liquid is more intimately connected with the specific properties of the molecules than is the viscosity of a gas. It is known that the mechanism of viscous resistance in a liquid must be different from that in a gas, but there are no theories at present capable of giving an adequate account of the phenomena. It is to be expected that the information given by the change of viscosity produced by pressure will be especially significant, since under pressure the molecules are brought into closer contact, and any peculiarities in their specific relations accentuated. There have been, however, only a few measurements of the effect of pressure on the viscosity of pure liquids, and these over only a comparatively small range of pressure.[1] The effect of pressure on the viscosity of a number of lubricating oils has been recently determined over a range sufficient to change the viscosity many fold,[2] but oils are not simple substances, and the results obtained with them cannot be easily interpreted, although information of this character is evidently of much importance technically.

In this paper the effect of pressure has been studied on some forty-three pure liquids, most of them organic, but including water. In addition, petroleum-ether and kerosene have been studied because of their interest in connection with high pressure apparatus. The method used has demanded that all the liquids be electrical non-conductors. The range of the measurements is 12000 kg./cm.[2] (unless the liquid freezes at a lower pressure, as is often the case), and measurements have been made at 30° and 75°, thus giving both the effects of pressure and temperature. It may be said in general that the effects of pressure on viscosity are greater than on any other physical property hitherto measured, and vary very widely with the nature of the liquid. The increase of viscosity produced by 12000 kg. varies from two or three fold to millions of fold for the liquids investigated here, whereas such properties as the volume decrease under 12000 seldom vary by as much as a factor of two from substance to substance.

The Method and the Apparatus.

None of the methods previously used for the measurement of the effect of pressure on viscosity was adapted to the greater pressure range of this work, in general the apparatus being too bulky or too cumbersome in operation, requiring for example that the pressure chamber be reopened and the apparatus refilled for every new determination at a new pressure. The apparatus adopted does not give the absolute viscosity, but does give the relative viscosity, which is all we are interested in here, and is shown in Figure 1. The general idea of the method is very simple; in a steel cylinder of approximately 6 mm. internal diameter, filled with the liquid under investigation, there is a steel cylindrical weight separated from the walls of the cylinder by a narrow annular space. The time of vertical fall of the weight from one end of the cylinder to the other is determined; the time is a measure of the viscosity.

The pressure producing apparatus is so mounted and connected with the viscosity cylinder that it may be rotated through 180°, so that after the time of fall of the weight has been determined the apparatus may be reversed, and the time of fall in the opposite direction determined. This may be repeated as often as desired, allowing an indefinite number of readings at any pressure with the same set-up.

The time of fall is determined electrically. At each end of the cylinder there is an electrically insulated terminal D against which the weight rests at the end of its fall. In this position electrical connection is made from the walls of the cylinder through the weight to the terminal, and the completion of this connection may be made to operate a suitable timing device. It is necessary, therefore, that the weight be in close enough contact with the cylinder to permit

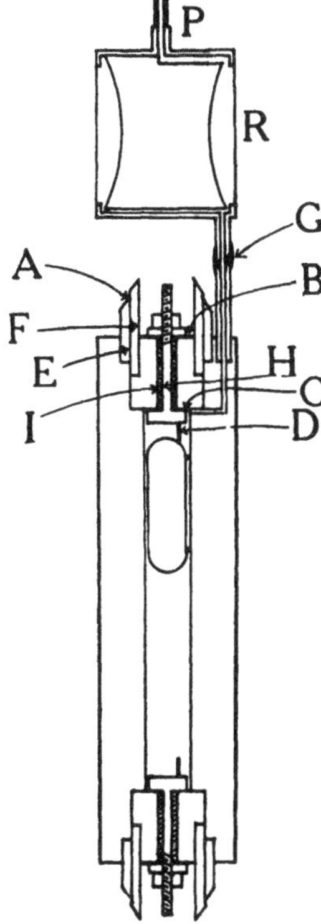

FIGURE 1. Section of viscosity apparatus.

passage of current, but at the same time be capable of free fall. With the electrical arrangements used it was necessary that the contact

FIGURE 2. Detail of falling weight of viscosity apparatus.

between weight and cylinder be closer than 0.0025 cm. The weight, Figure 2, was provided with three small projecting lugs* at the top

* Shown as two in the figure for convenience of drawing.

and bottom ends, which act as guides to keep the weight concentric as it falls, and through which electrical contact is made with the cylinder. By putting the stop D against which the weight rests at the end of its fall off center, it was ensured that the lugs on one side of the weight were pressed more firmly against the cylinder, thus making better contact. In order to keep the time of fall within reasonable limits, different weights were used according to the absolute viscosity of the liquid, the annular space between weight and cylinder varying between 0.0125 cm. for water to 0.075 cm. for the more viscous liquids such as glycerine. There was further provision for variability in the viscosity of the liquid by making the weight hollow (see detail in Figure 2), and changing its total weight by placing within the cavity appropriate weights of tungsten or gold. The hollow weight was made in two pieces, of the same material as the cylinder. The main body of the weight, with the lugs, was milled from one piece, and a cap to close the cavity was made a tight push fit for the body of the weight. The external shape of the cap was made as much like that of the other end as possible, but the two ends were never exactly alike, because for one thing it was necessary to provide a small hole in one end by which the interior might be filled. The times of fall in the two directions were, therefore, never exactly the same; the difference was not often more than one per cent.

The insulating plugs at the two ends of the cylinder are shown in the figure. It is the function of these plugs not only to provide a terminal electrically insulated from the cylinder, but to prevent the liquid under investigation in the cylinder from mixing with the pump liquid in which the cylinder is immersed. The central stem of the the plug, H, is of brass, with the projecting terminal of platinum, D, insulated by a sleeve of pipestone, I, and flat washers of mica B and C. The mechanical separation of the liquid inside from that outside is provided entirely by the washer C. The seats on which the mica washers rest are ground flat; the washers are made as tight as possible initially with the nut on the stem, and become still tighter under pressure because of the differential compressibility of the brass stem and the steel of which the surrounding parts of the apparatus are made. The insulating plug was soldered into the end of the cylinder; in this way the liquid under investigation came in contact only with metals or mica, and was kept pure. The soldering and unsoldering were facilitated by two german silver sleeves E and F permanently soldered to the cylinder and the plug respectively. The soldered connection between cylinder and plug was made and broken at A. This could be easily done since in the first place german silver is easy to solder, and in the second place is a poor conductor of heat, so that the soldering could be done rapidly with a soldering copper without seriously heating the rest of the apparatus. The unsoldering was facilitated by a special soldering copper, of cylindrical shape to fit the sleeve.

The liquid under investigation, which fills the inside of the cylinder, must be kept from contact with the surrounding liquid by which pressure is transmitted, but at the same time must freely receive the pressure. This was done by means of a reservoir, R, of very thin pure tin, attached to the upper end of the cylinder as shown. The pressure is transmitted from the liquid on the outside to that on the inside through the tin walls of the reservoir, which offer so little resistance that the pressures outside and inside are the same within less than 1 kg./cm.[2] This device of the reservoir has already been used in connection with measurements of the effect of pressure on the thermal conductivity of liquids, and will be found described in greater detail in that paper.[3] The connection from the reservoir to the cylinder was through a tube of german silver; this connection, G, was resoldered and a new reservoir made at each new filling of the apparatus with a new liquid. The apparatus was filled through the fine tube, P, of german silver at the top of the reservoir. The filling liquid was placed in a glass thistle connected in such a way with rubber tubing to the pipe at the top of the reservoir that practically no liquid came in contact with the rubber, and then the whole combination was placed bodily under the receiver of an air pump. It was filled by exhausting the receiver, the air inside the cylinder bubbling out through the liquid in the thistle, and then readmitting air to the receiver, driving the liquid into the reservoir and the cylinder. The receiver was exhausted to such a point that the liquid boiled vigorously, and this was repeated a number of times, so that the liquid was nearly air free. The reservoir was then closed by soldering into the top of the pipe, P, a brass plug, the outside of the pipe being kept cool in a simple way with water, so that the liquid inside could not boil during the soldering. The cylinder and the reservoir, filled with the liquid, were now attached with screws to an insulating plug, not shown, but essentially like the one described in the paper on thermal conductivity. Soldered connection was made from the insulated terminals of the cylinder to the insulated terminals of the plug, and the plug, with cylinder and reservoir as one self-contained unit, were screwed into the large pressure cylinder.

The large pressure cylinder was connected with a pipe to the pressure generating apparatus and the pressure gauge, which were the same as that used in previous work. They were now mounted horizontally, instead of vertically, and were so arranged that they could be rotated about the connecting pipe as an axis. During this rotation the hydraulic press with which pressure was produced had to be disconnected from the hand pumps which were the source of pressure. This was made possible by valves mounted to rotate with the press, so that the pressure produced by the pump could be maintained after the transmitting pipe has been disconnected. The large pressure cylinder which contained the viscosity apparatus was T-shaped; there was a side connection at which entered the pipe connecting to the pressure producing apparatus, which acted as the long arm of the *T*, which was horizontal and about which rotation took place. The main body of the cylinder acted as the cross arm of the T, and by rotation was changed from one vertical position to the inverse. The falling weight within the viscosity apparatus fell along the axis of the cross arm of the T. The viscosity cylinder was kept at constant temperature by the conventional stirred bath and regulators; the horizontal connecting pipe entered the bath through a simple stuffing box.

The timing apparatus was a comparatively simple affair. A Warren clock, which is essentially a small motor running synchronously with the commercial 60 cycle alternating current source, was connected to the second hand shaft of an ordinary clock movement. The Warren clock made one revolution per second, so that the clock movement rotated at 60 times the normal speed. The circuit of the Warren clock was made and broken through a relay operated by the circuit in which was the falling weight of the viscosity apparatus. Since the Warren clock did not stop immediately when the circuit was broken, it was necessary to provide a mechanical stop; this was done by an arm with a pointed end pulled by a spring into the teeth of one of the gears of the clock movement. This was operated by a relay in series with the other relay. To provide against too great mechanical shock to the clock movement on stopping, there was a spring in the connection between the Warren clock and the clock movement. Since there were 60 teeth in the gear used for stopping purposes, the smallest time interval capable of being distinguished was 1/60 second, so that the necessary error in the determination of a single time interval was not over 1/120 second.

Considerable difficulty was found in getting a proper source of current to operate the relay through the contact made by the falling weight. If the weight is to fall freely, there is considerable necessary contact resistance between the lugs and the walls of the cylinder. Direct current is not suitable, for if the voltage is high enough to jump the gap an arc follows, which may decompose the liquid or make the weight stick to the walls of the cylinder. After some trial, the alternating current delivered by a small bell ringing magneto of the type used in insulation testing was found suitable. The magneto was driven by a variable speed motor, connected through a belt and cone pulleys in such a way as to give a wide variation of speed. The speed was chosen as low as possible to give positive contacts, thus avoiding as much as possible contamination of the liquid by sparking. The use of such a source made it necessary to use in operating the Warren clock a relay sensitive to alternating current; an old telephone relay was found suitable.

There were various possible sources of error in the starting and stopping of the clock and the operation of the various relays, so that a calibration of the timing device was necessary. This was done by means of a second Warren clock, also rotating at one revolution per second. To the axle of this was attached a copper disc, one quarter of the periphery of which was insulated from the rest. Contacts made from the axle to the periphery gave a current which during every second was interrupted for exactly one-quarter second. The calibrating device could be connected to the timing device instead of the weight of the viscosity apparatus, and thus the clock tested by measuring known intervals of ¼ second. The calibration was made by the measurement of 50 such intervals. The error in the timing apparatus depends on the speed of the magneto; in by far the majority of cases the error in the average of 50 readings was only a few thousandths of a second, and in the worst case it never exceeded 0.050 second. Of course the error in a single interval could be considerably more. Since the error in the timing device is an absolute and not a relative matter, it is much more serious at the short intervals. The procedure was adopted of calibrating the clock with 50 ¼ second intervals after every readjustment of pressure when the absolute time of fall was less than 5 seconds, and at less frequent intervals for longer falls. If the time of fall was less than 5 seconds, 50 fall times of the weight were measured. The error in the finally corrected time of fall could not have been more than 0.002 or 0.003 seconds.

In addition to the corrections of the timing device there were a number of other corrections. What is desired is to so correct the measured time of fall that the finally corrected value shall be proportional to the viscosity of the liquid at different pressures and temperatures. This demands, among other things, that the weight be falling for the entire time under a constant force with a constant velocity. Under the actual conditions the weight did not fall for the entire distance under a constant force, because while the cylinder was being rotated to start a new fall the weight started to fall before a completely vertical postition was reached, and therefore was falling under a diminished effective gravity until the completely vertical position was reached. The correction for this effect could be determined by measuring the time interval between the weight starting to fall and the cylinder reaching a completely vertical position. This was done electrically in a simple way by measuring, with the same timing apparatus as was used to time the falling weight, the interval between breaking of contact by the weight as it began to fall, and the reaching of a vertical position, which could be determined by making the rotating part come against a stop at the end of its rotation. Since the rate of rotation during the last part of the 180° inversion was approximately constant, the time of fall was corrected by subtracting from the measured time interval one-half the interval between starting to fall and reaching the completely vertical position. This correction so determined was 0.05 second.

Another correction is that due to the inertia effect at the beginning of fall. Because of the inertia of the weight a certain amount of time is needed to build up the final velocity. The correction may be determined theoretically from the equation of motion of the falling weight in a vscous liquid, and with the dimensions used was very small. The distance of free fall was 3 cm. In only one or two cases in these experiments was a time of fall used of less than one second. The correction for the acceleration effect is — 0.003 second for a fall time of 1 second, — 0.0015 at 2 seconds, and inappreciable above this.

The corrections so far discussed are important only at short times of fall. The largest correction is for the change of buoyancy of the liquid on the falling weight as the density of the liquid changes under pressure or with temperature. This correction is a percentage correction and is naturally greatest at the highest pressures. It may be shown that the exact expression for the correction, which is somewhat more complicated, reduces to the simpler form:

$$\frac{\rho_p - \rho_0}{\rho_0}\left[\frac{\dfrac{W_1}{D_1}+\dfrac{W_2}{D_2}}{\dfrac{W_1+W_2}{\rho_0}-\left(\dfrac{W_1}{D_1}+\dfrac{W_2}{D_2}\right)}\right],$$

with an error of less than 1% of the correction. This is accurate enough, since the total correction does not rise to over 5%. In this formula W_1 is the weight of the hollow steel shell, and D_1 the density of the steel, W_2 the weight and D_2 the density of the core of the shell (either of tungsten or gold), ρ_0 is the atmospheric density of the liquid and ρ_p its density under pressure. The compressibility of the metal of the falling weight does not enter within the limits of error.

In the following, the relative viscosities for any liquid are expressed in terms of the viscosity at 30° at atmospheric pressure as unity. The results of the run at 75° were corrected by a double application of the above formula; a first application corrected the viscosity at atmospheric pressure at 75° for the change in buoyancy of the liquid due to thermal expansion at atmospheric pressure (this correction was very small), and a second application then corrected the readings obtained under pressure for the change of density of the liquid produced by pressure at 75°. Within the limits of error, the percentage correction for the pressure effect is the same at 30° and 75°.

The compressibility of about a dozen of the liquids of this investigation have already been measured,[4] so that for these liquids the pressure correction could be exactly computed. The compressibility of the other liquids has not been measured to high pressures, so that for them some sort of estimate had to be made of the correction. This estimate could be made with considerable confidence because it is known that $(\rho_p - \rho_0)/\rho_0$ as a function of pressure does not vary greatly from liquid to liquid among such organic liquids as those used here. The procedure in calculating the correction for a liquid whose compressibility had not been measured was to substitute into the formula above the value of ρ_0 of the liquid (obtained from the tables) and to use for $(\rho_p - \rho_0)/\rho_0$ the corresponding value for that one of the twelve liquids whose compressibility had been measured which was most nearly like the liquid in question. In selecting the most similar liquid, weight was given both to the chemical constitution and to the compressibility at low pressures in those cases in which this had been determined by other observers. It was not difficult to select a similar liquid in practically every case except that of glycerine. The low pressure compressibility of glycerine is so much less than that of any other of the liquids that an estimate of the correction did not seem safe, and the compressibility was therefore determined by special experiment, and will be described in detail later.

The correction for buoyancy rises to as much as 5% in only a few extreme cases. It was found by trial that the correction could be determined with an error of less than 0.1% in the final result over the entire range by actually computing it at only 2000, 6000, and 12000 kg., and then passing a smooth curve through the four points (the correction is zero at atmospheric pressure). It must be recognized that in some cases the error in the estimated correction may, because of an incorrect assumption about the compressibility, be over 0.1%, although I do not believe that this is often the case. For this reason, the corrections assumed for buoyancy are listed with the final data for each liquid together with the falling weights, so that at some future time, when the compressibility has been exactly determined, a more exact correction may be applied if necessary.

In computing the temperature correction for buoyancy at atmospheric pressure the effect of the thermal expansion of the steel is negligible, and only the thermal expansion of the liquid affects the result. The correction tends to become large when the density of the liquid is high, and small when the falling weight is of high density, as when a gold or tungsten core were used. The volume expansion of most of the liquids of this paper, with the exception of water, is about 5% between 30° and 75°, and this value was assumed for all except water. The correction was calculated by combining this value for the expansion with the particular weight and initial density of each liquid. The correction is 0.5% for nearly all the liquids, but for a substance with so extreme a density as ethylene dibromide rose to a maximum of 1.8%.

Finally a correction has to be applied for the change of dimensions of the apparatus under pressure. Since the cylinder and the weight were made of the same material (it is only the change of the *external* shape of the falling weight which is effective and so the core need not be considered), and since the pressure is hydrostatic, only the absolute dimensions of the apparatus change under pressure but the geometrical proportions remain unaltered. The way in which the time of fall varies with the absolute dimensions may be found by a dimensional argument. The time of fall, t, is a function of the viscosity of the liquid, η, the absolute linear dimensions, l, various dimensionless shape factors, and the total force, f, with which gravity pulls on the falling weight. Writing out the dimensions of η and f in terms of mass, length, and time, we see that

$$t \propto \eta f l^2.$$

The time of fall is therefore proportional to the square of the linear dimensions. Under pressure the linear dimensions become smaller, and so the time of fall becomes smaller. Hence the observed time of fall under pressure is to be corrected by adding a percentage amount equal to twice the linear compression of the steel of the cylinder at the pressure in question. This correction is linear with pressure, independent of the temperature, and at 12000 kg. is 0.46%.

In addition to the corrections described above peculiar to these measurements of viscosity, there are other corrections common to all high pressure measurements, which have been described in sufficient detail previously.[5] All the corrections together do not amount to over 10% in the extreme case, and more often were of the order of 3 or 4%.

Experimental Procedure.

The first operation of a pressure run was filling the viscosity cylinder with the liquid. This had to be done with extreme care to avoid the introduction of bits of mechanical dirt which might hinder the free fall of the weight. In the preliminary work a good many unsuccessful attempts were made before the appropriate procedure was found. The apparatus was first disassembled from the preceding run by unsoldering the insulating plugs and the tin reservoir. The inside of the cylinder, which is brightly polished, was scoured with a linen rag and whiting powder, and then all loose particles wiped out with a clean linen rag and a camel's hair brush. The brass terminals and the flat faces of the insulating plug were polished bright with the finest French emery paper. The reservoir was remade by unsoldering the thin tin wall, partially deformed from the previous run, and brightly scouring with French emery paper the top and bottom plates and the central core. The reservoir was now resoldered with a new tin wall, using no flux in the soldering, which might partially penetrate into the interior of the reservoir, contaminating the liquid and depositing a gum which would interfere with the free fall of the weight. To facilitate soldering, the various parts which were to come in contact were previously tinned, using a flux in this preliminary tinning, because any excess could be readily wiped away. A weight of the dimensions judged to be most appropriate for the liquid under investigation was now selected, using as a criterion the known viscosity of the liquid at atmospheric pressure, and the apparatus was assembled by soldering in the plugs and the reservoir. During this assembly, which was made in a very bright light, preferably sunlight, the most rigorous scrutiny was made for minute particles of dirt, which were removed with the camel's hair brush.

The cylinder was now filled by exhausting as already described, and was then connected to the insulating plug which closes the pressure cylinder and was assembled into the pressure apparatus. This was usually done the last thing at night. In the morning a viscosity reading was made at once at the prevailing temperature of the room, which was 15° to 20°. The temperature of the bath was then raised to 30°, and after temperature equilibrium was reached another viscosity determination was made at atmospheric pressure. Pressure was now applied and a run made at 30° to 12000, or to the maximum pressure allowed by the freezing of the liquid. The freezing pressure was known in those cases in which the liquid had been previously investigated, but in a number of cases the freezing had not been previously investigated, and in the following will be found a number of new freezing points under pressure approximately determined. To have made an exact determination would have taken a great deal of time for the apparatus is not well adapted to this sort of measurement. Freezing is, of course, shown by the weight refusing to fall. There is no previous warning of the approach of freezing, but the viscosity curve of the liquid runs without change into the subcooled region. In the following the melting pressures are recorded as for example "freezes between 8000 and 10000 kg." or "melts between 7000 and 6000 kg." The first means that the substance was liquid at 8000 and on increasing pressure to 10000 it froze. The second means that at 7000 it was solid, and on releasing pressure to 6000 it melted. Because of the possibility of the supercooling of the liquid, the second method of determination is of course much the more accurate.

The number of pressure readings depended to some extent on the character of the liquid, more points being taken for a liquid whose viscosity changes rapidly with pressure or which could be measured over only a small range because of freezing. For those liquids which could be investigated over the entire pressure range of 12000 kg., readings were made at 0, 100, 500, 1000, 2000, 4000, 6000, 8000, 10000, and 12000 kg. If log viscosity is plotted against pressure a curve will be obtained nearly straight above 4000, but below this there is more or less curvature; this is the reason that the readings were multiplied at low rather than high pressures. In making the runs with the first few substances, pressure was released after the reading at the maximum of 12000, and check readings were made on the way down at 6000 and 0. There is no reason why a perfect check should not be obtained, unless the reservoir leaked or some similar accident took place, because we are here dealing with the properties of a liquid which never have hysteresis. As a matter of fact, a perfect check was always obtained. The check readings having shown themselves unnecessary, in the case of the rest of the liquids, after the reading at 12000 and 30°, the temperature of the bath was changed to 75°, and the viscosity determined at 75° in decreasing pressure steps, retracing the steps of the ascending run to atmospheric pressure, if the boiling point of the liquid at atmospheric pressure was above 75°, or making the last reading at a pressure of a few atmospheres if the atmospheric boiling point was below 75°.

If the time of fall of the weight was less than 5 seconds, the time of 50 falls was taken, grouping the falls into five groups of ten. Above 5 seconds fall time, the mean of 10 fall times was taken, until the time got to be 40 or 50 seconds, when a smaller number of readings was taken, but never less than two even in the extreme case of one and a half hours. It has already been explained that the time of fall in the two directions might be slightly different because of lack of perfect symmetry in the two ends of the weight.

In the preliminary experiments a number of questions about the correct functioning of the apparatus were examined. In the first place it was established that the results could be repeated in that the apparatus could be taken apart and reassembled with the same liquid and the same readings obtained. If the falling weight was changed, either by changing the outside dimensions of the weight, or by putting a weight in the cavity keeping the outside dimensions the same, consistent results were obtained in that the ratio of the fall times at two different pressures was a constant.

Additional important checks are obtained by comparing the relative viscosities for different substances with those found by other observers. Nearly three-quarters of the readings were made with the same hollow weight, varying the total weight by the use of different cores. The relative times, corrected for the buoyancy of the liquid and the difference of total weight, of all these runs should be the same as the relative absolute viscosities determined by other observers. In Table I are given the corrected times of fall at 30° at atmospheric pressure, the absolute viscosity at 30° taken from the Smithsonian Tables, and the ratio of these two numbers for each liquid. The ratio should be constant, and the table shows that it is, except in one or two cases, with an error not greater than the usual discrepancy between viscosity determinations by different observers.

Another check is afforded by comparing the temperature coefficient of viscosity at atmospheric pressure given by these experiments with the values of others. In Table II the ratio of the viscosity at 30° to that at 75° for some of the liquids of this investigation is compared with all the values given in the Smithsonian Tables for the same liquids. Again the agreement is satisfactory.

TABLE I.

Comparison of Relative Viscosities Obtained with the Pressure Apparatus with the Absolute Viscosities of Other Observers.

Liquid	Corrected Time of Fall at 30° (seconds)	Viscosity at 30° $\times 10^2$	Ratio $\frac{\text{Time}}{\text{Viscosity}}$ $\times 10^3$
Ethyl Alcohol	10.4	1.003*	1.04
n-propyl Alcohol	17.4	1.779	.98
n-butyl Alcohol	22.2	2.24	.99
i-propyl Alcohol	17.2	1.757	.98
i-butyl Alcohol	28.9	2.864	1.01
n-pentane	2.17	.220	.99
n-hexane	2.79	.296	.94
n-octane	5.00	.483	1.04
i-pentane	2.18	.200	1.09
Ethyl Bromide	3.74	.368	1.02
Ethyl Iodide	5.52	.540	1.02
Ethyl Acetate	4.23	.407	1.04
CCl_4	8.65	.845	1.02
Chloroform	5.48	.519	1.06
CS_2	3.43	.352	.97
Ether	2.19	.223	.98
Benzene	5.72	.566	1.01
Toluene	5.53	.523	1.06
o-xylene	7.25	.709	1.02
m-xylene	5.92	.552	1.07
p-xylene	5.74	.568	1.01

The satisfactory functioning of the apparatus having been established, runs were made in the great majority of cases with only a single filling of the apparatus for a single substance.

It is not easy to estimate the probable accuracy of the relative

* The numbers in this column were obtained mostly from Smithsonian Tables or from Table III of this paper.

TABLE II.

COMPARISON OF TEMPERATURE COEFFICIENT OF VISCOSITY AT ATMOSPHERIC PRESSURE GIVEN BY THE PRESSURE APPARATUS WITH VALUES OF OTHER OBSERVERS.

Liquid	η_{30}/η_{75} Pressure Apparatus	Other Observers
Ethyl Alcohol	2.20	2.13*
n-propyl Alcohol	2.52	2.65
n-butyl Alcohol	2.84	2.81
i-propyl Alcohol	3.14	3.14
n-octane	1.55	1.58
Ethyl Iodide	1.46	1.44
CCl_4	1.74	1.69
Benzene	1.72	1.68
o-xylene	1.71	1.64
m-xylene	1.59	1.55
p-xylene	1.60	1.54
Toluene	1.60	1.56

viscosities under pressure yielded by these experiments. The maximum discrepancy in the time intervals was of the order of 0.5 seconds (10 intervals for fall times below 5 seconds, and single intervals above 5 seconds) giving a probable error in the averaged time of fall of something of the order of 0.05 second. The percentage error which this introduced into the final result depends on the total time of fall. As given in the following, the relative viscosities are the ratios of the corrected times of fall under the pressure and temperature in question to t_0, the corrected time of fall at atmospheric pressure at 30°. The percentage accuracy of these ratios obviously depends on the absolute value of t_0. In order to permit an estimate of this accuracy the value of t_0 is tabulated for each of the liquids. In Table V, to be given later, the logarithm to the base 10 of the relative viscosity is given as a function of pressure and temperature; one need expect no appreciable error in the shapes and relative positions of the log viscosity curves for times above 10 seconds. In addition to the accidental time errors, there may be consistent errors due to the various corrections; the more uncertain of the corrections are absolute and not percentage corrections, and so also are more important at the low

* Taken from Smithsonian Tables.

pressure end of the curves. Errors of this kind I believe to be considerably less than 0.01 seconds.

Some internal evidence as to probable accuracy is given by the closeness of the experimental points to a smooth curve. In constructing the table, the $\log_{10}$ of the corrected time of fall was plotted against pressure, smooth curves were drawn through the points, and the values at regular pressure intervals tabulated. The average numerical deviation from a smooth curve of $\log_{10}$ time is given in the table. The smoothness and self-consistency of the results was such that the theory of probable errors shows the justifiability of keeping more significant figures than are given here (logs to only three places are given), an uncertainty of 0.001 in the log corresponding to a percentage error of about 0.2%. The reason for not giving more significant figures was the very great increase of labor in computing the results which would have been involved. Three figures could be obtained fairly easily by direct graphical methods; another significant figure would have demanded computational methods at an enormous increase of time, which does not seem justified by any use which I can at present foresee will probably be made of these data.

In the case of water a great many experimental attempts were made, but an accuracy equal to that obtained with the other liquids was not achieved. The difficulty with water was caused by electrical conductivity of the water interfering with the operation of the timer through the electric contact arrangement. It was not possible to use the cylinder of bessemer steel used with the other liquids, because after the water had stood in contact with the steel for a while there was enough chemical action to produce a short-circuit in all positions of the weight. This was in spite of starting with water distilled in tin, from which all the air had been boiled out. Another viscosity cylinder was therefore made from one of the various grades of "rustless" iron recently put on the market. Considerable mechanical difficulty was found in machining this because of hard and soft spots, and the hole, although made as carefully as possible in the usual way with an ordinary reamer, was so much out of round as to introduce great irregularity, the weight falling freely in some positions, and in others sticking tight. A third apparatus was now made with another grade of rustless iron, this time the hole being *ground* true to within 0.00025 cm. To grind so long a hole as this to such an accuracy requires a high degree of mechanical skill, and I am much indebted to the firm of West and Dodge, who did the work. The falling weight was made of the same steel as the cylinder. But now a great deal of trouble, not given by the first cylinder of rustless iron, was found with the electrical part of the apparatus; the apparatus would work perfectly in preliminary trials, but after running for a while the contact would fail. This was finally traced to the steel itself; after a number of sparks had passed between the cylinder and the weight a film was deposited of such highly insulating properties that it could not be broken down with any voltage feasible with the magneto. The steel of which the weight was formed was uneven in quality, and the contaminating film formed on the cap end in preference to the other. The difficulty was avoided at the low temperatures by making another cap of the original grade of rustless iron. With this, successful runs were made at 0° and 10°, but at 30° the insulation resistance became so great that the weight had to be discarded entirely, and a new one made of pure nickel. This worked well at 30°, but at 75° there was again trouble, both because of high insulation resistance to the falling weight, and electrolytic conductivity through the water. Only readings at pressures below 9000 were obtained at 75°, and these are in considerable doubt.

In reducing the readings to give correct relative viscosities over the entire range, correction had to be made for the change of weight between 10° and 30°. I made this correction so as to give between 10° and 30° the relative change of viscosity given by Bingham and Jackson.[6] For the relative changes between 0° and 10° and between 30° and 75°, the values given by these experiments were used. This gives a temperature coefficient of viscosity agreeing fairly well with that of Bingham between 30° and 75°, but materially larger between 0° and 10°.

Description of the Liquids.

There follows now a description of the various liquids used. These were obtained from various sources and were of varying degrees of purity. Fourteen of them were obtained through Professor J. Timmermans of the Bureau Belge d'Étalons Chimiques, for whose courtesy in providing me with the liquids I am much indebted; these were of an unusually high degree of purity. They were furnished in sealed glass tubes of approximately 10 cc. capacity. The tubes were not opened until immediately before filling the viscosity apparatus. The total time from breaking the glass seal to finally sealing the liquid into the viscosity apparatus was of the order of 5 minutes. Some of the physical constants of these liquids were determined by Professor Timmermans, and are given in the following table.

To Professor F. Keyes of the Mass. Institute of Technology I am indebted for CS_2, methyl and ethyl alcohol, ethyl ether, and normal pentane, also in sealed glass receptacles. These are also of a high degree of purity; the preparation and properties have been previously described by Professor Keyes.[7]

TABLE III.

PHYSICAL CONSTANTS OF LIQUIDS FROM BUREAU BELGE D'ÉTALONS CHIMIQUES.

Compound	Boiling Point	Densities 0°/4°	15°/4°	30°/4°	Viscosities$\times 10^5$ 15°	30°
n-butyl alcohol	118°.10 (762 mm.)	0.8245	0.8134		3309	2237
n-amyl alcohol	137.27±.02 (751 mm.)	0.8395	0.8245			
i-pentane	27.95±.02	0.6393	0.62485			
Ethylene dibromide	132.00±.02		2.1911	2.1596	1880	1490
Ethyl acetate	77.15±.01	0.9245	0.9066			
n-butyl bromide	101.65±.02	1.3041	1.2830		626	537
CCl_4	75.75±.01	1.6326	1.6035		1040	845
Chloroform	61.20±.01	1.5264	1.49845			
n-amyl ether	185.26±.02 (750 mm.)	0.7990	0.7869			
Cyclohexane	80.75±.01		0.7830	0.7696	1056	828
Methyl cyclohexane	101.18±.01	0.7865	0.7734	0.7603	777	639
Benzene	80.15±.01		0.8841	0.8685	696	566
Chlorobenzene	132.00±.01	1.1280	1.1117	1.0954	844	711
Bromobenzene	156.15±.01	1.5200	1.5009		1196	985
Toluene	110.80±.01	0.88545	0.8716	0.8577	623	523
p-xylene	138.16±.01 (757 mm.)		0.8642	0.8526	682	568

The i-propyl alcohol I owe to Professor R. F. Brunel. This is from the same lot as that which I used previously in determining the effect of pressure on thermal conductivity. It was freshly distilled for this work in a Hempel column; and all came over at a temperature constant within 0.1°, the error of the thermometer.

Glycerine was Kahlbaum's purest, in a glass stoppered bottle.

The water was obtained from the Chemical Laboratory of Harvard University, and had been distilled in tin. There would have been no point in attempting to start with water of greater purity because of the impurities immediately absorbed by the water from the walls of the cylinder.

The other chemicals were obtained from the Eastman Kodak Co. and were of the best grades of those described in their catalogue of chemicals. The boiling points given in the catalogue for the liquids used here are reproduced.* In some cases the liquids were provided in glass stoppered bottles, but more often a cork stopper was used, separated from the liquid by a piece of paper. In such cases there was likely to be more or less mechanical dirt mixed with the liquid. Inclusion of any of this dirt in the viscosity apparatus was in almost all cases avoided by letting the bottle settle for some time, and then removing the liquid with a pipette, but in one or two cases an additional filtering was made through glass wool. Evidence of the purity of the liquid was in some cases obtained from the sharpness of freezing under pressure. If the liquid freezes sharply there are no pressures at which the weight will perceptibly leave one end of the cylinder (break contact), without falling completely to the other end, whereas if freezing is not sharp (liquid becomes mushy) the weight may break contact without falling completely to the other end.

Numerical Results.

The numerical results for all the substances except water are now given in Table V. The table gives $\log_{10}$ of the relative viscosity as a function of pressure and temperature, the viscosity at 30° and atmospheric pressure being taken as unity. The logarithm of the viscosity instead of the viscosity itself is given because the variation with pressure of the viscosity is very rapid, and the curve of viscosity against pressure has rapidly varying curvature, whereas the curve of log viscosity against pressure approaches a straight line at high pressure and is not too much curved at the low pressures. The pressures tabulated in the table are 0, 500, 1000, 2000, 4000, 6000, 8000, 10000, and 12000 kg./cm.^{2}, the intervals being shorter at the lower end of the range because of the much more rapidly varying curvature. From the values of the logarithms the ratio of the viscosity at atmospheric pressure at 30° to that at 75° may be found, and is tabulated. In the table are given the average numerical deviations from smooth curves of the observed logarithms. The deviation is usually not as much as 0.001 on the logarithm, and in general the observed points lay on a smooth curve within 0.1%. Since the accuracy of the time measurements depends on the absolute value of the time interval, the time of fall at 30° at atmospheric pressure (t_0) is given. For those substances with a relatively low value of t_0 the lower ends of the curves are relatively inaccurate, but there is

* See Table on p. 99.

TABLE V.

Substance		Pressure, kg./cm.²									Average Deviation Smooth Curve	η_{30}	t_0	Wt.	Buoyancy Correction
		1	500	1000	2000	4000	6000	8000	10000	12000					
Methyl Alcohol	$\log \frac{\eta}{\eta_0}$ 30°	.000	.094	.167	.286	.471	.616	.750	.874	.998	.001			B	.0048
	$\log \frac{\eta}{\eta_0}$ 75°	9.769	9.862	9.933	.043	.208	.334	.448	.555	.655	.001	.00520	1.73	C	.0130
	η_{30}/η_{75}	1.702	1.706	1.714	1.750	1.832	1.914	2.004	2.084	2.203					.0204
Ethyl Alcohol	$\log \frac{\eta}{\eta_0}$ 30°	.000	.107	.200	.363	.617	.829	1.023	1.211	1.390	.000			A	.0093
	$\log \frac{\eta}{\eta_0}$ 75°	9.657	9.772	9.873	.045	.289	.473	.634	.778	.919	.000	.01003	3.95	E	.0236
	η_{30}/η_{75}	2.203	2.163	2.123	2.080	2.128	2.270	2.449	2.710	2.958					.0359
n-propyl Alcohol	$\log \frac{\eta}{\eta_0}$ 30°	.000	.151	.283	.494	.836	1.131	1.402	1.667	1.915	.000			A	.0051
	$\log \frac{\eta}{\eta_0}$ 75°	9.598	9.754	9.880	.074	.368	.610	.827	1.033	1.223	.000	.01779	6.19	C	.0121
	η_{30}/η_{75}	2.523	2.495	2.529	2.630	2.938	3.319	3.758	4.305	4.920					.0179
n-butyl Alcohol	$\log \frac{\eta}{\eta_0}$ 30°	.000	.175	.321	.554	.934	1.289	1.609	1.912	2.208	.000			A	.004
	$\log \frac{\eta}{\eta_0}$ 75°	9.548	9.724	9.867	.089	.312	.690	.941	1.172	1.396	.000	.02237	2.90	D	.010
	$\eta_{30/75}$	2.845	2.838	2.858	2.932	3.343	3.991	4.679	5.521	6.518					.015
n-amyl Alcohol	$\log \frac{\eta}{\eta_0}$ 30°	.000	.188	.341	.607	1.060	1.448	1.811	2.164	2.495	.000				.0087
	$\log \frac{\eta}{\eta_0}$ 75°	9.540	9.723	9.871	.105	.466	.772	1.049	1.313	1.562	.000		2.29	F	.0212
	η_{30}/η_{75}	2.884	2.917	2.951	3.177	3.926	4.742	5.781	7.096	8.570					.0320
i-propyl Alcohol	$\log \frac{\eta}{\eta_0}$ 30°	.000	.193	.343	.591	.982	1.318	1.640	1.977	2.311	.000			A	.004
	$\log \frac{\eta}{\eta_0}$ 75°	9.505	9.695	9.851	.087	.425	.701	.957	1.191	1.424	.002	.01757	2.24	D	.011
	η_{30}/η_{75}	3.141	3.164	3.120	3.208	3.624	4.161	4.844	6.140	7.748					.015
i-butyl Alcohol	$\log \frac{\eta}{\eta_0}$ 30°	.000	.210	.388	.696	1.203	1.655	2.075	2.483	2.898	.000			A	.0042
	$\log \frac{\eta}{\eta_0}$ 75°	9.444	9.659	9.824	.075	.488	.838	1.158	1.459	1.747	.000	.02864	3.78	D	.0103
	η_{30}/η_{75}	3.597	3.556	3.664	4.178	5.188	6.561	8.260	10.57	14.16					.0153

TABLE V.—*Continued.*

Substance		Pressure, kg./cm.² 1	500	1000	2000	4000	6000	8000	10000	12000	Average Deviation Smooth Curve	η_{30}	t_e	Wt.	Buoyancy Correction
i-amyl Alcohol	$\log \frac{\eta}{\eta_0}$ 30°	.000	.209	.386	.686	1.185	1.636	2.069	2.505	2.952	.000				.0087
	$\log \frac{\eta}{\eta_0}$ 75°	9.424	9.618	9.787	.065	.492	.848	1.168	1.483	1.780	.000		.70	G	.0211
	η_{30}/η_{75}	3.805	3.939	4.012	4.221	4.970	6.200	8.042	11.41	15.76					.0316
n-pentane	$\log \frac{\eta}{\eta_0}$ 30°	.000	.181	.315	.524	.847	1.112	1.360	1.615	1.846	.002				.011
	$\log \frac{\eta}{\eta_0}$ 75°	9.811	.014	.163	.380	.676	.908	1.119	1.313	1.493	.000	.00220	2.32	A	.027
	η_{30}/η_{75}	1.545	1.469	1.419	1.393	1.483	1.600	1.742	2.004	2.254					.040
n-hexane	$\log \frac{\eta}{\eta_0}$ 30°	.000	.184	.332	.561	.914	1.224	1.514	1.803		.001				.0105
	$\log \frac{\eta}{\eta_0}$ 75°	9.803	.028	.171	.379	.701	.961	1.198	1.426	1.646	.000	.00296	2.99	A	.027
	η_{30}/η_{75}	1.574	1.432	1.449	1.521	1.633	1.832	2.070	2.382						.040
n-octane	$\log \frac{\eta}{\eta_0}$ 30°	.000	.196	.327	.641	1.088	1.487				.000			A	.0043
	$\log \frac{\eta}{\eta_0}$ 75°	9.810	.003	.153	.390	.763	1.080	1.363	1.630		.000	.00483	1.75	C	.0119
	η_{30}/η_{75}	1.549	1.560	1.493	1.782	2.113	2.553								.0144
i-pentane	$\log \frac{\eta}{\eta_0}$ 30°	.000	.202	.344	.559	.894	1.175	1.431	1.687	1.947	.000				.0101
	$\log \frac{\eta}{\eta_0}$ 75°	9.821	.040	.193	.408	.715	.960	1.179	1.381	1.586	.000	.00198	2.34	A	.0258
	η_{30}/η_{75}	1.510	1.452	1.416	1.416	1.510	1.641	1.786	2.023	2.296					.0398
i-amyl Decane	$\log \frac{\eta}{\eta_0}$ 30°	.000	.237	.435	.772	1.354					.001			A	.0084
	$\log \frac{\eta}{\eta_0}$ 75°	9.772	9.989	.178	.463	.925	1.334	1.727			.000		2.65	E	.0210
	η_{30}/η_{75}	1.690	1.770	1.807	2.037	2.685									.0319
Ethylene Di-bromide	$\log \frac{\eta}{\eta_0}$ 30°	.000	.138								.001				.0296
	$\log \frac{\eta}{\eta_0}$ 75°	9.756	9.885	.003	.203	.354 (3000)					.000 1 discard	.01490	20.6	A	.0739
	η_{30}/η_{75}	1.754	1.791												.0837

TABLE V.—*Continued.*

Substance		Pressure, kg./cm.² 1	500	1000	2000	4000	6000	8000	10000	12000	Average Deviation Smooth Curve	η_{30}	t_0	Wt.	Buoyancy Correction
Ethyl	$\log \frac{\eta}{\eta_0}$ 30°	.000	.134	.242	.405	.649	.837	1.008	1.172	1.323	.001				.0150
Chloride	$\log \frac{\eta}{\eta_0}$ 75°	9.850	.017	.131	.285	.514	.683	.834	.977	1.111	.000		2.54	A	.0341
	η_{30}/η_{75}	1.413	1.309	1.291	1.318	1.365	1.426	1.493	1.567	1.633					.0443
Ethyl	$\log \frac{\eta}{\eta_0}$ 30°	.000	.121	.222	.387	.631	.854	1.043	1.223	1.400	.000				.0195
Bromide	$\log \frac{\eta}{\eta_0}$ 75°	9.806	9,959	.072	.235	.472	.653	.816	.978	1.123	.000	.00368	4.54	A	.0486
	η_{30}/η_{75}	1.567	1.452	1.413	1.419	1.442	1.589	1.687	1.758	1.892					.0602
Ethyl	$\log \frac{\eta}{\eta_0}$ 30°	.000	.115	.218	.385	.656	.888	1.108	1.330	1.549	.001				.0144
Iodide	$\log \frac{\eta}{\eta_0}$ 75°	9.837	9.954	.057	.227	.467	.672	.854	1.030	1.200	.004	.00540	2.15	A C	.0363
	η_{30}/η_{75}	1.455	1.449	1.445	1.439	1.545	1.644	1.795	1.995	2.234					.0551
Acetone	$\log \frac{\eta}{\eta_0}$ 30°	.000	.135	.226	.373	.605	.804	.987	1.160		.000				.0051
	$\log \frac{\eta}{\eta_0}$ 75°	9.895	.017	.113	.245	.445	.610	.762	.898	1.031	.002	.00285	1.16	A C	.0139
	η_{30}/η_{75}	1.274	1.312	1.297	1.343	1.445	1.563	1.679	1.828						.0213
Glycerine	$\log \frac{\eta}{\eta_0}$ 30°	.000	.134	.260	.497	.936	1.346	1.741	2.133		.000				.0088
	$\log \frac{\eta}{\eta_0}$ 75°	8.810	8.920	9.023	9.204	9.529	9.818	.094	.369	.628	.001	3.8	9.42	H	.0186
	η_{30}/η_{75}	15.49	16.37	17.26	19.63	25.53	33.73	44.36	58.08						.0291
Ethyl	$\log \frac{\eta}{\eta_0}$ 30°	.000	.142	.258	.463	.818	1.120	1.393	1.686	1.974	.003				.0060
Acetate	$\log \frac{\eta}{\eta_0}$ 75°	9.836	9.976	.081	.253	.517	.761	.992	1.213	1.416	.000	.0039	1.50	A C	.0151
	η_{30}/η_{75}	1.459	1.466	1.503	1.622	2.000	2.286	2.518	2.972	3.614					.0229
n-butyl	$\log \frac{\eta}{\eta_0}$ 30°	.000	.143	.269	.474	.816	1.115	1.408	1.715	2.018	.000				.0088
Bromide	$\log \frac{\eta}{\eta_0}$ 75°	9.832	9.975	.090	.273	.564	.811	1.040	1.264	1.484	.002	.00537	2.21	A C	.0214
	η_{30}/η_{75}	1.472	1.472	1.510	1.589	1.786	2.014	2.333	2.825	3.420					.0319

TABLE V.—*Continued.*

Substance		Pressure, kg./cm.[2] 1	500	1000	2000	4000	6000	8000	10000	12000	Average Deviation Smooth Curve	η_{30}	t_0	Wt.	Buoyancy Correction
Cineole	$\log \frac{\eta}{\eta_0}$ 30°	.000	.315								.002				.0099
	$\log \frac{\eta}{\eta_0}$ 75°	9.654	9.905	.142	.575						.000		2.03	F	.0241
	η_{30}/η_{75}	2.218	2.570												.0364
Oleic Acid	$\log \frac{\eta}{\eta_0}$ 30°	.000	.306	.616							.000				.0103
	$\log \frac{\eta}{\eta_0}$ 75°	9.419	9.671	9.989	.255	.843					.001		2.72	I	.0265
	η_{30}/η_{75}	3.811	4.315	4.236											.0402
CCl_4	$\log \frac{\eta}{\eta_0}$ 30°	.000	.190	.351	.493 (1500)						.000			A	.0095
	$\log \frac{\eta}{\eta_0}$ 75°	9.760	9.949	.100	.349	.542					.002	.00845	3.26	C	.0267
	η_{30}/η_{75}	1.738	1.742	1.782											.0529
Chloroform	$\log \frac{\eta}{\eta_0}$ 30°	.000	.110	.211	.386	.660	.884				.002			A	.0104
	$\log \frac{\eta}{\eta_0}$ 75°	9.858	9.985	.094	.251	.480	.691	.914	1.141		.000	.00519	2.05	C	.0261
	η_{30}/η_{75}	1.387	1.334	1.309	1.365	1.514	1.560								.0396
CS_2	$\log \frac{\eta}{\eta_0}$ 30°	.000	.090	.150	.307	.509	.674	.840	1.010	1.189	.002				.0133
	$\log \frac{\eta}{\eta_0}$ 75°	9.875	9.972	.051	.180	.372	.527	.671	.808	.946	.000	.00352	4.05	A	.0375
	η_{30}/η_{75}	1.334	1.312	1.235	1.340	1.371	1.403	1.476	1.592	1.750					.0491
Ether	$\log \frac{\eta}{\eta_0}$ 30°	.000	.189	.324	.514	.792	1.042	1.261	1.469	1.670	.002				.0109
	$\log \frac{\eta}{\eta_0}$ 75°	9.878	.024	.149	.344	.601	.806	.986	1.155	1.311	.000	.00212	2.38	A	.0258
	η_{30}/η_{75}	1.324	1.462	1.496	1.479	1.552	1.722	1.884	2.061	2.286					.0398
n-amyl Ether	$\log \frac{\eta}{\eta_0}$ 30°	.000	.218	.401	.708	1.230	1.685	2.091			.002			A	.0048
	$\log \frac{\eta}{\eta_0}$ 75°	9.736	9.943	.107	.364	.776	1.125	1.437	1.728	2.007	.000		3.44	C	.0114
	η_{30}/η_{75}	1.837	1.884	1.968	2.208	2.844	3.631	4.508							.0169

TABLE V.—*Continued.*

Substance		Pressure, kg./cm.² 1	500	1000	2000	4000	6000	8000	10000	12000	Average Deviation Smooth Curve	η_{30}	t_0	Wt.	Buoyancy Correction
Cyclohexane	$\log \frac{\eta}{\eta_0}$ 30°	.000	.261								.000				.0063
	75°	9.723	9.975	.169	.341 (1500)						.002	.00828	9.18	A	.0169
	η_{30}/η_{75}	1.892	1.932												
Methyl Cyclohexane	$\log \frac{\eta}{\eta_0}$ 30°	.000	.220	.388	.710	1.274	1.804	2.318			.003			B	.0037
	75°	9.747	9.976	.154	.434	.900	1.335	1.756	2.167	2.582	.000	.00639	2.14	C	.0100
	η_{30}/η_{75}	1.791	1.754	1.714	1.888	2.366	2.944	3.648							.0163
Benzene	$\log \frac{\eta}{\eta_0}$ 30°	.000	.173	.347							.001				.0087
	75°	9.765	9.938	.081	.308	.498 (3000)					.003	.00566	6.34	A	.0202
	η_{30}/η_{75}	1.718	1.718	1.845											
Chlorobenzene	$\log \frac{\eta}{\eta_0}$ 30°	.000	.133	.253	.478	.867	1.223				.002			B	.0099
	75°	9.814	9.936	.053	.245	.563	.852	1.146	1.465		.000	.00711	2.38	C	.0208
	η_{30}/η_{75}	1.535	1.574	1.585	1.710	2.014	2.350								.0312
Bromobenzene	$\log \frac{\eta}{\eta_0}$ 30°	.000	.138	.262	.486	.897					.000				.0099
	75°	9.801	9.930	.044	.228	.558	.874	1.029 (7000)			.000	.00985	3.29	A	.0234
	η_{30}/η_{75}	1.581	1.614	1.652	1.811	2.183								C	.0346
Aniline	$\log \frac{\eta}{\eta_0}$ 30°	.000	.195	.376	.709						.001			A	.0037
	75°	9.551	9.703	9.847	.102	.560					.000	.0319	4.08	D	.0101
	η_{30}/η_{75}	2.812	3.105	3.381	4.046										.0162

TABLE V.—*Continued.*

Substance		Pressure, kg./cm.² 1	500	1000	2000	4000	6000	8000	10000	12000	Average Deviation Smooth Curve	η_{30}	t_0	Wt.	Buoyancy Correction
Diethylaniline	$\log \frac{\eta}{\eta_0}$ 30°	.000	.201	.394	.761	1.070 (3000)					.002				.0110
	$\log \frac{\eta}{\eta_0}$ 75°	9.690	9.839	9.984	.259	.758	1.250	1.775			.003		1.32	F	.0279
	η_{30}/η_{75}	2.042	2.301	2.570	3.177										.0424
Nitrobenzene	$\log \frac{\eta}{\eta_0}$ 30°	.000	.134	.264							.001				.0045
	$\log \frac{\eta}{\eta_0}$ 75°	Decomposes											2.23	A	.0121
	η_{30}/η_{75}													D	.0194
Toluene	$\log \frac{\eta}{\eta_0}$ 30°	.000	.145	.274	.497	.897	1.285	1.699	2.177		.000				.0049
	$\log \frac{\eta}{\eta_0}$ 75°	9.796	9.939	.065	.267	.597	.896	1.186	1.504	1.832	.000	.00523	1.96	A	.0137
	η_{30}/η_{75}	1.600	1.607	1.618	1.698	1.995	2.449	3.258	4.710					C	.0213
o-xylene	$\log \frac{\eta}{\eta_0}$ 30°	.000	.165	.311	.577						.000				.0054
	$\log \frac{\eta}{\eta_0}$ 75°	9.767	9.925	.057	.292	.689	1.087				.001	.00709	2.57	A	.0131
	η_{30}/η_{75}	1.710	1.738	1.795	1.928									C	.0196
m-xylene	$\log \frac{\eta}{\eta_0}$ 30°	.000	.154	.290	.529	.967					.001				.0092
	$\log \frac{\eta}{\eta_0}$ 75°	9.799	9.959	.079	.286	.637	.983	1.333			.001	.00552	6.55	A	.0226
	η_{30}/η_{75}	1.589	1.567	1.626	1.750	2.138									.0336
p-xylene	$\log \frac{\eta}{\eta_0}$ 30°	.000	.152								.000				.0054
	$\log \frac{\eta}{\eta_0}$ 75°	9.797	9.957	.092	.315						.003	.00568	2.04	A	.0130
	η_{30}/η_{75}	1.596	1.567											C	.0196
p-cymene	$\log \frac{\eta}{\eta_0}$ 30°	.000	.172	.333	.626	1.194	1.859				.000				.0059
	$\log \frac{\eta}{\eta_0}$ 75°	9.800	9.948	.087	.335	.749	1.168	1.612	2.164		.000		2.43	B	.0149
	η_{30}/η_{75}	1.585	1.675	1.762	1.954	2.786	4.977							C	.0226

TABLE V.—*Continued.*

Substance		Pressure, kg./cm.² 1	500	1000	2000	4000	6000	8000	10000	12000	Average Deviation Smooth Curve	η_{30}	t_0	Wt.	Buoyancy Correction
Eugenol	$\log \frac{\eta}{\eta_0}$ 30°	.000	.288	.541	1.081	2.273	3.007				.003				.0138
					1.652	(3000)	(5000)								
	$\log \frac{\eta}{\eta_0}$ 75°	9.429	9.616	9.810	.143	.805	1.520	2.343			.001		1.70	G	.0326
	η_{30}/η_{75}	3.724	4.699	5.383	8.670	29.38									.0496
Petroleum Ether	$\log \frac{\eta}{\eta_0}$ 30°	.00	.16	.30	.54	.93	1.28	1.59	1.88	2.18					
	$\log \frac{\eta}{\eta_0}$ 80°					.56	.83	1.06	1.28	1.49			2.75		(*)
	η_{30}/η_{80}					2.34	2.82	3.39	3.98	4.90					
Kerosene	$\log \frac{\eta}{\eta_0}$ 30°	.00	.24	.46	.88	1.71									
	$\log \frac{\eta}{\eta_0}$ 80°					.91	1.41	1.88	2.34	2.80			3.80		(*)
	η_{30}/η_{80}					6.3									

(*) No buoyancy correction applied.

Designation of Weights

A	1.014 gm. Fe	D	7.041 gm. Au	G	3.821 gm. Fe
B	.997 gm. Fe	E	1.921 gm. Fe	H	3.134 gm. Fe
C	2.008 gm. W	F	4.108 gm. Fe	I	3.470 gm. Fe

no error arising from this source in the relative values at the higher pressures. The falling weights are tabulated and in the last column the corrections assumed for the buoyancy of the liquid at 2000, 6000, and 12000 kg., respectively. For those liquids whose compressibility has not been directly measured a small correction may have to be made at some future time when the compressibility is measured, permitting a more accurate buoyancy correction. Also, for convenience in the discussion, the absolute viscosities at atmospheric pressure at 30° are listed when these are known; many of these values were taken from the Smithsonian Tables, and others were given by Timmermans.

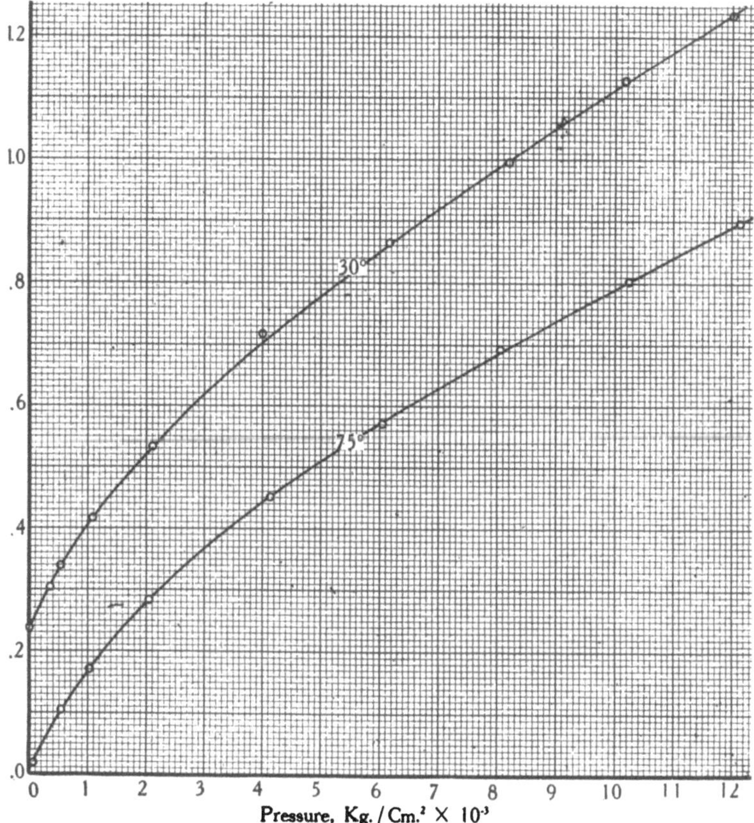

FIGURE 3. The common logarithm of the corrected time of fall of the weight against pressure for methyl alcohol. The time of fall is proportional to viscosity.

Sample curves for two substances, giving observed logarithm of the relative viscosity, are shown in Figures 3 and 4.

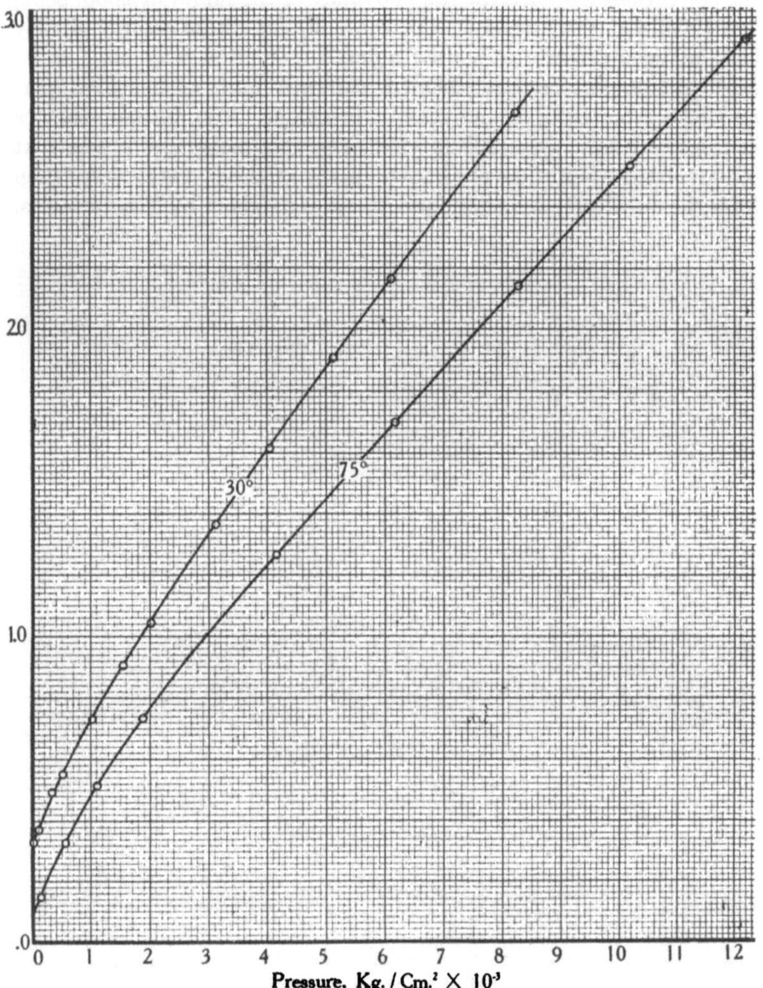

FIGURE 4. The common logarithm of the corrected time of fall of the weight against pressure for methyl-cyclohexane. The time of fall is proportional to viscosity.

Because of its comparatively small variation of viscosity with pressure, water is treated separately. In Table VI. are given the relative viscosities (not log relative viscosity) as a function of pressure at 0°, 10°, 30°, and 75°. The pressure range at the two lowest temperatures was terminated by freezing, and at 75° the limit was set by the experimental difficulties already described. The experimental points are plotted in Figure 5. The experimental irregularity is much greater than for the other liquids which are better insulators, but still is not great enough to leave any doubt as to the essential character of the facts.

TABLE VI.

Relative Viscosity of Water.

Pressure kg./cm.[2]	Relative Viscosity			
	0°	10.3°	30°	75°
1	1.000	.779	.488	.222
500	.938	.755	.500	.230
1000	.921	.743	.514	.239
1500	.932	.745	.530	.247
2000	.957	.754	.550	.258
3000	1.024	.791	.599	.278
4000	1.111	.842	.658	.302
5000	1.218	.908	.720	.333
6000	1.347	.981	.786	.367
7000		1.064	.854	.404
8000		1.152	.923	.445
9000			.989	.494
10000			1.058	
11000			1.126	

Some of the liquids require special comment.

Methyl Alcohol. Two attempts were necessary with this. At first a portion of the same sample was used as had been used for the thermal conductivity measurements under pressure. This had been sealed in glass since extracting the thermal conductivity sample, but nevertheless had absorbed sufficient moisture from the air during that brief handling to make it so conducting that the method failed. A second fresh sample was obtained from Professor Keyes, also prepared in 1923, but sealed in glass ever since. This was a sufficiently good insulator to give good readings.

Ethyl Alcohol. At 75°, at a pressure of 2000 kg. and below, the electrical conductivity became so great that the electrical timing device ceased to function, and a telephone and stop-watch had to be substituted. This gives less accurate results, the smallest division of the stop-watch being 0.2 seconds, so that the lower end of the 75° is much more in error than usual. The upper end of the 75° curve is not affected by this error, neither is any part of the 30° curve.

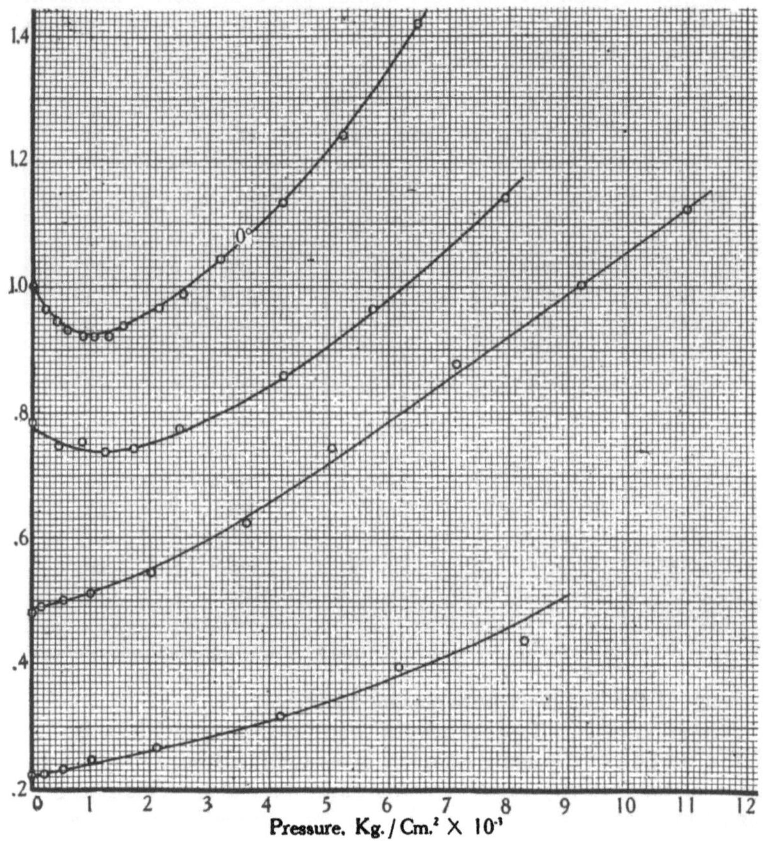

FIGURE 5. The relative viscosity of water at 0°, 10°, 30°, and 75° as a function of pressure.

i-Butyl Alcohol. Two fillings were necessary, the connecting pipe to the pressure generating apparatus having split on the first application of 10000 kg. at 30°. The second filling gave all the points at 75°, and the points at 10000 and 12000 at 30°, and several check points at lower pressures, which agreed within 0.1%.

n-Octane. Two fillings were necessary; after a number of readings at 30°, the weight stuck because of mechanical dirt. A heavier weight was used with the second filling. The results of the two fillings agreed after correction for the difference of weight.

Ethyl Chloride. The boiling point at atmospheric pressure is about 12°. The experiment was done in winter; during the filling of the apparatus the room was cooled to 0° C. by opening the windows.

Acetone. This was cleaned from mechanical dirt before filling the apparatus by filtering through glass wool.

Ethyl Ether. Two fillings of the apparatus was necessary, the weight sticking because of mechanical dirt after the run at 30°. A check reading was made at 30° with the second filling, agreeing within a few thousandths of a second.

Nitro Benzene. Readings were obtained only at 30°. At 75°, at 1200 kg., the liquid short circuited, as it did also at 2200. The apparatus was then taken apart, and the nitro benzene found completely decomposed, with a heavy deposit of lamp black.

Incidental Data.

In addition to the viscosity, a certain amount of incidental data was also obtained, as already mentioned. The approximate values of the freezing pressures are collected in Table VII.

The compressibility of glycerine was also determined approximately. This was done by a new method, which is to be applied in the future to a large number of liquids, and will be described in detail then. Briefly, the liquid is placed in a piezometer with freely moving piston, and immersed in the liquid by which pressure is transmitted. Under pressure, the piston moves in, equalizing pressure inside and out, and the motion of the piston is measured by a sliding wire scheme with potentiometer contacts much like the lever arrangement for measuring compressibility. Since the compressibility of glycerine was needed only in a correction term, no attempt was made to apply all the corrections in working up the results, and in Table VIII the change of volume at 30° is tabulated as a function of pressure to only two significant figures. The particular interest of this substance, glycerine, lies in its extremely low compressibility for a non-metallic substance.

Comparison with Previous Results.

For comparison there are practically only the results of Faust[1] to 3000 kg. on ethyl alcohol, ether, and CS_2. He finds for ethyl alcohol an increase of viscosity at 30° under 3000 kg. of 2.94 fold against my 2.31, for ether 3.96 fold against my 3.27, and for CS_2 3.43 against 2.03. The agreement certainly ought to be much closer. Faust used a flow method through a glass capillary under a varying head of mercury. It is not evident from his paper whether all the corrections were applied, but it is hard to see how any corrections could be responsible for so large a difference.

TABLE VII.

ROUGH FREEZING DATA UNDER PRESSURE.

Substance	Freezing Data 30°	75°
n-hexane	10600 kg./cm.^{2}	
n-octane	6100	10000
i-amyl Decane	Freezes between 4000 and 6100	Melts between 8200 and 7130
Ethylene Dibromide	Freezes between 600 and 800	Freezes between 2500 and 2800
Cineole	Freezes between 900 and 1000	Freezes between 2100 and 2500
Oleic Acid	Freezes between 100 and 1600	Melts below 5000
n-amyl ether	Freezes between 8200 and 9500	
Cyclohexane	Freezes between 400 and 500	Freezes between 1400 and 1600
Methyl Cyclohexane	Freezes between 8000 and 10000	
Diethyl Aniline	Melts between 3000 and 2600	Freezes between 7200 and 7800
o-xylene	2400	5500
m-xylene	4000	above 5200
p-xylene	490	1900

GENERAL CHARACTER OF THE RESULTS.

In certain qualitative features, the behavior of all the liquids investigated here, except water, is alike, although there are very large quantitative differences. The viscosity increases with pressure at a rapidly increasing rate, so that if viscosity is plotted against pressure a curve of very rapid upward curvature is obtained. This is unusual;

TABLE VIII.

COMPRESSIBILITY OF GLYCERINE.

Pressure kg./cm.^{2}	$\frac{\Delta V}{V_0}$ at 30°	Pressure kg./cm.^{2}	$\frac{\Delta V}{V_0}$ at 30°
2000	.042	8000	.107
4000	.068	10000	.121
6000	.089	12000	.134

most pressure effects become relatively less at high pressure by a sort of law of diminishing returns. In Figure 6 is shown viscosity against

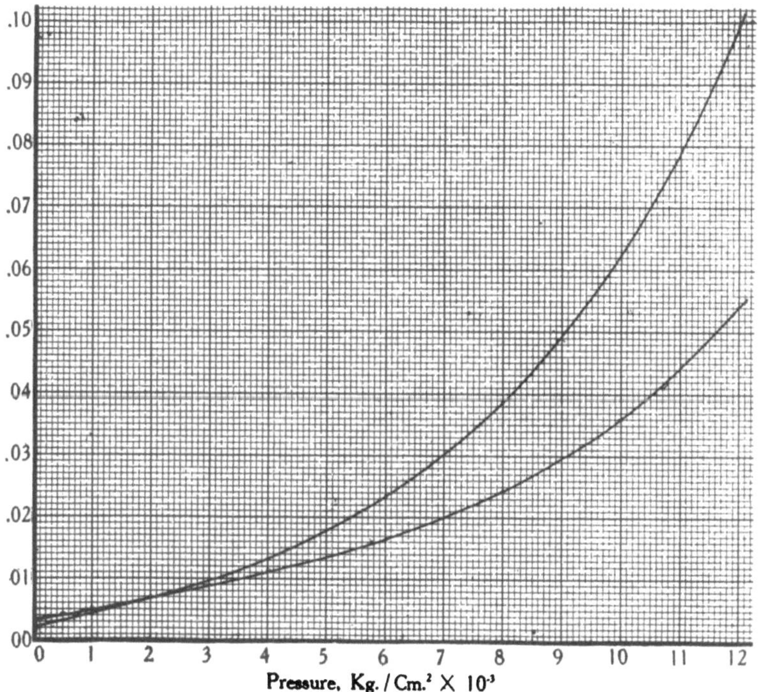

FIGURE 6. The viscosity in Abs C.G.S. units at 30° C. of CS_2 and ether as a function of pressure. The curve for ether starts below that for CS_2 and rises above it beyond 2000 kg.

pressure at 30° for CS_2 and ether, two substances with comparatively small pressure effect. It is seen that over the first two or three thousand kilograms the relation between pressure and viscosity is nearly linear, but above this the departure is extreme. If logarithm viscosity is plotted against pressure a curve is obtained in general concave toward the pressure axis. The curvature is much the greatest at low pressures; above 2000 or 3000 kg. the curve approximates to a straight line, or indeed in a number of cases reverses curvature. This means that above 3000 viscosity increases approximately geometrically as pressure increases arithmetically.

A number of substances show reverse curvature, log viscosity being convex toward the pressure axis at high pressures, which means that viscosity increases more rapidly than geometrically when pressure increases arithmetically. Eugenol and p-cymene show the effect more markedly than do any other substances. The following list

TABLE IX.

List of Substances for which log η is Convex to Pressure Axis at High Pressures.

Substance	Temperature	
i-propyl Alcohol	30°	
i-butyl Alcohol	30°	
i-amyl Alcohol	30°	
i-pentane	30°	75°
Ethyl Acetate	30°	
n-butyl Bromide	30°	75°
Chloroform		75°
CS_2	30°	75°
Methyl Cyclohexane	30°	75°
Chlorobenzene		75°
Diethyl Aniline		75°
Toluene	30°	75°
o-xylene		75°
m-xylene	30°	75°
p-cymene	30°	75°
Eugenol	30°	75°
Petroleum Ether	30°	

contains those substances for which this effect seems certain beyond experimental error.

The temperature coefficient of viscosity is shown by the rows in Table V. giving η_{30}/η_{75}. Here again the effect is abnormal; most temperature effects become less at high pressures, which is to be expected if the modification in the structure produced by temperature agitation becomes less under the greater constraints imposed by the high pressure. But here the relative change of viscosity with temperature becomes very markedly greater at high pressure, the ratio η_{30}/η_{75} changing under 12000 kg. by a factor of as much as 4.

Apart from these qualitative resemblances, the most varied quantitative behavior is shown by the various substances. In fact, viscosity is a unique property in regard to the magnitude of the pressure effect and its variation from substance to substance. The compressibility at atmospheric pressure, for example, varies by a factor of not more than four or five fold for the substances investigated here, and under 12000 kg. the compressibility of any one substance diminishes by not over 15 fold. The thermal expansion changes by a factor of two or three under 12000 for these liquids, and the specific heats and thermal conductivities do not vary more. Excepting water, the smallest effect of pressure on viscosity found above is that on methyl alcohol, which increases 10 fold under 12000, and the largest is by over 10^7 for eugenol (obtained by linear extrapolation, which gives too low a value).

In general the largest pressure effects are for those substances with the most complicated molecules. This is very plainly shown by the series of the alcohols, or by the various compounds derived from benzene; the relative pressure effect is greater the more complicated the group substituted for hydrogen. There is also a very marked constitutive effect, the iso-compounds having a larger effect than the normal compounds, and a similar effect is seen in the three xylenes. A heavier atom substituted into a molecule produces in general a larger pressure effect, as is shown in the series C_2H_5Cl, — Br, — I. or by chloro- and bromo-benzene. There appears, however, to be a tendency working in the other direction at low pressures. In the ethyl halogen series, the increase of viscosity produced by 500 kg. is in the order Cl, Br, I, whereas the increase under 12000 is in the order I, Br, Cl. It may well be that the abnormal effect at low pressure is due to the abnormally large compressibility of C_2H_5Cl, due to the neighborhood of the critical point, at which the compressibility is infinite. Abnormal behavior is also shown by methyl cyclohexane, the pressure effect being larger than for cyclohexane.

Water is quite different in character from the other liquids. Previous investigations have been made to 400 kg. by Hauser.[1] He found that, below 30°, viscosity decreases with increasing pressure, and above 30°, increases. At higher pressures we now find that at 0° and 10° there is a minimum viscosity, at a pressure roughly 1000 kg., the minimum being less pronounced at 10° than at 0°. At 30° and 75° there is a regular increase of viscosity with pressure over the entire range. (Not much weight must be placed on the precise numerical values given for 75°, there being much experimental uncertainty here because of electrical conductivity by the water.) It is natural to see in this abnormal behavior of water an association effect; at low pressures and temperatures, water is strongly associated with large molecules and a large viscosity, but as pressure increases the association decreases, and the average size of the molecules decreases, giving a term in the viscosity which diminishes fast enough to more than compensate the normal increase of viscosity under pressure. At higher pressures the association effect is exhausted, and the behavior becomes normal.

Theoretical Discussion.

It is generally recognized that the mechanism of viscosity is very different in a gas and a liquid. In a gas, viscosity increases with rising temperature, and is constant with increasing pressure, whereas in a liquid, viscosity decreases with rising temperature and increases with rising pressure. A further essential difference is shown by the connection between viscosity and thermal conductivity. In a gas, momentum and energy are conveyed by the same mechanism, so that reciprocal viscosity and thermal conductivity are proportional, whereas in a liquid the mechanisms are different, as is shown by the increase of thermal conductivity with increasing pressure and the very great decrease of reciprocal viscosity.

The mechanism of viscosity and thermal conductivity is well understood for a gas, but we have little understanding of the mechanism for a liquid, and there have been few attempts at theoretical explanation. In the following I shall not attempt to offer a theory of liquid viscosity, but shall try to show that the pressure phenomena make it pretty evident that there is a very important element in liquid viscosity which has been overlooked in previous theoretical attempts.

The few theoretical discussions of liquid viscosity have attached especial significance to the relation between viscosity and volume. Faust[1] drew from his measurements the conclusion that at high pressures viscosity tends to a limiting behavior, in which it is a linear function of the volume only, so that the viscosity is constant at a given volume independent of temperature (pressure varying). This behavior he found for CS_2 and ether in the pressure range of 3000 kg., but he found that ethyl alcohol did not approach such a behavior. The new data of this paper show that at high pressures the viscosity of all substances departs very far indeed from being a linear function of volume at a definite temperature, or from being a pure volume function of any kind. Faust saw a connection between the supposed linear relation between volume and viscosity and Tammann's theory that in the domain of high pressures the forces of molecular attraction are constant.

The most elaborate of the theories of liquid viscosity is that of L. Brillouin,[8] whose guiding idea is that part of the viscosity arises from a transfer of momentum through the liquid by elastic waves, in much the same way as Debye has a thermal conduction (energy transfer) by elastic waves. The mathematical working out of the idea is elaborate and reference must be made to the original paper; the important result is found that the dissipation of the continuity of the waves by temperature agitation produces a viscosity decreasing with rising temperature. No formula is given for the effect of pressure on viscosity at constant temperature, and indeed the theory must be extended to be capable of giving such a formula, but an exact expression is found for the change of viscosity with temperature at constant volume. The expression is:

$$\left(\frac{\partial \eta}{\partial \tau}\right)_v = \frac{5}{3}\frac{\chi}{V^2},$$

where χ is the thermal conductivity in absolute mechanical units and V is the velocity of sound. The value for the temperature coefficient of viscosity at constant volume given by the formula turns out to be about 100 times smaller than the coefficient as ordinarily measured at constant pressure (atmospheric). Brillouin tried to get an exact numerical check of his expression for $(\partial \eta/\partial \tau)_v$ from the data of Faust, and drew the conclusion that there was agreement within experimental error, which, however, he admitted might be very large.

The new data of this paper are sufficiently accurate to allow a check of the formula. It turns out that the formula fails by so large a factor as to constitute another order of magnitude. Thus for ether at a volume of 0.80 my value of viscosity at 30° and 75° gives

$$\left(\frac{\partial \eta}{\partial \tau}\right)_v = -.00106,$$

against a value of $(5/3)(\chi/V^2)$ of

$$2.18 \times 10^{-7}\left[= \frac{5}{3} \times \frac{1.38 \times 10^4}{(3.24)^2 \times 10^{10}}\right].$$

The values of χ, V, and volume involved in this calculation were taken from previous papers on compressibility and thermal conductivity.[3, 4] The difference between the calculated and observed value is so great that experimental error seems absolutely incapable of explaining it.

Another theory of viscosity has been developed by Phillips,[9] who has momentum transferred from part to part of the liquid by a quantum mechanism. He draws the conclusion that viscosity is a volume function only, so that his theory does not fit with these experiments even as well as that of Brillouin, who gives the right sign for $(\partial\eta/\partial\tau)_v$. Just how far η fails of being a pure volume function may be inferred from Figures 7 and 8, plotting log η against volume at 30° and 75° for i-amyl alcohol, ether, and CS_2. I-amyl alcohol, being associated, might be expected to be exceptional, but there is no such reason for the failure of ether and CS_2.

It seems fairly evident from the failure of the previous theories that there is some very important element in the situation not hitherto considered. This I believe to be an interlocking effect between the molecules which prevents the free motion of one layer of molecules over another. Slipping of two interlocking molecules past each other can take place only when haphazard temperature agitation has so far separated them that the interlocking parts are free. According to such a picture, viscosity would be expected to decrease with rising temperature both at constant pressure and constant volume. When the volume is decreased at constant temperature by increasing pressure, a comparatively small decrease of total volume may evidently produce a very large increase of interlocking, and the effect would be expected to increase more and more rapidly as the pressure increases. Furthermore, it is evident that the magnitude of the effect may be very different for different substances.

Such an interlocking effect would be expected to be most important in the most complicated molecules, and strong evidence in favor of such a picture is the very marked tendency found experimentally for the pressure effect to increase as the molecule becomes more complicated. In order to show this I have plotted the increase in log η produced by the first 500 kg. against a number which attempts to give

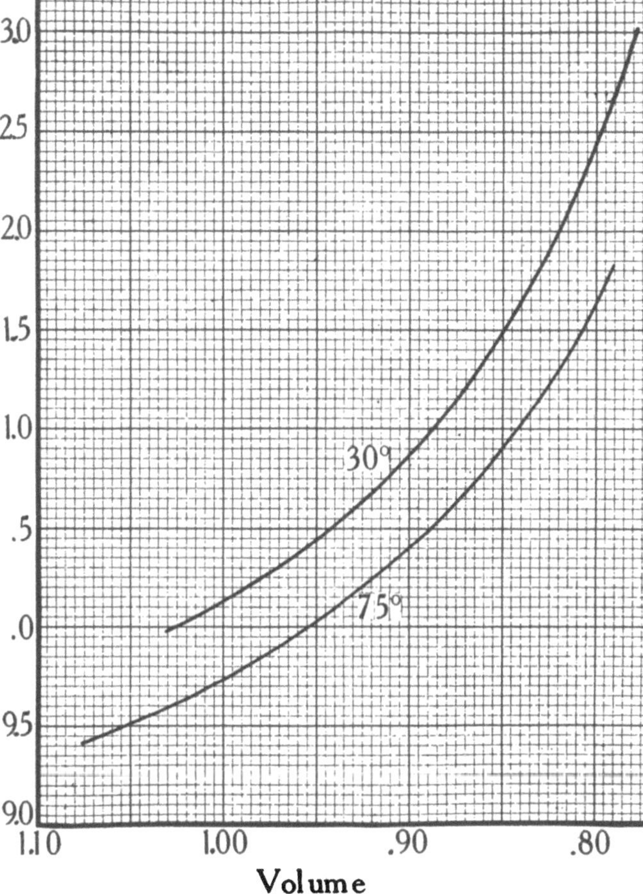

FIGURE 7. The common logarithm of relative viscosity at 30° and 75° of i-amyl alcohol as a function of volume.

some measure of the complication of the molecule. Evidently "complication of the molecule" is a very hazy concept, and any numerical measure of it can be only very crude. It is evident, I think, that the molecule may be more complicated either because it contains more atoms or because the atoms themselves are more complicated. As a measure of the complication of the atom I have taken the number of extra-nuclear electrons (the atomic number), and to measure the complication of the molecule I have multiplied the total number of extra nuclear electrons in all the atoms which the molecule contains by the total number of atoms in the molecule. We may call such a number the "complexity number." It evidently neglects many factors which we would like to include: for example, no distinction is made between an iso- and a normal alcohol, although our ordinary structural formulas would suggest that the molecule of an iso-alcohol is more likely to interlock with others of its kind than is a molecule of a normal alcohol.

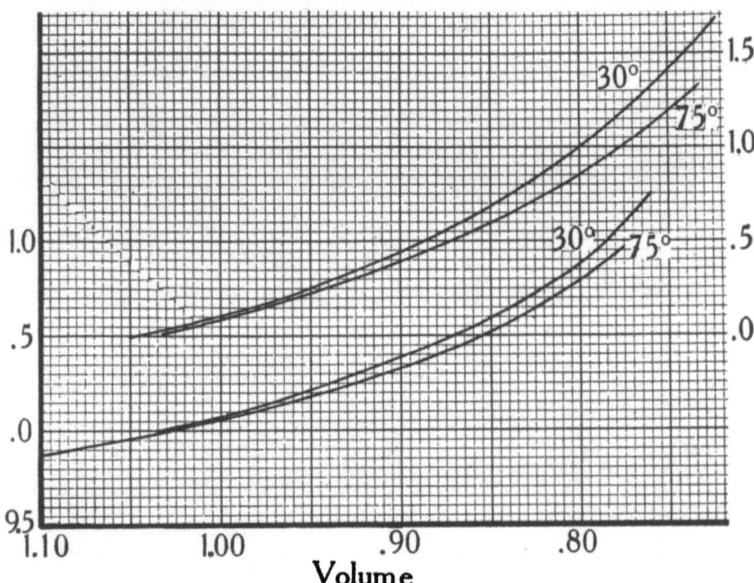

FIGURE 8. The common logarithm of relative viscosity at 30° and 75° of ether and CS_2 as a function of volume. The curves for ether are the upper curves with the scale of ordinates at the right.

In Figure 9 is plotted the effect of 500 kg. on log η against the logarithm of the complexity number. It is evident that there is a correlation which is perhaps as close as could be expected when the extremely arbitrary character of the complexity number is considered and the arbitrariness of comparing the pressure effect for all liquids at the same temperature, 30°, and at low pressures. It must be admitted that the correlation coefficient shown by the figure is not very high, but it will appear more significant if one will take the trouble to make similar plots against other properties which might be significant. I have not been able to find any other properties of the liquid which show nearly as close a correlation as the complexity number. For instance it has been suspected by some that viscosity in a liquid is due to the attraction between the molecules, which are prevented from moving freely with respect to each other by the attractive forces. If the molecular latent heat of vaporization is taken as a measure of the attractive force, and log η is plotted against it, no correlation whatever will be found. Or one might expect a correlation between absolute viscosity at atmospheric pressure and the pressure effect, but on plotting these two against each other much less correlation will be found than with the complexity number.

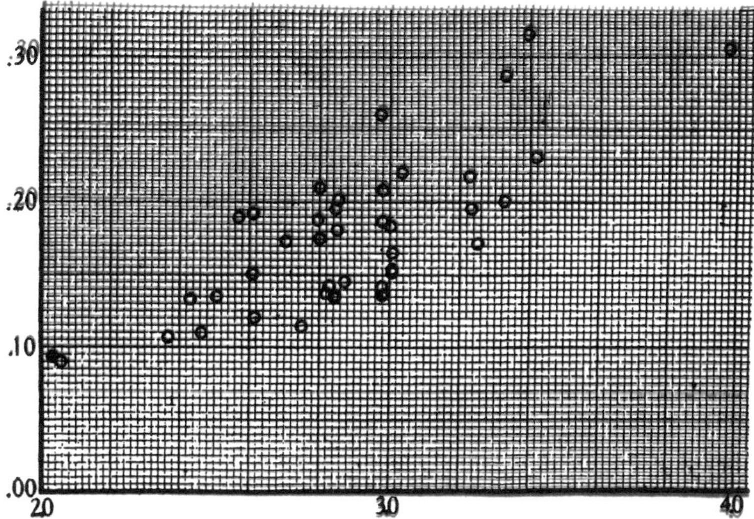

FIGURE 9. The common logarithm of the ratio of the viscosity at 500 kg. to that at atmospheric pressure at 30° C. of all the substances of this paper (except water) plotted as ordinates against the common logarithm of the "complexity number." The complexity number is defined as the product of the number of atoms in the molecule by the number of all the extra-nuclear electrons of the molecule.

If the correlation shown in the figure does not seem sufficiently real, perhaps still stronger evidence of the importance of molecular complexity may be found by confining oneself to a single related series of chemical compounds and noticing the effect of making the molecule more complicated in the series. Striking examples of this are the increase of pressure effect with molecular weight in the series of alcohols, or of hydrocarbons C_nH_{2n+2}, or the ethyl halogens.

If some such interlocking effect as suggested above is an important part of the viscosity mechanism, this means that so far as viscosity phenomena are concerned the molecule preserves its inviolability and continues to function as a unit when the volume is greatly decreased by high pressure. There are other phenomena, such as compressibility, in which the molecule seems to lose its significance at small volumes, and the atom becomes more significant. A reason for the difference may be seen in the different sorts of relative motion involved. If a molecule ceased to function as a whole during viscous shear it would be torn apart by the relative motion of the parts of the liquid, but the relative motion involved in a hydrostatic compression is not such as to destroy the molecule, even under comparatively large changes of volume.

Summary.

A method has been devised for the measurement of the viscosity of liquids under pressure and applied to 43 pure liquids up to a pressure of 12000 kg./cm.^{2} at 30° and 75° C.

Viscosity increases rapidly with increasing pressure; at low pressures the relation between viscosity and pressure is approximately linear, but at higher pressures viscosity increases much more rapidly. The effect of 12000 kg. may vary from 10 fold to 10^7 fold (excepting water). Log viscosity against pressure is at first concave toward the pressure axis, but above 3000 becomes nearly straight, and for a number of liquids even becomes convex toward the pressure axis. The temperature coefficient of viscosity increases under 12000 kg. by a factor of several fold, varying much with the liquid. Water is exceptional; at low temperature its viscosity decreases with rising pressure, but there is a minimum at about 1000 kg., and from here on the viscosity increases. At higher temperatures its viscosity increases at all pressures, but the increase is very much less than for any of the other liquids measured.

Incidentally a number of new freezing points have been determined under pressure, and the pressure-volume relation for glycerine roughly measured.

No theory proposed hitherto is adequate to account for these pressure phenomena. It is suggested that an interlocking effect between the molecules is an important element in the situation, an idea which is supported by the rapid increase of the pressure effect as the molecule becomes more complicated.

It is a pleasure to acknowledge the assistance given by my mechanic Mr. E. T. Richardson in setting up the apparatus and making many of the readings.

The Jefferson Physical Laboratory,
Harvard University, Cambridge, Mass.

References.

[1] W. Roentgen, Wied. Ann. 22, 510, 1884; Ann. Phys. Chem. (3) 45, 98, 1892.
R. Cohen, Wied. Ann. 45, 666, 1892.
L. Hauser, Ann. Phys. 5, 597, 1901.
O. Faust, ZS. Phys. Chem. 86, 479, 1914.

[2] J. H. Hyde, Proc. Roy. Soc. 97, 240, 1920.
Special Research Committee on Lubrication of the American Society of Mechanical Engineers, Progress Report in Mech. Eng. 45, May, 1923.

[3] P. W. Bridgman, Proc. Amer. Acad. 59, 139–169, 1923.

[4] P. W. Bridgman, Proc. Amer. Acad. 48, 310–360, 1912; 49, 1–114, 1913.

[5] P. W. Bridgman, Proc. Amer. Acad. 49, 627–643, 1914.

[6] E. C. Bingham and R. F. Jackson, Bull. U. S. Bur. Stds. 14, No. 298, 1917.

[7] F. G. Keyes, Jour. Math. and Phys. M. I. T. 1, 306–309, 1922.

[8] L. Brillouin, Jour. Phys. et Rad. 3, 326–340, 362–383, 1922.

[9] H. B. Phillips, Proc. Nat. Acad. Sci. 7, 172–177, 1921.

Boiling Points of Liquids from Eastman Kodak Co.,
see page 73

Liquid	Boiling point
n-propyl alcohol	96–98 C.
i-butyl alcohol	106–108
i-amyl alcohol	130–132
n-hexane	68–69
n-octane	124–126
i-amyl-decane	156–158
Ethyl chloride	12.5–13
Ethyl bromide	38–40
Ethyl iodide	71–72
Acetone	55.5–55.8
Cineole	53–54 at 8 mm.
Oleic Acid	U.S.P.
Aniline	mpt. —6
Diethyl aniline	102–104 at 10 mm.
Nitro benzene	mpt. 5
O-xylene	143.5–144.5
m-xylene	138–139
p-cymene	176.5–177.5
Eugenol	135–140 at 15 mm.

THERMAL CONDUCTIVITY AND THERMAL E.M.F. OF SINGLE CRYSTALS OF SEVERAL NON-CUBIC METALS.

By P. W. Bridgman.

Presented October 14, 1925. Received November 16, 1925.

TABLE OF CONTENTS.

Introduction.

In this paper previous investigations of the properties of several metals crystallizing in non-cubic systems[1] are extended to thermal conductivity and thermal e.m.f. The thermal conductivity is measured at room temperature and the thermal e.m.f. in the range between room temperature and 100° C. Fairly complete results are presented here for Zn, Cd, and Sn in all possible orientations in the crystal and for Bi over a restricted range of orientations, whereas for Sb and Te only measurements of thermal e.m.f. have been obtained and only for those orientations in which the cleavage plane is nearly parallel to the length. The mechanical difficulty of obtaining suitable specimens of the proper orientations is responsible for the lack of completeness.

66 — 2087

Apart from the exact numerical results, there are several interesting questions involved in the data presented here. With regard to thermal conductivity we have to ask whether the symmetry relations deduced by Voigt hold, and also whether the Wiedemann-Franz proportionality between electrical resistance and thermal conductivity holds in detail for all directions in the crystal. It will appear that the symmetry relations of Voigt do hold, but that the Wiedemann-Franz ratio does not hold in detail. With regard to the thermo-electric effects, we have to discuss the Thomson heat and the Peltier heat. It will appear that the analysis of Voigt considers only the Thomson heat, the Peltier heat at surfaces of discontinuity, which is numerically by far the more important effect, having been completely neglected. It is probable that the symmetry relations of Voigt apply to the Thomson heat, but that the symmetry relations of the Peltier heat are different. It is shown in the discussion that a Peltier heat varying with the orientation in the crystal involves an internal Peltier heat when the direction of current flow in the crystal changes.

Part I. Thermal Conductivity.

Method of Measuring Thermal Conductivity.

The methods available were much restricted by the requirement that the parts of the crystal subjected to measurement should not be subjected to any machining operation. Previous work with single crystals[1] has shown the very great difficulty of machining without seriously upsetting the crystal structure. The general scheme of the method adopted was as direct as possible; the temperature difference between two points at a known distance apart on a rod was measured when a known heat current passes along the rod. The heat current was produced electrically by a heating coil attached to one end of the rod, the other end of the rod being maintained at constant temperature in a temperature bath and so acting as a sink. The temperature difference was measured by a differential thermo-couple attached to the rod at intermediate points.

The experimental arrangements for measuring the thermal conductivity are shown in Figure 1. The metal is a unicrystalline casting, *A*, made by methods already fully described,[2] about 10 cm. long and very nearly 6 mm. in diameter. It is mounted by soldering with low melting solder into the massive copper block *B*. A special jig used during the setting of the solder ensured that the rod was central in the block. The heating coil *C* at the upper end was contained in a

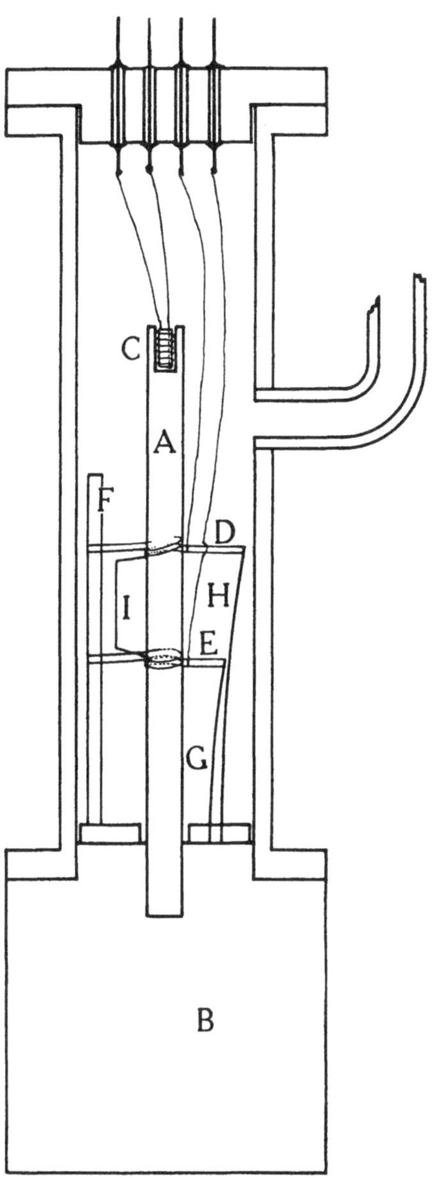

FIGURE 1. Section of apparatus for measuring thermal conductivity.

recess, 3.3 mm. in diameter and 6 mm. deep, drilled into the rod. Any damage to the crystal structure in drilling the hole was confined to the immediate neighborhood of the hole, and could not affect the lines of heat flow near the thermo-couples. The heating coil was of high resistance wire 0.009 cm. in diameter, wound in a thread cut in a miniature cylinder of pipestone, the total resistance being about 35 ohms. The coil was insulated externally with a wrapping of paper 0.0025 cm. thick; before an experiment the insulation resistance between coil and rod was checked and required to be over 10^8 ohms. The leads to the heating coil were of copper 0.013 cm. in diameter. To insure good thermal contact between coil and rod the hole in the upper end of the rod was filled with vaseline before inserting the coil, and the excess vaseline was afterward wiped away.

The thermo-couple was of copper-constantan ribbon, made by rolling to a thickness of less than 0.0025 cm. wire originally 0.015 cm. thick. The couple was held in contact with the rod with springs. The details of construction of the couple were as follows: Two ribbons of copper, *D* and *E*, attached at one end by soldered loops to the insulating pillar *F*, were passed once around the rod, and at the other ends were held taut by the piano wire springs *G* and *H* from which they were insulated by glass sleeves. Underneath the two loops of copper was passed the constantan ribbon *I*, which was then soldered to the copper with the minimum amount of solder. As shown in the figure, the constantan ribbon was bent at right angles to the rod where it made contact with the copper. Copper connections 0.012 cm. in diameter, soldered to the ribbons *D* and *E* near the rod, led to a potentiometer and provided means of measuring the potential difference and so the difference of temperature between the two ends of the constantan ribbon. The thermo-couple was insulated from the rod by a single wrapping of oiled paper 0.0025 cm. thick, stuck to the rod with a very thin coating of air drying lacquer. The insulation resistance between couple and rod was always required to be greater than 10^8 ohms. To ensure better thermal contact between rod and couple the loop around the rod was painted with lacquer above the paper after assembling. The distance between the junctions constantan to copper was read for each specimen with a telescope mounted on a comparator, and was always nearly 2 cm.

The springs holding the thermo-couple taut were so mounted that they could be slacked after a run, and the crystal rod slipped out through the loosened loops. The couple could conversely be slipped over a freshly mounted rod and the loops tightened into place. Similarly the heating coil at the upper end was removable from specimen to specimen. All the measurements on thermal conductivity on all the specimens were made with the same thermo-couple and same heating coil, and for all except Bi with the same heat input, so that all the results are comparable, apart from any question of absolute values.

The rod with the thermo-couple attached was placed inside a brass cylinder 3 cm. in inside diameter and 20 cm. long. This cylinder was provided with flanges at the top and bottom by which vacuum tight connection was made to the copper block *B* and to a brass plate at the upper end through which passed terminals for the thermocouple and the heating coil insulated with glass tubing and deKhotinski cement. Connection to the vacuum pump was made with a brass tube soldered into the side of the cylinder. The vacuum was produced with a mercury diffusion pump with the conventional forepumps. There was provision for a liquid air trap, but this was never used. Trial showed no difference in the results when the trap was packed in ice or when left at room temperature. Since this difference of temperature corresponds to a difference of vapor pressure of the mercury of about eight fold, it was concluded that the heat dissipated by convection or conduction by the gas surrounding the rod was negligible, and that the diffusion pump vacuum at ordinary temperatures was sufficient. The importance of the vacuum became at once obvious if one attempted readings at full atmospheric pressure. Here there were fluctuations so violent that no consistent readings could be obtained; what rough readings could be obtained indicated a conductivity materially higher than given with the vacuum, showing that an appreciable part of the heat input was carried away by the gas at atmospheric pressure.

The electrical measurements do not need detailed description The thermal e.m.f. of the couple and the heating current were both measured with the same potentiometer that was previously used in measuring the effect of pressure on the thermal conductivity of metals,[3] and a full description will be found in that paper. The heating current was provided by a storage battery of 12 volts in series with appropriate resistances and a ballast lamp to maintain constancy. The maximum heating current was about 0.4 amp. The sensitiveness of the measurements was very much better than 0.1%, and very much better than the consistency of the measurements with different specimens.

The experimental procedure was as follows: The rod was soldered into the copper block *B*, paper strips 6 mm. wide were lacquered on at the mean position of the thermo-couples, the thermo-couples adjusted, the heating coil inserted, soldered connection made between the couple and the heating coil leads and the fixed copper terminals in the upper brass plate, clamps applied to the flanges to make them vacuum tight (the conventional vacuum wax was used underneath the flanges), the assembly was then placed in the temperature bath of water at room temperature with the upper brass plate projecting a couple of cm. above the surface of the water, connections were made to the vacuum pump through a conical joint between glass and brass sealed in deKhotinski cement (the diffusion pump and the brass assembly were slung from counterpoised arms to allow free relative motion), the insulation resistance was tested between the thermocouple and the heating coil and the grounded rod, soldered connection made between the outer terminals in the upper brass plate and the potentiometer and the source of heating current, and the vacuum pump and heating current started. The vacuum was tested by noting the character of the discharge through a small discharge tube excited by a transformer on the commercial 60 cycle 110 volt circuit. No final readings were made until ½ hour had elapsed after the last previous change of heating current, this interval having been proved sufficient by trial, and after reaching a vacuum so high that all discharge had ceased. Readings were made for two different heat inputs, one about 4 times the other; the readings with the smaller heat input were made first and were merely by way of check. These were sometimes made before complete equilibrium was reached. In general the conductivity calculated from the small heat input was a few per cent higher than that with the larger input; the results with the larger input are to be preferred, and are the only ones retained in the final results.

Detailed Results.

The theoretical connection between thermal conductivity and direction in the crystal has been worked out by Voigt; the connection is the same as for the electrical conductivity. This means that for the crystals investigated here, which have rotational symmetry, the behavior of the conductivity in all directions is completely characterized by two constants. These two constants are the thermal conductivity parallel and perpendicular to the axis of rotational symmetry. Denote these conductivities by κ_{11} and $\kappa_{\perp}$, and introduce the reciprocal conductivity or thermal resistance, namely

$$\lambda_{11} = \frac{1}{\kappa_{11}}, \quad \text{and} \quad \lambda_{\perp} = \frac{1}{\kappa_{\perp}}.$$

The relation of Voigt is now

$$\lambda_\theta = \lambda_\perp + (\lambda_{11} - \lambda_\perp)\cos^2\theta,$$

where λ_θ denotes the reciprocal conductivity in the direction inclined at the angle θ to the axis. The relation for electrical resistance is the same, namely

$$\rho_\theta = \rho_\perp + (\rho_{11} - \rho_\perp)\cos^2\theta.$$

Now eliminate θ between the equations, giving

$$\lambda_\theta = \frac{\rho_\theta(\lambda_{11} - \lambda_\perp) + \rho_{11}\lambda_\perp - \rho_\perp\lambda_{11}}{\rho_{11} - \rho_\perp}.$$

Hence λ_θ is a linear function of ρ_θ, in virtue merely of the symmetry relations, and without any special hypothesis connecting electrical and thermal conductivity.

One of the points of chief interest in this investigation of thermal conductivity is whether the Wiedemann-Franz ratio continues to hold for the individual directions in a crystal. If this is the case, λ in any direction is proportional to ρ in that direction, which is at once seen to be consistent with the equation above, for on putting $\lambda_{11} = \alpha\rho_{11}$, and $\lambda_\perp = \alpha\rho_\perp$, the equation collapses to

$$\lambda_\theta = \alpha\rho_\theta.$$

Hence if the Wiedemann-Franz ratio holds for all directions, not only is λ_θ a linear function of ρ_θ, but the straight line passes through the origin. In the following this criterion is applied graphically to the thermal resistance.

In computing the results, the data needed are the heat input, given by the heating current and resistance of the heating coil, the temperature difference, given by the thermal e.m.f. of the couple and the constant of the couple, the cross section of the rod, and the distance between the junctions of the thermo-couple. In terms of all these data the absolute thermal conductivity can be at once obtained, provided that heat losses be neglected. The heat losses are of two kinds; losses by radiation or by convection and conduction through the small amount of gas remaining, and loss by conduction along the leads of the heating coil or an effective loss by conduction along the wires of the thermo-couple. The precise amount of these various losses would be difficult to determine experimentally; calculation shows most of them to be very small, and direct experiment by varying the gas pressure showed that loss through the gas must be small. In the calculation of conductivity such losses were entirely neglected, and it must be recognized that there is here a possible source of error. The heating was so small that the rise of temperature above the surroundings at the heating coil end of the rod was only 2° for Sn, less for Zn and Cd, and not over 7.5° in the extreme case of Bi. The heat loss should be proportional to the temperature rise and therefore inversely proportional to the thermal conductivity. There is another possible source of error in the value of the constant assumed for the thermo-couple. It would have been difficult to calibrate the couple directly, and the value (100×10^{-6} volts = 2.48° C.) given by L. H. Adams[4] in his Table for copper—" Ideal " was assumed. It is questionable, however, whether the very great mechanical deformation involved in rolling the wire flat (it was not annealed afterward) may not have seriously affected the constant. However, since all specimens were measured with the same couple, any error in the constant can produce no error in the relative conductivities of the various samples, and can have no effect on any conclusions made as to connections between electrical and thermal conductivity. The absolute thermal conductivities calculated, neglecting heat losses and with the assumed value for the thermo-couple constant, are uniformly too high to be consistent with the values given in the literature for cast rods in which the crystal grains are at haphazard. The inconsistency amounts to about 20% for Cd, Zn, and Sn, and for Bi to about two fold. I do not believe that nearly all of this discrepancy can be due to the factors discussed (heat leak and constant of the couple) but that an important part must be real, which means that the thermal conductivity in my single crystals is high, arising perhaps from a combination of unusual purity with a perfectly regular atomic arrangement. It is to be expected that this effect will be particularly large in Bi, in which the crystalline character is most pronounced, and cleavage most easy.

The Specimens. The rods were cast and examined to ensure that they contained only one crystal grain by methods already described.[1] A simple method of controlling the orientation of the crystalline axis with respect to the axis of the rod has not yet been developed, and I did as before, selecting those best oriented from a number of castings. Because of the small diameter of the casting, only 6 mm., large angles of inclination between the axis of the crystal and the casting were more common than in the larger size castings.

Zinc was Kahlbaum's best; measurements were made on 8 rods, of angles varying from 86.5° to 33°. The angle was determined from the position of the cleavage plane.

Bismuth was electrolytic metal from the U. S. Metals Refining Co., the same as the Bi of the previous measurements. Before casting into the 6 mm. rods it had been cast by the regular procedure for making unicrystalline castings into rods 2.5 cm. in diameter. These were not one grain. The lower part of these rods, which thus had experienced an additional purification, was used for the 6 mm. rods. In the previous paper it was mentioned that it is particularly hard to get Bi in unicrystalline rods, there being a very strong tendency to form several grains of very nearly the same orientation. This difficulty was again found in high degree, and in many trials only a few satisfactory rods, of angles varying from 90° to 68.5° were obtained. Since these measurements were completed, however, a much better method has been developed for obtaining unicrystalline rods of Bi. The difficulty is apparently connected with the expansion on solidifying. The crystal that separates in the bottom of the container is lighter than the liquid and is therefore mechanically unstable. Portions of the crystal tend to detach themselves after being laid down, and act as new centers of crystallization. This tendency may in large part be avoided by drawing the mold up through the top of the furnace instead of lowering through the bottom, so that the Bi crystal is in a mechanically stable position at the top.

The angle between the axis of the casting and the crystal for these Bi rods was determined from the cleavage plane.

Cadmium was Kahlbaum's best; successful measurements were made on 9 rods. The singleness of the castings was determined in the regular way by the appearance of the castings in reflected light. There was a difficulty not met before, however, in determining from the reflection pattern the location of the crystal axis, since for some reason under these special conditions not all the faces of the reflection pattern were developed. It turned out to be much simpler to obtain the location of the axis from measurement of the specific electrical resistance, the specific resistance parallel to and perpendicular to the axis having been previously determined. In measuring the resistance the apparatus of the previous paper was used. The values found for the specfic resistance varied from 6.92×10^{-6} to 8.31, the previous values for $\rho_{\perp}$ and $\rho_{\parallel}$ being respectively 6.80 and 8.30. It thus appears that the 9 rods used ranged through nearly all the possible orientations.

Tin was measured in 5 specimens; 3 of these were Bureau of Standards melting point samples, and 2 were Kahlbaum's purest. Again, although the reflection patterns were sufficiently developed to determine the uniqueness of the grains, all the faces were not present, so that the orientation of the axis had to be determined from the specific electrical resistance. This ranged from 9.92×10^{-6} to 13.20; the previous values are $\rho_{\perp} = 9.9$, $\rho_{11} = 14.3 \times 10^{-6}$, so that about two-thirds of the total possible range of orientations is here represented.

Numerical Results. The accuracy of the results is so low that graphical representation is adequate. In Figure 2 is plotted for all

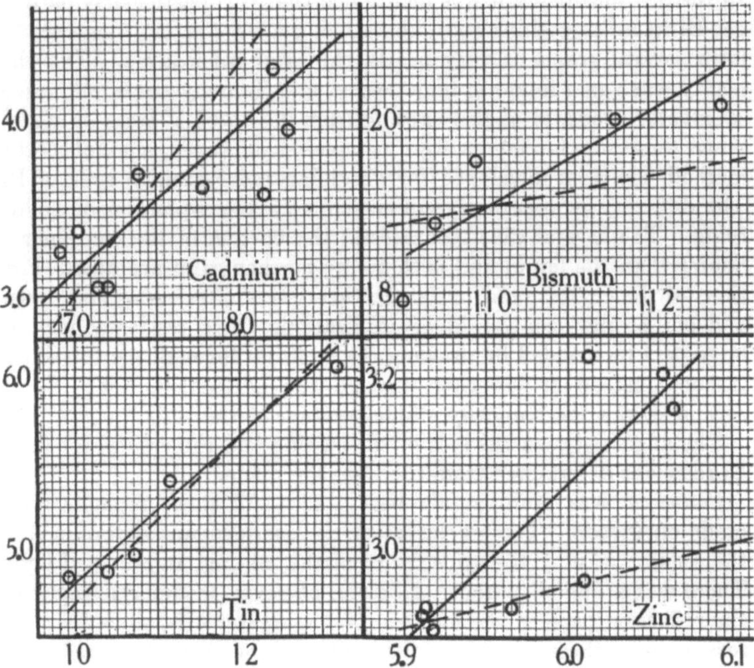

FIGURE 2. Reciprocal of thermal conductivity (ordinate) against specific resistance $\times 10^6$ (abscissa).

the metals the reciprocal of thermal conductivity (calculated as explained) against specific electrical resistance of the same sample. According to the symmetry relations developed by Voigt[5] the relation between these two should be linear. The heavy lines in the diagrams appear to be the best straight lines connecting the observed points. The experimental error is seen to be considerable. If, further, the Wiedemann-Franz ratio holds for the individual directions in a crystal, the straight line should pass through the origin. The dotted lines in the diagrams are the best lines drawn through the origin.

In the case of Sn and perhaps in that of Bi, this line connecting with the origin may also within experimental error be the line on which the points actually lie, so that for Sn and possibly for Bi the proportionality between electrical and thermal resistance holds for different directions in a single crystal. It has already been found by Kaye and Roberts[5a] that in Bi the ratio of the thermal conductivities in the perpendicular and parallel directions is 1.39, which is to be compared with my value[1] 1.27 for the ratio of the electrical conductivities. The discrepancy is in the same direction as indicated in Fig. 2.

But with Cd and Zn there seems no room for doubt that the experimental points do not lie within experimental error on the line through the origin, so that for these metals the Wiedemann-Franz proportionality does not hold in detail for different directions in the crystal. In the case of cadmium, electrical resistance varies more with direction than thermal conductivity, and for Zinc thermal conductivity varies more than electrical resistance. It would seem that these conclusions cannot be affected by any uncertainties in the values given for the absolute conductivities.

Whether or not the Wiedemann-Franz ratio holds, it is evident that for all four metals the electrical conductivity increases in the same direction in which the thermal conductivity increases.

Part II. Thermal E.M.F.

Methods of Measuring Thermal E. M. F.

The single crystal rods on which the thermal e.m.f. measurements were made were 6 mm. in diameter and about 15 cm. long. In point of time the thermal e.m.f. measurements were made before the measurements of thermal conductivity; after the thermal e.m.f. measurements some of the rods were selected for the measurements of thermal conductivity, being cut down to 10 cm. in length. The number of specimens used was as follows. Zn, 8 thermal e.m.f. specimens were measured and all 8 were again used for thermal conductivity; Sn, 14 thermal e.m.f. specimens were used from which 5 were selected for thermal conductivity; Cd, there were 14 thermal e.m.f. specimens from which 9 were selected for thermal conductivity; Bi, there were 10 thermal e.m.f. specimens of which 5 were used for thermal conductivity; Sb, there was only 1 thermal e.m.f. specimen which was not suitable for thermal conductivity; Te, there were 3 thermal e.m.f. specimens of which none were suitable for thermal conductivity. With regard to Sb and Te it is to be noticed that geometrical imperfections are not a source of inaccuracy in measurements of thermal e.m.f., whereas geometrical perfection is necessary for a good measurement of thermal conductivity.

The two ends of the specimen were maintained at different temperatures by two oil baths. These baths were contained in rectangular copper boxes placed about 2.5 cm. apart, and were provided with stuffing boxes through which ran the crystal rods, projecting about 6 cm. into the bath and 10 cm. below the surface. The stuffing boxes were carefully designed so as not to exert an appreciable mechanical stress on the rods, since a mechanical stress at the stuffing boxes, which is the region of rapid change of temperature, would introduce extraneous e.m.f.'s. Both baths were vigorously stirred. One bath was provided with an electric heater with which temperature was varied in the range between room temperature and 100° C. No regulator was used; in changing temperature a heavy heating current was used, and after the desired temperature was reached it was kept very nearly constant by adjusting the heating current to that value which experiment showed was appropriate to that particular temperature. The exact temperature of each bath was read on calibrated thermometers to 0.01°. Since temperature equilibrium was attained between the crystal rods and the bath almost immediately, the error due to drift in the bath temperature was negligible. Readings were made with both ascending and descending temperature; the usual temperatures of the hot end were: room temperature, 40°, 60°, 80°, 100°, 75°, and 55°. The cold end, at the temperature of the cold bath, started at room temperature and gradually increased by conduction through the rod until at the end of the run it was usually about 28°. In order to decrease the total time of the run, the bath was cooled for the decreasing readings by drawing out hot oil and pouring in cold. The total time of a run was about 2 hours.

The thermal e.m.f. measured was that between the crystal rod and commercial copper wire. Connections were made to the crystal rods by soft soldering to each end a copper wire 0.030 cm. in diameter. The identical copper wire was used with all the specimens so that the relative results are not affected by any peculiar properties which this particular copper wire may have had.

It is evident from the dimensions of the rods and the bath that the temperature was perfectly uniform in the neighborhood of the soldered connections, and that therefore a true measure was obtained of the thermal e.m.f. between the crystal and copper.

The thermal e.m.f. was measured by a null method on the same potentiometer which was previously used in measuring the effect of pressure on thermo-electric quality,[6] and it has already been sufficiently described in detail. The thermal e.m.f. of Bi was so large that it was measured with a Siemens and Halske millivoltmeter, which was especially calibrated for this work by a simple method with a standard cell and high resistances.

In calculating the results, the observed e.m.f.'s had to be first corrected for drift of the temperature of the cold end, reducing all readings to a cold end temperature of 20°. This was easily done graphically from the readings at room temperature, 40°, and 60°, the curvature in no case being high enough to introduce perceptible error into the correction. The corrected readings were then plotted on large scale plotting paper, and a smooth curve drawn through the points. In almost all cases the ascending and descending readings agreed within experimental error. This is an important point, and is evidence of freedom from internal changes in the crystal produced by changes of temperature, such as might occur if there were stresses at the stuffing boxes or if there were incipient cleavages. This condition has not been attained in considerable of the work previously done with Bi.

The smooth curve was in all cases within experimental error a curve of the second degree. From these curves two data were now taken, the total e.m.f. of the rod between 20° and 100° (E_0), and the deviation from a linear relation of the observed e.m.f. at the mean temperature of 60° (Δ). These two data are sufficient to determine the two constants of the second degree relation between temperature and e.m.f. For each of the metals E_0 and Δ were now plotted against a parameter determining the orientation of the rod with respect to the crystal axes. In the case of Zn and Bi this parameter was the angle between the basal plane and the length of the rod; in the cases of Sn and Cd it was the specific electrical resistance at 20° C. of the rod, which was especially measured. It has already been explained that for these samples of Sn and Cd the specific resistance gave the most reliable determination of orientation. Through the observed values of E_0 and Δ smooth curves were now passed, and from these smooth curves the values of E_0 and Δ were taken at regular intervals. From these values the corresponding values of Peltier heat and Thomson heat against copper were calculated by well known methods as follows:

The relation between total e.m.f. and temperature is of the form

$$E_{Cu-M} = a(t-20) + b(t-20)^2,$$

where t is temperature in degrees Centigrade. The usual sign convention is employed; a positive E means that current flows from copper to the crystal at the hot junction. Now for the Peltier heat between the metal under investigation and copper we have the familiar thermodynamic relation

$$\begin{aligned} P_{Cu-M} &= \tau \frac{dE}{d\tau} \\ &= \tau[a - 40b + 2bt] \\ &= \tau[a' + 2bt] \end{aligned}$$

where τ is absolute Centigrade temperature and $a' = a - 40b$. For the Thomson heat we have

$$\sigma_M - \sigma_{Cu} = \tau \frac{d^2E}{d\tau^2} = 2b\tau.$$

Now by definition,

$$E_0 = a \cdot 80 + b \cdot \overline{80}^2,$$

and

$$\begin{aligned} \Delta &= \frac{80}{2}[a + b \cdot 80] - [a \cdot 40 + b \cdot \overline{40}^2] \\ &= b\frac{\overline{80}^2}{4}. \end{aligned}$$

These equations, solved for a' and b give

$$a' = \frac{E_0 - 6\Delta}{80}, \qquad b = \frac{4\Delta}{\overline{80}^2}.$$

The values of a' and b calculated in this way were now plotted against the orientation parameter. For the sake of uniformity and for theoretical reasons to be described later, the orientation parameter was in all cases now chosen as the specific resistance. For Zn and Bi the specific resistances were calculated from data given in a previous paper[1] from the location of the cleavage plane. It will be noticed that except for the factor τ, a' and $2b$ are equal to P_{Cu-M} and $\sigma_M - \sigma_{Cu}$ at 0° C.

There now follows the detailed data for the various metals.

Detailed Data.

Zinc. The 8 samples used have already been described under the measurements of thermal conductivity. In Figure 3 are shown the observed e.m.f.'s corrected for the drift of the lower temperature for

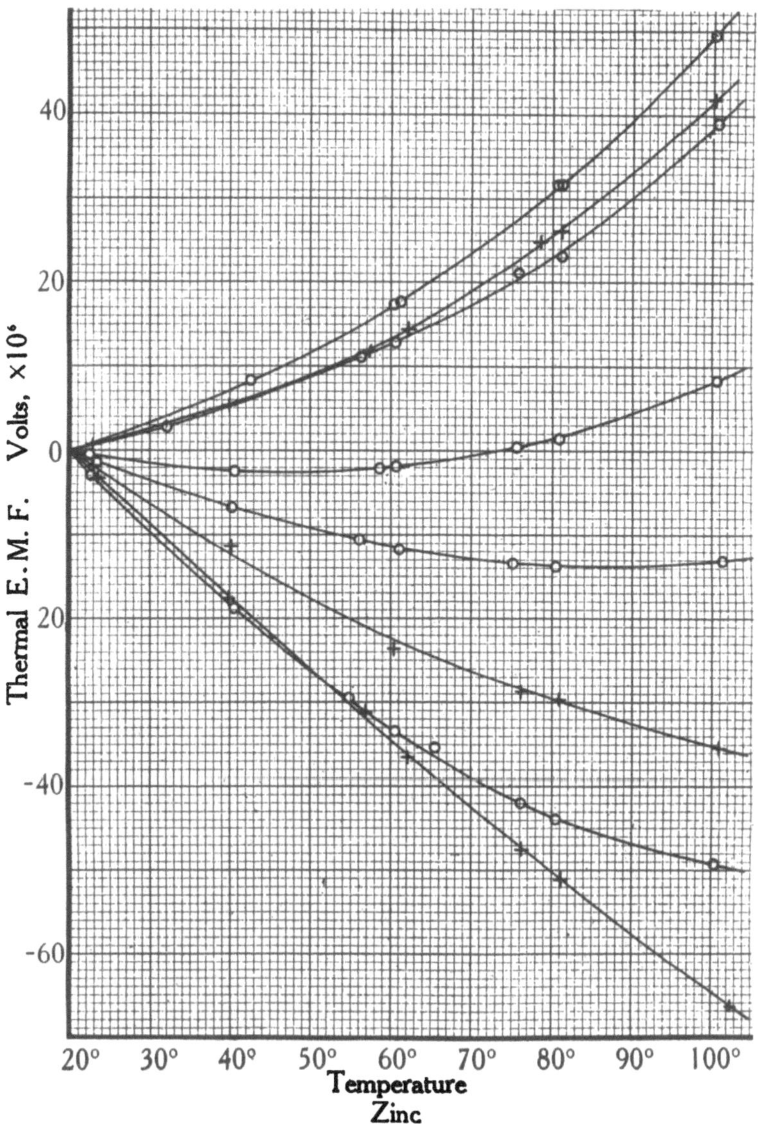

FIGURE 3. Thermal E.M.F. against copper as ordinate against temperature of the hot junction as abscissa for zinc rods of different orientations. Reading from the top down, the axis of the crystal makes the following angles with the axis of the rods: 86.5°, 83°, 80°, 60°, 57°, 48°, 46.5°, 35°.

all the specimens of zinc of various orientations. Notice that the thermal e.m.f. changes sign with the orientation. This series for Zn gives a fair idea of the experimental accuracy reached with the other metals also, and detailed data will not be given for them. The experimental curves for Zn are of the second degree within experimental error. The simplest criterion of this is that at 60° the difference between the actual curve and the straight line connecting the 20° with the 100° point is 4/3 as great as the difference at 40° and 80°. From the family of curves of Figure 3 the values of E_0 and Δ were obtained graphically, and are shown in Figure 4, plotted against the angle between the hexagonal axis and the length of the rod. In Figure 4 are also drawn the smooth curves which seemed to best

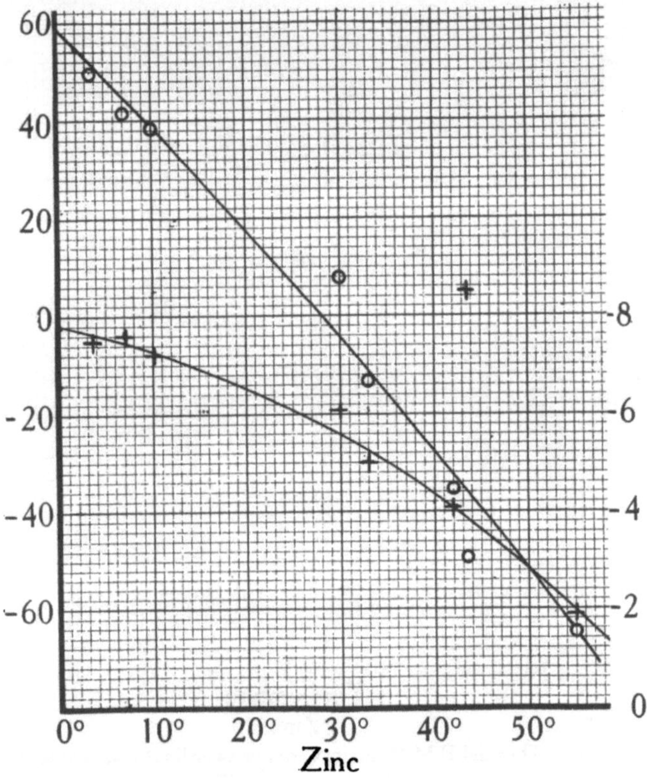

FIGURE 4. Values of E_0 in micro-volts (circles, scale to the left) and Δ in micro-volts (crosses, scale to the right) against the angle between the crystal axis and the length of the rod as abscissa for zinc.

summarize the experimental results. There does not seem to be much question possible as to the best way to draw these curves; only one experimental point is badly off, the value for Δ of the 43.5° specimen. It is of course evident that the percentage error in Δ is much larger than in E_0.

Finally from the smooth curves for E_0 and Δ, the values of a' and b were calculated, and are plotted in Figure 5 against specific resistance.

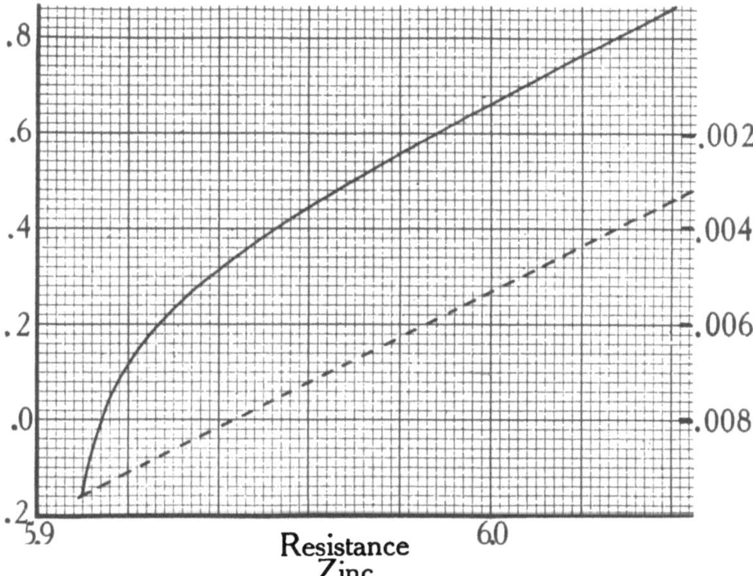

FIGURE 5. Values of a' (full line, scale to left) and of b (dotted line, scale to right) against specific resistance of zinc, a' is closely related to the Peltier heat, and b to the Thomson heat; see equation on page 114.

In computing the specific resistances from the angles of the axis the values $\rho_{11} = 6.13 \times 10^{-6}$ and $\rho_{\perp} = 5.93 \times 10^{-6}$ were used. It appears from Figure 5 that the relation between b and resistance is linear within experimental error, but that the relation between a' and specific resistance cannot well be linear.

Bismuth. 10 rods were used, of the material already described under thermal conductivity, of orientations between the cleavage plane and length varying from 0° to 21.5°. The departure of e.m.f. from linearity with temperature is relatively much less for Bi than for

Zn, so that the value deduced from the curves for Δ is relatively much more uncertain. The following values were found (Table I.). In

TABLE I.

THERMAL E.M.F. DATA FOR BISMUTH.

Angle between Cleavage Plane and Length	Total E.M.F. against Cu 20° to 100°	Departure from Linearity at 60°	Average by Groups		
			Angle	Total E.M.F.	Departure
0°	4460×10^{-6}	50×10^{-6}	0°	4500	25
0°	4540	0			
4.5°	4610	0	5.5°	4610	20
6.5°	4610	40			
10°	5040	80	10°	5040	80
17.5°	5080	100	17.7°	5110	50
18.0°	5150	0			
20.3°	5200	0	21.1°	5170	70
21.5°	4900	100			
21.5°	5420	100			

plotting, rods of nearly the same orientation were grouped together. The results by groups are plotted in Figure 6, with the smooth curves.

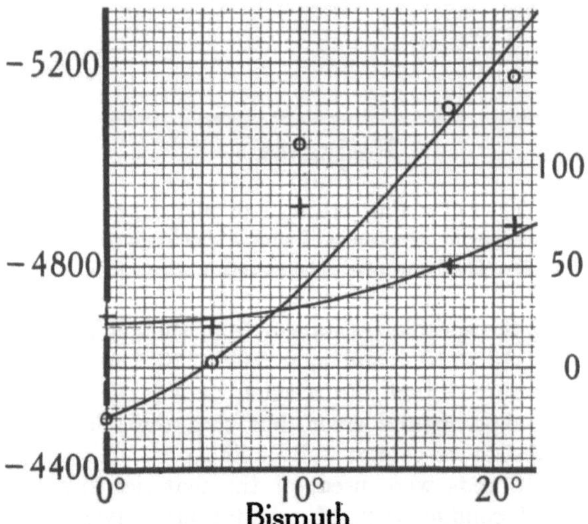

FIGURE 6. Values of E_0 in micro-volts (circles, scale to the left) and Δ in micro-volts (crosses, scale to the right) against the angle between the basal plane and the length of the rod as abscissa for bismuth.

The proper location of the curve for Δ is evidently in much greater doubt than that for E_0.

In Figure 7 are plotted a' and b calculated from these curves at intervals of 0°, 5°, 10°, 16°, and 22°. In converting orientation to specific resistance the values $\rho_\perp = 109 \times 10^{-6}$ and $\rho_{11} = 138 \times 10^{-6}$ were used. Again it is evident that within experimental error b is linear against specific resistance, and a' is probably not linear, although the departure is much less than for *Zn*.

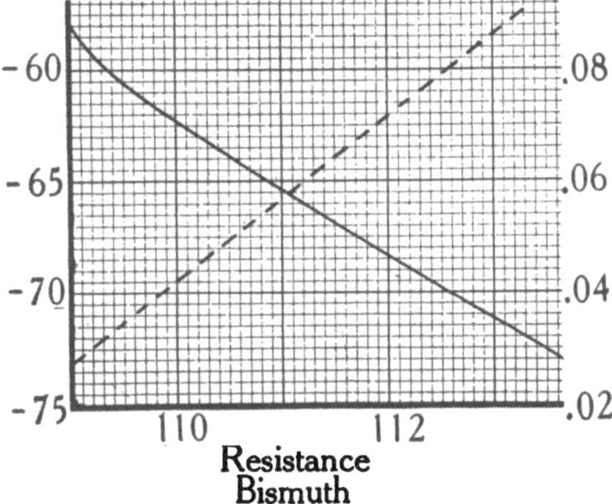

FIGURE 7. Values of a' (full line, scale to left) and of b (dotted line, scale to right) against specific resistance of bismuth. a' is closely related to the Peltier heat, and b to the Thomson heat; see equation on page 114.

Cadmium. 14 rods were used, of which 9 were later selected for the thermal conductivity measurements. These were all of Kahlbaum's best grade of Cd. The specific resistances of these rods were distributed with fair uniformity over the range of values from 6.88 to 8.36×10^{-6}. The values previously found for the specific resistances are: $\rho_\perp = 6.80$, $\rho_{11} = 8.30$, so that practically the entire range of possible orientations was represented here. (The high value 8.36 given above against 8.30, the maximum previously found, deviates no more than single samples sometimes do.)

In calculating the results, there was at first considerable question whether the relation between temperature and e.m.f. was really of the second degree or not. Several of the individual curves differed appreciably from the quadratic form, but the discrepancies were irregular, sometimes the greater curvature being found at the higher temperature end of the curve, and sometimes at the low temperature end. In order to see whether there was any systematic deviation from the second degree relation, the deviations from linearity of all the curves were found at 40°, 60°, and 80°, and the deviations at each temperature plotted against specific resistance. It was found that the irregularities were unsystematically distributed, so that the best smooth curve through the 40° points was identical within experimental error with that through the 80° points and the ordinates of this curve were ¾ of those of the smooth curve through the 60° points. But this is exactly the condition that the curve be of the second degree. All three of these deviation curves were linear in specific resistance.

In Figure 8 is shown E_0 and Δ as a function of specific resistance

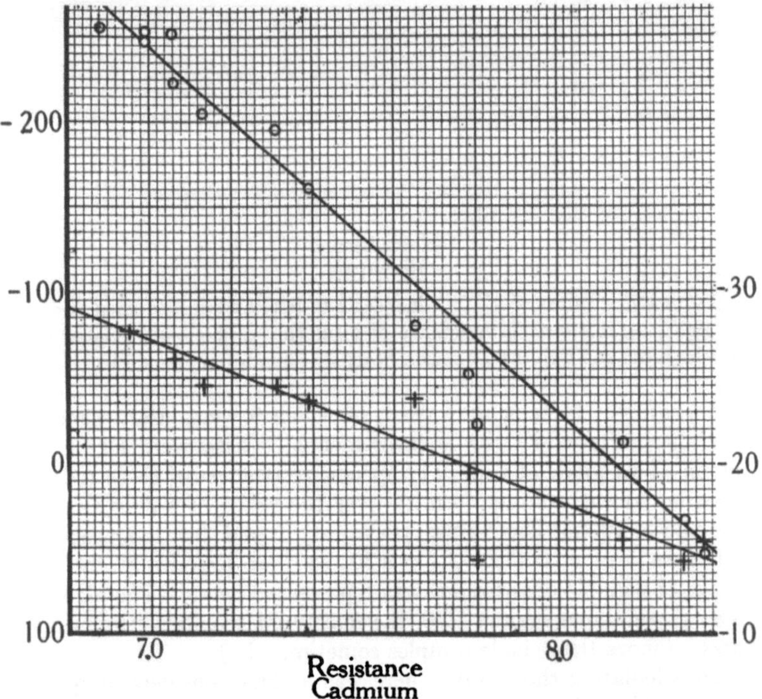

FIGURE 8. Values of E_0 in micro-volts (circles, scale to the left) and Δ in micro-volts (crosses, scale to the right) against the specific resistance in the direction in question as abscissa for cadmium.

(In the curve for Δ several of the values at low specific resistances are averaged into a single point, the values being so nearly coincident that it was not easy to separate them graphically.) In Figure 9 are plotted the values calculated from E_0 and Δ of a' and b. Since E_0

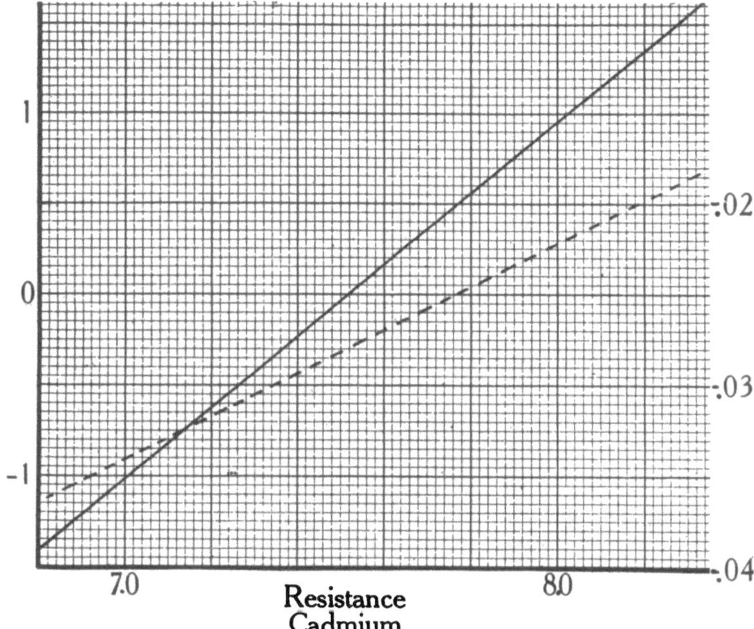

FIGURE 9. Values of a' (full line, scale to left) and of b (dotted line, scale to right) against specific resistance of cadmium. a' is closely related to the Peltier heat, and b to the Thomson heat; see equation on page 114.

and Δ are linear in specific resistance, a' and b must be linear also, so that it was necessary to calculate only two points on each of the curves. Notice that this is the first case yet found of an a' linear with specific resistance.

Tin. 14 rods were used; of these 14 rods, 8 were of Bureau of Standards melting point tin, and 6 were of Kahlbaum's purest grade. No difference whatever could be seen in the behavior of tin from these two sources, and the results obtained with both were averaged indiscriminately. The range of specific resistances of these rods was from 9.89 to 13.23 $\times 10^{-6}$ against 9.9 to 14.3 found previously for the entire range of orientations. However, the distribution of the rods over the range of orientations was not uniform, tin showing a particular disinclination to crystallize with the basal plane perpendicular to the length. Of the samples used 1 had a specific resistance of 13.23, 3 were between 11.1 and 11.2, 1 was 10.6, and the rest were between 9.9 and 10.3.

The relation between thermal e.m.f. and temperature was much more nearly linear than usual, so that the value of Δ is more in error. In Figure 10 are shown the experimental values of E_0 and Δ against

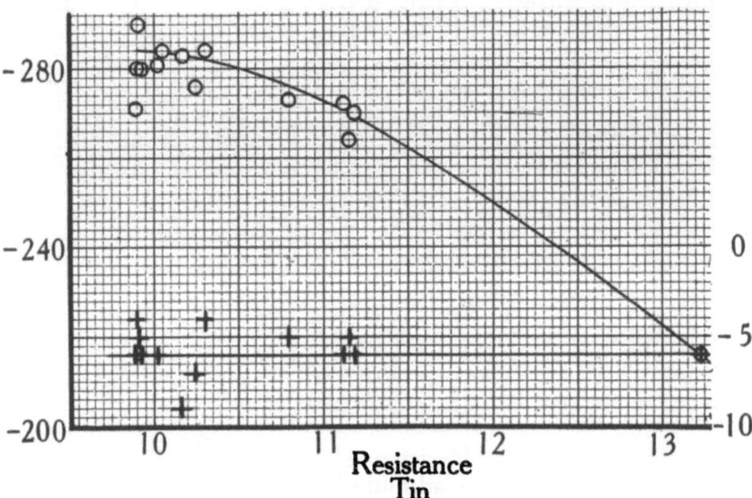

FIGURE 10. Values of E_0 in micro-volts (circles, scale to left) and Δ in micro-volts (crosses, scale to right) against specific resistance in the direction in question for tin.

specific resistance and in Figure 11 the values calculated therefrom for a' and b. Since Δ is constant against specific resistance, it is obvious that b is constant also. Taking as the best value for Δ -6×10^{-6}, the corresponding value of b is -0.0075×10^{-6}.

As in all the other cases, b is linear against specific resistance, but as was also the case with Zn and Bi, a' is not linear within experimental error.

Antimony. Measurements were made on only one specimen of antimony, so that the data are not at hand for a determination of the variation of thermo-electric quality with direction, but since this measurement was on a single crystal rod, and most previous measurements have been on haphazard aggregates, it seems of interest to record the results.

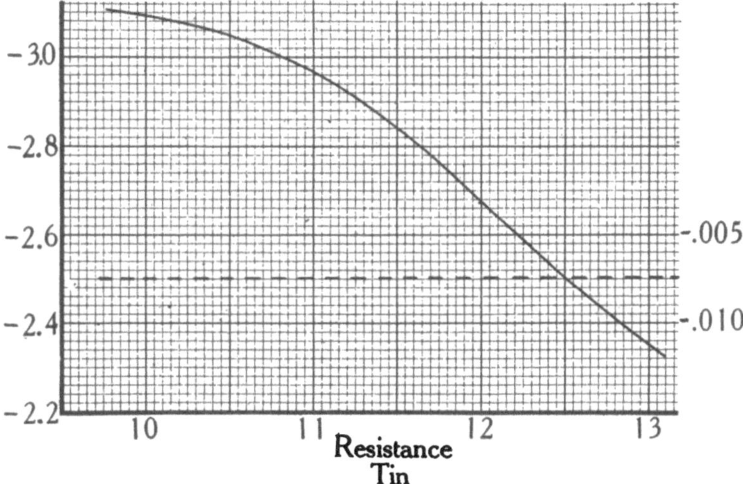

FIGURE 11. Values of a' (full line, scale to left) and of b (dotted line, scale to right) against specific resistance of tin. a' is closely related to the Peltier heat, and b to the Thomson heat; see equation on page 114.

The specimen was of Kahlbaum's purest grade, with the cleavage plane parallel to the length. This was the only one of the specimens investigated for which there was hysteresis between readings with increasing and decreasing temperature of more than the experimental error. The e.m.f.'s with decreasing temperature were about 2.5% lower than those with increasing temperature, the difference therefore being in a direction the reverse of ordinary hysteresis. The mean of the increasing and decreasing readings was used in calculating the results which were as follows:

$$E_0 = +3850 \times 10^{-6} \text{ volts}, \qquad \Delta = +135 \times 10^{-6},$$

whence

$$P_{Cu-Sb} = \tau[58.25 + .169t] \times 10^{-6}.$$

This is more than twice as high as the value found by Matthiesen[7] for crystalline Sb.

Tellurium. Results, not very satisfactory, were obtained for 3 specimens. The difficulty of making castings which are truly one grain is even greater than for Bi, and I have not yet attained success. The requirements here are much more exacting than in the previous work because the specimens must be much longer. The specimens investigated here were doubtless not all one grain, but the grains of which they were composed all had very nearly the orientation given. The tellurium used was from the Raritan Copper Works, to whom I am much indebted for specially refining it to remove all selenium. The properties of this Te are reported in greater detail in a previous paper.

The measured e.m.f.'s against temperature are shown in Figure 12. There is no consistent variation of e.m.f. with direction. The curvature is large, corresponding to a Thomson heat much larger than for any of the other metals.

The specific resistance of 2 of the 3 samples was measured; that of the 7° sample was 0.0597 ohms per cm. cube, and that of the 12.5° sample 0.0613.

Theoretical Discussion.

Consider a thermo-couple composed of a straight crystal rod running from a hot region in which the temperature is uniform to a cold region in which the temperature is also uniform. The circuit is to be completed by a copper wire (Figure 13). Then it is a matter of experiment that the e.m.f. of the couple depends only on the terminal temperatures and the orientation of the rod with respect to the crystal, and not at all on how the copper wire is attached at the ends, or on any other modification which may be made in the two regions at constant temperature. In particular, the e.m.f. is the same whether the copper wire is attached as at A to a surface perpendicular to the length of the rod, or whether it is attached as at B to an oblique surface. Within the hot region, heat is absorbed in the one case at the surface A, and in the other case at the surface B, and since the total e.m.f. is the same and no modification has been made in any other part of the circuit, the heat absorbed at the two surfaces A and B must be the same. But if the orientation of the rod with respect to the crystal were changed, the heat absorbed locally at the surface of attachment of the copper wire would in general change, because the total e.m.f. changes with the orientation. Hence it follows that the heat absorbed when a current passes out of a crystal rod into an isotropic medium depends on the direction of flow of the current in the rod, and not on the orientation of the surface by which it leaves (of course it also depends on the nature of the isotropic medium).

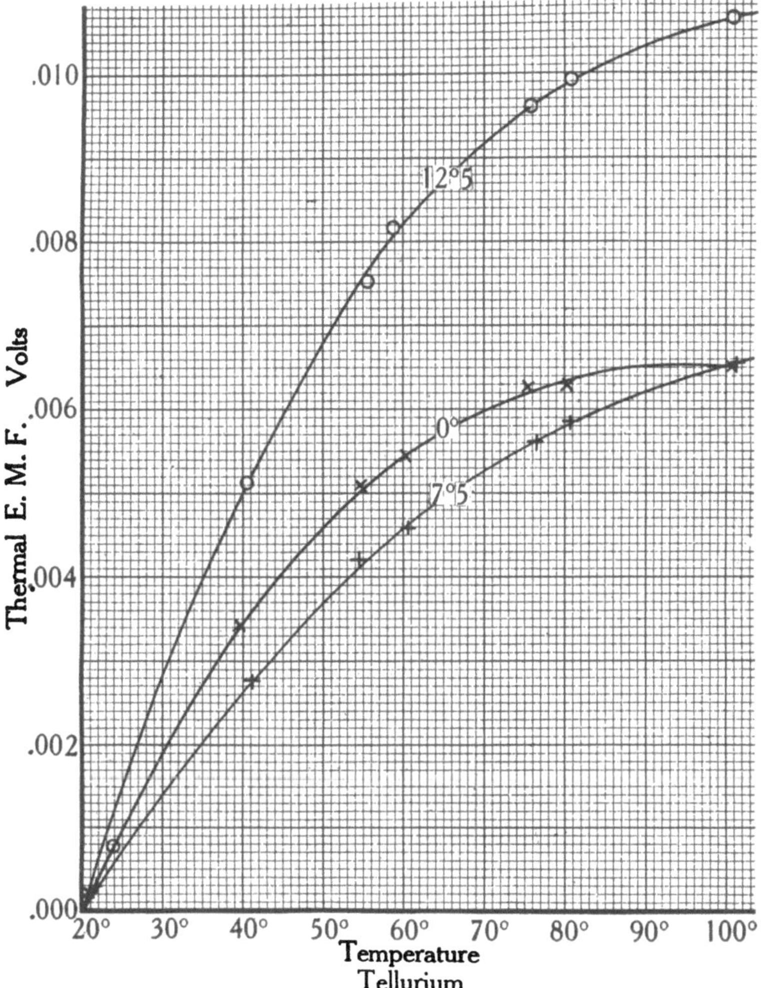

FIGURE 12. Thermal E.M.F. against copper in volts as ordinates plotted against the temperature of the hot junction as abscissa for rods of tellurium of the indicated angles between the crystal axis and the length of the rod.

Let us now suppose that the copper terminal is attached at the end of a right angled arm as shown in Figure 14, the crystalline orientation being uniform throughout the entire specimen. The total e.m.f. is the same as in the first case, because no modification made in

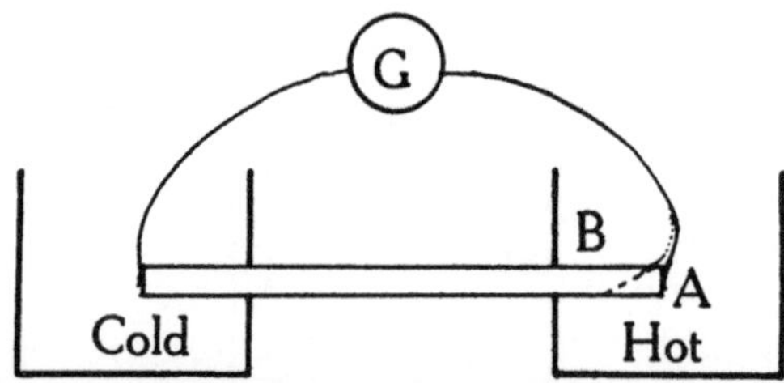

FIGURE 13. Schematical representation of the connections in measuring the thermal E.M.F. of a crystal rod against copper. The E.M.F. is the same whether the crystal rod is cut off perpendicularly as at *A* or obliquely as at *B*.

the region at constant temperature affects the e.m.f., and therefore the total heat absorbed in the hot region is unaltered. But the heat absorbed on leaving the surface *C* is now that appropriate to a flow in the crystal at right angles to the original direction. This is in general different from the original heat. There must therefore be an absorption of heat somewhere else in the hot region, and the principle of sufficient reason suggests that this can be only where the direction of the current flow changes. The amount of heat so absorbed where the direction changes is determined by the requirement that together with the surface heat at *C* it add to the right amount.

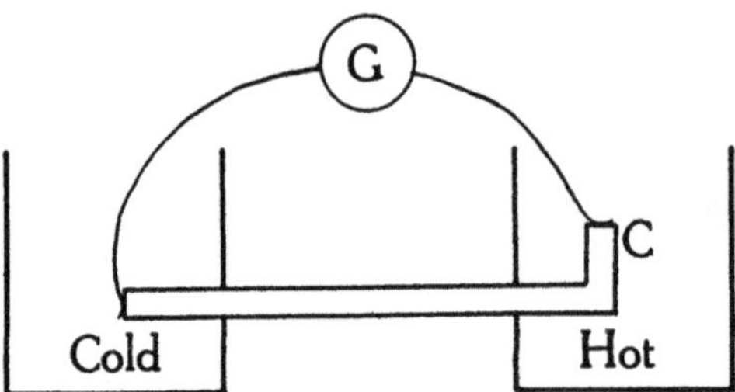

FIGURE 14. The thermal E.M.F. is the same in this figure as in figure 13.

We are thus led to the concept of an internal Peltier heat in a crystal which occurs when the direction of current flow changes at constant temperature. The magnitude of this effect is obviously determined when the ordinary Peltier heat at emergence is known as a function of direction, for it is determined by the simple requirement that the internal Peltier heat on changing from one direction of flow to another is equal to the difference of the emergent Peltier heats for these two directions of flow.

The necessity for the existence of this internal Peltier heat does not seem to have been previously noticed; it is not mentioned by either Voigt or Thomson.[8] The inclusion of it will essentially modify Voigt's analysis. It would seem that the mere existence of this effect should have important consequences on our theoretical views as to the nature of electrical conduction; it is hard to see according to any of the usual pictures why there should be any such effect at all.

For theoretical considerations the dimensions of P are important. P is heat absorbed per coulomb, or it may also be expressed as heat absorbed per electron. This means that the effect cannot be a momentum or a kinetic energy effect arising from a difference of natural velocity of the electrons in different directions in the crystal, for such effects would be proportional to the velocity of each electron, so that the resultant effect per electron would be proportional to the current, and therefore not a constant. In explanation, it almost seems necessary to ascribe to the electron a polar character, and to suppose that in the crystal there are different orienting forces depending on the direction of motion in the crystal. One is reminded of the magnetic doublet character which Compton[9] would ascribe to the electron.

The total e.m.f. in such a circuit as we have been considering is maintained by the Peltier heats at the two ends, and by the Thomson heat in the region where there is a temperature gradient. The relations between the Peltier heats and the Thomson heats for a long slender rod of any orientation with respect to the crystal we assume to be given by the thermodynamic analysis of Thomson, so that for any definite direction we have the relations:

$$P_{\mathrm{Cu-Crys}} = \tau \frac{dE}{d\tau},$$

$$\sigma_{\mathrm{Crys}} - \sigma_{\mathrm{Cu}} = \tau \frac{d^2E}{d\tau^2},$$

where E is the total e.m.f. for the direction considered, and σ is the Thomson heat for current flow in the crystal in that particular direction. Now it seems highly probable that the analysis of Voigt for the heat absorbed internally when the current flow is straight may be accepted without change. Voigt puts the Thomson heat absorbed reversibly equal to the work received by the current in flowing through a Thomson "e.m.f." This Thomson e.m.f. is in nature a vector, and as suggested by the adequacy of the ordinary thermodynamic analysis for an isotropic substance, we assume it is a linear vector function of the temperature gradient. The coefficients of the linear vector function are the Thomson coefficients of the crystal. Precisely the same symmetry considerations apply to them as to the coefficients of other linear vector functions, for instance, to the electrical resistance in different directions, or to the thermal resistance; the symmetry relations are the same and the manner of variation with direction is the same. In particular, the Thomson coefficients for the crystals considered here belonging to the trigonal, tetragonal, and hexagonal systems have rotational symmetry about the principal axis and the variation with direction is given by the equation

$$\sigma_\theta = \sigma_{11} \cos^2 \theta + \sigma_\perp \sin^2 \theta,$$

where θ is the angle between the direction considered and the principal axis, and σ_{11} and $\sigma_\perp$ are the Thomson coefficients for flow along and at right angles to the axis.

It is a consequence that σ is a linear function of the specific resistance, just as was the reciprocal thermal conductivity.

The Peltier heat at the surface, as well as the internal Peltier heat, was not at all considered by Voigt, so that his formulas cannot be applied without further examination to any actual thermo-couple. (The inadequacy of the analysis of Voigt may be inferred from his formula 433 on page 550, according to which the total heat absorbed in a complete thermo-electric circuit must be zero, so that there is no provision made for the source of the integral e.m.f. around the circuit.) The question that we now have to consider in supplementing Voigt's analysis is what are the necessary symmetry relations on P? The symmetry relations on σ involved essentially the assumption that it was a linear vector function of the temperature gradient. Now P is a function of direction, but there seems to be no natural argument for the symmetry relations of a *general* direction function such as there is for a *vector* function. In particular there would seem to be no reason for expecting P to be a linear function of the direction cosines—why might it not as well be a linear function of the direction tangents? As far as I can see, there is no reason to expect any special manner of variation of P with direction, and in particular no reason at all to expect that P should be the same function of direction as σ. It would follow that the total e.m.f. of a circuit would not necessarily be expected to follow Voigt's formula, and in particular there is no reason to think that the total e.m.f. ought to be a linear function of the resistance.

It does, however, seem natural to expect that P should have rotational symmetry about the principal axis, although I can see no mathematical necessity for such. This is strongly suggested by the experiments. When I made the experiments, I had not examined Voigt's analysis carefully, and expected his results to apply. Therefore in determining the orientation of the crystal rods I determined only the angle between the length of the rod and the principal axis. If P has rotational symmetry, this is adequate. That it is adequate is suggested by the experimental results, since with only a few exceptions, smooth curves are obtained on plotting total e.m.f. against orientation with respect to the axis. If two parameters were required, the curves in one parameter would not be expected to be smooth. In the case of Bi, I have since determined that, in the two specimens whose cleavage planes were parallel to the length, the secondary cleavage planes in one rod made an angle of 66° with the length, and in the other an angle of 81°, with corresponding values for E_0 of 4460 and 4540×10^{-6} volts, which probable differ no more than the experimental error. One may at least draw the conclusion that for these four metals P does not depend importantly on orientation with respect to the secondary crystal axes.

That P does not have the symmetry relations of Voigt has already been brought out in the detailed presentation of data. In all cases the Thomson heat is within experimental error a linear function of electrical resistance, but for Bi, Zn, and Sn, P is not such a function.

An examination of the signs shows that there is no universal relation between heat absorption and the change of direction of flow. Thus for Sn and Bi heat is absorbed when the direction of current flow changes from parallel to perpendicular to the axis, whereas for Zn and Cd heat is given out for the same change of direction.

The fact that thermal e.m.f. varies with direction in a crystal has an interesting bearing on a question which has been much discussed, namely the law of Magnus. This law states that in a circuit composed of a single perfectly homogeneous metal no circuital e.m.f.'s can be generated by any distribution of temperature. It has been claimed a number of times that the law is not true, but the concensus of opinion seems to be that such apparent exceptions are to be explained by lack of perfect homogeneity in the metal. Now it is at once obvious that this law can have no application to a crystal, for high e.m.f.'s may be generated in such circuits as that shown in Figure 15 cut from a single crystal. Now of course all metals of practise are multi-crystalline, the grains being presumably oriented at haphazard.

Experiments made on a large scale would be expected to verify the law of Magnus, but when any part of the circuit becomes so small that the individual crystals are comparable in size with the cross section of the circuit, one might expect these characteristically crystalline properties to produce an appreciable net effect, and the law of Magnus to fail. It is not quite evident from the experimental evidence now in hand whether to expect such small scale effects with metals crystallizing in the cubic system. If the Peltier heat satisfies the same symmetry relations in a cubic crystal as does the Thomson Heat, then we should not expect such an effect, but we have seen that there seems to be no simple necessary connection between P and direction, so that we need not expect P necessarily to have spherical symmetry

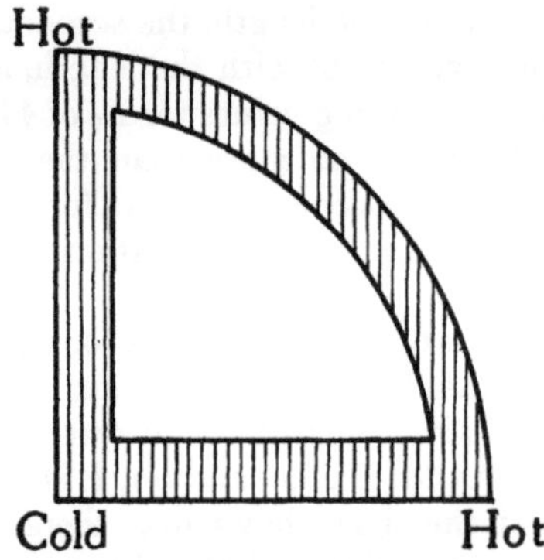

FIGURE 15. To illustrate conditions under which current may flow in an unequally heated circuit cut from a single crystal. If the materials were amorphous, no current would flow.

in a cubic crystal. If P has not spherical symmetry, it need not average to zero when the cross section of the circuit is composed of comparatively few grains. Entirely apart, however, from these crystalline effects, there is the consideration that non-uniform temperature gradients cannot be maintained in a solid without producing mechanical stresses, and so making the material non-homogeneous.

It seems then that interest in the law of Magnus is reduced to a somewhat academic position. It is a result of experiment that any such effects are small, but under properly chosen conditions of experiment we may expect such effects to be found. Two interesting questions remain, however: does the law of Magnus hold for a liquid metal such as mercury? And in a single crystal rod, in which there is a uniform temperature gradient and no mechanical stress, is the Thomson heat absorbed by the current a function only of the temperature difference between two points and not a function of the temperature gradient?

Average Values. There are no previous results on single crystals to compare with these (except for Bi), but it is natural to expect that previous results on polycrystalline metals should agree with some sort of an average, throughout the crystal, of the results found here. The theoretical discussion indicates what sort of agreement to expect. The Thomson heat satisfies the symmetry relations of Voigt, so that his value for the average should apply. The Peltier heat seems to have no such symmetry relations, so that Voigt's expression for the average does not apply, but we expect, nevertheless, an agreement as to sign and order of magnitude between the observed P, for polycrystalline metals, and the average calculated for the crystal by Voigt's formula.

In Table II. are collected the values for P and σ at 0° C. against copper taken from my previous paper on extruded wires,[6] and the values for P and σ in different directions of the crystals of this paper. In the case of Bi, which has a very large P, there is agreement in both sign and order of magnitude. σ is also of the right sign, but when averaged by Voigt's formula over all directions in the crystal it is evident that σ for the crystal will be much larger than for the extruded wire. But it must be remembered that there is considerable error possible in the σ given for the crystal, and it is also possible that in the extruded wire all orientations were not uniformly represented, so that one cannot be perfectly sure that the data for Bi are inconsistent with some sort of average law. But the other metals seem to make this impossible. Both P and σ for all directions in Sn have the wrong sign to agree with previous values for Sn wire. Cd crystals have the wrong sign for σ and a value for P also of the wrong sign and at least 10 fold too small numerically. With Zn it is not impossible that some sort of an average of the P of the crystal should agree with the P of the wire; σ furthermore has the right sign, but the average for the crystal is about 4 times too small. One may try to improve the agreement, on remembering that the copper of these experiments may have been different from that used previously, by correcting the values given for the extruded wire by a constant additive correction applied to all the P's, and another additive correction to all the σ's. But a little trial will show that this is not possible; a change that improves the agreement in one place makes it worse in another.

TABLE II.

Comparison of the Thermal E.M.F. against Cu of Isotropic and Unicrystalline Metals.

Metal	Isotropic		Unicrystalline		
	Peltier Heat at 0° C Against Cu.	Thomson Heat at 0° C Against Cu.	Specific Resis.	Peltier Heat at 0° C Against Cu.	Thomson Heat at 0° C Against Cu.
Sn	$-2.547\times10^{-6}\tau$	$-.0110\times10^{-6}\tau$	9.9×10^{-6} 13.0	$3.10\times10^{-6}\tau$ 2.35	$.0075\times10^{-6}\tau$ .0075
Cd	9.225	.227	6.8 8.3	−1.41 +1.54	−.036 −.019
Zn	.270	−.0196	5.910 6.039	− .147 + .854	−.0097 −.0035
Bi	−77.20	.0223	109 113	−57.8 −71.5	.0275 .0875

The conclusion seems to be forced that the thermo-electric properties of a polycrystalline aggregate (at least of non-cubic metals) cannot be computed from the average of the properties of a single crystal in the same way that the electrical resistance or the thermal conductivity can be calculated. In explanation it is to be noticed that, in any polycrystalline aggregate of non-cubic metals formed in the usual way by casting, there must be intense internal stresses, because of the unequal thermal expansion of the grains in different directions, and that thermo-electric properties are much more sensitive to such stresses than resistance or thermal conductivity. If this is correct, one will in the future attach less theoretical significance than hitherto to thermo-electric properties measured on polycrystalline aggregates of these metals.

Summary.

1. Thermal Conductivity. The thermal conductivity of Bi, Zn, Cd, and Sn is measured at room temperature for different directions in the crystal. It is found that Voigt's symmetry relations are satisfied in that in any single crystal the reciprocal of thermal conductivity varies linearly with electrical resistance as the orientation changes. The generalized Wiedemann-Franz law, however, does not hold for all directions in a crystal in that the ratio of thermal to electrical conductivity is in general a function of the orientation. Within experimental error it is not impossible that the Wiedemann-Franz ratio is constant for Bi and Sn, but in Zn the thermal conductivity varies more rapidly with direction than does the electrical conductivity, and Cd the variation of thermal conductivity is less rapid.

2. Thermal E.M.F. The thermal e.m.f. against copper is measured between 20° and 100° C. as a function of direction in the crystal for Bi, Zn, Cd, and Sn, and isolated results are also found for Sb and Te. In general the thermo-electric force varies greatly with direction. From the thermal e.m.f. data the Peltier and the Thomson heats are calculated. Voigt's symmetry relations are shown to be satisfied by the Thomson heat, which is a linear function of electrical resistance. The Peltier heat, on the other hand, was apparently not considered in Voigt's analysis, and these experiments show that the symmetry relations to which it is subject are not the same as those for the Thomson heat. It is probable that the Peltier heat has rotational symmetry about the principal crystal axis, but there appears to be no necessary restriction as to its variation in directions inclined to the axis. It is pointed out that the Peltier heat depends only on the

direction of current flow within the crystal just before emergence, and not on the orientation of the surface through which it emerges. It is a consequence that there is an internal Peltier heat, i. e., a reversible absorption or generation of heat where the direction of current flow changes within the crystal. The mere existence of this effect should be most important for theories of electrical conduction. It is also a consequence of the difference of thermal properties of a crystal in different directions that the generalized law of Magnus cannot hold for a crystal, but in general thermal currents will flow in an unequally heated closed circuit cut from a single crystal. Finally, the thermo-elective properties of an ordinary polycrystalline aggregate of non-cubic crystals cannot be found by averaging the properties of the single crystals in different directions.

It is a pleasure to acknowledge the assistance received from Mr. E. T. Lane in making nearly all the readings of thermal e.m.f. It is also a pleasure to acknowledge financial assistance received from the Rumford Fund of the American Academy.

The Jefferson Physical Laboratory,
Harvard University, Cambridge, Mass.

References.

[1] P. W. Bridgman, Proc. Nat. Acad. Sci. 10, 411–415, 1924; Proc. Amer. Acad., 60, 305–383, 1925.

[2] Second reference under 1.

[3] P. W. Bridgman, Proc. Amer. Acad. 57, 77–126, 1922.

[4] L. H. Adams, Jour. Amer. Chem. Soc. 36, 65–72, 1914.

[5] W. Voigt, Lehrbuch der Kristallphysik, Teubner, Leipzig, 1910.

[5a] G. W. C. Kaye and J. K. Roberts. Proc. Roy. Soc., 104, 98-114, 1923.

[6] P. W. Bridgman, Proc. Amer. Acad. 53, 269–384, 1918.

[7] A. Matthiesen, Pogg. Ann. 103, 412–428, 1858.

[8] W. Voigt, Reference 5, page 534.
W. Thomson, Edin. Proc. 3, 255, 1854.
W. Thomson, Edin. Trans. 21, 153, 1857.
W. Thomson, Phil. Mag. 11, 379 and 433, 1856.

[9] A. Compton, Jour. Frank. Inst. Aug. 1921, 45–155.

Dimensional Analysis Again.

By P. W. BRIDGMAN.

THE subject of dimensional analysis has been so much discussed in the pages of the Philosophical Magazine in the last few years that one would like to see it die a natural death, were it not that in the last two communications † things have taken such a turn that an unchallenged acceptance of the views there presented may easily affect the use which the physicist makes of this analysis.

My position on this general subject has already been expounded in considerable length and detail in my book 'Dimensional Analysis,' Yale University Press, 1922. This book was written primarily because there was nowhere any previous attempt to give a systematic examination of the arguments on which the dimensional method rests, nor

† Mrs. T. Ehrenfest-Afanassjewa, Phil. Mag. [7] i. pp. 257–272 (1926); Norman Campbell, Phil. Mag. [7] i. pp. 1145–1151 (1926).

to show how it may be applied in many of the cases of practice, although the dimensional method was used to a greater or less degree by nearly every physicist, and there were incomplete and incidental references to the theory in a great many places. Now in these previous incomplete references were contained many misconceptions and fallacies; it was the first task of the book to show what these fallacies were. Among these fallacies was the idea that an equation expressing a connexion between physical things must be dimensionally homogeneous. I then showed that, in spite of the unsatisfactory form in which the dimensional argument had been often stated, it was possible to put it on a thoroughly sound basis, and apply the method with confidence to the solution of many important problems. I particularly stressed the point that the method, when properly understood, has nothing esoteric about it, that it gives no information which we did not have already in a less available form, and that therefore an adequate physical grasp of each individual problem is necessary. No refinement of mathematical argument will give correct results if our physical grasp of the broad features of the problem is inadequate. Just here is the feature in which my treatment differed most from those preceding; I emphasized that the broad physical grasp necessary to an application of the method finds expression in a knowledge of the *general character* of the equations which govern the motion of the system. Only in this way could the puzzling matter of dimensional constants be satisfactorily treated. Thus if the problem is one involving fluid motion, as in discussing airplane problems, the governing equations are those of hydrodynamics, into which no dimensional constants enter, so that no such constant enters the final result; but if the problem is one of electrodynamics, the field equations of electrodynamics govern, which do contain a dimensional constant (velocity of light, or ratio of units), and the dimensional constant may be expected in the final result.

For the method, purged from many previous fallacies and made (comparatively) rigorous, I used the name "Dimensional Analysis." Now it is gratifying that the name has apparently been accepted, as shown by the titles of recent papers, but, unfortunately, not in the sense in which I used it, but with a more or less indefinite blend of many former misconceptions. Thus Mrs. Ehrenfest-Afanassjewa in particular, although generously attributing the name to me, ascribes to the "dimensional analysist" a miscellaneous assortment of errors, which even I, in my enthusiasm in my book for putting the worst light on previous misconceptions, would hardly have ventured to attach to any individual. Thus she begins by committing the dimensional analysist to the thesis that all equations expressing physical relations are dimensionally homogeneous, which is supposed to be equivalent to the statement that dimensional constants cannot occur. She then allows him to talk the most meaningless nonsense about other possible universes in which the gravitational constant might be different from ours, and she ends by making him so blind in his resolution to use only a pure dimensional method that he forgets that he knows that the gravitational attraction of the earth varies inversely as the square of the distance, and so is unable to find how the period of a pendulum varies with the distance from the earth's centre.

Of course Mrs. Ehrenfest-Afanassjewa has no trouble in making merry with her poor analysist. Now comes the important point for the physicist. Because of these difficulties, she would have us abandon the dimensional method altogether, and substitute for it another, which she calls the theory of similitudes, much more mathematical in character, and much more difficult for the physicist to apply, as one may see on reading her treatment of the simple pendulum problem. There are many formal resemblances between the methods, as she points out, but that they are not the same is shown by the fact that they do not always give the same solutions. The method of similitudes resembles that of dimensional analysis as expounded in my book in that the important role of the fundamental equations governing the motion of the system is recognized, but there is a vital difference in that her method demands a knowledge of the equations in *detail*, whereas the dimensional method needs to know only the general character of the equations. It is well known that the dimensional method has its most important practical application in such complicated problems as that of the airplane, where it is hopeless to attempt to write the equations in detail. I do not see how Mrs. Ehrenfest-Afanassjewa's method can be applied to this problem at all. On the other hand, dimensional analysis, in the sense expounded in my book, can certainly solve all the problems which she gives to show the limitations of dimensional analysis, and she will further find in my book an explanation, on purely dimensional grounds, of the questions raised in the correspondence between Lord Rayleigh and Riabouchinsky which she discusses in her paper from the standpoint of her theory.

The main purpose of this note is then to dispute the conclusion that might be drawn from an unchallenged acceptance of Mrs. Ehrenfest-Afanassjewa's paper. I believe that the dimensional method *can* be made rigorous, that it is *not* necessary to replace it by the mathematical method of similitudes which she offers, that the dimensional method is much easier for the physicist to apply, and in many important practical cases is more powerful.

Opposed to Mrs. Ehrenfest-Afanassjewa is Norman Campbell, who will have none of her method of similitude, but prefers that of dimensions. However, I cannot agree with Campbell, as it seems to me that his argument is put on an incorrect basis. It is certainly not necessary to assume that physically similar systems are possible in order to apply the dimensional method. For example, the deduction of the relation between the mass of an electron, its charge, radius, and velocity of light (dimensional constant of the electrodynamic equations) does not involve the assumption at all that more than one kind of electron is physically possible. Neither is it necessary in deducing the pendulum formula to so far analyse the situation as to recognize that "the period of the pendulum seems to be the physical sum of elementary periods, during each of which it may be supposed to move with constant acceleration." I fear that if as deep an analysis as this were necessary, few physicists would ever have the courage to attempt a dimensional analysis at all.

The Jefferson Physical Laboratory,
Harvard University, Cambridge, Mass.

THE BREAKDOWN OF ATOMS AT HIGH PRESSURES

THE BREAKDOWN OF ATOMS AT HIGH PRESSURES

By P. W. Bridgman

Abstract

Thermodynamic evidence supports the experimental suggestion of a previous paper that at ordinary temperatures sufficiently high pressures are capable of breaking down the quantum structure of atoms, reducing matter to an electrical gas of electrons and protons. We may, therefore look for atomic dissociation under two sorts of conditions: high temperatures and comparatively low pressures, such as we have in the stellar atmospheres, and high pressures and comparatively low temperatures, which we may surmize we have in the interiors of stars, possibly in stars like the sun, and almost certainly in stars of the enormous density of the dark Sirius type. The possibility of two sorts of dissociation, together with the more rapid increase of pressure than density when the diameter of a star is reduced, offers the possibility of a critical condition determining whether a star is of the dark Sirius type or not.

IN THE Physical Review for January 1926 I called attention to a reversal in the behavior of certain properties of potassium (the atom of which has an abnormally loose structure) at high pressures and room temperature, which I suggested might indicate the initiation of an ultimate breakdown of the atom at much higher pressures and an approach to a gas of electrons and protons. Supporting this idea that a breakdown of the atoms is possible at high pressures, there is an argument from a theorem of Schottky's which I presented in the previous paper; furthermore we have the physical feeling that the quantum orbits to which the atom owes its structure ought not to be able to resist an indefinitely great force, and also the fact that there are stars of enormous densities. Nevertheless, the assumption of this sort of atomic disintegration involves certain apparent inconsistencies, for in the atmospheres of the stars we have direct spectroscopic evidence of atomic disintegration, as was first extensively shown by Saha, and simple thermodynamics shows that this decomposition *increases* with rising temperature and *decreases* with rising pressure. Since such decomposition at the high temperatures and greatly reduced pressures of the stellar atmospheres is only partial, it would appear at a first glance that there is no reason to expect any decomposition at all at ordinary temperatures and pressures of tens of thousands of atmospheres. It is the purpose of this note to present additional thermodynamic evidence suggesting that decomposition may nevertheless occur at high pressures and low temperatures, and to resolve the apparent inconsistency.

Let us examine the consequences of assuming that it is possible to apply sufficient pressure to a substance at room temperature to break the atoms down into a perfect gas of electrons and protons. Under such a pressure the thermal expansion must assume the value appropriate to a perfect gas, and the specific heat must also become very much larger than that of a normal solid, because each electron makes its full individual contribution to specific heat. Now thermal expansion and specific heat cannot vary independently, but there is a thermodynamic connection, namely:

$$(\partial C_p/\partial p)_\tau = -\tau(\partial^2 v/\partial \tau^2)_p$$

Hence if C_p is to increase with pressure, $(\partial^2 v/\partial \tau^2)_p$ must be negative, which is the reverse of its usual behavior, because the thermal expansion at constant pressure of normal substances increases with rising temperature instead of decreasing. Now since the thermal expansion of a gas is much higher than that of a normal solid, a higher thermal expansion at low temperatures means a closer approach to the perfect gas condition at low temperatures. This indicates therefore that if a solid is decomposed by high pressure and made to approach the behavior of a gas, the approach to this condition will be most rapid at low temperatures. My experiments on potassium were made at low temperature.

This state of affairs is also consistent with other thermodynamic evidence. In the atmospheres of the stars atomic decomposition decreases with rising pressure; here it increases. Now a homogeneous reaction is driven by pressure in such a direction as to decrease the volume. In the stellar atmospheres, therefore, the volume of the neutral atoms is less that that of the ionized atoms and the detached electrons; this is a consequence of the comparatively low pressures. At high pressures, on the other hand, simple calculation shows that the electrical gas has a smaller volume than that of the undissociated atoms from which it comes. In normal substances under high pressures it appears therefore that the quantum orbits act like skeleton frameworks *distending* the structure; if these frameworks are destroyed, the substance collapses. It was shown in the previous paper that at 300°K the pressure at which the volume of the electrical gas is equal to that of the neutral atoms is of the order of a few 10,000 atmospheres, which is certainly a negligible pressure compared with cosmic possibilities. As temperature increases, the dissociated volume gains relatively to the undissociated volume in consequence of the high thermal expansion of the gas, thus bearing out the evidence above that the approach to gaseous decomposition is closest at low temperature.

In the stellar atmospheres decomposition increases with rising temperature at constant pressure; under our conditions it decreases. This means that in the stellar atmospheres heat must be absorbed by the dissociation, whereas under our conditions heat is given out. The reason for this difference is evident. Under atmospheric conditions the volumes are so large that the electron must be removed to a great distance against the electrostatic forces of the core during decomposition. There is a compensating effect arising from setting free the kinetic energy of the electrons in the quantum orbits, but this is only half the electrostatic effect. Under our conditions however the volume is small, and the electrons on the average are no further away from the core after decomposition than before. The electrical effect vanishes, therefore, leaving the kinetic effect outstanding. Now the kinetic energy of the electrons in the quantum orbits is much higher than the equipartition temperature energy at ordinary temperatures, so that heat must be given out rather than absorbed when an atom decomposes at high pressures. (Has this been considered as a source of stellar energy?) It is also evident, since the kinetic energy of the electrical gas is higher at the higher temperature, that this heat of dissociation decreases with rising temperature, again demanding that decomposition be greater at lower temperatures than high. Further, there is another effect tending to accentuate the reversal of sign of the heat of dissociation at high pressures. After the first few 10,000 atmospheres the internal energy of a solid increases when pressure increases at constant temperature. This increase of internal energy is divided in the ratio of two to one between energy of position and increased kinetic energy of the electrons in their orbits. This latter part is set free during dissociation, so that the heat of dissociation under high pressures is greater by this amount than would be indicated by our argument above applied to normal atoms at atmospheric pressure.

All the lines of evidence converge, therefore, to indicate a pure pressure decomposition of atoms at high pressures, and this decomposition is favored by low temperature. We must then visualize the condition of matter over extreme ranges of pressure and temperature somewhat as follows. The pressure-temperature plane is crossed by a diagonal band rising from low pressures and temperatures to high pressures and temperatures, within which matter exists in the normal form of neutral undissociated atoms as we know them. To one side of this band is the region of high temperature and low pressure in which the atoms are dissociated into a gas of electrons and protons, and it is in this region that we find matter in the stellar atmospheres. On the other side of the band, at high pressures and low temperatures, we also have matter dissociated into an electrical gas. We now have to ask to which of these two regions the apparently perfectly gaseous *interiors* of the stars belong. If to the first, we have decomposition in spite of high pressure, if to the second, in spite of high temperature. It seems almost certain that stars of densities of

50,000, like the dark companion of Sirius, indicate the second region. If the density is of the order of magnitude of unity, and nevertheless the star acts like a perfect gas (as does our sun, according to Eddington), we may suspect the first region, but perhaps even under these conditions the second region is not impossible.

Perhaps the following calculation is worth recording as suggesting possible orders of magnitude. Imagine an atom of atomic number 40, with the negative electricity all concentrated in a uniform spherical shell of radius 1.5×10^{-8} cm. Due to the mutual repulsion of its parts this shell is exposed to a distending pressure of approximately 3×10^{14} dynes/cm^2. The positive nucleus exerts an inward pressure of twice this; the difference, or 3×10^{14} dynes/cm^2, is the effective distending pressure of the quantum structure of the atom preventing collapse. The pressure in the interior of the sun is of the order of ten times higher, so that possibly we may expect pressure dissociation in the sun, although it seems more likely that we have the first sort of dissociation.

As far as I know, no adequate physical difference has been suggested to account for some stars with densities of 1.5 and others with 50,000. In view of the comparatively small range of mass of the stars this seems to demand some explanation. The possibility of two regions of dissociation seems to offer a clue. If the mass of a star is concentrated in spheres of decreasing radius, the pressure rises faster than the density. Thus if the sun were concentrated in a small sphere of density 50,000 times the present density, the pressure would be nearly 2,000,000 times greater (in general pressure varies as (density)$^{4/3}$). This sort of thing suggests instability and critical conditions, with the possibility of some stars in the second state of dissociation. If, however, the sun should turn out to be in the second condition, then we may perhaps recognize the possibility of two different stable states in the second condition, not unlike the two amorphous phases of ordinary matter of van der Waals.

A complete description of the state of affairs involves many complicated considerations. It is evident that the atoms in a star are in varying degrees of dissociation; according to the nature of the atom perhaps some of these are in the first condition, while others may be in the second. We need a detailed mathematical treatment of the comparatively simple problem of the equation of state of matter dissociated into electrons and protons under very high pressures without quantum conditions; this would show how nearly our assumption is realized of perfect gas behavior.

THE JEFFERSON PHYSICAL LABORATORY,
HARVARD UNIVERSITY, CAMBRIDGE, MASS.
October 4, 1926.

68 — 2129

THE TRANSVERSE THERMO-ELECTRIC EFFECT IN METAL CRYSTALS

By P. W. Bridgman

Jefferson Physical Laboratory, Harvard University

Communicated January 17, 1927

Nearly seventy-five years ago Lord Kelvin[1] made the following prediction: "If a bar of crystalline substance possessing an axis of thermo-electric symmetry has its length oblique to this axis, a current of electricity sustained in it longitudinally will cause evolution of heat at one side and absorption of heat at the opposite side, all along the bar, when the whole substance is kept at one temperature."

Much later Voigt,[2] in developing his theory of thermo-electric action in crystals, finds the same effect in his equations, and says of it: "This transverse Peltier heat, peculiar to crystals, has not yet been investigated. But there is no reason to doubt its existence."

So far as I can find, the effect has not been reported since Voigt wrote. I have just found the effect in bismuth, zinc, tin, and cadmium. As is to be expected, the effect is by far the largest in bismuth; in fact, it is so large that the experiment may be easily arranged to throw a sensitive galvanometer off the scale. The experiments were made on single crystal bars of approximately 10 cm. length and 6 mm. diameter; the same bars were previously used in measuring thermal conductivity.[3] Three bars of bismuth were used; in two of these the basal plane was inclined at about 20° to the length, and in the other it was parallel. The simplest possible experimental arrangement suffices to show the effect. Current connections were made at the two ends of the bar with clamps; there was an ammeter and reversing switch in the current circuit. The temperature difference between the two sides of the bar was measured with a copper-constantan thermo-element of two junctions, made of ribbon 0.002 cm. thick rolled from wire 0.010 cm. in diameter, so arranged that the two junctions were pressed by springs, one against each side of the bar, from which it was insulated by thin cellophane paper. The longitudinal positions of the two couples on the bar were adjustable independently of each other. The bar could be rotated about its longitudinal axis.

With currents of the order of 1 amp., thermal e. m. f.'s of the order of 15×10^{-4} volts were found, corresponding to a temperature difference between the two sides of the bar of 0.4°C.

A number of different sorts of checks were made to verify that the effect is the one expected. The effect reverses sign when the direction of the current is reversed; it is proportional to the current within the range investigated up to 1 amp.; it is a maximum at the two ends of that diameter which lies in the plane perpendicular to the basal plane and passing through the axis of the bar; it reverses sign if the bar is rotated through 180° beneath the couples and vanishes half way between the two orientations of maximum effect; and finally it is independent of the relative positions of the two junctions longitudinally on the bar, thus showing that the whole of one side of the bar is warmed and the other cooled, and incidentally eliminating the possibility that the effect may be due to defective insulation between the couples and the bar. The effect was of about equal magnitude in the two bismuth bars with basal plane 20° to the length, and zero in the bar with basal plane parallel to the length.

The effect was also detected in bars of zinc, cadmium, and tin, in which the axis was at approximately 45° to the length. In these metals the effect is very much less, and the disturbances from chance fluctuations in the air of the room were so great that not much more could be stated with certainty than the existence of the effect. It is of the order of 500 times smaller in these metals than in bismuth.

The existence of the effect being established, let us examine the significance of it. The theories of both Kelvin and Voigt not only demand the existence of the effect, but also give an explicit expression for its dependence on the direction of the bar with respect to the crystal axis, in terms of the difference of Peltier heat for a current flowing parallel and perpendicular to the axis. These two theories indicate a close connection between the transverse thermoelectric effect and the precise way in which the Peltier heat at the end of a bar inclined at an arbitrary angle with the crystal axis varies with the angle. The expression which Kelvin and Voigt find for this latter is:

$$P_\theta = P_\parallel \cos^2\theta + P_\perp \sin^2\theta.$$

Now the argument by which Voigt derives this expression for P_θ I have already shown to be incorrect,[4] because it takes into account only the body e. m. f.'s, leaving out the surface e. m. f.'s, and in so doing failing to provide for any net e. m. f. in a complete thermo-electric circuit. The same considerations show also that Voigt's analysis for the transverse effect cannot be correct. Experimentally, I have previously found that the relation of Voigt, which is also that of Kelvin, for the dependence of P_θ on direction is not true.[5] Granting the correctness of my experimental result, the implication must then be that Kelvin's theory, as well as Voigt's, must be incorrect. Only since writing the previous paper have I found the argument of Kelvin, buried in his long paper on the "Dynamical Theory of Heat."[6] Kelvin's argument is so simple and apparently straightforward compared with that of Voigt that it is of interest to examine it in detail. The argument is made to depend on the following axiom: "Each of any number of coexisting systems of electric currents produces the same reversible thermal effect in any locality as if it existed alone." Kelvin applies the axiom to the case in point as follows. The crystal bar, cut with its length at an angle θ with the axis of rotational thermo-electric symmetry, and carrying a longitudinal current of density q, is imagined to be imbedded in some standard non-crystalline metal. The longitudinal current is now resolved into two components, one along the crystal axis and the other at right angles. Each of these virtual components flows across the transverse surface into the standard metal, and hence there are Peltier heats at this surface, which, because of the crystalline character of the metal, are different for the two components, so that there is a net effect. Now, by the condition of continuity, the components of these two components normal to the transverse surface of the bar cancel each other, so that in the metal outside only a longitudinal current flows. This current in the metal outside, parallel to the surface, may now be suppressed, as well as the external conductor itself, as having no physical connection with the phenomena inside, leaving a net Peltier heat at the transverse insulated surface of the bar. The argument easily gives for the Peltier heat per unit lateral surface the numerical expression:

$$q(P_\parallel - P_\perp) \sin\theta \cos\theta.$$

My experiments were not accurate enough to show whether Kelvin's precise expression for the variation of this transverse heat with direction holds exactly or not. The effect observed was, however, of the order of magnitude to be expected according to the formula, using in the calculation the probable value of $P_{\parallel} - P_{\perp}$ and the known thermal conductivity.

The argument and axiom of Kelvin certainly make a strong appeal, but that they cannot possibly be right may be shown by resolving the longitudinal current into another pair of directions. The two directions at 45° to the crystal axis are a pair of perpendicular directions for which the Peltier heats are equal, so that if Kelvin had resolved along these two directions instead of the two that he chose, his same argument would have given zero transverse Peltier heat. It appears in general, therefore, that Kelvin's "Principle of the Superposition of Thermo-Electric Action" cannot be maintained, and that *in general it is not allowed to resolve the current in a crystal into perpendicular components.*

We thus have the amusing situation that the transverse effect exists actually, although the theoretical prediction of it seems to be based on inadequate reasoning.

At first sight a surface heat, proportional to the volume density of current, produced by a current parallel to the surface, appears most paradoxical. In fact, if the current were physical entirely parallel to the surface, such an effect would be quite incomprehensible. But the explanation is apparent when one considers the electron structure of an electric current. In the stream of electrons which constitutes the current, and which on the whole is parallel to the surface, there are always some electrons moving obliquely. These oblique electrons are reflected on striking the surface, with change of direction with respect to the crystal axis. I have already shown[7] that in a metal crystal there is a reversible absorption or evolution of heat when a current changes its direction. Hence we are to expect a similar effect when the electrons are reflected at the surface. It is easy to see that in general, if the crystal axis is inclined to the surface there is, after reflection, a change of net direction with respect to the crystal axis of the average of all the reflected electrons and, therefore, a net heating effect, which constitutes the effect observed.

These considerations show that the precise way in which the transverse heat varies with the inclination of the bar to the crystal axis is determined by the way in which the current-carrying-electrons are distributed about the crystal axes. If the electrons move only along the axis and at right angles, then Kelvin's argument applies, and his result holds, but with the most important restriction that it is allowable to resolve the current into two rectangular components *only* along the crystal axis and at right angles. It is probable that there is a good deal of physical correctness in such a procedure. My measurements show that the Peltier heat at the end of a bar at least does not depart by a large amount from the relation of Kelvin and Voigt, and entirely apart from such considerations, it has been suggested several times recently[8] that in a metal the electrons may flow along definite guided paths, which it is natural to suppose are simply connected with the crystal axes. A more accurate measurement of the variation of these effects with direction, which I hope to be able to make on bismuth in which the effects are largest, may be expected to throw important light on these questions.

[1] Lord Kelvin, *Mathematical and Physical Papers*, Vol. I, p. 267; *Trans. Roy. Soc. Edin.*, March, **1851.**

[2] W. Voigt,, *Lehrbuch der Kristallphysik, Teubner*, **1910,** p. 551.

[3] P. W. Bridgman, *Proc. Amer. Acad., Boston*, **61,** 1926 (101–134).

[4] Reference 3, p. 128.

[5] Reference 3, p. 129.

[6] I owe this reference to a forthcoming paper by Linder in the *Physical Review* on the t. e. m. f. of single zinc crystals.

[7] Reference 3, p. 126.

[8] H. Kamerlingh Onnes, *Rapport au IV[e] Conseil Solvay;* H. A. Lorentz, *Ibid.*, also *J. Inst. Met., London*, **33,** 1925 (257–278); P. W. Bridgman, *Proc. Amer. Acad., Boston*, **60,** 1925 (442).

Some Mechanical Properties of Matter under High Pressure

By P. W. Bridgman, Cambridge U. S. A.

(For figures 8 and 9 see plate II.)

One does not ordinarily think of hydrostatic pressure as very important in affecting the properties of the solid and liquid bodies of our commun experience. Thus it is only when special precision is sought that we specify that the measurements must be made under standard atmospheric pressure in, for example, defining the zero of the Centigrade scale, or in defining the calorie in terms of the thermal properties of water. The justification for customarily neglecting the effect of pressure is that the pressure range of ordinary experience is small. If, however, pressure is sufficiently increased, effects may be produced large enough to give useful information about the constitution of matter. For a good many years I have been experimenting with pressure high enough to produce such significant effects: pressure of the order of 10,000 to 20,000 kg/cm², and sometimes even considerably higher. The effect of such pressure on a large number of the physical properties of matter has been determined; in this lecture I propose to briefly summarize a few of these facts which are more distinctly mechanical in character.

Perhaps the simplest effect of hydrostatic pressure is to decrease volume. Ordinarily, changes of volume under pressure are not considered important—thus the error still persists in some books that water is incompressible. However, important changes of volume are produced by the pressures with which we are concerned, as may be realized from the fact that 12,000 kg/cm² will decrease the volume of many metals by an amount several fold greater than the volume contraction produced by cooling to 0° Abs. at atmospheric pressure.

It is perhaps worth while to mention in the first place a point on which there is sometimes considerable misapprehension; pressures of this order of magnitude never produce the slightest permanent change of volume, but volume changes under pressure are perfectly elastic. There is no reason to think that this is not also the case at pressures indefinitely higher than those yet reached; effects which at first seem to indicate a permanent change of volume can always be traced to flaws in the original material, or failure of the stress to be truly hydrostatic, so that plastic flow of the material results.

With respect to the effect of pressure on volume we may roughly divide matter into the two groups of solids and liquids, since under these high pressures there is no essential difference between a liquid and a gas. All the ordinary liquids, leaving out liquid metals, behave roughly alike at high pressure; the volume decrease under 12,000 kg/cm² varies from about 20 % for water to 30 % for ether, the most compressible organic liquid. The greatest volume decrease yet found in any substance is that of hydrogen and helium; the volume of these at 12,000 kg/cm² at ordinary temperature is less than one half the volume of the corresponding solids at atmospheric pressure at a temperature close to 0° Abs.

The compressibility of ordinary liquids is a very strong pressure function; typical organic liquids decrease in compressibility by a factor of about 15 from atmospheric pressure to 12,000 kg/cm²; for water, on the other hand, the corresponding decrease is by a factor of only 5. By far the largest part of the decrease of compressibility occurs in the first two or three thousand kg/cm². This we may take to indicate that initially the effect of pressure is mainly to squeeze out the empty spaces between the molecules. This effect is exhausted comparatively soon, thus accounting for the very rapid initial decrease of compressibility; after this effect is exhausted, further change of volume is provided by the actual change of volume of the molecules under pressure, and this effect is much smaller than the initial effect and persists with comparatively little change over a wide range. This change of volume of the molecules under pressure is not usually thought of as important, and is not suggested by experiments at low pressures. Thus practically all the equations of state that have been proposed for liquids on the basis of experiments to 2000 or 3000 kg/cm² do not take account of this phenomenon, so that these equations uniformly indicate too large a volume, that is, too low a compressibility, at high pressures.

Initially, different liquids show much greater individual differences than they do at high pressure. In fact, ordinary liquids approach so close to a common behavior at high pressures that it is possible to set up a sort of ideal liquid at high pressures, roughly typical of all liquids, in much the same way that we have an ideal gas at ordinary pressure.

The effect of pressure on the volume of solids (metallic elements) is markedly different from that on liquids. The volume changes are in general much less, being of the order of only a few per cent under 12,000 kg/cm². More important, the decrease of compressibility under pressure is much less than for liquids, so that the decrease of volume is much more nearly a linear function of pressure. This property is easy to connect with the difference of structure, for, of course, all these solid metals are crystalline and remain so at high pressure. This means that the atoms are always arranged on the same space lattice, so that the spaces between the corners of the atoms are never removed by pressure, and therefore the initial phase in the compressibility of liquids is absent. The compressibility is then from the first due only to a uniform distortion of the space lattice, and this in turn is doubtless connected with a distortion of the atoms themselves.

Although the compressibility of the average solid metal is much less than that of the average organic liquid, there are important exceptions. The alkali metals are by far the most compressible of the metals. In the series of the alkalis, compressibility rapidly increases with increasing atomic weight, until in caesium we have a metal whose volume decrease under 12,000 kg/cm² is as great as that of ether, one of the most compressible organic liquids. Initially the compressibility of caesium is much less than that of ether, but at 12,000 kg/cm² it is much higher, the average over the entire pressure range being about the same. This means that the compressibility of the atom of caesium is much higher than that of the molecule of ether. Among the alkali metals the most interesting is potassium. Under ordinary conditions its atomic volume is abnormally high, indicating a very open electronic structure of the atom. We would expect such a structure to have an abnormally high compressibility, and this in fact is the case. Under high pressure the abnormally high compressibility compared with the other alkali metals becomes very much accentuated, and all the evidence points to a great deformation of the potassium atom. In fact this deformation is so high as apparently to endanger the integrity of the atomic structure, for at the highest pressures there is a reversal in direction of some of the physical properties of such a character as to suggest the beginning of an atomic disintegration, such as we suppose we have in the stars, where the atoms are broken down into a gas of protons and electrons.

We have said that compressibility decreases with rising pressure. This is exactly what one could expect; it means thas as the atoms or the parts of the atoms are pushed closer together the forces which resist further compression becoming greater. In fact this behavior seems so natural that it is sometimes assumed that compressibility must necessarily decrease when pressure increases, but that this is not necessary is shown by at least one exception.

The compressibility of amorphous quartz increases with rising pressure by a marked amount. This is a genuine effect; there is no hysteresis in the connection between volume and pressure, the effect is perfectly reversible, and the experiment may be repeated an indefinite number of times. One is at first inclined to see a connection between this and the amorphous condition, since quartz may also exist in the crystalline form with a smaller volume. But that the effect may be more intimately connected with the law of force between the molecules is suggested by the fact that crystalline quartz shows the same effect at high pressures along one of the directions in the crystal.

After compressibility, perhaps the most purely mechanical property of a liquid is its viscosity. I have measured the effect of pressure up to 12,000 kg/cm² on the viscosity of a large number of liquids, most of them organic, but including water and measurements not yet complete on mercury. In general, the effects of pressure on viscosity are higher than on any other physical property yet measured, and vary by a far greater amount with the character of the liquid. In every case except that of water, viscosity increases under pressure. Roughly, viscosity increases geometrically when pressure increases arithmetically, so that log (viscosity) plotted against pressure is approximately linear. Fig. 1 shows a typical curve of

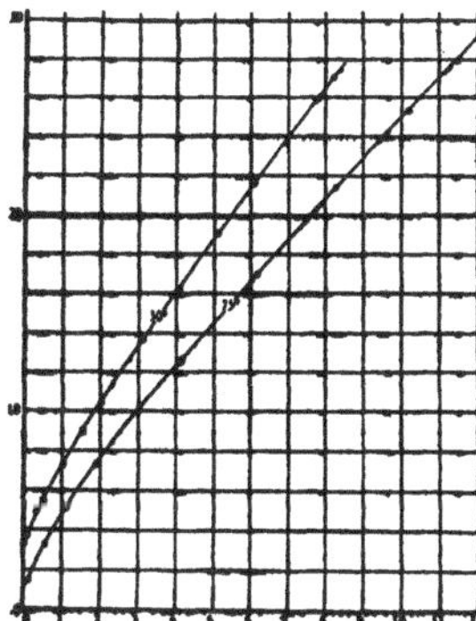

Fig. 1.
The common logarithms of the relative viscosity of methylcyclohexane at 30° and 75° C plotted as ordinates against pressure in thousands of kilograms per square cm.

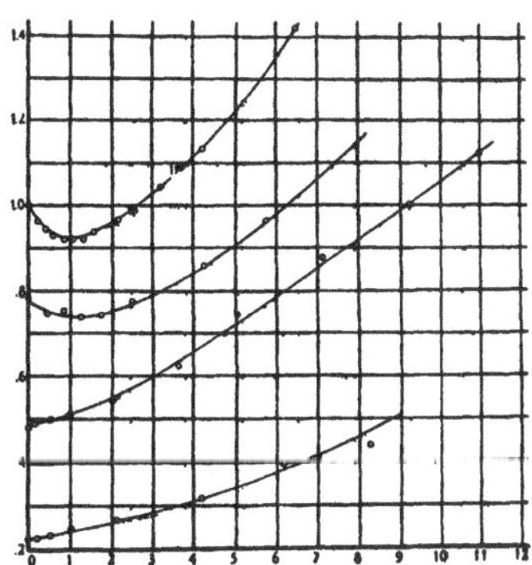

Fig. 2.
The relative viscosities of water at 0°, 10°, 30°, and 75° (reading downward) plotted as ordinates against pressure in thousands of kilograms per square cm.

this sort. At low pressures, however, there is usually a departure from this rule in that the initial increase of log (viscosity) is less rapid than the increase of pressure, the graph being initially concave toward the pressure axis. There are also departures at high pressures from this rule; often log (viscosity) of those liquids which show the highest pressure effects increases more rapidly than linearly with pressure. The increase of viscosity produced by 12,000 kg/cm² varies from something of the order of 20 % from mercury to more than 10^7 fold for a complicated organic liquid. It seems then that some substances ordinarily liquid may be made to approach the behavior of a glass by high enough pressures. The change of viscosity under pressure is always perfectly continous; there are no phenomena indicating freezing under pressure.

Water is an exception, as it is with respect to so many of its other properties. The results are shown in fig. 2. At low temperatures, the initial effect of increasing pressure is to decrease viscosity instead of to increase it. This abnormal decrease does not continue indefinitely, however, but presently the viscosity has a minimum and beyond this the normal increase. As temperature increases, the initial abnormal negative pressure coefficient becomes smaller, until at 30° C. it reaches zero, and above this the sign changes to positive, as is normal. In general the effect of pressure on the viscosity of water is like that of the other properties, that is, initial abnormalities are wiped out at high pressures, and at pressures high enough water tends to become completely normal. The fact that under ordinary conditions the properties of water are abnormal, which is so essential to phenomena of life, we owe to the fact that atmospheric pressures are so extremely low.

The relations found between viscosity and pressure indicate that the mechanism of viscosity cannot be as simple as has been supposed in most of the theories hitherto proposed for the viscosity of liquids. There are not many such theories, but all are similar in supposing that the viscosity of a liquid is due to the same sort of momentum transfer that it is in a gas. It is a consequence of such a mechanism that viscosity must be a function of volume only, or at any rate must change only very slightly when volume changes. If, then, the temperature of a liquid is raised, and at the same time the pressure is increased so as to keep the volume constant, there should at most be a very small change of viscosity. That this is not the case is shown by fig. 3; viscosity decreases when temperature is increased at constant

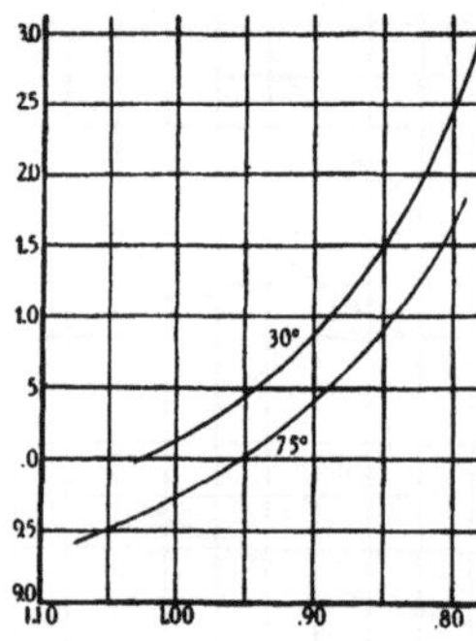

Fig. 3.
The common logarithms of the relative viscosity of i-amyl alcohol at 30° and 75° C plotted as ordinates against relative volume.

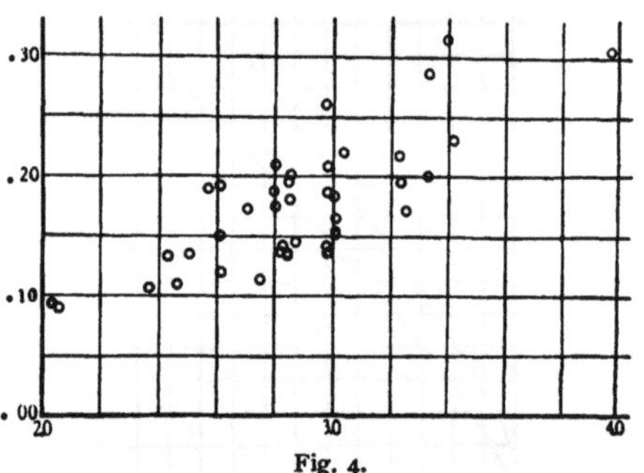

Fig. 4.
The common logarithm of the ratio of the viscosity at a pressure of 500 kg/cm² to that at atmospheric pressure at 30° for a large number of liquids plotted as ordinates against the common logarithm of the « complexity number ». See the text for a definition of this number.

volume. In seeking for an alternative or supplementary explanation of the viscosity of liquids, the enormous variation of the pressure coefficient with the nature of the liquid seems significant. There is a rough parallelism between the magnitude of the coefficient and the complexity of the molecule. Thus the effect of 12,000 kg/cm² on the viscosity of mercury is 20 %, on water perhaps twofold, on methyl alcohol four fold, up to 10^7 on the most complicated organic liquid. This is shown in fig. 4 in which the initial effect of a pressure of 500 kg/cm² is plotted against a number measuring roughly the complexity of the molecule. «Complexity of the molecule» is obviously a very rough concept. It may be thought to involve both the number of atoms in the molecule and the complexity of the individual atoms; we have arbitrarily chosen as a measure of complexity the product of the total number of electrons in all the atoms of the molecule by the number of atoms. The diagram now shows a very marked parallelism between the pressure coefficient of viscosity and molecular complexity. Such a thing of course is not suggested at all by any kinetic momentum transfer mechanism. The simplest interpretation seems to be that in a liquid there is some sort of interlocking mechanism which prevents the molecules freely sliding past each other. This view is consistent with the fact that at constant volume viscosity decreases with rising temperature. Such an interlocking effect may perhaps be important in determining the properties of liquids as mechanical lubricants. In a liquid in which the interlocking is great we would expect that it would be comparatively easy to tear apart a film which has become very thin, as in the journal of a machine, and thus destroy the lubrication. Now it does seem to be an experimental fact that, other things being equal, a liquid is a better lubricant if it has a *small* pressure coefficient of viscosity, that is, a small interlocking effect.

The two effects hitherto described (compressibility and viscosity) are homogeneous effects, in that the substance is submitted to a stress the same in every direction, and unless we are dealing with a non-cubic crystal, the strain is homogeneous also. Besides simple effects of this kind, there are more complicated ones produced when part, not the entire, surface of a body is exposed to hydrostatic pressure, so that non-homogeneous conditions are produced. In this way, solids may be subjected to permanent deformations and eventually rupture. Results are to be obtained in this way which cannot be otherwise obtained, because the extreme uniformity of external conditions when stress is applied by a liquid makes it possible to reach higher stresses than can ordinarily be reached without rupture, and so to observe the deformation and rupture under extreme and instructive conditions. In my early pressure work I have observed many cases of this sort, and I have thought a description of them would be particularly appropriate to a Congress of Applied Mechanics.

We first discuss phenomena where there is no rupture, but only permanent deformation. The results here differ from those of ordinary experience chiefly with respect to stress-strain hysteresis, which presents many anomalies, such as abnormal directions and abnormally large values. Interesting results are found sometimes in minerals. A mineral need not be a homogeneous substance, so that even when the surface is exposed to a hydrostatic pressure all over so as to produce in the mineral a hydrostatic stress uniform from the macroscopic point of view, the strain on a microscopic scale may be non-homogeneous. Under these conditions there may be hysteresis between pressure and deformation. In fig. 5 is shown such a hysteresis

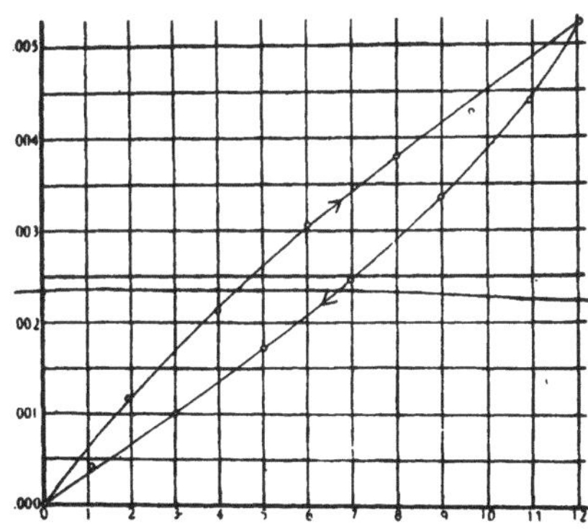

Fig. 5.

Hysteresis loop in pipe stone when subjected to hydrostatic pressure. The fractional decrease of length is plotted against pressure in thousands of kilograms per square cm.

loop for pipe-stone. The most surprising feature is the abnormal direction of the loop. Properly we do not have hysteresis here at all, for the deformation, instead of *lagging behind* the pressure, is in advance of it. Evidently a highly unusual type of mechanism must be involved in such an effect as this, but there is nothing impossible in such a situation and mechanisms can be invented which show effects of just this kind.

Other unusual hysteresis effects are found when a hollow cylinder, closed on the ends, is exposed to hydrostatic pressure over the entire external surface so as to decrease the internal volume. The results of one such test are shown in fig. 6. Most of the features of this test are shown also by simple tension tests, such as the raising of the yield point by overstrain to the previous maximum stress, hardening by resting after overstrain, and hysteresis. But the particular character of the stress applied here, which does not tend to pull the metal apart and so ultimately produce rupture, but rather compacts the metal more closely as stress

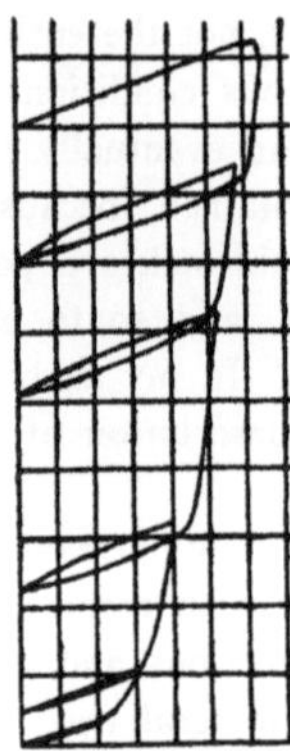

Fig. 6.

The relation between fractional decrease of internal volume (plotted as ordinates) of a hollow steel cylinder when subjected to cyclic applications of external pressure (plotted as abscissa, one division corresponding to 1000 kg/cm²).

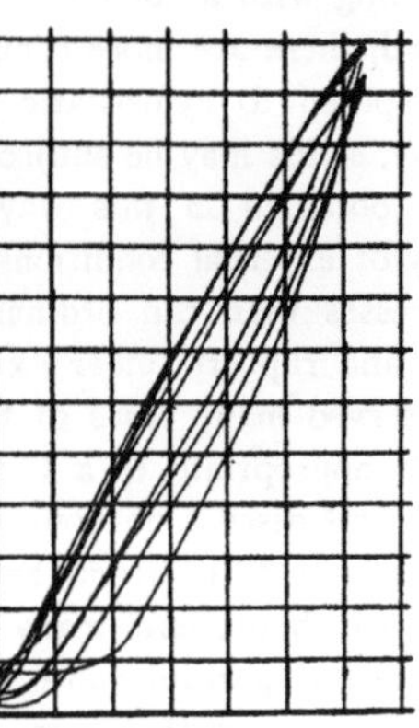

Fig. 7.

The same data as shown in Fig. 6 for another and thicker-walled cylinder. Notice that when the external pressure is at first increased the internal volume may *increase* also.

increases, permits reaching stresses after the initial yield point very much higher than is possible in other kinds of test, and so gives much exaggerated effects. Thus the yield point in collapsing tests of this sort may be raised by overstrain to ten times the initial value, and the width of the hysteresis loop may rise to one half that of the maximum deformation.

In addition to these effects, similar in character but different in magnitude from those of the usual tests, sometimes unique effects are obtained as shown in fig. 7. At one stage in the cycle of operations it was found that an increase of external pressure was followed by an *increase* of internal volume over a short range. Evidently the very great distortion of the metal far beyond the yield point has stored up in it a most complicated system of internal strains, which are responsible for this highly anomalous effect. Similar anomalous effects have been found also in thick cylinders stretched by internal pressure. In general, such effects are produced more readily in thick walled cylinders, in which there is a large mass of metal to store up a system of internal strains varying greatly from point to point. If the cylinder has thin walls, the internal strains are more nearly uniform from point to point, and such effects do not appear. The effects are also larger in a substance of complicated molecular structure, as for example such effects are larger in a high carbon tool steel than a steel of lower carbon content.

What now happens when the external pressure on such a hollow cylinder is increased indefinitely? Of course it is well known that a thin walled cylinder eventually becomes unstable, and folds into a star shaped figure. If however the walls are above a certain critical thickness, instability does not occur, but instead the internal volume decreases regularly with increasing pressure until it vanishes and the hole is squeezed shut. The internal diameter is very nearly a *linear* function of pressure; the pressure at which the cavity is closed is a function only of the material of the cylinder, and is independent of the ratio of internal to external diameter. This pressure is about 10,000 kg/cm² for copper, and 20,000 kg/cm² for a mild steel. One important conclusion can be drawn from the linearity of the relation between internal diameter and pressure, namely that plastic flow of the metal is not determined by the maximum stress difference. It is easy to show that if the maximum stress difference is constant after flow begins, such a cylinder would support an indefinitely great pressure without completely closing, the internal diameter decreasing logarithmically as pressure increases. The fact that the cavity eventually closes shows that in the later stages of flow the maximum stress difference is less than in the initial stages; in other words we have here a weakening by excessive flow instead of a strengthening.

Suppose now we experiment with a brittle substance like glass, whose breaking stress in ordinary tensile tests is perhaps 500 kg/cm², instead of a metal, and subject a thick walled hollow cylinder of glass to hydrostatic pressure all over its external surface. Will the glass flow like the metal? Experiments show that nothing happens—pressures of 25,000 to 30,000 kg/cm² produce in the glass neither flow nor rupture, and such a cylinder recovers from the application of pressure unscathed. Now this failure to break is most instructive in what it tells us about conditions of rupture. At the internal surface of such a hollow cylinder, when exposed to 25,000 kg/cm² external pressure there is a maximum stress (compressive) of 25,000 kg/cm², a maximum stress difference of 25,000 kg/cm², a shearing *strain* corresponding to a stress difference of 25,000 kg/cm², and an elongation of the fibres in the radial direction equal to that produced by an ordinary tension of 12,500 kg/cm² (assuming Poissons ratio = ¼). This single example is sufficient to show, therefore, that rupture is not determined either by maximum stress, maximum stress difference (or strain), or maximum elongation, the three criteria which are most usually considered. Furthermore, these criteria fail to hold by enormous amounts. We notice, however, that the stresses in this case are never tensile in character, but are everywhere compressive, or at most zero. May it not be possible then that rupture occurs when the maximum tensile stress reaches a critical value? Imagine now a solid cylinder of glass, exposed to hydrostatic pressure over its curved external surface, but with no pressure on the ends, so that there is no stress across planes perpendicular to the axis. Experimentally we may realize such a stress distribution by leading a rod of glass into and out of a pressure chamber through two stuffing boxes. What now happens when pressure is increased indefinitely? The experiment is easily tried. The rod parts on some plane perpendicular to the axis in the region exposed to pressure, and the two pieces are expelled from the pressure chamber through the two stuffing boxes, exactly as if the rod where pulled lengthwise by a tensile force. This I call the «pinching off» effect. Here then we have rupture across planes on which there is approximately no stress (actually there is a slight *compressive* stress across this plane arising from the friction of the packing), so that it is perfectly evident that maximum tensile stress does not determine rupture. We would be inclined in this case to say that it is the maximum elongation which determines rupture, but the experiment above with the hollow cylinder disposes of this criterion. That maximum tensile stress is not intimately concerned in rupture we can show still further by experiments on hollow glass cylinders exposed to *internal* pressure. It is possible to raise the internal pressure on such a cylinder to a point where the maximum circumferential fibre stress is more than twice the fibre stress at rupture under ordinary tensile conditions.

These results just described with glass depend essentially on its amorphous character, and results entirely different in character may be obtained with a brittle crystalline material, such as quartz.

Thus a hollow cylinder of quartz, exposed to external pressure, is ruptured when pressure is pushed to high. In fig. 8 (Plate II) is shown one such cylinder ruptured by external. pressure. Rupture takes the form of a microscopic disintegration of the inner surface. There are obviously interesting geological applications here.

Tests similar to those on glass may be made on metals capable of plastic deformation. Such metals show the pinching-off effect just like brittle materials. There is a difference of appearance in the ruptured surface, however, in that instead of a clean rupture on a single plane we get a drawing out and necking down, in appearance like an ordinary tensile specimen. This however is not an important difference, and the essential fact remains that we have rupture by simple parting of the fibres across planes on which there is no stress. The results of rupture tests on hollow cylinders subjected to internal pressure are, however, different for brittle and plastic materials. The rupture of a brittle material is sudden, and I have not been able to follow the details of it; there is every reason to think, however, that rupture begins at the internal surface where we would expect, where all stresses and strains have their maximum values. With plastic metals on the other hand, if the cylinder is sufficiently thick so that high pressures can be reached before rupture occurs, it will be found that the crack starts on the *outside* and travels inwards. In fig. 9 (Plate II) is shown a section of one such cylinder of steel which was burst by internal prssure. Besides the fact that rupture starts at the exterior surface, there are two important features to be emphasized. In the first place, the magnitude of the internal pressures which such a cylinder can support are much higher than ordinary considerations would lead one to expect; for example 40,000 kg/cm^2 was reached in the steel cylinder of fig. 9 although the normal tensile strength was only 10,000 kg/cm^2. In fact it is only this unexpected behavior which has made possible these high pressure experiments. In the second place, the very high deformations reached at the inner surface, where rupture does not occur, are significant. Circumferential elongations up to 300 % have been found, against 25 % in ordinary tensile tests. This very great elongation is accompanied by profuse slipping of the metal on itself at the inner surface, as careful examination of the diagrams will show. This slipping at the inner surface is, of course, a sort of rupture. The experiment shows that although rupture may *start* at the inside, the conditions are such that it can not propagate itself from the inside, and the ultimate destructive rupture has to start at the outer surface.

Finally, we describe another most amusing type of rupture. Around a cylindrical steel core place a closely fitting ring of hard rubber and subject the steel and rubber together to hydrostatic pressure over the entire external surface. When the pressure is raised high enough it will be found that the rubber ring is broken exactly as when a wedge is driven into the ring under ordinary conditions. In fact, reflection shows that the effect is much like driving in a wedge, because the volume compressibility of rubber is much higher than that of steel, so that under pressure the rubber is distorted by the steel core to a diameter greater than that of free rubber under the given pressure. But this explanation amounts to assuming that maximum shearing strain, or stress, determines rupture, and this we have seen is not true. Whatever the explanation, the fact is that here we have rupture in a substance whose every stress is compressive and every fibre is shortened.

These examples are sufficient to show that none of the usual criteria of rupture can apply, either to brittle or plastic materials. In comment on this situation, it seems that at least one consideration not usually entertained is pertinent. It is evident that the hollow cylinder of glass under external pressure did not break for the reason that there was nowhere for the fragments to go if rupture did occur, because of the perfect symmetry, and similarly in the case of the metal cylinders under internal pressure, rupture did not occur at the inside because rupture could not propagate itself after starting. We must therefore add to our usual conditions geometrical considerations, such as those of the familiar problems of instability, which shall state that rupture can occur only if it is of such a nature as to *relieve* the applied stress and bring about a decrease of potential energy of the whole system. In these geometrical considerations symmetry relations must be involved, so that the conditions of rupture will in general be different for crystalline and amorphous substances.

But a much more serious consideration is whether properly speaking there is any such thing as a criterion of rupture. I believe we have no right to expect any general criterion in view of the extremely varied structure of different sorts of matter. It is sufficient in dealing with many phenomena to think of the molecules which compose matter as being to a certain degree like small rigid bodies, in which are located, according to more or less complicated patterns, centers of electrical and magnetic forces which hold the substance together and give it its properties. When such a substance is subjected to stress, the molecules must readjust themselves to each others irregularities in the most complicated ways, and according as one or another pair of the local centers of force are separated by more than the critical amount, we may have rupture under the most varied conditions.

The theoretical considerations, as well as the experimental facts summarized above, indicate therefore that we should attempt to establish general criteria of rupture only as a matter of practical convenience, as for engineering purposes, and that we should expect any such criteria to be valid only in a narrow range of conditions, both of stress and material.

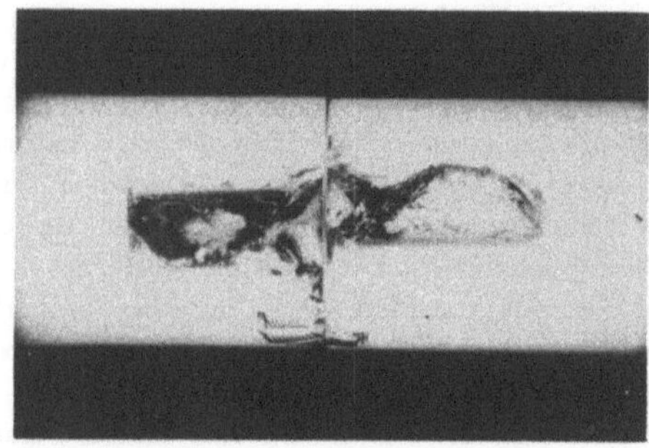

Fig. 8.
Photograph of a hollow cylinder of quartz cut from a single crystal, showing how the internal cavity has been eroded by the application of a pressure of 12 000 kg/cm² to the external surface.

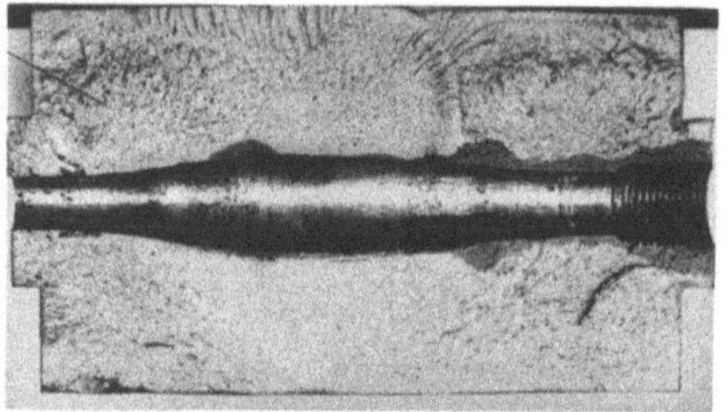

Fig. 9
One of the halves of a cylinder of tool steel split by the application of 40 000 kg/cm² internal pressure

Electrical Properties of Single Metal Crystals

by

P. W. Bridgman - Cambridge (U. S. A.)

Within the last few years improvements in technique have made it possible to produce single metal crystals large enough to allow the experimental determination of a number of their electrical properties, of which we hitherto had knowledge only through theoretical analysis. These experiments have brought to light facts which have far-reaching implications, not only on our pictures of the mechanism of conduction in solid bodies, but also on the very concepts in terms of which we describe electrical phenomena.

Electrical Resistance. In general, the electrical resistance of a single crystal is different in different directions. There is a connection between the variation of resistance with direction and other properties; the resistance is greater in those directions in which the planes of the lattice are more widely spaced, and is greatest across the plane of easiest cleavage. The mere fact of this difference of resistance in different directions is important for our pictures of the mechanism; the classical free electron theory would give a resistance the same in all directions.

The symmetry relations of electrical resistance have been worked out by Voigt, and as far as our knowledge at present goes, are verified by experiment. Voigt treats the current and the applied field (taken as the gradient of a scalar potential function) as having the symmetry of ordinary polar vectors. The fact that the theoretical symmetry relations are verified is presumptive evidence that our ordinary concepts of the vector character of the current and the field are adequate.

Thermo-Electric Phenomena. The termo-electric properties of a crystal are different in different directions. This means that if a straight bar is cut from a crystal, the two ends connected to an electrically isotropic metal such as copper, and the two junctions brought to different temperatures, the thermal e.m.f. will be found to depend on the orientation of the bar. The symmetry relations have been discussed by Lord Kelvin and by Voigt. The basis of Kelvin's analysis is his axiom of the superposition of thermo-electric action: « It may be assumed as an axiom that each of any number of coexisting systems of electric currents produces the same reversible heating effect in any locality as if it existed alone ». Voigt has obtained a result similar to that of Kelvin. His analysis rests on the treatment of the Thomson e.m.f. as a quantity of the same vector symmetry as the electric vector. He finds that the thermal e.m.f. varies with direction in the same way as the electrical resistance, so that one should be a linear function of the other; this gives a convenient method of checking the equation experimentally.

The difference of thermo-electric quality in different directions may be large, and is easy to show experimentally; in Zn, for example, the thermo-electric power of a bar parallel to the axis against one perpendicular to it is 1.2×10^{-6} volts/C°, and in Bi the difference is about 50×10^{-6}. The precise symmetry relations of t.e.m.f. are difficult to get experimentally with great accuracy. The present evidence makes it probable that the symmetry is *not* that predicted by Kelvin and Voigt; it is probable that the Thomson heat does have the predicted symmetry, but that the Peltier heat is somewhat different.

What now is the meaning of the disagreement between experiment and the predictions of Kelvin and Voigt? A reexamination of the argument of Voigt will show that he has considered only the action within the body of the crystal, that is, the Thomson heats, and has neglected the surface effects, that is, the Peltier heats. We would expect, therefore, that Voigt's symmetry relations would apply only to the Thomson heat, and we have just seen that this is consistent with experiment. The argument of Kelvin is also faulty; in his application of his axiom of superposition he resolved the current into directions parallel and perpendicular to the crystal axes. It is easy to show, however, that if the resolution were made into another pair of perpendicular directions, the same result would not be obtained.

It is interesting that Kelvin predicted on the basis of his incorrect analysis an unknown thermo-electric effect peculiar to crystals. Namely, if a current passes lengthwise in a bar inclined at some angle neither 0° nor 90° to the crystal axis, then a transverse difference of temperature will be set up in the bar, one entire side becoming warmer and the other colder. This effect I have recently found; in bars of Bi of the order of 6 mm. diameter with currents of the order of 1 amp., temperature differences of several tenths of a degree may be established. This is approximately the magnitude to be expected according to Kelvin's theory.

The facts that the transverse effect exists, and also the fact that the symmetry of the Peltier heat does not differ greatly from that predicted by Kelvin, lends a certain probability of physical correctness to his actual process of analysis, although we must recognize that his general argument was fallacious. That is, Kelvin's resolution of current into components parallel and perpendicular to the axis may correspond to the physical fact that in a crystal the conduction electrons actually move mostly along only these two directions. This idea that the electrons may be confined to definite guided paths is also one which has been suggested by several other lines of evidence.

Leaving now this question of symmetry, there are most important implications contained in the mere fact that thermo-electric quality is different in different directions. I pointed out some time ago that this mere fact, in conjunction with the thermodynamic requirement that no work can be got out of a cyclic operation in a system at one uniform temperature, demands that there be a reversible internal Peltier heat in a crystal where a current changes its direction of flow. The existence of this heat has important bearings both on our picture of the mechanism of conduction, and on our electrical concepts.

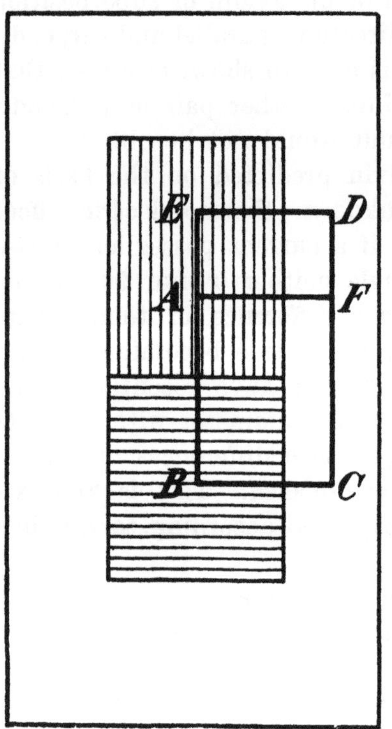

Fig. 1

Figure 1 shows two pieces of the same crystal joined so that the axis in one part is perpendicular to that in the other, and imbedded in an isotropic conductor. Consider the two closed paths *ABCFA* and *ABCDEA* by which an electron may be carried from *A* back to *A* again. The only place in this path where heat can be absorbed is at a surface of discontinuity, or where the direction of motion changes. Now the heat absorbed by the electron in the second path differs from that in the first path by the internal Peltier heat at the corner *E*, and therefore the total heat absorbed in these two paths cannot be the same. But thermodynamics demands that the total heat absorbed in any closed path in a system at constant temperature be zero. Hence at least one of these paths cannot be a closed path in the sense of thermodynamics, or in other words we have not returned the electron to its initial condition. The only way of saving the situation is to recognize that the condition of the electron depends not only on its position in the body, but also on its direction of motion. The electron in the cycle *ABCFA* was not returned to its initial condition because it left *A* in the direction *AB*, and returned in the direction *AF*.

The physical explanation of this most surprising fact would seem to demand the existence of a fine structure in the metal, of which ordinary phenomena give no hint. The guided path idea is again suggestive. It is as if the electron could move in one direction only when it is in one kind of channel, and in the perpendicular direction only if it is in another kind of channel, and energy is required to change the electron from one channel to another. The magnitude of the energy difference between the channels can be easily calculated. In the extreme case of Bi, 1 coulomb absorbs at 300° Abs. $300 \times 50 \times 10^{-6}$ joules $= 1.5 \times 10^5$ ergs, in changing direction by 90°. Hence one electron absorbs $1.5 \times 10^5 \times 1.59 \times 10^{-19} = 2.4 \times 10^{-14}$ ergs. The energy of a single degree of temperature agitation at this temperature is 6×10^{-14}. If we suppose the work in getting from one channel to another is electrical work against a small scale difference of potential, then this potential difference is 1.5×10^{-2} volts, which, in a distance of 10^{-8} cm., means a gradient of 1.5×10^6 volts/cm. Or possibly the effect may be magnetic. If we suppose that there is an internal magnetic field in the channels along which the electrons move, and if we suppose that the electron has a magnetic moment of one Bohr magneton, and that the axis of the magneton must set itself in the direction of motion, then work will be required to change the orientation in the magnetic field, and hence to change the direction of motion. It will be found that the intensity of the internal magnetic field required for this is 2.7×10^6 Gauss.

Whatever the nature of this small scale structure, we cannot suppose it to be confined only to non-cubic crystals, but it must be present in all crystals. In a cubic crystal the structure along the three axes has the same properties, so that no work is required to change the direction of motion of an electron from one axis to another.

What now are the bearings of this directional energy on the concepts by which we describe the situation? By what means is energy transmitted to the current when it changes direction? If we keep our ordinary concept of electrical potential at points inside a metal, as I believe we may and must, then, since potential is determined by the position of the distant electrical charges, there can be no abrupt changes in potential in atomic distances, and we must recognize that this work cannot come from a change of potential energy in the field in the ordinary sense. We must say, therefore, that the current acquires its energy through the action of an impressed e.m.f. This is what we mean by an impressed e.m.f., namely some agency, other than that covered by the electromagnetic field equations, by whose action energy is conveyed to the current. We must recognize the existence of such an impressed e.m.f. wherever there is a reversible absorption of heat, that is, either at a surface of discontinuity, or where the direction of current flow changes. In addition to this energy aspect of an e.m.f., we always think of a second aspect, in that an e.m.f. tends to produce a current. The current flowing between two points whose electromagnetic difference of potential is ΔV and between which an impressed e.m.f. E acts, is given by Ohm's law:

$$i = \frac{-\Delta V + E}{R}.$$

Do the impressed e.m.f.'s in a crystal act in this way? It is evident at once that the directional e.m.f. does not, because for two points infinitely close together $\Delta V = 0$, and $R = 0$, so that if E were finite, we would have an infinite current. It is also true that the surface e.m.f. cannot produce a current. In Figure 2 convey an electron from A to B by the path indicated. There can be no jumps of potential in the homogeneous metal, either crystalline or isotropic.

The only two surfaces of discontinuity in the path are identical, except that they are crossed in opposite directions. Hence if there are any jumps of potential at these surfaces, they must cancel, and the potential at B must be the same as at A. Hence there is no double layer at the interface. But heat is absorbed in passing through the surface from B to A, and hence according to our definition there is an impressed e.m.f. in the surface. The fact that there is no accompanying double layer shows that this impressed e.m.f. has only one of the two usual aspects of an e.m.f.; it delivers energy to the current, but it does not act like an ordinary body force, and cannot be put into Ohm's law.

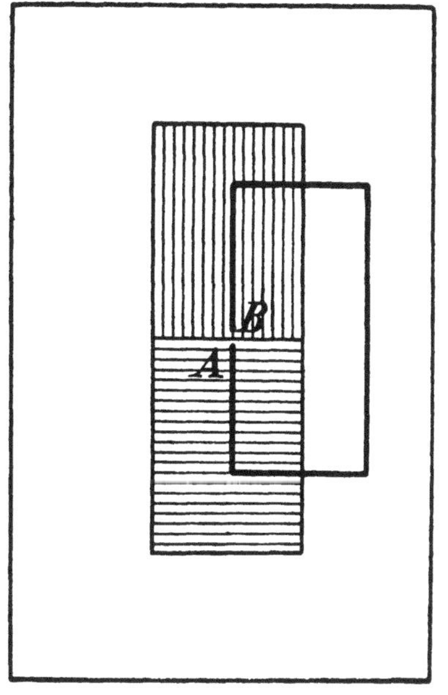

Fig. 2

The fact that in crystals we may have a Peltier heat but no double layer has at once an application to the ordinary thermo-electric analysis. Leaving out of account the Volta effect, which is of a different order of magnitude, it is almost always stated that at a surface of separation of two unlike metals there is a jump of potential corresponding to at least the Peltier heat; the example just considered shows that there is no necessity in such a state of affairs, and in fact it is in general not true.

The existence of an impressed e.m.f. which delivers energy to the current, but which does not enter into Ohm's law, demands that the current carry energy with it. Consider a current i flowing in a crystal with a bend, so that there is an internal Peltier heat Pi at the bend. This means that there is a reversible in-put of heat at the bend in addition to the Joulean heat developed in overcoming resistance. Let us write down the equation of energy balance for the region included between the dotted lines and the planes A and B in Figure 3. Assume for the moment that unit current carries into the region across the plane A the energy E_1, and carries E_2 out across B, call the resistance between A and B, R, and the difference of potential ΔV. Then:

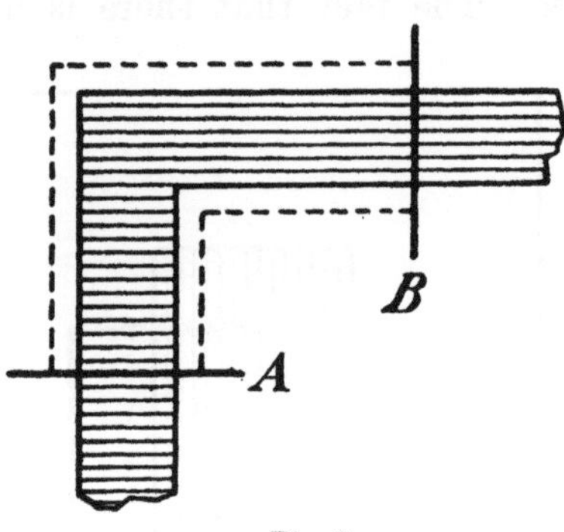

Fig. 3

$$(E_1 + P)i = E_2 i + i^2 R + i\Delta V,$$

where the five terms in this equation are, respectively: the energy carried into the region by the current, the reversible heat flowing in to provide the heat absorbed by the current at the bend, the energy carried out of the region by the current, Joulean heat flowing out of the region, and electromagnetic energy carried from the region by the Poynting vector. But now Ohm's law gives $i = \frac{-\Delta V}{R}$, whence $i^2 R = -i\Delta V$, and

$$E_1 + P = E_2.$$

This means that the energy carried out differs from that carried in by P, and therefore in general we must recognize that energy is convected by the current. It can be shown that this must be ***heat*** energy; the proof is too long to give here. This conclusion can be avoided only if $E_1 = E_2$, and this demands that $i = \frac{-\Delta V + P}{R}$. But this we have seen to be impossible; it would demand a difference of potential between the two ends of a crystal, which is inconsistent with crystal symmetry, and is furthermore opposed to experiment, because I have found to a sufficient degree of accuracy that i is proportional to ΔV for small ΔV's.

The assumption that an electric current carries heat with it was of course made in the classical electron theory, and was the basis of its explanation of the Thomson heat. But there were great difficulties with the classical idea that the electrons carry heat energy, and the conclusion to be drawn from the classical explanation of the Thomson heat has usually been disregarded. It is, in fact, justifiable to disregard the classical conclusion, because it is possible to give another account of the matter. It is usual to speak of a Thomson e.m.f., which not only delivers energy to the current, but which is in nature like an ordinary body force, entering Ohm's law, and producing a difference of potential between the ends of an open-circuited conductor in which there is a temperature gradient. It is perhaps not generally recognized that such an account of the Thomson e.m.f. is inconsistent with an energy convected with the current. It is easy to apply to an unequally heated bar an analysis like that applied above to a crystal, and it will be found that the conservation of energy demands either that the current convect energy with it ($\sigma\tau$), and that the e.m.f. do not enter into Ohm's law, or else that the current convect no energy which varies with the temperature, but that the Thomson e.m.f. does enter Ohm's law and produce a difference of potential between the ends of an unequally heated bar on open circuit.

But now the positive proof from the properties of a crystal that at least under some conditions the current *must* convect heat energy, makes it highly probable that the Thomson e.m.f. is not in nature like an ordinary vector, and therefore eliminates the second of the two possibilities just mentioned. This removes the basis of Voigt's discussion of the symmetry of the Thomson heat, so that apparently we have no adequate theoretical basis left for the discussion of thermo-electric symmetry.

In addition to these comparatively simple effects, other

more complicated effects in crystals not yet investigated experimentally may be expected to throw much additional light on the nature of electrical action in metals. For example, we must recognize the possibility of Volta differences of potential between different faces of the same metal crystal; important conclusions as to the nature of the Volta effect, which has not yet been adequately explained, might be drawn from a knowledge of the facts in crystals. The Hall effect, and the other transverse effects in a magnetic field, have been barely touched experimentally: in fact the symmetry relations are not known. An experimental knowledge of the symmetry of the Hall effect would give important information as to the nature of the Hall e.m.f.; whether it is a body force like the electrostatic vector, or whether it is something quite different.

THE VISCOSITY OF MERCURY UNDER PRESSURE.

By P. W. Bridgman.

Presented October 19, 1927. Received October 23, 1927.

CONTENTS.

Introduction.

In a previous communication,[1] results have been given for the effect of pressure up to 12000 kg/cm² on the viscosity of 43 liquids. These liquids varied greatly in complexity, from CS_2 and H_2O on the one hand, to Cineole ($C_{10}H_{16}O_5$) and Oleic acid ($C_{15}H_{33}COOH$) on the other. A simple generalization suggested by those experiments was that the percentage effect of pressure on viscosity is greater for those liquids which have a more complicated molecule. Hence the interest of extending the measurements to a substance with as simple a molecule as possible is obvious, and mercury, which is monatomic, at once suggests itself. In this paper the results of such measurements on mercury are recorded; these measurements were made at 30° and 75° up to 12000 kg.

The viscosity of mercury under pressure has been previously measured by Cohen and Bruins,[2] who determined the effect of 1500 kg at 20°.

The Method.

The method previously used for 43 liquids demanded that the liquid be an electric insulator, and so was not applicable to mercury. It proved unexpectedly difficult to devise another method, and a number of attempts were made without success. The first attempt was with a modification of the method used with other liquids, in which a magnetized steel weight rose through a mass of mercury, the motion of the weight being indicated outside the pressure apparatus by the current induced in a solenoid. This device failed because of the great irregularities arising from the failure of the mercury to wet the walls of the containing vessel, so that solid friction entered. The performance was somewhat improved by amalgamating the walls of the vessel, but the amalgamation of the steel was only temporary and partial and the improvement was not sufficient. Next, various flow schemes through glass capillaries were tried. These failed either because of my inability to seal platinum contacts into glass in such a way that the glass did not crack under high pressure because of the differential compressibility, or because the glass itself cracked when made in at all complicated shapes, or because of irregular surface tension effects due to the necessary small size of the apparatus.

Somewhat more successful was a capillary flow apparatus constructed of steel, in which about 50 gm of mercury ran of its own weight from one end of a vessel with two compartments to the other through a steel capillary coil 15 cm. in length and 0.05 cm. in diameter. A great many measurements were made with various modificatons of this general scheme. The individual results were of a high degree of irregularity, but by making a great many readings a rough value for the effect of pressure was found. The method did not work well at 75°, however, the various irregularities becoming much greater, and I was particularly anxious to get the effect of temperature, in order to find whether there is a simple connection between viscosity and volume, a matter of considerable theoretical importance.

The difficulty with all these methods arose primarily from irregular surface effects because of the failure of the mercury to wet the containing vessel. It presently became obvious that the mercury must wet the walls of the vessel if regular results were to be obtained, and I could think of no other way of doing this than making the vessel of copper, or some similar metal, and amalgamating it. I was long reluctant to do this because of my dislike to introduce impurities into the mercury, but I decided finally that this was the lesser of two evils. Professor Richards[3] had found by careful measurements that mercury dissolves less than 0.02% of copper, so that very little error is to be expected from the impurity of copper, and doubtless my early scruples were without foundation.

The general scheme of the apparatus finally adopted is shown in Figure 1. It consists essentially of two copper reservoirs partly filled with mercury, and connected by a capillary tube of copper. The apparatus could be tipped back and forth through a definite angle, and the time of flow to a definite height in either reservoir determined by the closing of an electric circuit through a platinum contact. Success was not immediate, however, even when the apparatus was constructed of amalgamated copper. Great trouble was always experienced from the contacts. In use, the surface of

FIGURE 1. Section of the flow part of the apparatus for determining the viscosity of mercury. The reservoirs may be tipped in one direction or the other through a fixed angle, and the time of flow determined by means of the platinum contact.

the mercury rapidly becomes covered with a coating of dirt, so that a high potential is necessary to overcome the contact resistance. For this purpose the current from the same magneto was used that had been used in the previous work with 43 liquids. Such a high potential may introduce serious errors, however, because the dirt, whatever it is, has conducting properties, so that the contact may be formed by a spark jumping to the dirt on the surface of the mercury, or dirt may actually hang in whiskers to the platinum point, again closing the circuit too soon. In the effort to minimize this effect, the top of the reservoir was narrowed down to a small neck, as shown in Figure 2, so that the contact was made with the surface of mercury

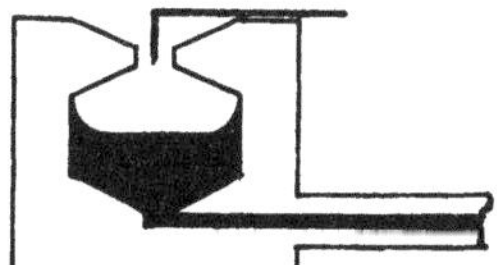

FIGURE 2. An early form of reservoir, designed to give greater sensitivity. It was discarded because of the error from the unknown effect of pressure on the capillary action.

when it was moving comparatively rapidly, thus reducing the time error resulting from a conducting film of constant depth on the surface of the mercury, or from whiskers of definite length. This device did not, however, produce markedly more regular results, and it was given up because of a serious theoretical objection to it, namely the head of mercury in the two reservoirs differs from that calculated from the difference of height by the amount contributed by the capillary forces in the surfaces in the two reservoirs. This capillary effect does not vanish if the cross section of the two reservoirs is not the same. Furthermore, it varies with pressure in an entirely unknown way, and hence introduces an entirely unknown error into the pressure coefficient of viscosity. It seemed, therefore, that the only safe course was to eliminate the effect altogether by making the cross sections of the two reservoirs everywhere the same, so that there is no differential capillary action, and the driving pressure can be calculated simply from the difference of height.

The liquid by which pressure is transmitted is a matter of some importance, because obviously its compressibility must be known, as the effective head driving the mercury from one reservoir to the other is determined by the difference of density between mercury and the surrounding liquid. The transmitting liquid must also be of low viscosity at every pressure, so as not to effectively increase the total viscosity by the act of completing the return flow from one reservoir to the other. Ether has these properties, and it was used in a number of the experiments. Its use had to be given up, however, because the short-circuiting effects were so bad. Apparently ether decomposes under the passage of the spark at high pressures, giving some decomposition product of comparatively high conductivity, perhaps water. This seems to be a purely high pressure phenomenon; attempts to reproduce it by sparking under ether to mercury at atmospheric pressure have failed. The performance was improved by replacing ether by petroleum ether. This latter, being a hydrocarbon, can give no water on decomposition, and this theoretical expectation was justified. The compressibility of petroleum ether was not known however, and it was therefore necessary to make a special determination of it, although not with a high degree of accuracy, since it was to be used only in a correction term.

The method by which the copper is amalgamated was also found to be of great importance. At first the copper was amalgamated with sodium amalgam, washing away the sodium in the form of hydroxide with water as carefully as possible. It is probable, however, that this washing was never complete. Always after amalgamation in this way, the dirt which formed on the surface after contact had been made a number of times, was of a disagreeable mechanical texture, difficult to break through, and very apt to form whiskers on the platinum points, particularly at 75°. The trouble disappeared on amalgamating with $HgNO_3$, again washing away the salts with water as thoroughly as possible. Under these conditions dirt still formed on the surface, but its mechanical texture was entirely different, being a light powder like lamp black, without the bad effects on the contacts of the other. The apparatus with which the final measurements were made was amalgamated in this way.

The experimental procedure with the final apparatus was as follows. The apparatus was first amalgamated by filling it with an $HgNO_3$ solution, and then washing with many changes of water. The capillary was amalgamated and washed by pulling through it a thread with a closely fitting knot. After amalgamating and drying, a weighed amount of mercury was placed in the reservoirs, enough to fill them about two-thirds full. The reservoirs with mercury were now attached to the insulating plug, connections made to the platinum contacts, and viscosity apparatus and plug placed together, in a horizontal position, as a single assembly, in the pressure cylinder. The pressure part of the apparatus was in all respects the same as that used formerly with 43 liquids. Because of the necessity of maintaining the mercury cups always in a vertical position, the details of the insulating plug formerly used had to be modified. The former plug, which carried in a single piece the insulation and the threaded part with which the hole in the pressure chamber was closed, was now made in two parts. One part carried the insulation, and together with the copper reservoirs constituted a single assembly which could be thrust without rotation into the pressure chamber. The other part carried the thread, and was turned up against the other part, compressing the packing and closing the hole. It is not necessary to go further into the details of this construction, which was sufficiently straightforward.

The pressure chamber could be rotated back and forth through fixed angles, which were determined by suitably placed stops. The mercury was run from one reservoir to the other while the apparatus was inclined to the horizontal at approximately 11°. The timing clock was stopped by the making of contact, through a system of relays actuated by the magneto in much the same way as in the previous work. Immediately when the clock stopped, the pressure chamber was turned back to a smaller angle, about 4°, which trial had shown was the angle at which the mercury just makes contact with the platinum point. In this position a sufficient time was allowed for the mercury to come to equilibrium, assuming the same level in the two reservoirs. The run was now started by rotating the pressure chamber from an inclination of 4° on one side to an inclination of 11° on the other side. The clock was so connected with various contacts that it started to run simultaneously with the rotation, the rotation itself occupying only a negligible time. In the early work the clock was started on breaking the contact with the platinum point, but the breaking contact proved much less reliable than the making contact, because of danger of dirt sticking to the point, and in all the later work the clock was started with the rotation.

At each pressure, ten readings of the time of flow were made, five in each direction, and the mean taken. No attempt was made to set up the apparatus perfectly symmetrically, so that the time of flow on the two sides was never exactly the same. In fact, it would not have been possible to attain symmetry at all pressures, even if

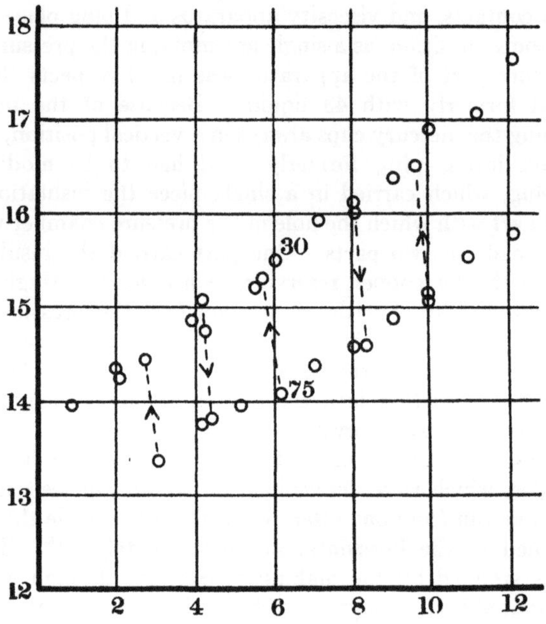

FIGURE 3. The observed times of flow in seconds plotted against pressure in thousands of kg/cm² at 30° and 75° C. The circles connected by dotted lines refer to readings obtained by changing the temperature at approximately constant pressure. The other readings were obtained during constant temperature runs by varying the pressure.

it had been attained at one, because unequal yielding of the packing at different pressures throws the apparatus out of allignment. The five readings for one direction of flow varied in consistency; when the apparatus was working well, the extreme variations were not more than 1 or 2 per cent, and from this they might rise to 3 or 4.

Readings were made at a number of pressures over the range from

0 to 12000 at 75°, and then at 30°, and then the temperature was changed from 30° to 75° and back a number of times at each of several different pressures, this giving direct measurements of the temperature coefficient of viscosity at a number of different pressures. This latter procedure was adopted to more certainly fix the temperature coefficient, and ensure that there had been no permanent changes produced in the apparatus during the readings at constant temperature.

The experimentally observed times are shown in Figure 3.

Corrections.

A number of corrections must be applied to the directly measured times in order to obtain the changes of viscosity. Unlike many of the other pressure effects which I have measured, these corrections are large, amounting altogether to the total magnitude of the measured effect, so that it will be necessary to give a careful discussion of them.

A determination of the corrections involves an accurate solution of the flow problem. The conditions of our problem differ in two important respects from the simple flow problem usually treated in connection with experimental determinations of viscosity, in that firstly, flow takes place under a pressure difference which varies so much from the beginning to the end of flow that the results cannot be simply expressed in terms of the average pressure difference, and secondly, the so-called kinetic energy correction is large, and varies in amount at different pressures. Allowance for variable head may be simply made by writing the equation of flow in differential form and integrating it, but the kinetic energy correction requires more discussion. The flow here is from one reservoir to another, the reservoirs themselves being so large that the kinetic energy of the rise and fall of the liquid in the reservoirs may be entirely neglected, so that we have to consider only the kinetic energy of the liquid in the capillary as it flows from one reservoir to the other. Now it is the opinion of Grüneisen[4] that the kinetic energy correction should not be applied under these conditions, the argument being that there is no net increase of kinetic energy of the liquid. Grüneisen considers that the correction for turbulence is much more important. Cohen,[5] in discussing his own experiments, has quoted Grüneisen with approval. It is, however, not the classical opinion that the kinetic energy effect may be neglected under such conditions, as may be found set forth at length in Bingham's book.[6] It is my belief that the classical point of view is correct in this case, as I shall try to show by the following argument. This argument is restricted by the assumption that there is no turbulence in the capillary itself, and that the rate of rise of the liquid in the reservoirs is not fast enough to disturb the equilibrium configuration of the surface. We also neglect end effects in the tube near its mouth. The length of the tube of these experiments was 160 times the diameter, so that doubtless this condition was satisfied.

We now consider the flow of the liquid in three parts; that in the higher reservoir, where the liquid gains kinetic energy, that in the capillary, where conditions of steady flow prevail, and that in the lower reservoir, where the kinetic energy of the flow in the capillary is dissipated. The energy condition evidently controls the situation. There is a decrease of potential gravitational energy during flow (or of pressure energy if the flow takes place between two reservoirs under a pressure difference) which is evidently all dissipated as heat. This dissipation takes place primarily in two places: in the capillary, where the viscous forces of the liquid are overcome, and in the lower reservoir, where the kinetic energy of the liquid emerging from the capillary is degenerated into heat energy. There is also some generation of heat in the higher reservoir, where the liquid acquires kinetic energy just before it enters the capillary, because in this region where velocity is gained, different parts of the liquid move with different relative velocities, and this necessarily involves viscous dissipation, but it seems safe to assume that by far the larger part of the gain of velocity in this region is produced by the reversible action of simple mechanical pressures, and that the irreversible aspects are relatively unimportant.

The amount of viscous generation of heat in the capillary may be calculated by the classical analysis by simply setting the heat equal to the difference of pv energy as the liquid flows through the tube, neglecting of course the compressibility, and the thermal expansion of the liquid under the heating. The rate of flow is determined by the pressure difference, so that the heat generation may be expressed in terms of the rate of flow. Of course in working out the exact expressions, integrations across the section of the tube are necessary because the velocity varies throughout the tube. In the lower reservoir the heat generated is equal to the kinetic energy of the entering liquid, and this again may be expressed in terms of the rate of flow. Finally, the sum of these two amounts of energy is equal to the loss of energy of position, and this again can be expressed in terms of the rate of flow. The resulting equation is a cubic in terms of the rate of flow ($= dV/dt$), from which dV/dt may be once cancelled, giving:

$$\frac{\rho}{\pi^2 R^4}\left(\frac{dV}{dt}\right)^2 + \frac{8l\mu}{\pi R^4}\frac{dV}{dt} - g\rho(h_2 - h_1) = 0, \qquad \text{(I)}$$

in which ρ is the density of the liquid (or the difference of two densities if the apparatus is immersed in another liquid as in the pressure experiments), R is the radius of the capillary, l its length, μ the viscosity in Abs C.G.S. units, dV/dt the rate of flow in cm^3/sec. and $h_2 - h_1$, the difference of height of the surfaces of the liquid in the two reservoirs. In this equation, the first term comes from the loss of kinetic energy, the second from the viscous dissipation in the capillary, and the third is the loss of gravitational potential energy.

This equation may now be compared with that given by the classical analysis, and will in fact be found to be exactly Bingham's equation (8) on page 18, except that the first term in Bingham's equation differs from ours through a numerical factor "m". In fact, the simple analysis given by Bingham makes $m = 1$, but a more elaborate analysis due to Boussinesq makes $m = 1.12$. The argument above shows that in fact m should be larger than 1 because of irreversible effects in the higher reservoir, where the liquid is gaining energy.

It seems to me, therefore, to be not open to question that the kinetic energy correction should be applied, under the conditions of these experiments, and in working out further the corrections I have assumed the flow equation (I), except that the first term is multiplied by Boussinesq's factor 1.12.

Entirely apart from this theoretical argument, we have the experimental fact that when the kinetic energy correction is applied, the experimentally determined temperature coefficient of viscosity at atmospheric pressure agrees with that of other observers, whereas without the correction, the experimental value is 40% too low.

Equation (I), with $m = 1.12$, may now be solved for dV/dt, and integrated for T. I find:

$$T = \frac{m\rho}{4\pi\mu l}\int_0^{V'} \frac{dV}{\sqrt{\alpha + \beta V} - 1}, \qquad \text{(II)}$$

where

$$\alpha = 1 + \frac{m\rho R^4}{16\mu^2 l^2} g\rho(h_2 - h_1)_0,$$

and

$$\beta = \frac{m\rho R^4}{16\mu^2 l^2} g\rho \frac{(h_2 - h_1)' - (h_2 - h_1)_0}{V'}.$$

The integration may be explicitly carried out, and a closed formula found, which involves the logarithms of expressions containing radicals. In practice, of course, T is the measured quantity; μ is the unknown, and the above relation is to be solved for μ in terms of T and other experimentally determinable quantities which enter the relation. Now an examination of the relation shows that μ enters in a most complicated way, and the only method of solution is by successive approximations in any special numerical case. It is natural therefore to seek for some method by which an approximate solution of the relation for μ may be found. An examination of the equation shows that the complication is introduced by the fact that the relative importance of the kinetic energy term compared with the principal viscous term varies as the head varies. But the effect of this is in the nature of a correction on a correction, and may therefore presumably be approximately treated. Let us in equation (I) substitute for dV/dt the average rate of flow, which may be written as V'/T. Equation (I), with $m = 1.12$, may now be solved for μ, giving:

$$\mu = \frac{\pi g R^4}{8V'l} g\rho (h_2 - h_1)_{av} T - \frac{1.12\rho V'}{8\pi l} \cdot \frac{1}{T} \qquad \text{(III)}$$

$$= AT - \frac{B}{T}. \qquad \text{(IV)}$$

In this equation, which is only approximate, everything on the right hand side may be determined except $(h_2 - h_1)_{av}$, which is evidently not a space average, whereas the space average is the only sort of average that can be determined experimentally under our conditions. Equation (IV) may be solved for A if μ, B, and T are known.

We now have to consider the question whether equation IV is a sufficiently good approximation over the range of these experiments. The answer can be found by comparing with the results of the exact equation. The test was made in the following way. Into the exact equation, I, was substituted the constants of the apparatus and the viscosity of mercury at atmospheric pressure, giving a value of T. With these values, the constant A of the approximate equation, IV, was calculated. Keeping now the same values of the constants of the apparatus, a 33% higher value for μ, which experiment and rough corrections had shown to be the approximate value at 12000 kg, was substituted into the exact equation, and the value of T again found. This value for T was next put back into the approximate equation, IV, in which we now know both constants A and B, and the value of μ computed. This value should be the same as that put into the exact equation, and as a matter of fact, it did agree with it within $\frac{1}{4}\%$, showing that the approximate equation IV is a sufficiently good approximation for μ over the range of these experiments.

Granting now the allowability of the approximate formula IV, the determination of the various corrections becomes much simplified, although it is still complicated enough. The coefficient A is what would be given by the classical analysis without the kinetic energy correction, taking account, however, of the variation of head during flow, which means integrating a differential equation. The classical equation for this is:

$$\frac{dV}{dt} = \frac{\pi g}{8} \frac{R^4}{\mu l} \Delta p.$$

Here of course Δp is a function of the volume that has flowed, and during flow at constant pressure and temperature is a linear function. Hence the equation reduces to the form:

$$\frac{dV}{a - bV} = \frac{\pi g}{8} \frac{R^4}{\mu l} dt.$$

This may be integrated and solved for μ. Since the μ obtained from this equation is uncorrected for the kinetic energy effect, we may denote it by μ_{unc}, and write $\mu_{\text{unc}} = AT$, where A has the value:

$$A = \frac{\frac{\pi g R^4}{8l} b}{\log \frac{a}{a - b(h_1' - h_{10})S}}$$

where

$$a = \rho\left[l \sin \theta_2 - \cos \theta_2 \left(2h_{10} - \frac{1}{S} V_{\text{Hg}}\right)\right]$$

and

$$b = \frac{\rho}{S} 2 \cos \theta_2.$$

In this formula the letters which have not been already explained are: S = cross section of the reservoirs, V_{Hg} = total volume of the mercury in the apparatus, h_{10} the height of the mercury in the reservoir toward which flow takes place at the instant of beginning of flow, and h_1', the final height at the instant of making contact in the same reservoir, and θ_2 is the angle of inclination with the horizontal of the apparatus during flow. h_{10} is determined by the constants of the apparatus and the angle, θ_1, from which flow was started.

The quantity A above, since it contains the dimensions of the apparatus and the density of the liquids, is obviously a function of pressure. If, as a matter of notation, we write

$$A = \frac{\pi g R^4}{8l} \frac{b}{\log I},$$

then we shall obviously have

$$\left(\frac{\mu_p}{\mu_0}\right)_{\text{unc}} = \frac{T_p}{T_0} \cdot \frac{A_p}{A_0},$$

where the subscripts 0 and p denote respectively the values at atmopheric pressure and pressure p. The change with pressure of the factor $(\pi g R^4/8l)b$ can at once be written down by inspection, so that we have:

$$\frac{A_p}{A_0} = \frac{(\rho_{\text{Hg}} - \rho_l)_p}{(\rho_{\text{Hg}} - \rho_l)_0} (1-\kappa_{\text{Cu}} p) \frac{\log I_0}{\log I_p}.$$

Here we have put explicitly $\rho = \rho_{\text{Hg}} - \rho_l$, where ρ_{Hg} is the density of mercury and ρ_l the density of the transmitting liquid. κ_{Cu} is the linear compressibility of copper.

The term $\log I_0/\log I_p$ is more complicated to work out as a function of pressure, but it is a straight forward job, in terms of the dimensions of the apparatus, and the known distortions under pressure, remembering the initial conditions, namely that the liquid is adjusted before flow starts by inclining the apparatus at a fixed angle independent of pressure, that flow takes place also at a fixed angle independent of pressure, and that flow terminates when the mercury strikes a fixed contact whose position is subject to the pressure distortion of the apparatus.

The dimensions of the apparatus were such that under 12000 kg. log I increases by nearly 5.5%. The density of mercury, which enters the formula, can be taken from previous experiment,[7] as can also the compressibility of copper.[8] The transmitting liquid, however, for which petroleum ether was chosen for reasons already described, had not had its compressibility previously determined, and a special determination was necessary. A rough determination was good enough for the present purpose. This was made in a new apparatus, which has been previously used and described in connection with a measurement of the compressibility of glycerine.[9] The following values for the density were found (Table I).

TABLE I.

DENSITY OF PETROLEUM ETHER AT 30° C.

Pressure Kg/cm²	Density
1	.638
2000	.742
4000	.792
6000	.825
8000	.849
10000	.872
12000	.889

The values of *A* so far discussed are for 30° C. The value at 75° may be found by a discussion exactly like that given above for the effect of pressure, since obviously the distortion produced by a change of temperature is exactly like that produced by a hydrostatic pressure. The results of the calculation are contained in Table II,

TABLE II

RELATIVE VALUES OF "*A*" (SEE TEXT) AS A FUNCTION OF PRESSURE AND TEMPERATURE

Pressure Kg/cm²	A	
	30°	75°
1	1.000	1.002
2000	.989	.991
4000	.983	.985
6000	.977	.979
8000	.971	.974
10000	.969	.971
12000	.966	.968

which shows the correction to be applied to the value of *A* at 30° and atmospheric pressure in order to obtain the value of *A* at higher pressures and at 75°.

If instead of integrating the equations of flow, one had used the less exact equation obtained by supposing flow to take place at a constant head equal to the mean of the initial and the final heads (that is a space average instead of a time average) the correction factor for A at 30° and 12000 kg. would have been found to be 1–0.046 instead of 1–0.034, as given by the more accurate calculation above.

We next have to determine the variation with pressure of the coefficient in the kinetic energy correction, that is, the correction of B ($= 1.12\rho V/8\pi l$). Since this term is comparatively small, it will be sufficient to assume that the variation of B with pressure is linear. The same data used in calculating the value af A at 12000 give for B at 12000 a 6.6% correction. The correction at 75° may be similarly found. Table III gives the results.

TABLE III.

Relative Values of "B" (see text) as a Function of Pressure and Temperature.

Pressure Kg/cm²	B 30°	B 75°
1	1.000	.988
2000	1.011	.999
4000	1.022	1.010
6000	1.033	1.021
8000	1.044	1.032
10000	1.055	1.043
12000	1.066	1.054

We are now ready to correct the observed values of time, except for the absolute values of A and B. An examination of the formulas for A and B shows that A is subject to considerable experimental error, through the factor R^4, whereas B involves only ρ, V, and l, which may be all easily measured. It appeared therefore that the most accurate method of getting A was to calculate B, and then substitute in equation IV the experimentally measured T and the absolute value of μ determined by other experimenters, and solve for A. The value adopted for μ at 30° and atmospheric pressure was 0.01514, as the best of all the values given in Landolt and Börnstein. The values so found for A and B were 0.001372, and 0.0497, respectively.

We are now in a position to calculate the viscosity at various pressures and at 30° and 75° in terms of the measured times of flow. In doing this, smooth curves were drawn through the experimentally determined times, and the times read from these curves at intervals of 2000 kg. These times were now corrected as above, and the results obtained which are shown in Table IV.

TABLE IV.

VISCOSITY OF MERCURY AS A FUNCTION OF PRESSURE AND TEMPERATURE.

Pressure Kg/cm²	Absolute Viscosity 30°	75	Percentage Increase of Viscosity 30°	75°
1	.01516	.01341	0.0	0.0
2000	.01588	.01399	5.1	4.6
4000	.01663	.01463	9.9	9.6
6000	.01742	.01528	15.2	14.4
8000	.01825	.01599	20.0	18.8
10000	.01913	.01675	26.3	24.7
12000	.02008	.01757	33.5	31.1

Checks on the Measurements.

The absolute value of viscosity cannot be calculated with much accuracy from these measurements because of the difficulty, already mentioned, of measuring the absolute dimensions of the capillary. There is, however, a check that may be applied, by comparing the temperature coefficient between 30° and 75° with that determined by other observers. In Figure 4 are plotted all the values of viscosity between 0° and 100° listed in Landolt and Börnstein's tables. In the same diagram are shown, as double circles, the values given above in Table IV; of these the value at 30° was assumed in determining one of the constants, and the 75° value was then calculated in terms of the experimental data. This value is seen to fall within the range of previous experimenters, and in so far gives presumptive evidence of the correct functioning of the apparatus.

It is perhaps worth while to mention the results obtained with the various forms of preliminary apparatus. It must be remembered that all these results were very irregular, that the apparatus would hardly function at all at 75°, and that we can expect from them only a correct order of magnitude for the effect. The first of the preliminary forms of apparatus which gave at all consistent results was

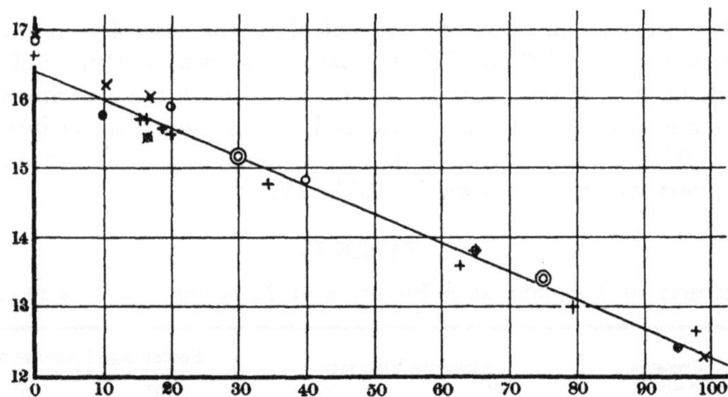

FIGURE 4. Collection of the data given in Landolt and Börnstein's Tables for the viscosity of mercury as a function of temperature at atmospheric pressure. Abscissae are Centigrade degrees, and ordinates 1000 times the absolute viscosity. The different symbols refer to data of different observers. The double circles refer to the data of this paper. The constants were so chosen that the point at 30° should fall on the curve of other observers. The closeness with which the double circle at 75° falls on the line measures the agreement of my temperature coefficient with that of other observers.

that of flow through a coiled steel capillary. The pressure cylinder was inverted through 180° in this experiment, with the result that the capillary was completely emptied at every reversal, and the walls wet with the transmitting liquid. During the flow, therefore, the mercury had to push away from the walls of the capillary a film of transmitting liquid. Since the viscosity of the transmitting liquid increases under pressure much more rapidly than that of mercury, this apparatus would be expected to give too high an increase of viscosity with pressure. The actual value found from a number of measurements with this apparatus, after applying a rough correction for the kinetic energy effect, was an increase of 40% at 30° and 12000 kg, against 33.5% found with the final apparatus above. This 40% was the mean of three separate runs, differing from the mean by 6% of the coefficient (that is varying from 42.4% to 37.6%). Beside the steel capillary coil apparatus, several forms of tipping apparatus, more or less like the final one, were used before success was attained. These preliminary attempts failed because of dirt on the surface of the mercury sticking to the platinum contacts, making the time of flow too short. Since the effect became larger at higher pressures, the pressure coefficients found with this apparatus were too low.

The first of these preliminary tipping apparatuses had a capillary smaller than the final one, so that the time of flow was three times greater. Under these conditions the kinetic energy correction may be neglected. This apparatus gave for the increase at 30° at 12000 the value 24.5% when the transmitting liquid was ether, and 25% with petroleum ether. A second form of apparatus, which was muck like the final one, so that the same corrections could be applied to it, gave an increase of 26%.

We now have to compare the results above in Table IV with the only previous determination of the effect of pressure, namely that of Cohen.[2] His measurements were made at 20° and at a single pressure, 1500 atmospheres. He found an increase of viscosity of 4.8% under these conditions. Correcting this from 20° to 30° by my results above, and changing from atmospheres to kg/cm^2, Cohen's result is equivalent to an increase of 6.1% for 2000 kg/cm^2 at 30°, against my value 5.1% above. We now seek to understand this difference. One small source of error in Cohen's work is that he has treated the flow as if it took place under a constant, instead of under a variable, head. Cohen gives a scale drawing of his apparatus, from which the dimensions may be obtained, and it appears that during flow the head varied from 6.3 to 2.1 cm. of mercury. The more rigorous equations of flow seem called for under these conditions. I find, on using the more rigorous equations that the value 6.1% given above for the effect of 2000 kg. should be reduced to 5.9%. This is not a large change, but is in the direction toward my value. There are also other possibilities of error in Cohen's work, the numerical magnitude of which are more difficult to estimate, but which work in the same direction. Thus the viscosity of the transmitting liquid, in Cohen's case, water, affects the measured results, because as mercury flows down from the high reservoir to the lower, water must flow back to take its place. Cohen's diagram does not show any channels specially made for the return flow of the water, so that the presumption is that they were small. Any such channels would become smaller at high pressure because of the compressibility of the apparatus, and furthermore water increases in viscosity at 20° under 1500 atmospheres, so that the net effect would be to give too large an apparent increase of viscosity of mercury. In my apparatus the return channel was at least 100 times greater in area than that of the capillary, and the distortion of the apparatus was in such a direction as to increase this with increasing pressure. Since the time of flow varies as the square of the section, no error from this effect is to be expected in the values of Table IV. Another possibility of error in Cohen's experiment is the surface tension effect. The electric contact in his apparatus occurred in a constriction of about 3 mm. diameter, the diameter of the main reservoir being 1.8 cm. While the mercury is in the constriction, and while it is approaching it, the difference of surface tension at the two surfaces of unequal area is in such a direction as to oppose the flow, thus simulating too high a viscosity. If the surface tension increases under pressure, as seems probable, this effect will become relatively larger at high pressures, resulting in too high a pressure coefficient. The kinetic energy correction,[5] which I believe theoretically should have been applied to Cohen's experiment, assumes a more or less academic position, because his time of flow was so long as to make this correction numerically negligible. It is unfortunate that Cohen's seven repetitions of his experiment were made under exactly the same conditions; if the quantity of mercury had been varied somewhat we would have some basis for an estimate of the magnitude of another possible correction, due to the almost complete emptying of the upper reservoir, for which there is at present no data. All of these suggested possibilities will probably, however, not have a large effect, so that it seems likely that Cohen's value for the pressure coefficient would remain at least 10% higher than mine.

Discussion of Results.

If the viscosity of mercury is plotted against pressure at 30° and 75°, (Figure 5) the results will be found to be of the same general character as those previously found for other liquids. Viscosity increases with increasing pressure at an accelerated rate, and the percentage increase at 75° is less than at 30°. It will be found on inspection of the diagram that the curve for 75° may be obtained by displacing bodily the 30° curve along the pressure axis by about 5750 kg. This means that the fractional change of viscosity produced by a given increment of pressure is a function only of the viscosity, no matter whether a given value of the viscosity is found at a low pressure and low temperature or at a high temperature and a high pressure. The experimental data are not sufficient to allow a similar statement about temperature coefficients at equal viscosities. The result for the pressure coefficient was not found to hold in the case of other liquids, so that possibly in the case of mercury the result is only approximate, holding only over a comparatively narrow range.

The upward curvature of viscosity against pressure is slight, but is certainly beyond experimental error.

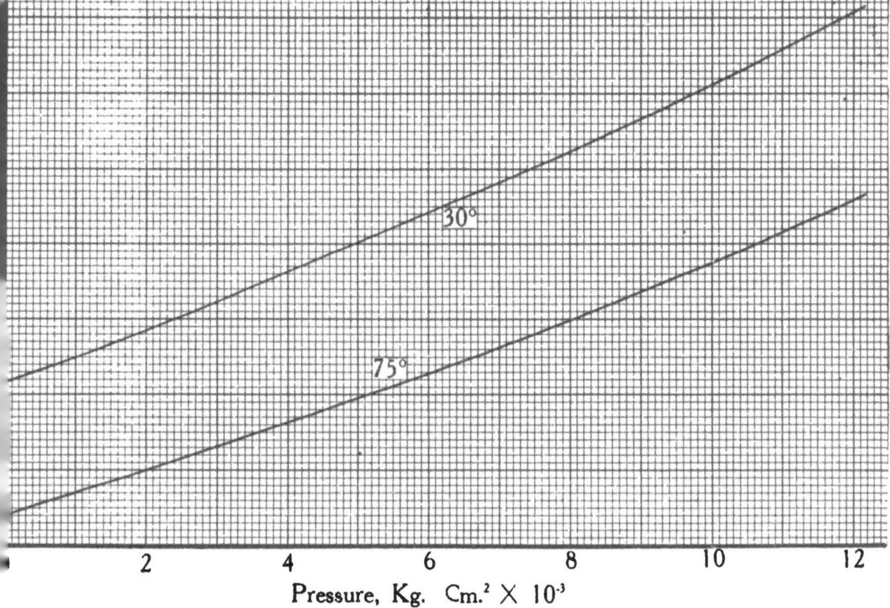

FIGURE 5. The absolute viscosity of mercury at 30° and 75° as a function of pressure.

As compared with the results for other liquids, the viscosity of mercury is striking regarded as a volume function. With other liquids, viscosity at constant volume decreased with rising temperature, but the decrease was not large. With mercury, on the other hand, the departure from this simple relation is very marked. The volume at 75° and 2000 kg. is the same as that at 30° and atmospheric pressure, but the viscosity at 75° and 2000 kg. is 0.0140, against 0.0152 at 30° at atmospheric pressure. The pressure coefficient of viscosity would have to be nearly three times as great as found above to make viscosity a function of volume only. Such a change is of course entirely outside the possibilities of experimental error; even if Cohen's value for the pressure coefficient is adopted, essentially the same conclusion is reached. Considerable theoretical significance has been attached to the idea that viscosity is a function of volume only; the fact that the largest departure from this relation has been found in mercury, a monatomic liquid and the simplest yet investigated, would seem to be of considerable significance.

The results bear out the expectation mentioned at the beginning of this paper that because of its simple monatomic structure the pressure coefficient would be found to be especially small in mercury. The smallest value previously found was for water, which increases 2.7 fold at 30° under 12000 kg., against an increase of only 33% for mercury. These results in general, therefore, bear out the view that in a liquid viscosity is intimately connected with some sort of interlocking mechanism between the molecules, instead of being primarily a kinetic matter, as in most theories hitherto. In our present state of knowledge it is exceedingly difficult to advance further with the picture of an interlocking mechanism and make it yield quantitative results, for the reason that we need detailed knowledge of the shape of the molecule and its deformation under pressure, which we do not have.

It is a pleasure to acknowledge the help received in most of the manipulations of this paper from my mechanician, Mr. E. T. Richardson.

The Jefferson Physical Laboratory,
Harvard University, Cambridge, Mass.

References.

[1] P. W. Bridgman, Proc. Amer. Acad. 61, 57–99, 1926.; Proc. Nat. Acad. 11, 603–606, 1925.

[2] E. Cohen and H. R. Bruins, Kon. Akad. Wet. Amst. Proc. 27, 9 and 10, 1924.

[3] T. W. Richards, Carnegie Inst. Wash. Pub. No. 118, 1909.

[4] E. Grüneisen, Wiss. Abh. Phys. Reichsanstalt, 4, 151–180, 1905.

[5] W. Cohen and H. R. Bruins, ZS, f. Phys. Chem. 103, 404–450, 1922.

[6] E. C. Bingham, Fluidity and Plasticity, McGraw Hill, New York, 1922.

[7] P. W. Bridgman, Proc. Amer. Acad. 47, 347–438, 1911.

[8] P. W. Bridgman, Proc. Amer. Acad. 58, 166–242, 1923.

[9] P. W. Bridgman, Proc. Amer. Acad. 61, 89, 1926.

THE COMPRESSIBILITY AND PRESSURE COEFFICIENT OF RESISTANCE OF TEN ELEMENTS.

By P. W. BRIDGMAN.

Presented Oct. 19. 1927. Received Oct. 23, 1927.

TABLE OF CONTENTS.

INTRODUCTION.

The data to be presented in this paper were obtained in continuation of the program of measuring the two properties above for all the elements available.[1] The range of these measurements is the same as that of the previous work, 0 to 12000 kg/cm^2, and 30° and 75° C, in most cases. The methods of measurement were also the same, so that no new discussion of them is needed. The compressibilities were all measured in the so-called "lever piezometer for small specimens." The measurements on Pr, La, Be, and Ce were made with the identical apparatus that was used in the previous measurements. Damage to the high resistance wire attached to the lever necessitated its replacement for the other substances. After this replacement the lever apparatus was recalibrated as previously described, both by measuring the magnification with a micrometer, and by measuring the apparent compressibility of pure iron at 30° and 75°. The constants of the apparatus were found to have changed only a few tenths of a per cent.

Numerical Data.

Praseodymium. I owe this metal to the kindness of Dr. H. C. Kremers, of the University of Illinois, who has long specialized in the preparation of the rare metals. The metal used in my measurements was some of the identical metal, the preparation and the properties of which have been described by Kremers.[2] The salts from which the metal was electrolyzed had a purity of 99.7%; the only known impurities being 0.02% Nd, and 0.01% La. Kremers states that the metal formed by electrolysis, which took place in a graphite crucible, was doubtless of high purity, but no special analysis was made to find exactly what the purity was. It is evident that there could have been no appreciable impurity of other rare metals, but it does not seem impossible that some of the more common impurities may have been absorbed from the crucible. The metal as furnished to me was in the form of small slugs, about 6 mm. in diameter, sealed into glass under oil. I formed the specimens for my measurements by extrusion of these slugs at a bright red heat through dies of Cr-Va steel, in an atmosphere of nitrogen. The compressibility sample was 3 mm. in diameter, and 1.3 cm. long. The extruded wire for the resistance measurements was 0.085 cm. in diameter, and about 15 cm. long. This metal extrudes to wire very easily under the proper conditions, and pieces were obtained much longer than necessary for these measurements. This wire has high mechanical strength, and in superficial appearance is much like soft iron wire.

Compressibility. This was measured in the regular way at 30° and 75°. The crystal system of Pr has apparently not yet been determined. Assuming that the compressibility of my sample is the same in every direction, which is probably not far from the truth because of the method of formation by extrusion, even if it should prove not to be cubic, the following results were computed from the measured linear compressibility:

At 30° $$\frac{\Delta V}{V_0} = -33.8 \times 10^{-7}p + 13. \times 10^{-12}p^2.$$

At 75° $$\frac{\Delta V}{V_0} = -34.6 \times 10^{-7}p + 13. \times 10^{-12}p^2.$$

The mean arithmetical deviation of a single reading from a smooth curve was 0.55% at 30°, making two discards, and also 0.55% at 75°, with one discard. The scattering of the readings was such that no great accuracy can be attached to the second degree term; it was certain, however, that the compressibility decreases with rising pressure, as is normal.

Electrical Resistance. This was measured with a potentiometer in the regular way. Contacts were made with spring clips, since it is not possible to solder this metal, taking pains that the contacts should be at only a single well defined point. The resistance decreases with pressure, as is normal, and within the limits of error, the decrease is linear. The mean coefficients to 12000 kg. were:

$$\text{At } 30°, \ -3.1 \times 10^{-7}$$

$$\text{At } 75°, \ -8.3 \times 10^{-7}$$

At 30° the mean deviation of a single reading from a smooth curve was 6.0%, and at 75°, 2.4% of the maximum effect. The pressure coefficients are seen to be unusually low, which accounts for the unusually large scattering of the individual readings. The very large change of the pressure coefficient with temperature is most unusual.

In addition to the pressure effect, the specific resistance at 30° was found to be 6.9×10^{-5}, and the temperature coefficient at 0° 0.00165. This last value is so low that it would seem probable that a rather large amount of impurity must have been absorbed during the electrolysis. Presumably this may have been gaseous in character; Kremers states that gases are easily absorbed.

By combining the values above for the pressure and the temperature coefficients, it will be found that the temperature coefficient at 12000 kg. is about 10% less than at atmospheric pressure.

Lanthanum. This I also owe to the kindness of Dr. Kremers. The methods of preparation have been described by him.[3] He states that the metal is free from other metals. He found it to melt sharply at 826°, which is much higher than the previously accepted value for the melting point, 810°, and therefore evidence of high purity. The samples on which my measurements were made were formed by extrusion, and were of approximately the same dimensions as those of Pr. La extrudes somewhat more readily than Pr, and at a somewhat lower temperature. Its mechanical properties are not markedly different.

Compressibility. The compressibility measurements went perfectly smoothly, without incident of any kind. Assuming that the sample was equally compressible in all directions (the crystal system of La seems not to have been yet determined), the following results were found for the cubic compressibility from the measured linear compressibility:

$$\text{At } 30^\circ \qquad \frac{\Delta V}{V_0} = -\,35.13 \times 10^{-7}p + 14.7 \times 10^{-12}p^2.$$

$$\text{At } 75^\circ \qquad \frac{\Delta V}{V_0} = -\,35.01 \times 10^{-7}p + 17.1 \times 10^{-12}p^2.$$

The average deviation of a single reading from a smooth curve was 0.45% at 30° and 0.42% at 75°.

The temperature coefficient of compressibility of La is abnormal in that it is negative; this effect seems larger than possible experimental error.

Resistance. The resistance was measured at 30° and 75°. The pressure coefficient is negative, as is normal. At 30°, the relation between pressure and resistance was linear within experimental error, the average deviation of a single reading from a straight line being 0.84% of the maximum change. At 75° the relation was obviously not linear within experimental error, but could be represented by a second degree equation with curvature in the normal direction. The average deviation from a smooth curve of a single reading at 75° was 0.54% of the maximum change. The results follow:

$$\text{At } 30^\circ \qquad \frac{\Delta R}{R(1\text{ kg}, 30^\circ)} = -\,1.199 \times 10^{-6}p.$$

$$\text{At } 75^\circ \qquad \frac{\Delta R}{R(1\text{ kg}, 75^\circ)} = -\,1.810 \times 10^{-6}p + 9.7 \times 10^{-12}p^2.$$

Again, as in the case of Pr, the temperature coefficient of the pressure coefficient is unusually large.

The absolute specific resistance of this sample was found to be 5.76×10^{-5} at 0°. The temperature coefficient of resistance at atmospheric pressure at 0° is 0.00213, reduced from readings at 20° and 75°, assuming a linear relation between temperature and resistance.

The pressure coefficient of resistance of another sample of La has been measured by me.[4] This sample had a temperature coefficient of 0.00148, considerably lower than that above, so that presumably the previous sample was less pure. The pressure coefficient of the previous sample was, on the other hand, about twice as great as that of the new sample, being $-\,3.9 \times 10^{-6}$, independent of temperature.

This is unusual; the pressure coefficient of the purer metal is usually greater numerically than that of the less pure metal, although the difference is not usually so pronounced as for the temperature coefficient.

Cerium. This metal I again owe to the kindness of Dr. Kremers, who has described the method of preparation and its properties.[5] The specimens were prepared for my measurements by extrusion in the same way as Pr and La. The extrusion temperature is lower, and the extrusion easier than for the other two metals. I have previously measured the compressibility of Ce[6], but on a sample whose temperature coefficient of resistance was so low, 0.001, that it did not seem worth while measuring the pressure coefficient of resistance, in view of the presumptively high impurity. The point in now repeating the measurements is that this sample is presumably of much higher purity; Kremers states that the method of preparation is such that the metal should be free from iron, which is known to be present in nearly all the Ce that has hitherto been available.

This pure Ce was found to be polymorphic at high pressures, a phenomenon of which no trace whatever was shown by the impurer sample. A transition at high pressures was first found during the compressibility measurements, and was shown by large irregularities at the upper end of the curve, and was verified on taking the apparatus apart by distortion and permanent changes of dimensions of the specimen. The existence of the transition was later verified by measurements of the electrical resistance. These measurements gave fairly good coordinates for the transition, which was found to take place at 7600 kg. at 30° and 9400 at 75°. The transition is slow and not perfectly sharp; perhaps if the metal were perfectly pure the transition would be found to be sharp. Unfortunately, there was not enough of the metal available to determine the volume change at the transition, or the other thermal parameters.

Compressibility. After the transition was found during the first attempted compressibility measurements, the sample was annealed by heating for six hours to a nearly red heat in a sealed glass tube. After annealing, the compressibility was determined, the pressure being restricted to a range of 4000 kg., in order to avoid all possible effects of the transition. It was not possible to determine the compressibility of the high pressure modification without changing the apparatus. The following results were found for the volume compressibility, assuming equal compressibility in all directions:

$$\text{At } 30^\circ \quad \frac{\Delta V}{V_0} = -45.63 \times 10^{-7}p - 161.4 \times 10^{-12}p^2.$$

$$\text{At } 75^\circ \quad \frac{\Delta V}{V_0} = -45.03 \times 10^{-7}p - 151.5 \times 10^{-12}p^2.$$

The average deviation of a single reading from a smooth curve was 0.51% at 30° and 0.54% at 75°.

These results for the compressibility of Ce are highly anomalous in two respects. In the first place, the temperature coefficient of compressibility is negative, in both the first and the second degree terms. Secondly, the direction of curvature is abnormal, in that the compressibility becomes greater at high pressures. This is the first pure substance in which this phenomenon has been found, the only previously known examples are some of the glasses with high SiO_2 content.[7] It is natural to see some connection between the anomalous effect here and the existence of a second modification. It will be noticed that both these anomalies are much too large to be accounted for by any possible experimental error.

The compressibility of this pure Ce is very much higher than that previously found for impure Ce, 35.74×10^{-7} at 30°. The difference is in a direction to be accounted for by a large impurity of iron in the first sample. The temperature coefficient and curvature of the impure Ce were both normal.

Resistance. The resistance phenomena under pressure are no less surprising than the compressibility phenomena. The low pressure modification is abnormal in that the resistance increases with pressure. The pressure coefficient of resistance of the low pressure modification was determined on a virgin sample, which had never experienced the transition. Within experimental error, the relation between pressure and resistance was linear at both 30° and 75°. The following results were found:

$$\text{At } 30^\circ \qquad \frac{\Delta R}{R(1\text{ kg}, 30^\circ)} = +4.42 \times 10^{-6}p, \qquad \text{range 3000 kg.}$$

$$\text{At } 75^\circ \qquad \frac{\Delta R}{R(1\text{ kg}, 75^\circ)} = +2.77 \times 10^{-6}p, \qquad \text{range 6000 kg.}$$

The average deviation of a single reading from a smooth curve was 0.7% of the maximum effect at 30°, and at 75°, 1.6%.

The high pressure modification is normal in that the resistance decreases with rising pressure. Within the limits of experimental error, which might perhaps have been as high as 10%, the relation between pressure and ΔR is linear between 9000 and 12000 kg. at 30°, and the pressure coefficient is -1.42×10^{-5}, in terms of the resistance at 9000 kg. This is also the coefficient, within somewhat larger limits of error, at 75°. It was not possible to compare the specific resistances of the two modifications, because of the unknown difference of dimensions.

The specific resistance of the low pressure modification at 30° was 7.48×10^{-5}. The temperature coefficient, reduced to 0° from readings at 19° and 75°, assuming linearity, was 0.00097. This is surprisingly low, lower even than found for the previous sample, which was known to be impure with iron. All the chemical evidence, as well as the fact that a transition has been found, points to the high purity of this sample. It is evident enough that the low pressure modification is highly anomalous. It may be that in addition to its other anomalies, we have here the first known example of a pure metal with a temperature coefficient of resistance very much less that $1/t$.

Beryllium. My particular interest in this metal was my expectation that it would be found to be one of the comparatively few metals with a positive pressure coefficient of resistance. The reason for this expectation was its proximity to Li at the left in the periodic table and to Ca and Sr below, all of which have positive coefficients. This expectation did not turn out to be correct.

Beryllium was obtained from two sources. The first I owe to the kindness of Professor A. Stock of Berlin, who sent me a nugget of beryllium prepared by electrolysis of the fused salts, the linear dimensions of which varied from 1 to 2 cm. The compressibility was determined on a sample of this, and I spent much time trying to get from it a sample suitable for the resistance measurements. Beryllium is very hard, and can be worked only by grinding. The interior of the piece unfortunately proved to have many inclusions of the original fused salt, so that it was not possible to grind from it a straight piece of sufficient length. I found that Be becomes soft when hot, and a pellet may be squeezed flat at a red heat between cold steel plates. This led to the attempt to extrude it hot, but unsuccessfully. I used Cr-Va steel dies, and also dies of a high speed steel, specially recommended for hot work, but at no temperature between a dull red and a bright yellow did the Be become softer than the steel, but the die was always deformed before extrusion started. An attempt to cast a slender rod in a vacuum electric furnace was also unsuccessful. I was finally fortunate enough to obtain a rod of cast Be, 4.6 mm. in diameter and 9 cm. long, made in this country by Dr. H. S. Cooper of the Kemet Laboratories, and I take this occasion to express my indebtedness. This rod was apparently entirely free from inclusions of any kind. It was rather larger than would have been desirable to give the best results for the change of resistance, but nevertheless the accuracy was high enough to leave absolutely no doubt as to the general character of the results.

Compressibility. The compressibility was measured, as usual, at 30° and 75°, with the following results, assuming equal compressibility in all directions. Be is known to be hexagonal, but the crystalline structure of this casting was very fine, so that the measured linear compressibility along the length of the rod doubtless gave a fair average of the linear compressibility in all directions.

$$\text{At } 30° \text{ and } 75°, \quad \frac{\Delta V}{V_0} = -\,8.55 \times 10^{-7}p + 3.88 \times 10^{-12}p^2.$$

Within experimental error, no difference could be found between the compressibility at 30° and 75°. The average deviation of a single reading from a smooth curve was 0.12% at 30° (two discards out of 13 observations) and 0.17% at 75° (three discards out of 14 observations). The compressibility of the German sample was 9.0×10^{-7}; this is somewhat higher than the other, as is to be expected because of the effect of the inclusions.

Resistance. The resistance of only the sample from Dr. Cooper was measured. This was determined with the potentiometer in the regular way, with four terminals. Connections were made by turning in the lathe with a diamond point four grooves of V section, two near each end of the rod, and snapping around the grooves a very fine helical spring. The resistance decreases under pressure, which is normal for the majority of metals. The results were accurate enough to establish that the curvature is also in the normal direction. The results were as follows:

$$\text{At } 30° \quad \frac{\Delta R}{R(1\,\text{kg}, 30°)} = -\,1.11 \times 10^{-6}p + 1.2 \times 10^{-11}p^2.$$

$$\text{At } 75° \quad \frac{\Delta R}{R(1\,\text{kg}, 75°)} = -\,1.58 \times 10^{-6}p + 2.6 \times 10^{-11}p^2.$$

At 30° the average deviation from a smooth curve of a single reading was 3.3% (13 readings with no discards), and at 75° was 3.4% (14 readings, with no discards).

The specific resistance at 30° was 10.6×10^{-6}. The temperature coefficient at 0°, reduced from readings at 30° and 75°, assuming linearity, was 0.00328. This suggests fairly high purity.

The density at room temperature, 20° ±, was found to be 1.820.

Barium. I owe this metal to the kindness of Dr. A. J. King of Syracuse University, who prepared it by the method of P. S. Danner.[8] Professor Saunders was kind enough to make a spectroscopic analysis, which showed only a trace of Sr. The spectroscopic examination was made with a quartz spectrograph, over a wide spectral range. Professor Saunders characterizes the Ba as "extraordinarily pure."

Two specimens were provided by Dr. King. One had been melted in an iron vessel to a coherent slug, and was furnished to me sealed into glass under oil. The other had not been melted since electrolysis, and was sealed into glass in an atmosphere of argon. This was presumably the purer of the two. The compressibility sample was taken from the first sample, the second not being large enough, and the resistance sample from the second. The specimens were formed by cold extrusion under oil through a steel die; this could be done very easily.

Compressibility. The measurements went smoothly, without incident. The following are the results, assuming equal compressibility in all directions. (The crystalline system has apparently not yet been determined).

$$\text{At } 30^\circ \qquad \frac{\Delta V}{V_0} = -\,101.9 \times 10^{-7}p + 129 \times 10^{-12}p^2.$$

$$\text{At } 75^\circ \qquad \frac{\Delta V}{V_0} = -\,106.3 \times 10^{-7}p + 149 \times 10^{-12}p^2.$$

The average deviation of a single reading from a smooth curve was 0.53% at 30°, and 0.56% at 75°. The results are represented above by a second degree curve, but it is probable that a strict regard for the magnitude of the experimental error would have justified the inclusion of another term. This extra term, if it had been given, would have been abnormal, in that the maximum deviation from linearity occurred at both temperatures at pressures somewhat higher than 6000, instead of exactly at 6000, as is the case with a second degree curve, or at less than 6000, as for all other substances so far studied whose compressibility could not be represented by a second degree curve, as for example the alkali metals.

Resistance. The resistance measurements were made with the potentiometer in the regular way. The sample was about 0.07 cm. in diameter and 15 cm. long. Current and potential connections were made near each end with springs resting in notches cut across the wire in such a way as to prevent motion. The results were not at all what was expected. A high positive pressure coefficient was expected, by analogy with Ca and Sr. Instead, the initial effect is negative, as is normal, but the resistance passes through a minimum, and above a pressure varying with temperature from 8000 to 10000, increases again. This is the behavior of Cs, which occupies the adjoining cell in the first column of the periodic table, and is the only other known example.

Measurements were made at 0°, 30°, and 75°, at the latter temperature first. At 75°, there was some chemical action, which resulted in a gradual increase of resistance during application of pressure, amounting in the aggregate to 1.1% of the total resistance. The mean of the results found with increasing and decreasing pressure was taken as the best results to be found from this run; the mean points lay on a smooth curve within the limit of the sensitiveness of the readings, which was about 0.14% of the maximum pressure effect. At 30°, there was no chemical action, and no point lay off a smooth curve by more than 0.2 mm. of the slide wire setting of the potentiometer, the maximum displacement of the slider being 4 cm. At 0°, again there was a small permanent change of resistance of 0.2% of the total resistance, probably not due in this case to chemical action, but to distortion of the wire brought about by the viscosity of the transmitting medium. This is to be expected because barium is a rather soft metal. The mean of the up and down points at 0° did not lie off a smooth curve by more than 0.1 mm. of the slide wire.

The results are contained in Table I. It will be seen, as already stated, that the resistance passes through a minimum, and that the pressure of the minimum is a temperature function. Within experimental error, the relation between temperature and the pressure of the minimum is linear, being 8100 kg. at 0° and 9600 kg. at 75°.

TABLE I.

RELATIVE RESISTANCE OF BARIUM AS A FUNCTION OF PRESSURE AND TEMPERATURE.

Pressure kg/cm²	Resistance 0°	30°	75°
1	1.0000	1.1520	1.4380
1000	.9922	1.1408	1.4183
2000	.9855	1.1319	1.4028
3000	.9803	1.1251	1.3904
4000	.9764	1.1198	1.3805
5000	.9736	1.1156	1.3726
6000	.9717	1.1127	1.3665
7000	.9707	1.1109	1.3613
8000	.9705	1.1100	1.3586
9000	.9710	1.1100	1.3569
10000	.9721	1.1107	1.3565
11000	.9739	1.1123	1.3581
12000	.9764	1.1144	1.3620

In addition to the measurements under pressure, careful measurements were made of the temperature coefficient of resistance at atmospheric pressure. The maximum temperature was limited to 75°, in order to avoid error from chemical action, which however did not prove troublesome after the first exposure to 75°. Temperature was varied in both directions from the mean temperature several times, and the results always lay on a smooth curve. A second degree equation in the temperature will not quite reproduce the results. A third degree curve through the points at 0°, 30°, 50° and 75° has the equation:

$$\frac{\Delta R}{R_0} = .004795t + 5.71 \times 10^{-5}t^2 + 1.119 \times 10^{-6}t^3.$$

The mean coefficient between 0° and 100°, given by extrapolation with this equation, is 0.00649, which is very high, and evidence of high purity.

Thorium. I owe this metal to the kindness of Dr. Rentschler of the Westinghouse Lamp Co. The compressibility sample was in the form of a rod about 3 mm. in diameter, and 2.7 cm. long. It had been prepared in the laboratory of the Westinghouse Co., and was known to be of high purity, as far as foreign metals are concerned, but the method of preparation did not ensure the absence of all oxides. The resistance sample was in the form of wire, 0.07 cm. in diameter, which was especially prepared for these measurements in the Westinghouse Laboratory, so as to be free from all oxides, as well as metallic impurities. I am much indebted to Dr. Rentschler for the pains which he took in the preparation of this sample.

Compressibility. Runs were made as usual at 30° and 75°. Thorium is known to be cubic. The volume compressibility, calculated from the linear compressibility, is:

$$\text{At } 30^\circ \quad \frac{\Delta V}{V_0} = -18.18 \times 10^{-7}p + 12.78 \times 10^{-12}p^2.$$

$$\text{At } 75^\circ \quad \frac{\Delta V}{V_0} = -18.46 \times 10^{-7}p + 13.29 \times 10^{-12}p^2.$$

At 30° the average departure of a single reading from a smooth curve was 0.61%, and at 75° 0.31%.

Resistance. The resistance was measured with the potentiometer in the regular way. Connections were made to the thorium wire with small springs, it not being possible to solder it. The results were:

$$\text{At } 30^\circ \quad \frac{\Delta R}{R(1\text{ kg}, 30^\circ)} = -2.787 \times 10^{-6}p + 1.89 \times 10^{-11}p^2.$$

$$\text{At } 75^\circ \quad \frac{\Delta R}{R(1\text{ kg}, 75^\circ)} = -2.966 \times 10^{-6}p + 2.18 \times 10^{-11}p^2.$$

The average numerical deviation of a single reading from a smooth curve was 0.07% at 30° and 0.11% at 75°.

At atmospheric pressure the resistance was measured at 20°, 55°, 75°, and 90°. Within this range, resistance varies linearly with temperature. The average temperature coefficient between 0° and 100° is 0.00239 at 0°. This is surprisingly low, considering the presumably high purity of this metal.

Chromium. The only sample available was a massive, strongly crystalline lump of unknown purity, which I owe to the courtesy of the Hoskins Co. of Detroit. This metal was not quite as hard as hardened steel tools, and by sacrificing many hack saw blades, a rod 2.7 cm. long and 6 mm. in diameter was obtained for the compressibility measurements. No attempt was made to measure the resistance; the probable purity did not justify the effort. Chromium is known to be cubic. The following results were found for the volume compressibility from the measurements of linear compressibility.

$$\text{At } 30° \quad \frac{\Delta V}{V_0} = -5.187 \times 10^{-7}p + 2.19 \times 10^{-12}p^2.$$

$$\text{At } 75° \quad \frac{\Delta V}{V_0} = -5.310 \times 10^{-7}p + 2.19 \times 10^{-12}p^2.$$

The average numerical deviation of a single reading from a smooth curve was 0.48% at 30°, and 0.93% at 75°.

The compressibility given above for Cr is very much less than that found by Richards,[9] who gives the value 9×10^{-7}. That there was very great doubt about the value of Richards is indicated by the fact that he gave only one significant figure.

Vanadium. This material I owe to the courtesy of Mr. B. D. Saklatwalla of the Vanadium Corporation of America. It was highly crystalline and brittle, and came in the form of small rectangular prisms bounded by the cleavage faces, so that the specimen was without doubt a single crystal. Since vanadium is cubic, the linear compressibility of a single crystal in a single direction gives the material for calculating the cubic compressibility. The Vanadium Corporation provided a partial analysis of the material. The best sample was described as 95.02% Va, and .07% C; this was the sample on which measurements were made.

The cubic compressibility, calculated from the linear compressibility was:

$$\text{At } 30° \quad \frac{\Delta V}{V_0} = -6.090 \times 10^{-7}p + 2.58 \times 10^{-12}p^2.$$

$$\text{At } 75° \quad \frac{\Delta V}{V_0} = -6.117 \times 10^{-7}p + 2.55 \times 10^{-12}p^2.$$

The average numerical deviation of a single reading from a smooth curve was 0.11% at 30° and 0.10% at 75°.

No attempt was made to measure the resistance under pressure.

Phosphorus. Measurements were made on both the red and the black modifications.

Red Phosphorus. Although this variety of phosphorus is technically described as "red," violet would more closely describe its appearance. This sample had been formed in 1915 from white phosphorus in the presence of a trace of sodium as a catalyzer by heating to 200° C. under a pressure of 4000 kg, and then increasing the pressure to 12000.[10] The density was 2.348 at room temperature. Since 1915 it has been kept sealed in glass. This specimen possessed no apparent structure. The fracture was conchoidal in character, and extremely fine grained; it probably is a microscopic aggregate of crystals arranged approximately at random. The crystal system of red phosphorus has apparently not yet been determined, but the probability is that it is not cubic. It is to be expected from the manner of formation that the linear compressibility of my sample would be approximately the same in all directions, and that therefore the cubic compressibility could be calculated from a single measurement of linear compressibility. This was checked, however, by making measurements in two different directions. The linear compressibility of a piece 1 cm. long was first measured, and then this piece was cut transversely, and two of the resulting pieces piled together to make a length of 0.5 cm. perpendicular to the original direction, and the linear compressibility of this also measured. The linear compressibilities in these two directions were not the same, but differed by a maximum of 14% at 30° and 12% at 75°, that of the longer single specimen being the greater. This would seem to show that red phosphorus cannot be cubic, and also that in my particular specimen there was a preferred direction of orientation.

The compressibility of neither specimen can be given by a two-power series in the pressure, but the maximum departure from linearity occurs at a pressure less than the mean pressure. This means that the initial rate of decrease of compressibility with increasing pressure is greater than the final rate of decrease. The two samples did not agree, however, with regard to the temperature behavior; the linear compressibility of the longer sample behaved abnormally in that it was less at 75° than at 30° by an amount varying from 4.5% at atmospheric pressure to 1.7% at 12000 kg, whereas that of the shorter sample was less at 30° than at 75° by an amount varying from 1.0% at atmospheric pressure to 5% at 12000. The accuracy of the measurements on the two samples was about the same, the average departure of a single reading from a smooth curve being about 0.3%.

The smoothed average of the results for the two specimens, converted from change of length to change of volume, is shown in Table II.

TABLE II.

VOLUME CHANGE OF RED PHOSPHORUS AS A FUNCTION OF PRESSURE AND TEMPERATURE.

Pressure kg/cm²	$\Delta V/V_0$ 30°	75°
2000	.0101	.0100
4000	.0190	.0189
6000	.0270	.0270
8000	.0342	.0344
10000	.0408	.0412
12000	.0469	.0476

The initial compressibility of this red phosphorus may be found by extrapolation to be about 5.45×10^{-6} at 30°. This is very much less than the value given by Richards[9] for red phosphorus, namely 9.0×10^{-6}. The red phosphorus of Richards was apparently the red phosphorus of commerce. The density of this is much less than the density of my violet phosphorus, so that it is probable that it was in a state of suspended equilibrium, and contained an appreciable proportion of untransformed white phosphorus molecules, which would account for the difference.

Black Phosphorus. This was the identical sample, the pressure coefficient of the resistance of which has been previously measured.[11] It has been kept sealed in glass since those measurements. The linear compressibility of this was measured in only a single direction. Black phosphorus is known not to be cubic; the crystal structure of this specimen was microscopically fine, however, and there is no doubt from the appearance of the specimen that all orientations contributed to the final result. The example of red phosphorus shows, however, that there may nevertheless be a preferred direction, so that the cubic compressibility can be calculated from the linear compressibility only with considerable uncertainty. For this reason I give in the following (Table III) only two significant figures for the initial cubic compressibility, calculated on the assumption of equal compressibility in all directions. The average deviation of a single reading from a smooth curve was 1.0% at 30°, and 1.1% at 75°.

The absolute compressibility is seen to be about one-half that of red phosphorus. The variation of compressibility with temperature is similar to that of red phosphorus, in that the temperature coefficient

TABLE III.

VOLUME CHANGE OF BLACK PHOSPHORUS AS A FUNCTION OF PRESSURE AND TEMPERATURE.

Pressure kg/cm²	$\Delta V/V_0$ 30°	$\Delta V/V_0$ 75°
2000	.0052	.0052
4000	.0095	.0095
6000	.0129	.0129
8000	.0158	.0158
10000	.0182	.0185
12000	.0205	.0209

does not assume an appreciable positive value until the highest pressures are reached. There is, however, a striking difference between the effect of pressure on the compressibility of red and black phosphorus. The compressibility of red phosphorus at 30° drops from 5.45×10^{-6} at atmospheric pressure to 0.54 of this value at 12000 kg., whereas the compressibility of black phosphorus at 30° drops from 2.9×10^{-6} at atmospheric pressure to 0.38 of this at 12000. That is, the relative drop in the compressibility of black phosphorus is greater in spite of the fact that its absolute compressibility is much less. This is not usual, and is directly opposed to a rule proposed by Professor Richards.[12] It would seem to indicate that structural differences in the intermolecular spaces furnish a comparatively greater contribution to the initial compressibility of black phosphorus, and that the compressibility of the *molecule* of black phosphorus is appreciably less than even one-half that of the molecule of red phosphorus.

Sulfur. Sulfur is not cubic, but orthorhombic, so that a complete description of the change of dimensions under pressure demands a measurement of the linear compressibility in the three crystallographically independent directions. Fortunately sulfur occurs in nature in large clear crystals, so that the material is comparatively easy to obtain. I am indebted to the Harvard University Museum for the single crystals of my measurements. Great care is needed in cutting the specimens, because of the great brittleness of this substance and its extreme sensitiveness to temperature changes, the heat of the hand being sufficient to crack it. The method finally adopted was to whittle the specimen with a razor blade from a large crystal, which was imbedded in plaster of paris. Care must be taken that the plaster is of the same temperature as the crystal before imbedding.

A special form of holder had to be used, because sulfur is soluble in the kerosene or petroleum-ether by which pressure is transmitted. A holder was designed which permitted the sulfur to be submerged beneath a surrounding mass of water, pressure being transmitted to the water by oil as usual. A further difficulty was that under pressure the sulfur attacks the steel of the holder, forming ferrous sulfate. This action was reduced to a harmless amount by gold plating the holder, a suggestion for which I am indebted to my assistant Mr. W. N. Tuttle. Since the compressibility of steel and gold are almost exactly the same, no error is introduced by the plating.

Measurements were made of the linear compressibility in three directions at 30° and 75°, in the regular way. The designation of these directions is that conventional with crystallographers. The "*a*" and "*b*" directions are in the plane of the rhombic table-like faces. "*a*" bisects the acute angle, and "*b*" the obtuse angle. The "*c*" direction is perpendicular to "*a*" and "*b*".

In no case could the results be represented by a two-power series in the pressure, but the compressibility drops initially faster than given by such a relation. The results are collected in Table IV,

TABLE IV.

THE COMPRESSIBILITY OF RHOMBIC SULFUR.

Pressure kg/cm²	$\Delta l/l_0$ Direction "a"		Direction "b"		Direction "c"		$\Delta V/V_0$	
	30°	75°	30°	75°	30°	75°	30°	75°
2000	.0099	.0104	.0093	.0113	.0042	.0043	.0233	.0258
4000	.0180	.0190	.0165	.0201	.0079	.0082	.0419	.0466
6000	.0246	.0261	.0224	.0273	.0113	.0116	.0571	.0638
8000	.0304	.0323	.0277	.0334	.0143	.0148	.0707	.0784
10000	.0357	.0372	.0323	.0388	.0171	.0177	.0839	.0908
12000	.0412	.0427	.0370	.0433	.0198	.0203	.0949	.1027

giving the change of linear dimensions at intervals of 2000 kg., and also the change of volume, calculated from the change of linear dimensions. The accuracy is indicated by the average departure of a single reading from a smooth curve, and was as follows: direction "*a*", at 30°, 0.47% deviation, and at 75°, 0.67% deviation; direction

"b", at 30°, 0.69% deviation, and at 75°, 0.63% deviation; direction "c", at 30°, 1.42% deviation, and at 75°, 0.55% deviation.

The initial compressibility, which may be found by extrapolation of the above values, is 13.2×10^{-6} at 30° and 14.4×10^{-6} at 75°. This gives 12.9×10^{-6} at 20° for the initial compressibility, or 12.7 $\times 10^{-6}$ as the average compressibility between 100 and 500 kg, at 20°. This latter agrees exactly with the value found by Richards.[9]

Discussion and Survey.

Most of these new results on compressibility can be dismissed with little further discussion. It will be found that the compressibilities fall entirely into line with what would be expected from the position of the metals in the periodic table. A convenient way to exhibit the relationship is to plot the logarithm of the compressibility to the base 10 against atomic number. This is shown in Figure 1.

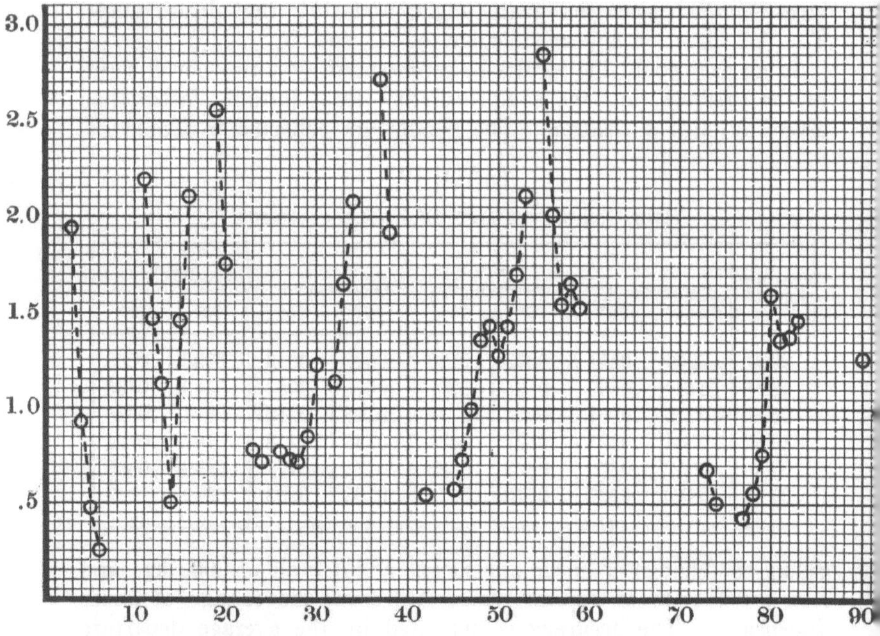

FIGURE 1. The logarithm to the base 10, plus 7, of the compressibility plotted as ordinate against the atomic number as abscissa.

Va and Cr come about where they would be expected, being nearly the same as Fe, Co, and Ni. A good value for manganese is needed to complete our knowledge of this portion of the periodic table; it is evident that there is some sort of minor irregularity in this group of elements. The compressibility of Be falls smoothly between that of Li on the one side, and that of *B* and *C* (diamond) on the other. Apparently black *P* must be considered the normal modification of *P*, since its compressibility falls more smoothly between that of Si and *S* than does that of the red modification, whereas the compressibility of white *P* falls entirely out of sequence. The compressibility of Ba falls smoothly between that of Cs and La, and also terminates the increasing sequence Ca, Sr, Ba, as would be expected. The compressibility of Th is very materially higher than that of Ur, but no other elements have been measured in this part of the table, so we cannot tell whether this was what was to be expected or not. The group of three elements, La, Ce, Pr, (atomic numbers 57, 58, and 59) perhaps offers the most unexpected feature shown by these ten elements, the compressibility of Ce being higher than that of either La or Pr, although the general situation on the falling branch of the curve would lead to the expectation of an intermediate value. But we have seen that the compressibility of Ce is by itself highly anomalous, the temperature coefficient being negative, and the curvature in the abnormal direction. Doubtless there is some connection with the new polymorphic form stable at high pressures.

The phenomena of electrical resistance under pressure are much less simple than those of compressibility. These new data only increase the sense of confusion in this domain that has been increasing with every further penetration into unusual parts of the periodic table. When I began my measurements on the effect of pressure on resistance, the facts seemed very simple, for the resistance of every known element decreased under pressure with the single exception of Bi, and this element is exceptional in so many of its properties that an exception in the pressure coefficient of resistance was more to be expected than not. But the list of elements whose resistance increases under pressure has rapidly grown to: Bi (solid only), Sb, Li (both liquid and solid), Ca, Sr, Ti (perhaps), Cs at high pressure, and now Ce (low pressure modification), and Ba (at high pressures).

The similar behavior of Cs and Ba is striking and perhaps significant, but that we cannot yet put our fingers on the precise feature which is responsible for the similarity is shown by the diametrically opposite behavior of the three pairs of elements above Cs and Ba in the periodic table, namely, Li and Be, K and Ca, and Rb and Sr.

It is interesting that in the group of four successive elements in the table, Cs, Ba, La, and Ce (numbers 55, 56, 57, and 58), three are highly anomalous, Cs and Ba being the only elements found thus far with a pressure minimum of resistance, and Ce having two modifications, of which the low pressure modification is highly anomalous in several respects, as we have seen. Why is it that La is entirely normal, as is also Pr, the next element, of atomic number 59?

The behavior of these new elements only strengthens the view to which I have been coming for some time, namely that the mechanism of electrical conductivity is not as simple as at first appeared, but that a single type of mechanism is not competent to explain the various behavior of different metals.

I am much indebted to my assistants Mr. R. L. Steinberger and Mr. W. N. Tuttle for making most of the readings of this paper.

THE JEFFERSON PHYSICAL LABORATORY,
Harvard University, Cambridge, Mass.

REFERENCES.

[1] P. W. Bridgman, Proc. Amer. Acad. 52, 573, 1917; 56, 61, 1921; 58, 151, 1923; 58, 166, 1923; 59, 109, 1923; 60, 305, 1925; 60, 385, 1925.
Proc. Nat. Acad. Sci. 5, 351, 1919.

[2] H. C. Kremers, Chattanooga Meeting of the American Electro-Chemical Society, Sept. 1925.

[3] H. C. Kremers, Jour. Amer. Chem. Soc. 45, 614, 1923.

[4] Second reference under 1 above.

[5] H. C. Kremers, Niagara Falls Meeting of the American Electro-Chemical Society, April 1925.

[6] Fourth reference under 1 above.

[7] P. W. Bridgman, Amer. Jour. Sci. 10, 359, 1925.

[8] P. S. Danner, Jour. Amer. Chem. Soc. 46, 2382, 1924.

[9] T. W. Richards, Carnegie Institution of Washington, Pub. No. 76, May, 1907.
Jour. Amer. Chem. Soc. 37, 1643, 1915.

[10] P. W. Bridgman, Jour. Amer. Chem. Soc. 38, 609, 1916.

[11] Second reference under 1 above.

[12] T. W. Richards, Jour. Amer. Chem. Soc. 48, 3063, 1926.

THE LINEAR COMPRESSIBILITY OF THIRTEEN NATURAL CRYSTALS.

P. W. BRIDGMAN.

INTRODUCTION.

In this paper the work of a former paper[1] is extended to thirteen more natural crystals. The method of measurement is the same as that employed in the former paper, and requires no further discussion. The new material of this investigation was obtained partly by a personal search through the resources of the National Museum in Washington, for which I am much indebted to the courtesy of Dr. Merrill, and also partly from the Harvard University Museum, for which I am again greatly indebted to Professor Palache. This paper covers practically all the available material in these two sources. As stated in the previous paper, specimens are not common which have the requisite perfection for measurements of this character.

DESCRIPTION OF THE SPECIMENS.

Cubic Crystals.

Andradite. This is a massive variety of garnet. The specimen, which was from British Columbia, and was obtained from the National Museum, was without flaws, greenish in color, and translucent. The composition of andradite is given by Dana as $3CaO.Fe_2O_3.3SiO_2$. A single sample was used for the measurements, 2.7 cm. long.

Garnet, from the National Museum, originally from Burke Co., N. C. This was in the form of a pebble 7 or 8 cm. in diameter, much rounded by the action of water, but with the original form of the natural crystal still traceable. The color was deep red, not transparent, but an appreciable amount of light was transmitted through a section 6 mm. thick. This

[1] Bridgman, P. W., this Journal, 10, 483, 1925.

was probably the variety pyrope, the composition of which is given by Dana as $3MgO.Al_2O_3.3SiO_2$. The specimen was profusely flawed but by careful cutting a single piece 2.6 cm. long was obtained sufficiently free from flaws.

Non-Cubic Crystals.

Beryl, hexagonal, two orientations required, one along and one perpendicular to the hexagonal axis. Two different samples were used, both light green and transparent—gem material. The longitudinal specimen was a single slender crystal 2.7 cm. long, from Mursinka, Siberia, from the National Museum. Two transverse pieces were used, piled together, 1.8 and 0.9 cm. long, cut from another single crystal, from the Harvard Museum, labelled "Tip of crystal with etched faces, Topsham, Maine, L. B. Merrill, 1923."

Apatite, hexagonal, two orientations required. The specimen was from Durango, Mexico, and was furnished by the National Museum. Two longitudinal samples, 1.1 and 1.6 cm. long, were used, piled together. Four transverse specimens were used, of lengths 0.5, 0.6, 0.7, and 0.8 cm., piled together.

Hanksite, hexagonal, two orientations required. The source of the material was several crystals, obtained from the National Museum, all from California, locality not specified, and all of the same general appearance. The crystals had numerous occlusions, but by cutting it was possible to obtain clear pieces large enough for the measurements. Three longitudinal pieces were used, of lengths 0.5, 0.6, and 0.9 cm., and two transverse pieces, of lengths 1.4 cm. each.

Tourmaline, rhombohedral, two orientations required. This had been previously measured, but only in the longitudinal direction, on a clear light pink and green specimen. The tourmaline of these measurements was black; the color is supposed to indicate a fairly large percentage of iron. Measurements of the compressibility parallel to the axis were made on two specimens. The first of these was a long slender crystal from the National Museum, obtained from Lost Valley, San Diego Co., California, and was 2.5 cm. long. The second longitudinal specimen was from a large crystal, not so clear, also from the National Museum, which obtained it from Pierrepont, N. Y.; it was 2 cm. long. The transverse specimen was cut from the same sample as the second longitudinal piece; one piece was used, 1.3 cm. long.

Quartz, trigonal, two specimens required, one parallel and one perpendicular to the trigonal axis. I am indebted for this specimen to Dr. L. H. Adams of the Geophysical Laboratory in Washington. It was cut from the identical specimen on which he had previously made measurements of the cubic compressibility.[2] Each piece was 2.7 cm. long.

Rutile, tetragonal, two orientations required, parallel and perpendicular to the tetragonal axis. This material had been previously measured in the longitudinal direction. A new specimen was now available, thick enough to allow transverse specimens. This was furnished by the National Museum, and originated in White Plains, N. C. Four pieces were used, piled together, each 0.6 cm. long.

Barite, orthorhombic, three orientations required. This substance crystallizes in the form of long prisms, with a section of the shape of an elongated hexagon. Following Dana, the A direction is taken as the axis of the prism, the B direction as the long axis of the hexagonal section, and the C direction as the short axis of the hexagon. Two large crystals were available. The A and B orientations were obtained from a single crystal from Dufton, England, from the National Museum. This crystal unfortunately had many flaws in such a direction as to make it impossible to obtain from it the C orientation. The A and B orientations were unflawed, the A piece 2.6 cm. long, and two B pieces piled together, 0.6 and 1.5 cm. long. The C orientation was obtained from another crystal, perfectly clear, from the Harvard Museum, which obtained it through Ward's from the Wotherton Mine, Chirbury, Shropshire, England. Two specimens were used, 1.3 and 1.4 cm. long.

Topaz, orthorhombic, three orientations required. Two specimens were used, furnished by the National Museum, obtained by them from Tanakami, Yami, Omi, Japan. The cleavage face, which is the 001 plane, contains the A and the B axes, the A axis being along the short diagonal of the parallelogram. The C axis is perpendicular to A and B. The following pieces were used: A direction, one specimen, 2.5 cm. long; B direction, one specimen, 2.7 cm. long; and C direction, one specimen, 1.6 cm. long. In addition to the measurements on these specimens, two sets of measurements were made on a sample which I owe to Professor Elihu Thomson of the

[2] Adams, L. H., and Williamson, E. D., Jour. Frank. Inst., 195, 475, 1923.

General Electric Co. Due to an erroneous identification, one of the samples was not cut in the correct direction, but the measurements on the other orientation afford a useful check. Three pieces were used, piled together, each 6 mm. long.

Jeffersonite, monoclinic, described by Dana as a manganese-zinc pyroxene. Three mutually perpendicular directions are required to give the change of volume under pressure, but four orientations are required to obtain all the independent changes of length, since the angle between the axes also changes under pressure. The following notation is that of Dana. The A and B axes are perpendicular to each other, and are in the natural cleavage face. The third axis is the C axis, which is perpendicular to the B axis, and makes an angle of 74° 10′ with the A axis. The fourth direction, which is not one of the crystal axes, but which, together with the B and C directions, gives the change of volume, I have called the Y direction, following Voigt. This is perpendicular to the C and B axes of Dana, or the X and Z axes of Voigt, respectively. The specimen was from the National Museum, from Franklin Furnace, N. J., and was a single piece 10 or 15 cm. on a side. It was badly flawed, and it was possible to find only small pieces for these measurements which were suitably unflawed. The A orientation was given by two specimens 1.7 and 1.0 cm. long piled together; the B orientation by one specimen 2.2 cm. long; the C orientation by one specimen 2.7 cm. long; and the D orientation by three specimens, piled together, 0.7, 0.9 and 1.1 cm. long.

Orthoclase, monoclinic, four specimens required. The orientations and the notation are the same as for Jeffersonite. The material was obtained from the National Museum and the Harvard Museum, both of which had obtained it through Ward's from Fianarantsoa, Madagascar. The specimens were light green in color and flawless, of the mineralogical variety known as Adularia. For the A direction one specimen 2.7 cm. long was used; for the B direction one specimen 2.5 cm. long; for the C direction one specimen 1.7 cm. long; and for the Y direction two pieces piled together, each 1.2 cm. long.

Spodumene, monoclinic, orientations and notation as above. The compressibility has been previously determined[3] for the A, B, and C directions. These are now completed by measurements of the Y direction. This was cut from the same crystal as the three previous specimens. A single piece was used, 1.7 cm. long.

[3] Op. cit., p. 488.

DISCUSSION OF NUMERICAL RESULTS.

The essential numerical results are contained in Table I.

It is interesting that the compressibility of the crystalline variety of garnet is materially less than that of the massive variety. This is what our general experience with liquids and solids would lead us to expect.

Beryl has a negative temperature coefficient of linear compressibility parallel to the hexagonal axis. The cubic compressibility, calculated from the linear compressibilities, is 5.33×10^{-7}, against the value 5.7×10^{-7} of Madelung and Fuchs[4] for a Brazilian beryl. The compressibility calculated from Voigt's determination of the other elastic constants of a Siberian beryl is 7.4×10^{-7}. Such indirect measurements have possibilities of considerable error.

The compressibilities of tourmaline listed in the Table were both measured on specimens cut from the same crystal of black tourmaline. This was flawed rather more than would have been desirable, and the results are somewhat more irregular than usual, as shown by the probable error of a single reading. The cubic compressibility of this specimen, extrapolated to 0° C, is 7.75×10^{-7}, which agrees fairly well with the values found by Madelung and Fuchs[4] for two different specimens, namely 8.2 and 7.9×10^{-7}, also at 0°.

In addition to the values tabulated, the longitudinal compressibility of the clear black tourmaline described above was measured and found to be:

$$\text{At } 30^\circ,\ \frac{\Delta l}{l_0} = -4.97 \times 10^{-7}p + 1.33 \times 10^{-12}p^2.$$

$$\text{At } 75^\circ,\ \frac{\Delta l}{l_0} = -5.03 \times 10^{-7}p + 1.58 \times 10^{-12}p^2.$$

These values are not greatly different from the values in the Table.

The values for the longitudinal compressibility of these two specimens of black tourmaline may be compared with the value previously found for the longitudinal compressibility of a clear green and pink tourmaline, namely:

$$\text{At } 30^\circ,\ \frac{\Delta l}{l_0} = -4.66 \times 10^{-7}p + 2.53 \times 10^{-12}p^2.$$

$$\text{At } 75^\circ,\ \frac{\Delta l}{l_0} = -4.61 \times 10^{-7}p + 2.39 \times 10^{-12}p^2.$$

[4] Madelung, E. and Fuchs, R., Ann. Phys., **65**, 289, 1921.

TABLE I.

Substance and System	Direction	Linear Compressibility, $\frac{\Delta l}{l_0} = -ap + bp^2$ 30° *a*	30° *b*	75° *a*	75° *b*	Volume Compressibility, $\frac{\Delta v}{v_0} = -ap + bp^2$ 30° *a*	30° *b*	75° *a*	75° *b*	Percentage Deviation of Single Reading from Smooth Curve 30°	75°
Andradite Cubic		2.210 x 10^{-7}	.70 x 10^{-12}	2.202 x 10^{-7}	.70 x 10^{-12}	6.630 x 10^{-7}	2.25 x 10^{-12}	6.606 x 10^{-7}	2.25 x 10^{-12}	.35 (0)	.41 (0)
Garnet Cubic		1.793	.73	1.813	.73	5.379	2.19	5.439	2.19	.17† (0)	.05† (0)
Beryl	∥ Heg. axis ...	2.046	.70	2.039	.70						
Hexagonal	⊥ Hex. axis ...	1.643	.70	1.648	.70	5.332	2.20	5.335	2.20	.59 (0)	.48 (2)
Apatite	∥ Hex. axis ...	2.410	.35	2.500	.06					.17 (2)	.20 (0)
Hexagonal	⊥ Hex. axis ...	4.160	2.31	4.205	2.31	10.730	5.34	10.910	5.07	.27 (0)	.37 (0)
Hanksite	∥ Hex. axis ...	11.651	12.34	12.345	13.70					.13 (0)	.13 (0)
Hexagonal	⊥ Hex. axis ...	6.240	5.31	6.374	5.65	24.131	24.80	25.093	26.98	.16 (0)	.06 (0)
Tourmaline (black)	∥ Trig. axis	4.78	1.59	4.85	1.76					.31 (0)	.27 (0)
Trigonal	⊥ Trig. axis ...	1.63	.70	1.82	.70	8.04	3.18	8.49	3.37	.8 (2)	1.1 (0)
Quartz	∥ Trig. axis	7.052	6.44	7.209	6.98					.14 (0)	.06 (0)
Trigonal	⊥ Trig. axis ...	9.764	7.79	9.923	7.81	26.58	24.4	27.05	25.0	.18 (0)	.07 (0)
Rutile	∥ Tetrag. axis* .	1.038	.70	1.090	.70					.35 (0)	.56 (0)
Tetragonal	⊥ Tetrag. axis .	1.871	.70	1.954	.70	4.780	2.18	4.998	2.18	.16 (0)	.07 (0)
Barite	A	4.940	3.57	5.045	3.51					.36 (0)	.26 (0)
Ortho-	B	6.695	4.34	6.758	4.72	17.295	12.70	17.604	13.44	.39 (0)	.37 (0)
rhombic	C	5.660	3.81	5.801	4.19					.41 (0)	.41 (0)
Topaz	A	2.145	.70	2.188	.70					.13 (0)	.10 (3)
Ortho-	B	1.486	.70	1.571	.70	6.024	2.32	5.991	2.32	.39 (0)	1.5 (0)
rhombic	C	2.393	.70	2.232	1.01					1.3 (0)	.11 (0)
	A	3.093	2.02	3.441	4.38					.52 (0)	.68 (0)
Jeffersonite	B	3.924	1.74	3.990	2.39					.35 (1)	.39 (0)
Monoclinic	C	3.078	2.51	3.384	3.41	8.947	5.21	9.400	6.79	.62 (0)	.67 (0)
	Y	1.945	.70	2.026	.70					2.0 (0)	3.0 (0)
	A	9.944	5.09	10.184	6.44					.10 (0)	.07 (0)
Orthoclase	B	5.490	5.15	5.394	4.20					.33 (0)	.43 (0)
Monoclinic	C	4.599	1.68	4.604	2.09	20.85	15.3	20.78	14.8	.48 (0)	.30 (0)
	Y	10.765	7.14	10.785	7.14					.65 (5)	.36 (5)
	A*	1.801	.70	1.826	.70					.69 (1)	1.48 (0)
Spodumene	B*	2.459	.70	2.587	1.46					.20 (0)	.14 (0)
Monoclinic	C*	1.997	.70	1.938	.70	6.930	2.84	6.969	3.50	.10 (0)	.11 (0)
	Y	2.474	1.28	2.444	1.28					.84 (0)	.29 (0)

* Copied from previous publication. † Maximum Deviation.

The black tourmaline does not show the anomalous temperature coefficient of the pink tourmaline. This, however, is a comparatively minor matter, and it appears in general that comparatively large differences in the appearance make comparatively small differences in the compressibility of tourmaline. The very large difference of compressibility parallel and perpendicular to the axis is a noteworthy feature of the elastic behavior of tourmaline, and is doubtless characteristic of all the varieties.

The results which I find for the new specimen of quartz do not agree with my results for the previous sample. The difference lies almost entirely in the direction perpendicular to the axis. The results found for the previous sample in this direction were anomalous in that initially there was a rapid drop of compressibility, and at 9,000 an inflection, so that above 9,000 the compressibility increased with increasing pressure, a highly abnormal result. The new specimen does not show this abnormal behavior at all; there is a very slight anomaly in the perpendicular direction, but of the opposite sign from that of the first specimen, in that now the decrease of compressibility with pressure is slightly greater at high pressure than at low, or in other words, the maximum of the curve of departure from linearity occurs at a pressure slightly above 6,000. I have not attempted to reproduce this slight effect in the final formulas, however, since probably other specimens would not show it, but have given in the Table the second degree equation which best represents the facts.

The high value, as compared with other observers, which the previous sample gave for the initial compressibility, is connected almost entirely with the anomaly in the perpendicular direction. The initial compressibility which I find for the new sample agrees almost exactly with the value found by Adams and Williamson.[5] My results for this sample, corrected to 20°, are:

$$\frac{\Delta V}{V_0} = -2.648 \times 10^{-6} p - 24.2 \times 10^{-12} p^2.$$

Their results, changed from megabars to kilograms, at 20°, and corrected by my value for the compressibility of iron, are:

$$\frac{\Delta V}{V_0} = -2.648 \times 10^{-6} p - 20.9 \times 10^{-12} p^2.$$

The first degree term, which determines the initial compres-

[5] Op. cit.

sibility, agrees exactly. Their second degree term is, however, lower than mine, which means that their changes of volume at high pressures are somewhat greater. At 12,000 kg. the difference is somewhat less than 2%. A comparison of our results is given in Table II. The agreement is perhaps

TABLE II. Comparison of Different Determinations of the Cubic Compressibility of Quartz at 20°.

	$\Delta V/V_0$		
Pressure kg/cm²	Adams and Williamson	Bridgman same sample	Bridgman previous sample
2,000	.00521	.00520	.00552
4,000	01026	1021	1077
6,000	1514	1502	1564
8,000	1984	1963	2019
10,000	2439	2406	2482
12,000	2877	2829	2939

to be considered satisfactory. It is to be remembered that their results were obtained on a large sample, of which my sample was only a small piece. If the material is not perfectly homogeneous, and my very different results with two different specimens would suggest this possibility, exact agreement between their results and mine is not to be expected.

A general conclusion to be drawn from these comparative measurements is that different samples of natural crystals of even so apparently well defined a substance as quartz may vary by quite measurable amounts. Further, too much emphasis should not be laid on the details of the differences of measurements by different observers; the most proper comparison of the results of Adams and Williamson with mine is given by the measured changes of volume over the entire pressure range, not by the initial slope of the curve.

In the previous paper, single measurements were made of the compressibility of rutile parallel to the tetragonal axis. A sample has now been obtained large enough to permit completing the description by measurements in the transverse direction. Since rutile has a simple chemical constitution, TiO_2, there should not be much danger in combining the constants obtained from two different specimens. The initial cubic compressibility calculated from my data is 4.78×10^{-7}, which is materially smaller than the direct value of Madelung and Fuchs, 5.8×10^{-7}.

The cubic compressibility of barite, calculated from the

linear compressibilities, is 17.3×10^{-7}, against direct values of Madelung and Fuchs, 17.4 and 17.8×10^{-7}, obtained from specimens from two different localities. The value calculated from the elastic constants given by Voigt is 18.5. The agreement of these figures is somewhat better than usual, doubtless because of the ease with which perfect crystals of this material can be obtained.

Topaz is noteworthy because of the very large negative temperature coefficient of compressibility in the C direction; this large value more than outweighs the smaller positive coefficients in the A and the B directions, so that the temperature coefficient of the volume compressibility is also negative. The cubic compressibility, calculated from Voigt's elastic constants, is 5.8×10^{-7}, against 6.02 above; again the agreement is better than usual. A further partial check of the constants given in the Table was obtained from the measurements parallel to the axis of the specimen obtained from Professor Elihu Thomson. This gives at 30°, $2.09 \times 10^{-7}p - .70 \times 10^{-12}p^2$, against $2.14 \times 10^{-7}p - .70 \times 10^{-12}p^2$ above. Considering that this sample was composed of three pieces piled together, the agreement is perhaps as close as could be expected.

There are no other values with which to compare the results found here for Jeffersonite. The linear compressibility of this was determined in four directions; the cubic compressibility may be found from the linear compressibilities in the B, C, and Y directions. From the linear compressibility in the A direction, another elastic constant may be found. The calculation follows, using the notation of Voigt. The initial compressibility in the C direction is the constant S_1 (3.078×10^{-7}); that in the Y direction, S_2 (1.945×10^{-7}); and that in the B direction, S_3 (3.924×10^{-7}). Then the following combination of these constants gives the linear compressibility in the A direction (3.093×10^{-7}).

$$S_1 \cos^2 74^\circ\, 10' + S_2 \sin^2 74^\circ\, 10' + S_6 \cos^2 74^\circ\, 10' \cos 15^\circ\, 50' = 3.093 \times 10^{-7}.$$

Substituting numerical values, S_6 is found to be 4.04×10^{-7}. S_6 is an abbreviation for the combination of constants $S_{16} + S_{26} + S_{36}$. If the stress system is a hydrostatic pressure, P, as here, then S_6 is connected with strain and stress by the equation:

$$e_{xy} = S_6 \frac{P}{3}$$

where e_{xy} gives the shearing strain between the X and Y (C and Y) axes. Substituting numerical values, we find that the acute angle between the C and the Y axes decreases by 0.095° under a pressure of 12,000 kg/cm².

Orthoclase is noteworthy in the negative temperature coefficient of volume compressibility, both in the first and the second degree terms. This is entirely due to the linear compressibility in the B direction, the temperature coefficients of the linear compressibility in the C and the Y directions being slightly positive, as is normal. The constant S_6 may be calculated by the same method as that just described for Jeffersonite. Assuming that the angle between the A and the C axes is 63°57′ (Dana), S_6 has the value 0.915×10^{-7} at 30°. This means that the acute angle between the C and the Y axes decreases by 0.021° under a pressure of 12,000 kg/cm².

As already explained, the compressibility of spodumene in the A, B and C directions was given in the previous paper. This data is now completed by the linear compressibility along the Y direction. The cubic compressibility can now be accurately calculated; its initial value of 30° is 6.93×10^{-7}. This is materially higher than the approximate value calculated in the previous paper, which was 6.26×10^{-7}. The reason for the discrepancy is that the assumption made in the approximate calculation, namely that under hydrostatic pressure the distortion is approximately equal in all directions in the plane perpendicular to the B axis, is not very close to the fact. The same calculation may be applied to spodumene that we have just applied to Jeffersonite and orthoclase. It will be found that S_6 has the value -1.84×10^{-7}, which means that under a pressure of 12,000 kg. the acute angle between the Y and the C axes *increases* by 0.043°.

It is a pleasure to acknowledge the assistance of Mr. W. N. Tuttle in making most of the readings of this paper.

THE JEFFERSON PHYSICAL LABORATORY,
HARVARD UNIVERSITY, CAMBRIDGE, MASS.

[On page 293, the signs of the second-degree terms in the two equations should be +, not −. Andradite (page 287) should be grossularite; garnet should be almandite.]

The pressure transitions of the rubidium halides.

By

P. W. Bridgman in Cambridge (Mass.).

(With 10 figures.)

Contents.

Introduction.

In the course of his measurements of the compressibilities of the alkali halides, Slater (1) found that *RbBr* and *RbJ* are stable at high pressures in a second polymorphic modification, not hitherto known. The method used by Slater was the »lever piezometer« method, which I had devised for the measurement of linear compressibilities (2). This method is not well adapted to the study of polymorphic transitions, and in fact is capable of establishing little more than the existence of a transition, and its temperature-pressure coordinates. In this paper I examine these transitions by a new method, designed especially to allow a determination of the volume difference between the two modifications, so that in this way all the thermodynamic parameters of the transition may be found. I have furthermore extended the measurements to *RbCl*, which also turns out to have a transition at high pressures.

In the attempt to understand better the nature of the transitions of these rubidium salts, I made unsuccessful efforts to find new modifications at room temperature of *KBr*, *KJ* and *CsF*, carrying the exploration to 18000, 20000 and 12000 kg./cm.^{2} respectively.

In connection with one of the necessary corrections, the linear compressibility of manganin was determined at 30° and 70° to 12000 kg./cm.^{2}.

The method.

The method used in previous measurements of polymorphic transitions under pressure (3) would not have worked well here because the quantity of material available was too small. The new method may be briefly described as a combination of the free piston type of piezo-

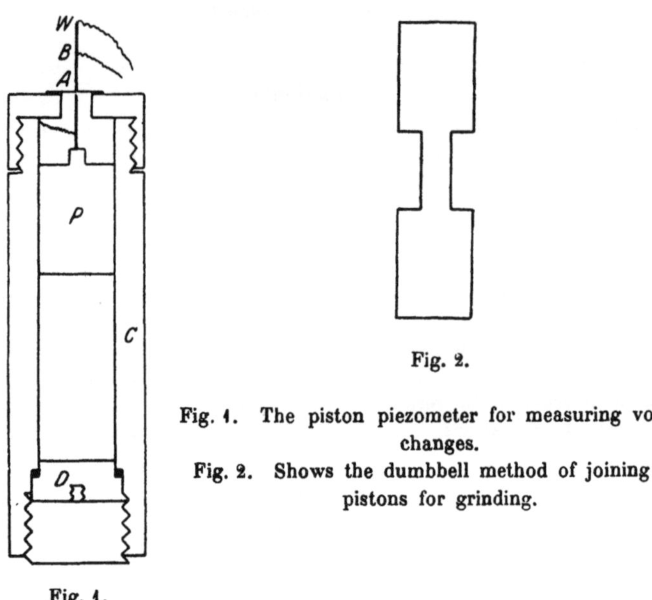

Fig. 1.

Fig. 2.

Fig. 1. The piston piezometer for measuring volume changes.

Fig. 2. Shows the dumbbell method of joining the pistons for grinding.

meter, which was apparently first used by Perkins (4), and which has been applied with modifications by me to the measurement of the compressibility of mercury (5), and lately to the compressibility of oils by Hyde (6), with the slide wire electrical measuring device, which I have used with various piezometers for measuring linear compressibility. By the use of the slide wire device the method may be made continuous reading; the chief inconvenience in the application of the original method was that it gave, for a single set-up of the apparatus, only a single reading, corresponding to the maximum pressure.

The new piezometer is shown in fig. 1. It consists essentially of a piston *P* sliding in a cylinder *C* closed at the lower end. The cylinder is filled with the substance whose volume change under pressure is to be determined, and then subjected, in the pressure chamber, to hydrostatic pressure exerted by a liquid over the entire external surface. The external pressure drives in the piston P until the pressures outside and inside are equal. The amount of motion is obtained from the change of resistance of the high resistance wire W, measured by a potentiometer, between the contact A attached to the cylinder, but insulated from it and over which the wire slides, and the contact B, attached to the wire.

It is evident that the most important detail of construction is the fit between the cylinder and the piston. I made a number of attempts before this was satisfactory. The cylinder and piston were made of hardened tool steel, of one of the brands made especially for die work, which warp very slightly on hardening. The cylinder was ground and lapped by the firm of West and Dodge. The piston was ground to fit the hole by Mr. H. L. van Keuren, Watertown, Mass. In the last few years a technique has been developed for the grinding of gauge plugs, by which results of high precision may be obtained. 18 pistons were ground at the same time, joined together two by two, in a sort of dumbbell, as shown in fig. 2. The grinder was set to grind a very slight taper, about 0,000025 cm. difference between the two ends of the dumbbell, and it was also so arranged that there was a progressive difference of diameter between the 9 dumbbells of about 0,00025 cm. altogether. The approximate size of the hole was already known, and it was not a difficult matter to select from the 18 plugs the best fit. The piston selected was only 0,000025 cm. smaller than the next size, which would not enter the hole.

The plug D at the bottom of the cylinder is also a matter of some importance, as it must fit without leak, and be freely removeable. The seat into which it fits was ground true, and the plug, which is of steel, was packed with a ring of pure gold. Gold has the advantage of being so soft that the joint may readily be made tight initially, and at the same time of being slightly less compressible than steel, so that the joint becomes tighter at high pressures.

The measuring wire does not require detailed description, as it was much like those already described in connection with previous piezometers. The diagram does not show various details of the construction; for example, there is a spring by which the wire is kept pressed against the contact A. The wire had best be made of manganin, if copper leads are used, in order to avoid thermal effects. Previously nichrome wire was used; this has the advantage of twice as high a specific resistance, but the t. e. m. f. against copper is much higher than that of manganin, and the motion of the piston is here so great that it is not necessary to give the highest attainable resistance to the wire in order to obtain the required sensitivity. The very fine helix shown in the diagram by which the piston is connected to the cylinder proved to be necessary, as otherwise the resistance of the film of oil between the piston and the cylinder was so high and so variable as to make measurements very difficult. I owe this suggestion to my assistant, Mr. W. N. Tuttle.

In use, the apparatus has proved entirely satisfactory in measuring changes of volume with pressure at constant temperature. After an excursion to 12000 kg./cm^2., the piston will return to its initial position within a small fraction of a per cent of the total motion. This apparatus was originally designed for the measurement of the compressibility of liquids, and I have already applied it to a number. Preliminary results obtained for the compressibility of glycerine were published some time ago (7). The apparatus has, however, up to date been disappointing in the results which it gives for thermal expansion under pressure. The temperature cannot be changed without a differential expansion between cylinder and piston, during which there is a small leak, so that the motion of the piston during a change of temperature at constant pressure does not correspond accurately enough to the change of volume of the contents. However, this objection does not apply to the measurements recorded here, which were all made at constant temperature.

The salt whose transition was to be studied was first formed into a coherent cylinder by compressing the dry powder with a steel plunger into a split steel mold with a pressure of several thousand atmospheres. The cylinder of salt was placed in the cylinder *C*, together with kerosene to fill the empty spaces. The amount of kerosene and the weight and dimensions of the cylinder of salt had to be so chosen as to allow free motion of the piston up to the maximum pressure. Pressure was then applied in steps, and from the resistance measurements the position of the piston was plotted as a function of pressure. The transition is shown by a change in the curve of piston displacement against pressure.

If the transition were one that takes place sharply, there would be an abrupt discontinuity at the pressure of transition. In the case of these three salts, the transition did not take place sharply, however, and the failure to do this constituted the particular difficulty of these experiments. Apparently the phenomena of the region of indifference (8) are particulary prominent with these salts, and the best that could be done in most cases was to shut the transition within limits.

In calculating the change of volume from the discontinuity in the motion of the piston, there are a number of small corrections that must be applied: for the change of resistance of the manganin wire under pressure, for the compressibility of the manganin wire, and for the compressibility of the steel of the piezometer. All these corrections are small, they all have the same sign, and altogether were less than 1,5% at 5000 kg./cm.², the approximate pressure of the transition.

The compressibility of manganin. The compressibility of manganin was determined by special direct experiment in the lever piezometer. To a sufficient approximation, the average linear compressibility of manganin between 1 and 12000 kg./cm². was found to be $2{,}51 \times 10^{-7}$ at 30° and $2{,}56 \times 10^{-7}$ at 75°.

The material.

The *RbJ* was Kahlbaum's, said to be of high purity. The *RbBr* and *RbCl* was obtained from Eimer and Amend, said also to be of high purity. It was evident, however, that the purity of these samples was not perfect, because there was some rounding of the corners of the transition; this rounding occurred at both corners, indicating, that the impurity formed mixed crystals with both phases. The nature of the impurity was not certain, but it seemed probable that it was one or more of the other alkali metals, which I found, on consultation with Professor T. W. Richards, are particularly difficult to completely remove. Fortunailly Professor Richards had a supply of very pure *Rb* material, which had been used in some of his atomic weight determinations, and he was so kind as to undertake personally the preparation of some pure *RbCl* from this material. The first specimen of *RbCl* which he furnished was known to have as impurity only a small amount of water. A transition measurement with this at 20° showed no rounding of the corners, and was a great improvement in this respect on the other sample, but the transition did appear to be unusually sluggish, since pressure had to be carried far beyond the equilibrium value to compel the transition to run to completion. In the attempt to remedy this, Professor Richards subjected the sample to a second purification, fusing it in a proper atmosphere, in order to cempletely remove the water. Measurements were then made on this at 30°, 60°, and 95°, and were by far the best obtained with any of the material, but it is difficult to be sure whether the improvement can be explained by the removal of the water, or whether it was due to the higher temperature, since the possibility of transgressing an equilibrium is always much less at higher temperatures.

In addition to the three rubidium salts, three others were subjected to an unsuccessful examination for new forms at high pressure. Of these, *KBr* and *KJ* were ordinary commercial chemically pure materials. The *CsF* I obtained from Dr. Hendricks of the Rockefeller Institute for Medical Research, who had had previous experience in the preparation of this somewhat difficult salt in connection with X-ray measurements of its structure, and who now prepared a fresh sample for this investigation at the request of Dr. Wyckoff. I am much indebted to both these gentlemen for their interest and trouble.

The results.

RbJ. The results found for the transition pressures at various temperatures are shown in fig. 3, and for ΔV, in cm³./gm., in fig. 4. The rounding of both corners, due to impurity soluble in both phases, was well marked at 100° and 75°, but was less, and in fact it is even not certain that it existed, at the lower temperatures. At low temperatures, on the other hand, the exact transition pressure was difficult to determine because of the sluggishness of the transition. A combination of these two effects resulted in a region of uncertainty which had about the same width at all temperatures, a rather unusual situation. The same rounding of the corners which makes uncertain the exact pressure of the transition introduces still larger error into ΔV, as is shown by the scattering of the points in fig. 4. It is fairly safe to draw the conclusion, however, that ΔV increases with rising temperature.

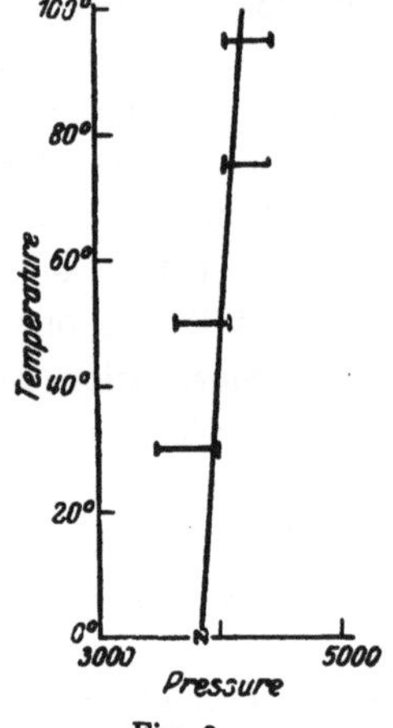

Fig. 3.

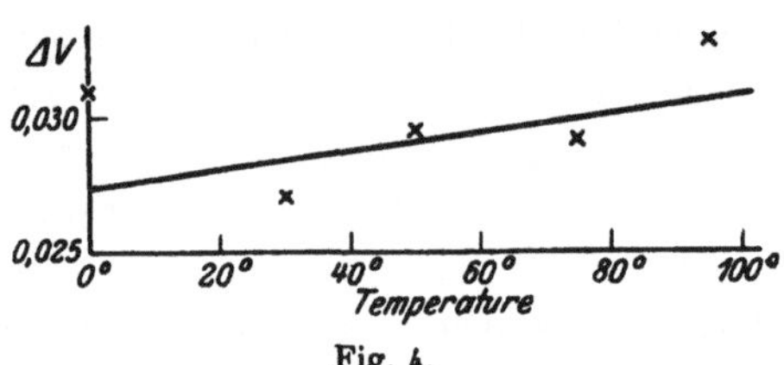

Fig. 4.

Fig. 3. The observed transition temperatures of *RbJ* as a function of pressure. The horizontal lines indicate the transition range.

Fig. 4. The observed volume decrements during the transition of *RbJ* as a function of temperature, in cm³./gm.

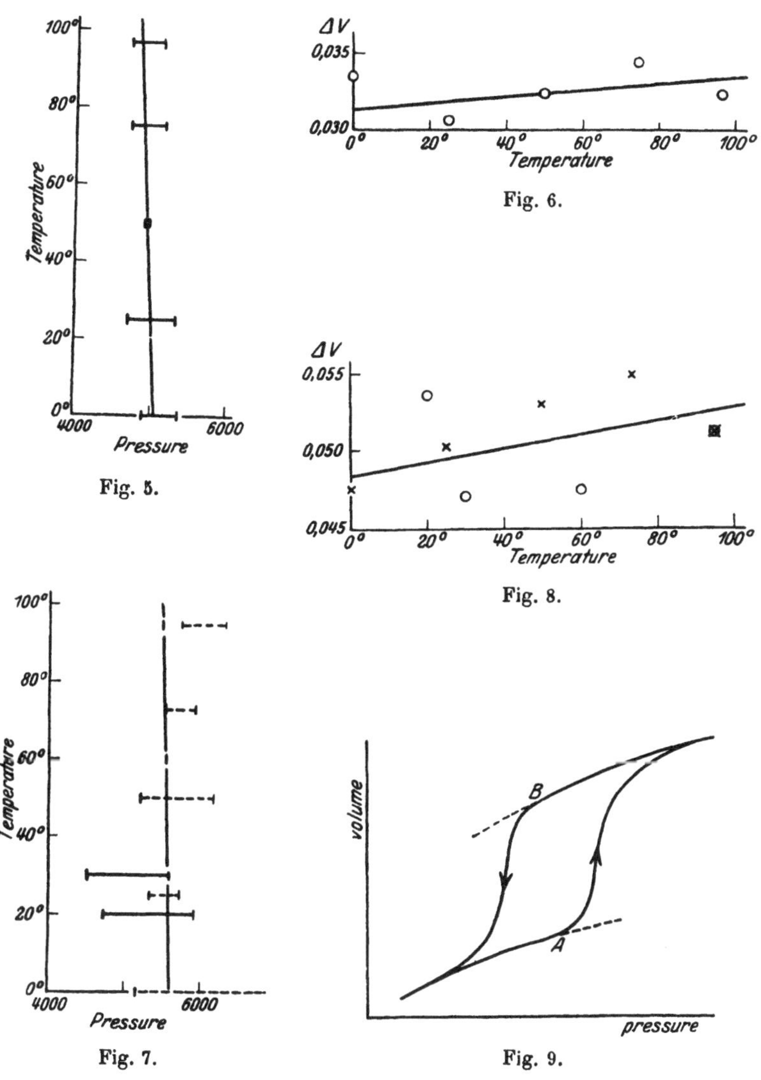

Fig. 5.

Fig. 6.

Fig. 8.

Fig. 7.

Fig. 9.

Fig. 5. Shows, for *RbBr*, the pressure range within which the transition takes place as a function of pressure and temperature.

Fig. 6. The change of volume during the transition of *RbBr* in cm^3./gm. as a function of temperature.

Fig. 7. Shows, for *RbCl*, the pressure range within which the transition takes place as a function of pressure and temperature. The dotted lines are for commercial material, and the full lines for the pure material of Professor Richards.

Fig. 8. The change of volume during the transition of *RbCl* in cm^3./gm. as a function of temperature. The crosses are for commercial material, and the circles for the pure material of Professor Richards.

Fig. 9. Shows the kind of hysteresis loop found during the transitions.

RbBr. The results for the transition pressure as a function of temperature are shown in fig. 5, and for ΔV in fig. 6. Here again the results are uncertain because of rounding of the corners, which is by far the most pronounced at the higher temperatures. Again ΔV appears to increase with rising temperature. The rounding of the corners, in the case of both *RbJ* and *RbBr*, was always greater at the upper corner, that is, when passing from the high pressure to the low pressure modification.

RbCl. The transition pressures are shown in fig. 7, those for the commercial sample dotted, and those for the material of Professor Richards in full. ΔV is shown in fig. 8, the commercial material as crosses, and the pure material as circles. Unlike *RbJ* and *RbBr*, it was never possible to certainly establish a rounding of the corners with this material, even with the commercial sample. The difficulty here was rather a hysteresis in the transition. This became much less at the higher temperatures, and was also much less in the purer sample, but the difficulty never completely disappeared. Fig. 9 shows the sort of hysteresis loops that were obtained. With rising pressure the transition was taken to have begun at the first point *A* which lay appreciably above the expected curve, and with decreasing pressure at the first point *B* lying appreciably below it. It will be seen from fig. 7 that these two points become identical at 60° and 95° for the pure sample. The explanation of this hysteresis is not at once obvious. The most probable explanation would seem to be that the material gets broken into lumps, which are so isolated that when the transition starts in one lump it will not spread by contact to the adjacent lumps. Further, we must suppose that the lumps are of different sizes, and that the transition is likely to run with less transgression of the equilibrium pressure in the larger lumps. This sort of explanation is in line with my previous experience of polymorphism under pressure, but the mechanical appearance of the salt after the application of pressure was not particularly favorable to this explanation.

Fig. 7 shows that the equilibrium pressures of the purer sample are probably slightly less than those of the impurer, but the difference is not important. Within experimental error there is no distinction between the ΔV values of the two samples. Again ΔV appears to increase with increasing temperature.

When the transition is sharp, it is possible to obtain the difference of compressibility between the two modifications from the difference of slope of the curve plotting piston displacement against pressure for the two modifications. This was possible only with the second of the two sambles provided by Professor Richards, and even then the results were only rough. It is certain that the high pressure modification is less compressible than the low pressure modification, and the difference is of the order of 3×10^{-6}. The initial compressibility of *RbCl* has been found by Saerens (9) to be $7{,}3 \times 10^{-6}$. Hence at the transition the high pressure modification has approximately half the compressibility of the low pressure modification. This sort of behavior appears natural enough, but it is nevertheless an exception, as I have previously found in studying a large number of transitions under pressure that in the majority of cases the phase of smaller volume has the higher compressibility (10).

Search for similar transitions of other salts. The search was made by the method of volume discontinuity.

Since the compressibility of *KBr* and *KJ* had been previously measured by Slater to 12000 kg./cm². , without the discovery of new forms, it was necessary to extend the pressure beyond this range in this new search. This was done in the small apparatus already used for the measurement of the compressibility of gases (11). The smaller size of this apparatus permits more thorough heat treatment of the steel and therefore higher pressure. Pressure was pushed to 20000 kg./cm². with *KJ* and to 18000 kg./cm². with *KBr*.

CsF has not been subjected to previous pressure measurements, so that results to only 12000 kg./cm². were of significance. Furthermore, the quantity available was smaller than that of *KJ* and *KBr*, so that so accurate results could not have been obtained with the gas apparatus. The examination of *CsF* was made with the same piston piezometer as that used for the rubidium salts, and gave no new modification to 12000 kg./cm².

In the search for new modifications of these three salts the accuracy was not pushed to its absolute limit, but the search was thorough enough to discover any transition with a volume change only a few per cent of the volume change of the rubidium salts.

Discussion.

Digest of experimental results. The most important of the thermodynamic transition parameters are collected in Table I, calculated for a mean temperature of 50° C. The material from which this table was calculated is contained in the diagrams already given, except for the column headed V_{1t}/V_{10}. Here V_{1t} is the volume of the low pressure modification at the pressure of the transition, and V_{10} is its volume at atmospheric pressure. ϱ_{1t} is the density of the low pressure modification at the transition pressure, and ϱ_{2t} is the corresponding value for the high pressure modification. In calculating V_{1t}, the values of Slater for the compressibility of *RbJ* and *RbBr* as a function of pressure were used. Slater did not measure the compressibility of *RbCl*. For this I have assumed the formula:

$$\frac{\Delta V}{V_v} = -7{,}04 \times 10^{-12} p + 120 \times 10^{-24} p^2, \text{ pressure in dynes/cm}^2.$$

The first degree term in this formula was taken directly from the work of Saerens (9). His pressure range was not sufficient to give the second degree term; this I have estimated by a sort of extrapolation of the data of Slater.

Most of the transition parameters vary regularly through the series, except the slope, and the latent heat calculated from the slope by Clapeyron's equation. The experimental accuracy of the slope is, however, not high. It is to be noticed that the transition lines are so nearly vertical that the latent heat makes a comparatively small contribution to the energy difference between the two modifications, and the sign of the energy difference is controlled by the relative volumes.

Table I.

Values of Various Transition Parameters at 50° C.

Salt	Mean Pressure kg./cm².	ΔV cm³/gm.	$\frac{V_{1t}}{V_{10}}$	Densities			Percentage volume Decrease at Transition	$\frac{d\tau}{dp}$	Latent Heat kg.cm/gm.	$p\Delta V$	Energy Differen[ce] kg.cm/g[m]
				ϱ_{10}	ϱ_{1t}	ϱ_{2t}					
RbCl	5525	.0505	.9654	2,81	2,91	3,41	14,6	− 67	− 24	280	304
RbBr	4925	.0325	.9643	3,37	3,49	3,93	11,3	− 40	− 26	160	186
RbJ	4050	.0290	.9655	3,57	3,69	4,135	10,7	+ 25	+ 37	118	81

A very rough idea can be obtained of the relative thermal expansion of the high and low pressure modifications from the variation of ΔV with temperature along the transition lines. ΔV changes in the 100° range by 0,0042, 0,0022, and 0,0034 cm³./gm. for *RbCl*, *RbBr*, and *RbJ* respectively, or by 0,0122, 0,0077, and 0,0127 measured fractionally. If we assume that the difference of compressibility of the two modifications is of the order of 3×10^{-6}, then in the extreme case, less than 10 % of the variation of ΔV along the transition line can be accounted for by the pressure variations. We accordingly neglect the pressure effect, and assume a mean variation of ΔV of 0,01 for the 100° along the transition line. This variation of 0,01 means a larger thermal expansion for the low pressure modification than for the high pressure modification, and the numerical value of this difference is at once seen to be 0,0001. But this is of the order of magnitude of the known total thermal expansion of the low pressure modifications. Hence we may draw the conclusion that the thermal expansion of the high pressure modification is very low, and may even be negative.

Theoretical implications. The crystal structure at atmospheric pressure of the 20 alkali halides, built of the alkali elements, *Li*, *Na*, *K*, *Rb*, *Cs*, and the halogens *F*, *Cl*, *Br*, and *J* is known to be of one or the other of two simple types, with the exception of *RbF*, which offers stability difficulties such that its structure has not yet been determined (11 a). Of the remaining 19 salts, all except three crystallize in a simple cubic lattice, composed of two interpenetrating face centered lattices of the positive and negative ions of the alkali and the halogen. The three exceptions are *CsCl*, *CsBr*, and *CsJ*; the lattice structure of these is body centered cubic, composed of two interpenetrating simple cubic lattices of the ions of *Cs* and the halogen. The volume relations throughout the entire series of 19 salts can be reproduced by representing the ions by spheres, which in the various lattices are in contact, and which have a radius which is always the same for the same ion and does not depend on the ion with which it is associated. The *CsCl* type of lattice is one of the close packed arrangements of spheres; the mean density of a lattice of equal spheres placed in this arrangement is 30% greater than when the same spheres are placed in the *NaCl* type of arrangement.

Hund (12) has indicated a possibility of explaining the different lattice types by energy considerations. The energy of the *CsCl* type is less than that of the *NaCl* type when the repulsive forces fall off as a high inverse power of the distance, that is, when the force field about the ions approaches that of an ordinary solid with rigid boundary, whereas the *NaCl* type is the more stable when the repulsive force varies more gradually. This difference corresponds to the difference indicated by the compressibilities of the various salts, but the parallelism between the two properties is very rough, and the discrepancies are high when expressed in numerical terms.

When I began work on the transitions of the rubidium salts I had no question but that the modification found by Slater at high pressures was of the *CsCl* type. All the a priori probability pointed in this direction: such as the approximately spherical symmetry and the presumable simplicity of the ions in this series of salts, and the fact that the volume relations are such that pressure will drive the transition in the suspected direction. It became at once abvious, however, on determining the change of volume that it is highly questionable whether the transition can be of this type, for the observed volume decrease is only 12 % on the average, against 30 % expected. The improbability that the transition is of this type was further much increased on studying the energy relations between the two modifications. It appears that in order to account for the observed energy change at the transition a value for the exponent in the repulsive potential as low as 1,5 is required, when the calculations are made by the method of Born (13). The exponent required to give the observed compressibility is 9 or 10.

If the transition is of the expected type, it seems plausible to expect the transition in other salts of the series. In particular, it is most natural to expect that *CsF*, the only *Cs* salt of the *NaCl* type, could

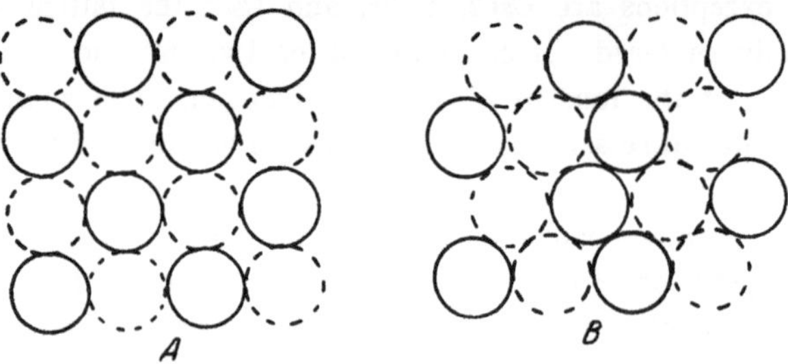

Fig. 10. Part A shows the normal *NaCl* type of arrangement viewed from the direction of one of the axes. Part B shows a possible rearrangement of the structure of part A which has approximately the volume ratio found experimentally for the rubidium salts.

be forced by pressure to assume the *CsCl* type of structure with its smaller volume. This transition does not take place, however, at least below 12000 kg./cm². Furthermore, if the ions are geometrically like simple spheres, one would certainly expect to find other transitions like those of the rubidium salts, whatever the type may be. Consider the two salts *KBr* and *RbCl*. The outer electron shells of the ions of which these salts are composed are supposed to have the configuration of the nearest inert gas. In *KBr* the *Br* is practically the same, except for the sign of its total charge, as the *Rb* of *RbCl*, and the *K* is practically, the same as the *Cl* of *RbCl*. It would be difficult to find two salts which, according to our present notions of atomic structure, should be more alike, and yet *RbCl* has a transition at 5500 kg./cm²., and *KBr* has none under 18000.

In speculating as to the probable crystal structure of the high pressure modifications, one may be guided by the volume relations. It does not appear that any of the commonly occurring simple structures have the correct volume ratio 1,12 to the volume of the *NaCl* type of structure, assuming the lattice to be built of spheres in contact. There is, however, a very simple modification of the *NaCl* structure which does give approximately the correct volume ratio. In figure 10 A the *NaCl* type of lattice is shown, looking down from above along the *Z* axis. Now displace every other plane parallel to the *X* and *Z* axes in the direction of *X* by a distance halt the distance apart of the unlike ions, and then push these planes together by a displacement along the *Y* axis until the spheres are again in contact, giving the arrangement of figure 10 B, again looking down along the *Z* axis. The volume of 10 B is less by 13,5 % than the volume of 10 A, which is very close to the observed difference of volume of the rubidium salts. Furthermore, a very rough calculation indicates that the energy relations of the two types of structure, making conventional assumptions about attractive and repulsive forces, are not widely different from those found experimentally. There are, however, difficulties in supposing that the actual structure is of the type of figure 10 B. No cases have ever been found, as far as I know, of this type of structure. Furthermore, the displacement which carries 10 A into 10 B treats the three axes unsymmetrically in a way inconsistent with the symmetry of the original structure, even if we accept the evidence of the etch figures that the symmetry of *NaCl* etc. is not holohedral.

The nature and the explanation of the transition of these rubidium salts therefore remains obscure. It is most disconcerting to find a situation of this sort in the series of salts which we had come to look upon as the simplest possible, and which we had felt we understood fairly well. We obviously can take little satisfaction in our explanation of various physical properties of these salts if we cannot foresee even the existence of two stable structures built from the same elements. The problem of understanding polymorphic transitions still remains most baffling, but nevertheless I feel that an important key to an understanding of the properties of solid matter in general will be found by further study in this field. It is perhaps not impossible that we shall be driven ultimately to recognize that in at least some cases of polymorphism the atoms have different electronic configurations in the different modifications.

I am much indebted to Mr. W. N. Tuttle for assistance in making most of the readings of this paper. It is also a pleasure to acknowledge financial assistance received from the Rumford Fund the Amerian Academy of Arts and Sciences.

References.

1. J. C. Slater, Proc. Amer. Acad. **61**, 135—150. 1926. Phys. Rev. **23**, 488. 1924.
2. P. W. Bridgman, Proc. Amer. Acad. **58**, 166—242. 1923.
3. P. W. Bridgman, Proc. Amer. Acad. **51**, 55—118. 1915; **52**, 91—187. 1916.
4. J. Perkins, Phil. Trans. Roy. Soc. Lon. 324. 1819—20; 541. 1926.
5. P. W. Bridgman, Proc. Amer. Acad. **44**, 255—280. 1909.
6. J. H. Hyde, Proc. Roy. Soc. London **97**, 240. 1920.
7. P. W. Bridgman, Proc. Amer. Acad. **61**, 89. 1926.
8. P. W. Bridgman, Proc. Amer. Acad. **52**, 57—88. 1916.
9. E. Saerens, Journ. chim. phys. **21**, 265. 1924.
10. P. W. Bridgman, Proc. Amer. Acad. **52**, 174. 1916.
11. P. W. Bridgman, Proc. Amer. Acad. **59**, 173—211. 1924.
11a. According to V. M. Goldschmidt, Geochemische Verteilungsgesetze VII and VIII, the structure of *RbF* is also of the *NaCl* type.
12. F. Hund, ZS. f. Phys. **34**, 833—857. 1926.
13. Max Born, Atomtheorie des festen Zustandes, B. G. Teubner, Leipzig 1923.

The Jefferson Physical Laboratory,
Harvard University, Cambridge, Mass.

Received January 30th 1928.

RESISTANCE AND THERMO-ELECTRIC PHENOMENA IN METAL CRYSTALS

By P. W. Bridgman

Jefferson Physical Laboratory, Harvard University

Communicated October 19, 1928

In a recent number of the *Physical Review* I discussed the significance of some of the more important qualitative features of the thermo-electric behavior of metal crystals, but had to leave unsettled the precise nature of some of the experimental facts until the completion of an experimental investigation then in progress. This investigation has now been completed, and the results are being published in full detail in *Proc. Amer. Acad.;* a brief summary of these results is the purpose of this note. In the extended paper will also be found a rediscussion of a couple of rather important theoretical matters, with regard to which I had previously reached erroneous conclusions because of my failure to take due account of the transverse thermal effect in crystals, a phenomenon at that time not known to me. A brief summary of these new conclusions is also given here.

An improved method of making single metal crystals has been developed by which it is possible to cast from the same melt a number of single crystal rods of a wide range of orientation. An apparatus has been constructed by which the thermal e.m.f. of 16 such rods may be measured simultaneously, so that the only variable factor in the results is the crystal orientation. The thermal e.m.f. between 20° and 88°C. of single crystal Zn, Cd, Sb, Sn and Bi against Cu has been measured, and from the results the thermal e.m.f., Peltier heat, and difference of Thomson heat between rods of the same metal in different orientations is calculated. The following results are found for thermal e.m.f.

Zn: $(\text{t.e.m.f.})_{\|-\perp} = 1.800 \times 10^{-6} \times (\tau - 293.1) + 4.27 \times 10^{-9} \times (\tau - 293.1)^2$

Cd: $(\text{t.e.m.f.})_{\|-\perp} = 2.973 \times 10^{-6} \times (\tau - 293.1) + 10.82 \times 10^{-9} \times (\tau - 293.1)^2$

Sb: $(\text{t.e.m.f.})_{\|-\perp} = 26.7 \times 10^{-6} \times (\tau - 293.1) + 11.1 \times 10^{-9} \times (\tau - 293.1)^2$

Sn: $(\text{t.e.m.f.})_{\|-45^\circ} = -0.259 \times 10^{-6} \times (\tau - 293.1) + 0.85 \times 10^{-9} \times (\tau - 293.1)^2$

$(\text{t.e.m.f.})_{\|-\perp} = -0.657 \times 10^{-6} \times (\tau - 293.1) + 1.82 \times 10^{-9} \times (\tau - 293.1)^2$

Bi: $(\text{t.e.m.f.})_{\|-45^\circ} = 21.6 \times 10^{-6} \times (\tau - 293.1) - 38.1 \times 10^{-9} \times (\tau - 293.1)^2$

$(\text{t.e.m.f.})_{\|-\perp} = 51.2 \times 10^{-6} \times (\tau - 293.1) - 75.3 \times 10^{-9} \times (\tau - 293.1)^2$

These formulas give the e.m.f. in volts in a thermocouple of two rods of the indicated orientations, one junction of which is maintained at 20°C. and the other junction of which is at the absolute temperature τ. A positive sign for $(\text{t.e.m.f.})_{\|-\perp}$ indicates that the positive current flows from the direction parallel to the crystal axis to the direction perpendicular to the axis at the hot junction.

It is emphasized that in the case of Sb and Bi the heat absorbed by an electron when its direction of motion changes from perpendicular to the axis to parallel to the axis is very large, being, respectively, 0.2 and 0.4 of the energy of a gas molecule at the same temperature. This is a difficult point for any theory like the recent one of Sommerfeld, in which that part of the energy of a conduction electron which varies with temperature is supposed small compared with the classical amount.

The Kelvin-Voigt law that thermal e.m.f. is a linear function of $\cos^2\theta$, θ being the angle between crystal axis and the length of the rod, is verified for Zn, Cd and Sb, but there are deviations for Sn and Bi which seem distinctly greater than possible experimental error. It was this that made it necessary to give separate formulas above for the 45° and perpendicular directions of these two metals. It would appear in general, therefore, that the Kelvin-Voigt symmetry law for thermal e.m.f. is only an approximation. This is perhaps the most important result of the new experimental work. I had previously come to the same conclusion, but Linder[2] had questioned whether this conclusion was justified by the experimental accuracy, and thought that his own data indicated the opposite conclusion over my temperature range, although over a wider range he also found departures. I felt that both the work of Linder and myself was capable of considerable improvement experimentally, whence this new investigation. The new results bear out the conclusion of Linder with respect to zinc in the temperature range below 100°. The new data for Sn and Bi are, however, very much more complete than the former data for these metals, so that I believe that my original conclusion with regard to the approximate character of the Kelvin-Voigt symmetry law must stand. It is shown in the detailed paper that the failure of the symmetry relation is due to the Peltier heat, as distinguished from the Thomson heat.

In the case of Bi a special examination was made of the Kelvin-Voigt assumption that thermal e.m.f. has rotational symmetry about the crystal axis, and no appreciable deviation was found.

The specific resistance of these metals as a function of orientation was subjected to a more complete examination than has been made hitherto. The possible error arising from distortion of the crystal rod is so great in the case of Cd that a special method of measurement had to be used. The Kelvin-Voigt symmetry relation for resistance is satisfied within experimental error. Especial attention was given to the resistance of Bi, and my previous low value perpendicular to the cleavage plane verified. The pressure coefficient of resistance of Sb and Bi was redetermined, so that now this quantity is known over the entire range of orientation.

In the theoretical discussion is shown that the third law of thermodynamics gives considerable plausibility, although not complete certainty, to the thesis that the symmetry of the Peltier and the Thomson heat must be the same; this would mean that the apparent experimental difference in the symmetry of the two heats is merely an effect of the much greater error in the Thomson heat. It is shown that, contrary to statements previously made by me, Kelvin's axiom of the superposition of the thermal effects of currents is not internally inconsistent, so that his proof of the symmetry relations and of the existence of a transverse temperature effect* are logically defensible. Experiment, however, seems opposed to the truth of the axiom. The revised discussion shows further that the local surface heat where a current leaves a metal crystal is a function both of the direction of flow with respect to the crystal axis within the crystal and of the orientation of the surface, contrary to my former statement. It is also shown that these crystal phenomena no longer offer a basis for the proof that the electrons must move in the crystal along something analogous to fixed channels.

* In *Proc. Nat. Acad.* for Jan., 1927, I announced the experimental detection of the transverse temperature effect, as I thought, for the first time. Recently, however, Dr. Borelius has called to my attention that this effect was experimentally established in a paper by Borelius and Lindh published during the war (*Ann. Phys.*, **53**, 97, 1917). The effect has also been independently announced by Terada and Tsutsui in Japan (*Proc. Imp. Acad.*, **3**, 132, 1927).

[1] P. W. Bridgman, *Phys. Rev.*, **31**, 221, 1928.

[2] E. S. Linder, *Phys. Rev.*, **29**, 554, 1927.

THE COMPRESSIBILITY AND PRESSURE COEFFICIENT OF RESISTANCE OF ZIRCONIUM AND HAFNIUM.

By P. W. Bridgman.

Presented Oct. 10, 1928. Received Oct. 15, 1928.

The data of this paper were obtained in continuation of my program to extend the measurement of compressibility and resistance under pressure to all the pure elements available. I have previously published results[1] for the effect of pressure on the resistance of a very impure specimen of zirconium, made by depositing a covering of zirconium on a core of tungsten. The impurity was so great that the temperature coefficient of resistance was only 0.00004, and the specific resistance was 200×10^{-6}. The pressure coefficient of resistance, in kilogram units, of this impure zirconium, was -4.0×10^{-7}. So far as I know, no previous measurements have been made on the compressibility of zirconium, or on either the compressibility or resistance under pressure of hafnium.

For the material of this investigation I am greatly indebted to Dr. G. Holst, of the Naturkundig Laboratorium, Eindhoven, Netherlands, Philips Gloeilampen Fabrieken. In connection with an investigation of possible commercial uses, both these metals have been prepared in a state of high purity by the Philips Lamp Works.

Two pieces of zirconium were available. The resistance sample was a wire, about 0.050 cm. in diameter and 14 cm. long. The potentiometer method of measurement was used, which demands two current and two potential terminals. It is not possible to solder zirconium by any easy means, and therefore the terminals were attached by spring clips, in a way which has been used for a number of other metals. The compressibility was measured on a piece 2.7 cm. long, about 3 mm. in diameter. It was measured in the same apparatus as that used for many previous measurements,[2] which I have described as "the lever apparatus for short specimens."

Only a single piece of hafnium was available, about 2.6 cm. long and 2 mm. in diameter. The outside surface was not perfectly regular, the appearance being as if the piece were composed of several rather large crystal grains which were slightly staggered with respect to each other. This irregularity of figure is perhaps connected with the fact that the compressibility measurements were much more irregular than usual. Because of the very low resistance of this specimen (about 0.00112 ohms), the pressure coefficient could be determined only with much less accuracy than usual. I did not attempt to find the variation of pressure coefficient with temperature; neither was it possible to find any departure from linearity between pressure and resistance. It is no more possible to solder hafnium than zirconium, so again connections had to be made with springs. The potential terminals were placed only 1.2 cm. apart, to avoid as far as possible end effects. The current connections at the ends were specially constructed to touch the specimen as nearly as possible at all points around the circumference, so that the lines of current flow inside the specimen should be straight.

In addition to compressibility and pressure coefficient of resistance, the specific resistance, and temperature coefficient of resistance at atmospheric pressure were measured. The temperature coefficient was obtained from measurements at 30° and 75°; the values listed are the average coefficients between 0° and 100° (*i. e.*, $(R_{100} - R_0)/R_0$), obtained by drawing a straight line through the measurements at 30° and 75°.

The results follow. The pressure range was up to 12000 kg./cm.², as usual.

Zirconium.

Compressibility:

At 30°, $-\Delta V/V_0 = 10.97 \times 10^{-7}p - 7.44 \times 10^{-12}p^2$.

At 75°, $-\Delta V/V_0 = 11.06 \times 10^{-7}p - 7.80 \times 10^{-12}p^2$, pressure in kg./cm.²

At 30°, the average deviation of a single reading from a smooth curve was 0.6% of the maximum pressure effect, and at 75°, 0.7%. The decrease of compressibility with increasing pressure was well marked and far beyond experimental error.

Resistance:

$$\text{At } 30°,\ \Delta R/R\ (0\text{ kg}, 30°) = -4.31 \times 10^{-7}p + 6.5 \times 10^{-12}p^2,$$

$$\text{At } 75°,\ \Delta R/R\ (0\text{ kg}, 75°) = -6.01 \times 10^{-7}p + 5.8 \times 10^{-12}p^2.$$

The resistance decreases under pressure, which is the normal behavior. The coefficient is, however, unusually small, which accounts for the somewhat larger experimental error than usual. At 30° the average deviation of a single reading from a smooth curve was 1.4% of the maximum pressure effect, and at 75°, 1.6%. Expressed in terms of the total resistance, this means that at 30° the average deviation from a smooth curve was 0.006% of the total resistance, and at 75°, 0.009%. Not much better can be expected, considering the diameter of the specimen, and the fact that the contacts were made with springs.

The specific resistance at 30° was found to be 49.2×10^{-6}, and the mean temperature coefficient between 0° and 100° 0.00403. The latter figure is evidence of the high purity of this material. It is remarkable that the pressure coefficient of the sample originally measured should have been so close to that found above, in spite of the fact that its temperature coefficient and specific resistance showed very high impurity.

Hafnium.

Compressibility:

$$\text{At } 30°,\ -\Delta V/V_0 = 9.01 \times 10^{-7}p - 2.37 \times 10^{-12}p^2,$$

$$\text{At } 75°,\ -\Delta V/V_0 = 8.81 \times 10^{-7}p - 2.37 \times 10^{-12}p^2.$$

At 30° the average deviation of a single reading from a smooth curve was 1.7% of the maximum pressure effect, and at 75° 2.0%. The accuracy was not great enough to establish any deviation from linearity in the relative compressibility of hafnium and iron, which is the quantity directly measured. The second degree term in the formulas above comes from the compressibility of iron. It is probable that the second degree term is actually materially larger than given above. Neither is it certain that the compressibility at 75° should be less than that at 30°.

Resistance:

$$\text{At } 30°,\ \Delta R/R\ (0 \text{ kg}, 30°) = -10.0 \times 10^{-7}p.$$

The average deviation of a single reading from a smooth curve was 6.0% of the total pressure effect, which means 0.09% in terms of the total resistance. Again I believe that this irregularity is as small as could be reasonably expected in view of the dimensions of the specimen, and the method of connection by spring clips.

The specific resistance at 30° was found to be 35.7×10^{-6}, and the mean temperature coefficient between 0° and 100°, 0.00398. Again the high value of temperature coefficient is presumptive evidence of high purity.

Discussion. The compressibilities listed above are cubic compressibilities, whereas the measured quantities were linear compressibilities. In converting the one into the other, the assumption was made that the compressibility is equal in all directions. Now this assumption is doubtless not strictly justified, because neither zirconium nor hafnium crystallizes in the cubic system, but instead in the close-packed hexagonal arrangement. However, the assumption of equal compressibility in all directions is probably approximately justified for these two metals, because the axial ratios are such (1.644 for hafnium and 1.593 for zirconium) as to correspond nearly to close packed *spheres*, for which the ratio is 1.63. It must nevertheless be recognized that the assumption of equal compressibility in all directions in a hexagonally close packed arrangement of spheres has not yet been justified by experiment, and should receive such a test.

In general comment on the results we may note in the first place that the compressibilities fit well into the vacant places in the periodic table, as is shown, for example, in Figure 1 of my paper in Proceedings American Academy 62, 207–226, 1927. It is perhaps not to be expected that the compressibility of hafnium should be less than that of zirconium, since in the majority of cases the heavier elements in a given column of the periodic table are more compressible. However, this observation is confined mostly to elements of lower atomic number than hafnium, and it is evident from the figure that there is some reverse tendency at higher atomic numbers.

The pressure effect on resistance has the normal sign in both these elements, but it is rather smaller for each than usual. It is to be noted that the pressure coefficient of resistance of hafnium is greater numerically than that of zirconium, although its compressibility is less. This emphasizes the fact frequently mentioned before that the pressure effect on resistance is not directly concerned with the volume changes produced by pressure.

I am indebted to my assistant, Mr. Stephen Stark, for making the readings of this paper.

The Jefferson Physical Laboratory,
Harvard University, Cambridge, Mass.

References.

[1] P. W. Bridgman, Proc. Amer. Acad. 56, 112, 1921.
[2] P. W. Bridgman, Proc. Amer. Acad. 58, 175, 1923.

THE EFFECT OF PRESSURE ON THE RESISTANCE OF THREE SERIES OF ALLOYS.

By P. W. Bridgman.

Presented Oct. 10, 1928. Received Oct. 22, 1928.

CONTENTS.

Introduction.

There are many phenomena connected with the electrical behavior of alloys which are difficult to understand. For several years one of the chief reasons for my dissatisfaction with the theory of metallic conduction on which I have been working has been the anomalous behavior of a number of alloys; the behavior of pure metals is more satisfactorily accounted for. It is therefore important to extend our knowledge of the electrical properties of alloys as much as possible; the measurements on the three series of alloys presented here are a contribution in this direction.

Materials and Experimental Methods.

For the series of Fe-Co alloys I am much indebted to Mr. W. C. Ellis,[1] who prepared the alloys and investigated many of their electrical properties as part of a thesis done at Rensselaer Polytechnic Institute under the direction of Dr. M. A. Hunter. The value of Ellis's work is much increased by an X-ray determination which he made of the crystal structure of all his alloys. The series consisted of eight different alloys. Six of these were in the form of drawn wires. Two of them, which could not be drawn, were in the form of swaged rods about 6 mm. in diameter; slender pieces suitable for the measurements under pressure were prepared from these rods by grinding. A special feature of the material prepared by Ellis is the high purity of his cobalt, which was electrolytic. Its high purity is evidenced by the low specific resistance and high temperature coefficient, as given by Ellis in his paper. A remarkable feature of the series is the fact, emphasized by Ellis, that the specific resistance of the alloys from 32 to 53 atomic per cent Fe have a lower specific resistance than either of the pure constituents.* So far as I know, this is the first example of such behavior in alloys; in fact one can frequently find the generalization made in the literature that the resistance of an alloy is greater than that of either component. As Ellis remarks, this effect may be expected to be of significance in theories of metallic conduction. Certainly such behavior lends great interest to as complete a knowledge as possible of the properties of this series of alloys.

The series of Fe-Ni alloys I owe to the kindness of Dr. L. W. McKeehan, at that time with the Bell Telephone Laboratories, Inc. He had a series of 15 alloys prepared and drawn to wire. The following description of the alloys is quoted from Dr. McKeehan's letter. "These wires are all hard drawn. In order to check the approximate composition and to get some idea of the variation of resistivity with composition, I have had the resistivity measured in the state furnished, and also on short samples that have been annealed at 1000° in vacuum for one hour and allowed to cool slowly, *i. e.*, requiring several hours to cool to room temperature. The results are given on the attached photostat (reproduced in Figure 1). In regard to the exact composition, I have not figures for all the alloys, but on those that have been analyzed the actual percentage of iron or nickel falls within less than ½% of the nominal composition."

---

* I am not sure that this result can be regarded as established beyond all possibility of doubt. It is evident that the resistance of Co is extraordinarily sensitive to slight impurities; thus Kalmus[2] found for the specific resistance of pure Co reduced in hydrogen the value 8.96×10^{-6} at 20°, against 6.24 found by Ellis. The value of Ellis might possibly be materially reduced if there were a very slight amount of remaining impurity. The reduction need not be great in order to change Ellis's conclusion, because he finds for the resistance at 0° of his pure Co 5.60×10^{-6}, whereas the two alloys with lowest resistance, containing 33.3% and 50% Fe respectively, had resistances of 5.15 and 5.13. It is significant in this connection that Grüneisen[3] has indicated as the most probable value for the resistance of pure Co, on the basis of computations from the resistance at very low temperatures, made by means of Matthiesen's rule, on samples known to be impure, the figure 5×10^{-6}. Furthermore, it would appear that purer cobalt than that of Ellis has been prepared, for Holborn[4] gives for the temperature coefficient between 0° and 100° of pure Co 0.00658, against Ellis's 0.00604. Holborn does not give the specific resistance of his material.

In addition to the more or less complete results on these two series of alloys, I am taking this opportunity to publish some partial results obtained with my high pressure apparatus by Mr. J. R. Oppenheimer in 1925 on the Cu-Ni series. These results cover a range of composition up to about 50% Ni. The material was electrolytic copper from the Bureau of Standards, and commercially pure nickel

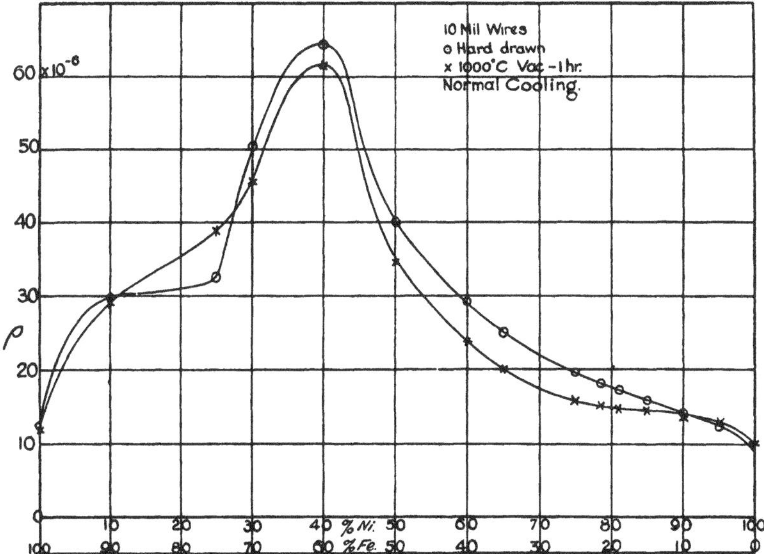

FIGURE 1. The specific resistance in the Fe-Ni series as a function of composition in the hard drawn condition, and after annealing.

of probably about 99% purity. The alloys were prepared by melting together in a vacuum in a quartz tube weighed amounts of the two constituents. Floating on the upper surface of the melted metal was a quartz capillary perhaps 8 cm. long and 0.5 mm. in diameter, closed at the upper end. When the melting was completed and the metals thoroughly mixed, air or nitrogen was suddenly admitted to the quartz tube, driving the molten metal up into the capillary, and thus producing a fine cast wire of the alloy. The wire so obtained was not subjected to any systematic heat treatment, but was used as it came from the tube, after cracking away the quartz. In general the capillary was always at a temperature lower than the melting point, so that the wire must have experienced a very rapid chilling from the melting temperature to several hundred degrees below.

The pressure measurements were made by the same method as that used in former work,[5] and no detailed description is necessary. The potentiometer method adapted to the measurement of small resistances was used. The description in the remainder of this paragraph applies to the Fe-Co and Fe-Ni series. The wire specimens were bent to the form of a hairpin, about 8 cm on either arm, and current and potential terminals were soft-soldered to either arm near the end. The changes of resistance listed in the following are measured changes; to convert to changes of specific resistance a correction equal to the linear compressibility must be applied. At a pressure of 10,000 kg. the change of resistance due to the change of dimensions is approximately 0.2%, or of the order of 1/10 of the measured change for the pure metals. The pressure range was 12000 kg./cm.^{2}; readings were made at 30° and 75°. At the lower temperature, pressure was transmitted with petroleum-ether to avoid distortion arising from the viscosity of kerosene under pressure. Before the pressure measurements, two seasoning applications of pressure were made, once to 2000 and once to 12000. The permanent changes of resistance produced by these preliminary applications were in almost all cases very small. This is evidence that no important change of internal composition was produced by pressure, such as an unmixing, and in fact inspection of the phase diagrams of these systems shows that no such effect is to be anticipated. In addition to the pressure coefficient of resistance, the mean temperature coefficient between 0° and 100° was obtained by linear extrapolation of the resistances at atmospheric pressure at 30° and 75°. Since the range 30°–75° is nearly in the middle of the range 0°–100°, linear extrapolation gives approximately the correct mean coefficient, although such an extrapolation would not in general be reliable because the relation between resistance and temperature is not linear. The dimensions of the specimens were also measured, and the specific resistance found. The chief error here is in the diameter of the wire, and should not be more than 1% or 2%. The diameter was measured with an ordinary micrometer, and varied from 0.025 cm. to 0.075 cm.

The measurements of the Cu-Ni series were less elaborate, and consisted of measurements by the potentiometer method of resistance as a function of pressure up to 12000 kg at 25°.

Experimental Results.

The numerical results of the measurements are given in the following tables. In all cases except one, the relation between pressure and resistance could be reproduced by a second degree formula in the pressure; the two constants of this formula are given in the tables. The one exception is the alloy 70% Fe, 30% Ni; for this the detailed dependence of resistance on pressure is shown in Table IV and in Table II only the mean coefficient to 12000 is given. The tables also contain the specific resistance at 30°, the mean temperature coefficient between 0° and 100°, and the mean deviation of a single pressure reading from a smooth curve in terms of the maximum pressure effect. This last gives an idea of the accuracy of the pressure measurements; of course the percentage deviation is much larger for those alloys with a small pressure coefficient.

Ellis's pure Co is materially purer than most of the Co whose electrical properties are listed in the literature, and much purer than the two samples for which I have previously measured the pressure coefficient.[6] The values given in the table should, therefore, supercede my previous values. Although the temperature coefficient of the new sample is much higher than that of my best previous sample, 0.0060 against 0.0044, the pressure coefficient is not materially different, the average coefficient between 0 and 12000 kg. of the new material being -0.950×10^{-6} against -0.934×10^{-6} found before. The difference is in the direction usually found to go with increasing purity.

The pressure effect on the resistance of Ellis's pure Fe was not measured, since I had previously measured[7] the effect on Fe of approximately the same grade of purity as that of Ellis. Ellis gives the impurity as 0.035%, and my American Ingot iron was stated to have about 0.03% impurity. Ellis's temperature coefficient is somewhat higher than mine, 0.00635 against 0.00621, so that on this ground Ellis's iron was probably somewhat purer than mine, but the difference is probably not sufficient to introduce an appreciable difference in the pressure coefficient. Neither the temperature coefficient of Ellis or myself is as high as the highest listed for iron, namely 0.00657 by Holborn.[8]

The values given in the table for the Fe-Ni series for pure Fe

TABLE I.

RESULTS FOR FE-CO SERIES.

Composition At. % Fe	Composition At. % Co	Effect of Pressure on Resistance $\frac{\Delta R}{R_0} = ap + bp^2$ 30° a $\times 10^6$	30° b $\times 10^{11}$	30° Deviation %	75° a $\times 10^6$	75° b $\times 10^{11}$	75° Deviation %	Specific* Resistance at 30° $\times 10^6$	Mean* Temperature Coefficient 0–100°
0	100	− .958	+ 0.68	1.1	− .960	+1.04	2.7	6.62	.00604
10	90	− .401	+ 0.69	1.7	− .305	+ .81	2.2	8.29	534
20	80	+4.32	+14.	.5	+6.29	+8.0	.6	9.91	563
33.3	66.7	− .796	+ 0.6	.7	− .643	0	.7	5.84	446
50	50	− .473	+ 2.65	10.2	− .467	+1.63	4.0	5.78	425
66.7	33.3	+1.153	+ 3.10	.1	+ .898	+4.61	1.4	10.37	241
80	20	+1.408	+ .73	.3	+1.346	+ .39	.6	19.34	218
90.9	9.1	−1.413	+ .62	.3	−1.360	+ .49	.5	18.12	275
100.	0	−2.427	+ 1.14		−2.450	+1.00		10.52	635

* Values of Ellis.

TABLE II.

RESULTS FOR FE-NI SERIES.

Composition Wt. % Fe	Composition Wt. % Ni	Effect of Pressure on Resistance $\frac{\Delta R}{R_0} = ap + bp^2$ 30° a × 10^6	30° b × 10^{11}	30° Deviation %	75° a × 10^6	75° b × 10^{11}	75° Deviation %	Specific Resistance at 30° × 10^6	Mean Temperature Coefficient 0–100°
100	0	−2.427	+1.14		−2.450	+1.00		10.5	00621
90	10	+0.036	−8.0	5.1	−0.091	−3.00	2.9	30.1	231
75	25	+2.822	− .92	.2	+2.315	− .92	1.9	32.4	262
70	30	+8.7	(See Table IV)	.1	7.7	(See Table IV)	.1	51.4	589
60	40	+9.100	−6.06	.1	+6.910	−7.08	.0	65.6	287
50	50	+5.904	+5.62	.2	+5.422	+3.68	.3	41.9	443
40	60	+3.002	+2.29	.9	+2.886	+1.70	....	30.7	474
35	65	+1.930	+1.09	.2	+1.776	+1.39	.4	25.7	473
25	75	+0.456		2.0	+0.394		3.1	18.9	628
21.5	78.5	−0.129		6.2	−0.181		12.2	18.2	462
19	81	−0.389		3.3	−0.461		4.2	17.1	446
0	100	−1.905	+0.50		−1.925	+0.56		(7.0)	634

TABLE III.

RESULTS FOR CU-NI SERIES.

Composition Wt. %		Effect of Pressure on Resistance at 25° $\frac{\Delta R}{R_0} = ap + bp^2$	
Ni	Cu	a	b
0	100	-2.03×10^{-6}	$+0.96 \times 10^{-11}$
0.145	99.855	−1.80	+0.71
1.50	98.50	−1.19	+0.38
4.58	95.42	−0.94	+0.21
8.23	91.77	−0.745	+0.10
12.9	87.1	−0.725	+0.10
24.3	75.7	−0.700	+0.12
50.8	49.2	−0.685	+0.10

TABLE IV.

RESISTANCE OF ALLOY 70% FE–30% NI AS FUNCTION OF PRESSURE.

Pressure kg/cm²	30° $\frac{\Delta R}{R(0.30°)}$	75° $\frac{\Delta R}{R(0.75°)}$
3000	.02478	.02114
6000	.05078	.04279
9000	.07751	.06473
12000	.10477	.08696

and pure Ni were taken from previous work.[7,9] It is probable that the end members of this series as supplied by Professor McKeehan had appreciable amounts of impurity. I measured the pressure and temperature coefficients of the nominally pure iron of this series, and found values considerably lower numerically than my best previous values for pure Fe: 0.00485 against 0.00621 for the temperature coefficient, and -2.09×10^{-6} against -2.43×10^{-6} for the pressure coefficient. It therefore seemed safer to use my previous values for both pure Fe and Ni. Slight impurities in the constitents produce by far the largest effects at the end of the series; there is no reason to expect any important error from impurities in the intermediate members of the series, between 10% and 90%, as given in the table.

Discussion of Results.

It is important to have before us the crystal structure of the various alloys. The data for the Fe-Ni and the Cu-Ni series will be found in International Critical Tables, Vol. I, p. 350–351, and for the series Fe-Co in the paper already quoted of Ellis.

In the Fe-Ni series, the two end terms have different structures, pure Fe being body centered cubic, and pure Ni face centered cubic (both at room temperature). The body centered structure of pure Fe persists up to between 25 and 30 At % Ni. In this range the effect of the addition of Ni is to increase slightly the side of the fundamental cube, that is to decrease the density slightly. The experimental error here is great; roughly the effect is linear with the amount of Ni, 30 At % Ni increasing the grating space from 2.84 to 2.88 Å, or 1.4%. Beyond 25 or 30% Ni the structure becomes face centered cubic, the same as that of pure Ni. The grating space decreases with increasing amount of Ni, roughly linearly from 3.61 at 25% to 3.53 Å at 100% Ni. In each case the effect of the addition of a slight amount of foreign metal is to distend the crystal structure, which is the opposite of the change produced by hydrostatic pressure. In the case of many dilute solutions of ordinary substances the effect of the addition of a small amount of solute is to produce the same change in physical properties as an extra hydrostatic pressure, that is, the effect of the dissolved substance may be likened to that of an increase of internal pressure. In the case of these Fe-Ni alloys, however, the effect is the exact opposite.

In the Fe-Co series the structure is body centered cubic from pure Fe up to about 70 A% Co. The ultimate effect of the addition of Co is to reduce the grating space, the value of which is 2.850 Å for pure Fe and 2.827 for 66.7% Co. The change of dimensions is not linear with composition, however, but up to about 33.3% Co the dimensions are independent of composition within an error of ± 0.001. The alloys with 80 and 90 A% Co are face centered cubic, with crystal parameters of 3.550 Å and 3.540 respectively. Finally, pure Co at room temperature is hexagonal close packed. As in the case of the Fe-Ni series, the effect of the addition of a small amount of foreign metal to pure Fe is not to produce the same change of dimensions as an external pressure.

The Cu-Ni alloys form an unbroken series of solid solutions, all of face centered cubic structure, the crystal unit varying from 3.605 for pure Cu to 3.527 Å for pure Ni.

The series of alloys measured here have in common the feature that at all compositions the crystal structure is that of a mixed crystal in which different kinds of atoms occur together, more or less haphazard, in the same crystal lattice. This is opposed to the structure of many binary alloys, in which over a range of composition there is a mechanical mixture in various proportions of mixed crystals of different fixed compositions, or of the pure components. In the range of mechanical mixture, the physical properties of such alloys are linear functions of the composition, the relation being that of the ordinary rule of mixtures. The alloys investigated by Ufford[10] had this structure throughout certain ranges, and within these ranges the pressure coefficient and other physical properties were linear functions of composition. Opposed to this, there is no range in which the pressure coefficient of these three series of alloys is a linear function.

The contents of the several tables are reproduced in Figures 2, 3, and 4 in order that the relations may be more easily grasped. The first and perhaps the most important result brought out by the figures is support of a generalization already made by Ufford,[10] namely that in all known cases the effect of adding a small amount of foreign metal to a pure metal is to increase algebraically its pressure coefficient of resistance. This is shown in the figures by the rise of the curve for pressure coefficient at either end on leaving the pure metal. This generalization is without known exception, and applies not only to these three alloys, but also to the three of very different type investigated by Ufford, and also to several series investigated by Lisell[11] and Beckman.[12] These new results are sufficient to show, however, that the explanation given by Ufford of this general effect cannot always be correct. Ufford pointed out that in the case of his alloys the addition of a small amount of foreign metal compresses the lattice; this is also the effect of hydrostatic pressure. This means, since in practically all cases hydrostatic pressure increases the pressure coefficient algebraically, that the effect of adding a foreign metal is equivalent, in the cases investigated by him, to an increase of external pressure. That this is not true in general is shown by the Fe-Ni series. Here the addition either of Ni to pure Fe or of Fe to pure Ni distends the structure, thus producing the same change of dimensions as that brought

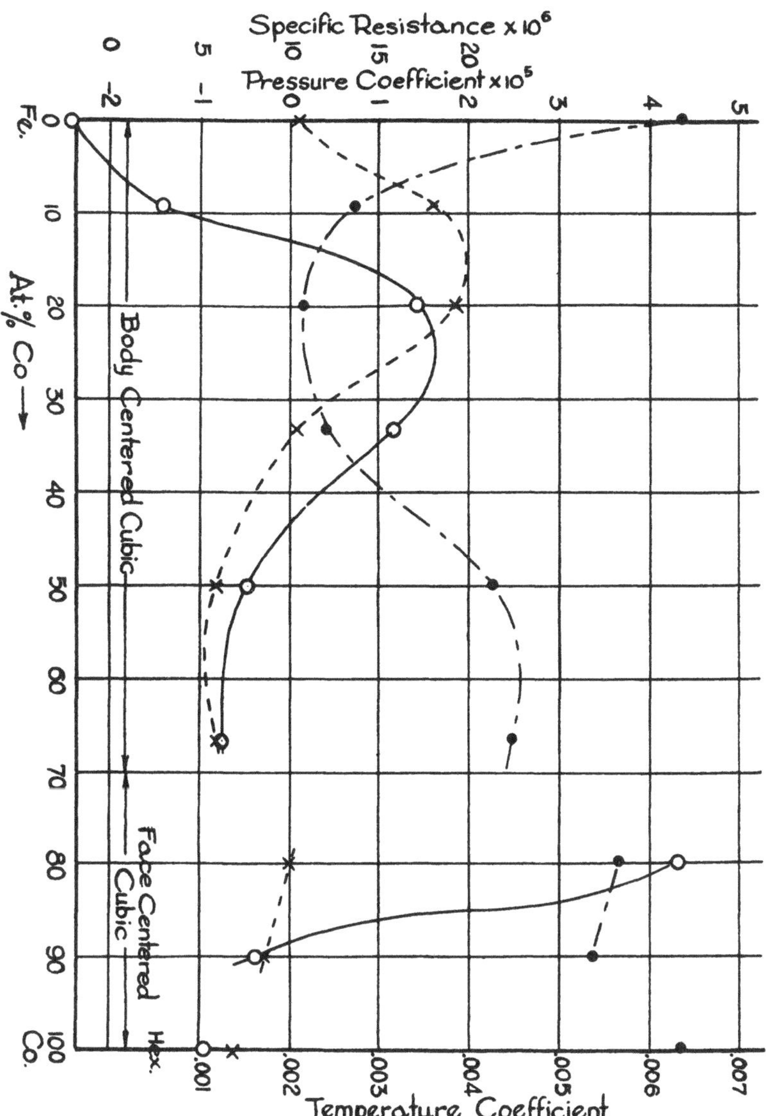

FIGURE 2. Specific resistance at 20° C (dotted line), pressure coefficient of resistance (full line), and temperature coefficient of resistance (dashed and dotted line), in the Fe-Co series as a function of composition.

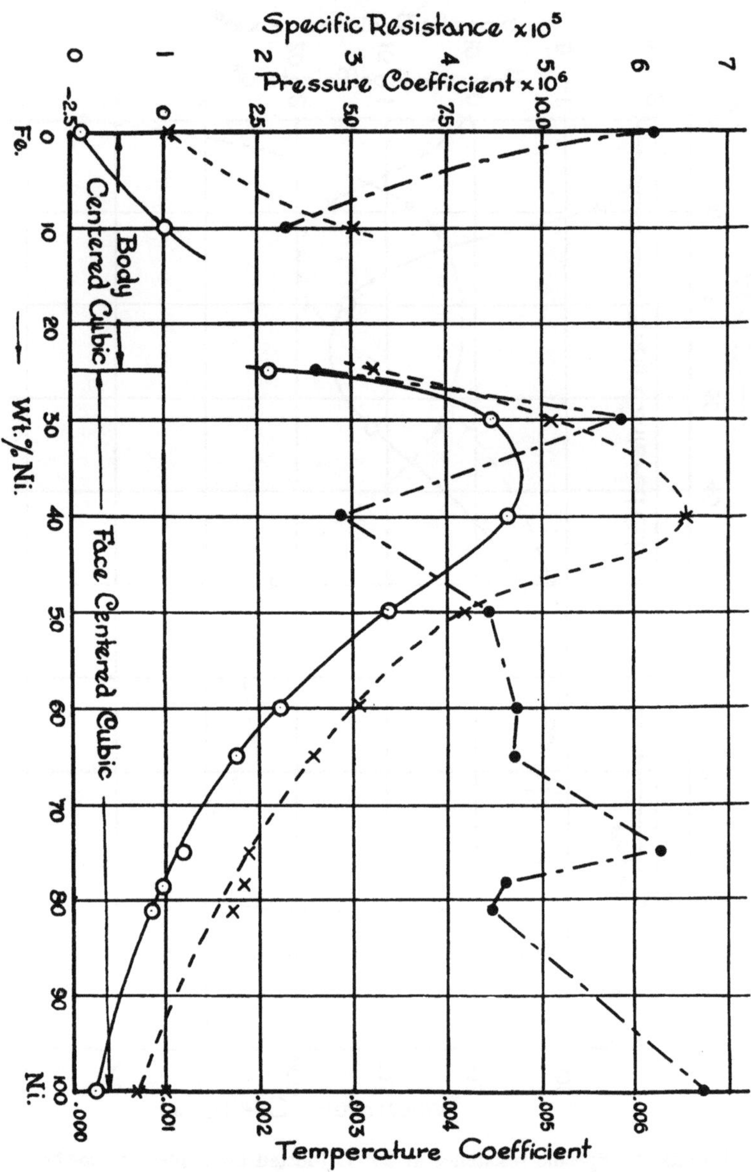

Figure 3. Specific resistance at 20° C (dotted line), pressure coefficient of resistance (full line), and temperature coefficient of resistance (dashed and dotted line), in the Fe-Ni series as a function of composition.

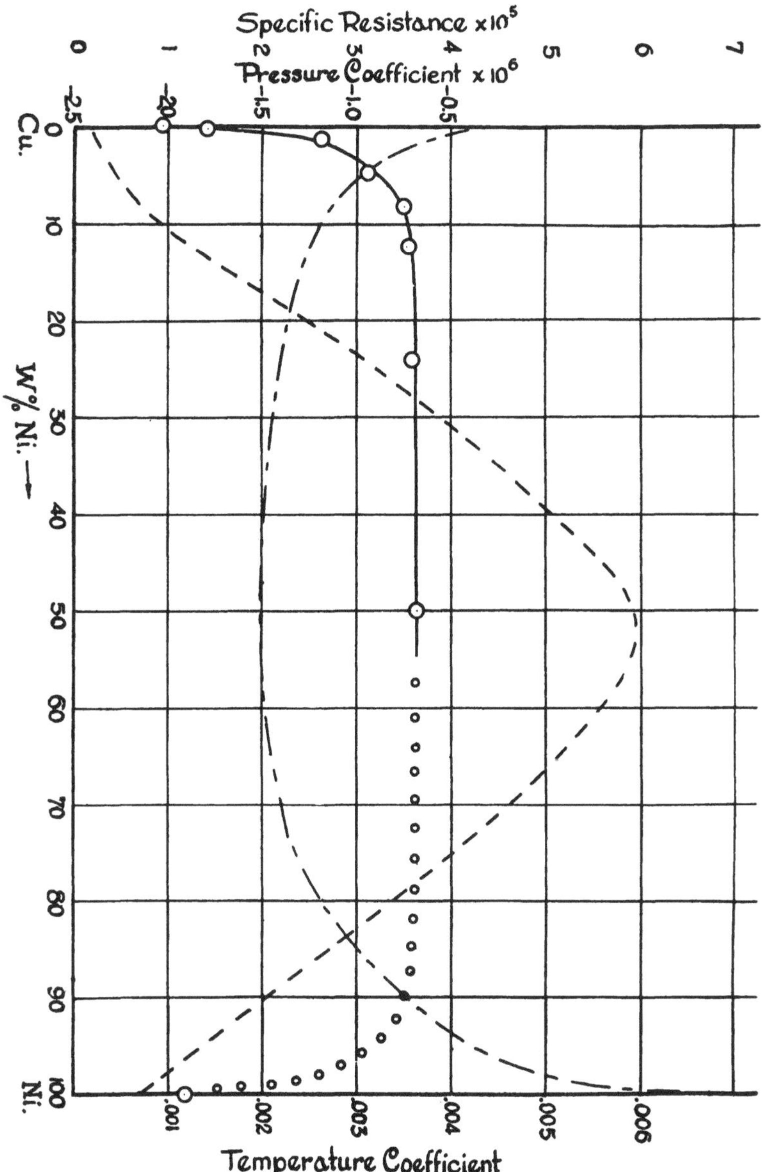

FIGURE 4. Specific resistance at 20° C (dotted line), pressure coefficient of resistance (full line), and temperature coefficient of resistance (dashed and dotted line), in the Cu-Ni series as a function of composition.. A hypothetical extension of the curve for pressure coefficient beyond the range of experiment is shown by the small circles.

about by a decrease of external pressure, whereas the change of pressure coefficient is in the direction brought about by an increase of pressure. The same is also true with regard to the addition of Co to Fe, although the numerical discrepancy is not so great, and is probably true, although this is not entirely certain, for the addition of Cu to Ni.

An inspection of the figures shows much greater regularity in the specific resistance and pressure coefficient of resistance than in the temperature coefficient. For example, in the face centered range of the Fe-Ni series, the behavior of the temperature coefficient is highly irregular, but that of specific resistance and pressure coefficient is comparatively regular, both rising from the ends of the range to a maximum at some intermediate composition. This is also the behavior in the body centered range of the Fe-Co series; here the maximum of resistance is more displaced with respect to the maximum of the pressure coefficient than in the Fe-Ni series. The same behavior is shown qualitatively by the Ni-Cu series; the range of variation of specific resistance with composition is here, however, much larger compared to the variation of pressure coefficient than in either the Fe-Ni or Fe-Co series. It would be interesting to know whether the same parallelism holds in other parts of the range, but there are only two points in the body centered range of Fe-Ni, and also only two points in the face centered range of Fe-Co, and two points are not enough to answer the question.

Although there is undoubtedly a parallelism between specific resistance and pressure coefficient, the example of Cu-Ni shows that the parallelism cannot be so close that one can be said to be a function of the other. Between 10 and 50% Ni the specific resistance of Cu-Ni changes six fold, whereas the pressure coefficient changes hardly at all.

An interesting question that cannot be answered at present is as to the discontinuity of resistance and pressure coefficient when the lattice type changes. If the relations previously found for all known cases of melting and polymorphic change hold here,[13] one would expect the resistance to increase, when the lattice type changes, in the same direction in which the volume increases. To answer this, many more alloys must be investigated in the range between 10 and 25% Ni for the Fe-Ni series, and between 66.7 and 80% Co in the Fe-Co series, than were available for the present work.

One of the most interesting results which I hoped to get from these measurements was some suggestion as to the mechanism which is responsible for a positive pressure coefficient. The conclusion seems to be justified that in these three series of alloys the same thing that tends to cause a high specific resistance also tends to make the pressure coefficient positive. According to the view of conductivity in which I have been interested for some time[14] this would be interpreted in terms of a failure of perfect fit of adjacent atoms. If the atoms do not fit closely together, as they do not in a lattice composed of different sorts of atoms of different sizes, the electrons encounter difficulty in passing from atom to atom and the resistance is high. This is the explanation of the high resistance in the middle of a series of mixed crystals. Now if pressure tends to accentuate the lack of fit, it is to be expected that the pressure coefficient of resistance will be positive. It is natural to expect this to be the case. At pressures of the order of magnitude of those with which we are here dealing, part of the compressibility of a solid comes from the closing of the empty spaces between the atoms, the compressibility of the atoms themselves contributing only partly to the total effect. Hence as pressure increases the relative part of the total volume occupied by the atoms themselves increases, which means that the relative difference between the diameters of the different sorts of atoms is accentuated. But this is merely another way of saying that the lack of fit is accentuated, so that the result to be expected is an increase of resistance, which is the effect actually found.

The general features of this somewhat gross and materialistic point of view can be taken over into the recent picture of conduction on the basis of the wave mechanics given by Houston.[15] According to this picture electrical resistance arises from the scattering by the atoms of the waves which constitute the electrons. If the atoms are spaced with perfect regularity, as in a crystal at 0° Abs., there is little or no scattering, and the resistance is very low. When the arrangement becomes more irregular, as it does when temperature is increased, the scattering increases and resistance increases. If there is a permanent additional cause of irregularity, as when atoms of different sizes are forced to occupy the same lattice in an alloy, there is an additional reason for scattering, which explains the increase of resistance when a foreign metal is added to either pure component. If the irregularity in the space arrangement is accentuated by pressure, as we have just seen is to be expected on geometrical grounds, the scattering is also increased, and the pressure coefficient of resistance tends to become positive.

The reason for the positive pressure coefficient of alloys given in the last paragraph is different from the reason given by Houston for the positive pressure coefficient of pure metals. Houston connects the positive pressure coefficient of pure metals with an abnormally small decrease of compressibility with pressure, or even possibly an abnormal increase of compressibility. It would seem that such a mechanism is not necessary to account for the positive pressure coefficient of alloys. In order to check this point, I have measured by the conventional method, which has now been applied to many substances,[16] the compressibility as a function of pressure of invar, a Fe-Ni alloy of 37.5% Ni content, corresponding approximately to the alloy of maximum resistance and pressure coefficient found above. The following was found for the compressibility at 30° between atmospheric pressure and 12000 kg./cm.^{2}:

$$-\frac{\Delta V}{V^0} = 9.99 \times 10^{-7}p - 6.18 \times 10^{-12}p^2.$$

The second degree term has the normal sign, and numerically is larger, rather than smaller, than that characteristic of pure metals of about the same compressibility. The compressibility itself is unexpectedly large, being 75% greater than the value calculated by the rule of mixtures from the pure components.

It would seem, then, that Houston's explanation of the positive pressure coefficient of pure metals need have no application to alloys. Here we have an adequate picture of the situation in a lack of fit in the crystal lattice, which is accentuated by pressure. The facts disclosed by the compressibility measurement of invar, namely a high compressibility combined with an abnormally rapid decrease of compressibility with pressure, are in complete agreement with this picture.

Houston's explanation of the positive pressure coefficient of pure metals must be taken, I believe, with considerable reservation. His explanation finds a correlation between the positive coefficient and an abnormally small decrease of compressibility with pressure; if examination is made of the behavior of a number of metals, instead of the one or two which Houston considered, it will be found that there is in general no such correlation. It may well be that under the proper conditions we have the same mechanism in pure metals as in alloys, namely a lack of fit accentuated by pressure. This would especially be expected if the atom has not spherical symmetry. It is perhaps the present tendency, suggested by the wave mechanics, to emphasize the spherical symmetry of the atom, but that the atom cannot always be spherically symmetrical is shown by the mere existence of non-cubic crystals.

I am indebted to my assistant, Mr. Stephen Stark, for making most of the readings in the Fe-Ni and Fe-Co series. I am also much indebted to Mr. Oppenheimer for permission to publish the results which he obtained.

THE JEFFERSON PHYSICAL LABORATORY,
Harvard University, Cambridge, Mass.

REFERENCES.

[1] W. C. Ellis, Paper presented at the Pittsfield, Mass., meeting on May 25–28, 1927, of A. I. E. E.

[2] H. T. Kalmus, Canadian Government publication, "The Physical Properties of the Metal Cobalt."

[3] E. Grüneisen, Handbuch der Physik, Vol. XIII, page 10.

[4] L. Holborn, Ann. d. Phys. 59, 145, 1919.
ZS. f. Phys. 8, 85, 1921.

[5] P. W. Bridgman, Proc. Amer. Acad. 52, 573, 1917; 56, 61, 1921.

[6] P. W. Bridgman, Proc. Amer. Acad. 58, 152, 1923.

[7] P. W. Bridgman, first reference under 5, page 609.

[8] L. Holborn, reference 4.

[9] P. W. Bridgman, reference 6, page 155.

[10] C. W. Ufford, Phys. Rev. 32, 505, 1928.
Proc. Amer. Acad. 63, 307, 1928.

[11] Erik Lisell, Om Tryckets Inflytande på det Elektriska Ledningsmotståndet hos Metaller, samt en Ny Metod att Mäta Höga Tryck, Upsala Universitets Årsskrift, 1903.

[12] Bengt Beckman, Arkiv för Matematik, Astronomi och Fysik, 7, No. 42, 1912.

[13] P. W. Bridgman, Rapports et Discussions du Quatriéme Conseil de Physique de l'Institut International de Physique Solvay, Paris, Gauthier-Villars, 1927, page 78.

[14] P. W. Bridgman, Phys. Rev. 9, 269, 1917; 17, 161, 1921.

[15] W. V. Houston, ZS. f. Phys. 48, 449, 1928.

[16] P. W. Bridgman, Proc. Amer. Acad. 58, 166, 1923; 60, 305, 1925.

THERMO-ELECTRIC PHENOMENA AND ELECTRICAL RESISTANCE IN SINGLE METAL CRYSTALS.

P. W. Bridgman.

Presented Oct. 10, 1928. Received Oct. 22, 1928.

CONTENTS.

Introduction.

In 1926 I published a paper[1] containing, in addition to some other matters, data for the thermal e.m.f. of single crystals of Zn, Bi, Cd, Sn, Sb, and Te as a function of the crystal orientation. The results thus obtained were not as complete or as accurate as could have been desired, the number of specimens being comparatively small, and in several cases the range of orientation being narrow. For example, the orientation of the specimens of Bi varied through only 22° instead of through the full 90°. A number of significant conclusions could be drawn from the data of that paper, but there were some important questions, as, for example, the precise way in which thermo-electric symmetry departs from Kelvin's relation (if indeed it departs at all), which could not receive a certain answer from the measurements. Furthermore, Linder[2] has also published thermo-electric data for single crystals, from which he draws conclusions not always the same as mine, so that a reëxamination was desirable.

In this paper improved results are obtained. In the first place, the method of producing single crystals has been much improved, so that it is now possible to cover the entire range of orientation, and in the second place, the apparatus by which thermo-electric force is measured has been improved so that now measurements can be made simultaneously with 16 different specimens, practically the only variable being the crystal orientation. This paper contains these new data, and a reëxamination, in their light, of the question of the symmetry relations. I have also remeasured, since the new specimens were better adapted to this purpose than the old ones, and since results disagreeing with my early ones have been since published by Schneider,[3] the specific resistance of most of these metals as a function of direction. I have further taken advantage of the new material to measure the effect of pressure on the resistance of Sb and Bi over the entire range of orientation.

The Method of Producing Single Crystals.

The new method is a modification of that used before;[4] this consisted in lowering a mold containing the molten metal slowly through the bottom of a furnace maintained at a constant temperature above the melting point. If the rate of lowering is enough less than the rate of crystallization, the liquid solidifies in a continuation of the crystal lattice already laid down, and if solidification can be forced to start as a single crystal grain, as by making the lower end of the mold taper to a sharp point, the mold becomes filled, after solidification, with a single crystal grain. In the previous work the largest furnace used had an interior diameter of about 2.5 cm., so that this was the greatest diameter of any crystal that could be made. The control of the orientation of the crystal was the difficult feature of the previous work. The crystals show a striking inclination to grow with the planes of easiest cleavage parallel to the vertical axis of the furnace; there is, however, considerable haphazard variation from this preferred direction, and by making a large number of castings, it was possible to cover a considerable range of orientation. The range depends to a marked degree on the nature of the metal. The direction of growth was much less strongly preferential in rods of small diameter than in the larger rods.

The new method depends on the use of a furnace of 10 cm. internal diameter, against the former 2.5. This makes it possible to use a mold consisting of criss-cross arms of glass tubing (a photograph of such a mold is shown in Figure 1), arranged at a variety of angles. If the rate of lowering is slow enough, the mold, as before, becomes filled with a single crystal grain. By breaking the mold at the corners of the arms, one has effectively a number of short crystal rods of a variety of orientations out of a single crystal. Of course this method does not give control of the orientation of any individual rod, but it does ensure that the orientation of the aggregate of rods is spaced over the entire range, which is the essential thing. The molds may be made of any desired degree of complication; my assistant, Mr. Philip Dalton, developed much skill in making them, and produced molds with as many as 50 arms. I myself was usually satisfied with more modest molds of 12 arms that could be made simply from a single length of tubing, bent first into a hairpin at the middle, and then the two arms of the hairpin bent to the desired angles.

Various new devices of manipulation were used. One of the most important matters is the deposition of some kind of a coating on the inside of the glass, to prevent the metal from sticking to the glass, and to allow the easy removal of the glass by cracking. If the mold is not coated in some way, the sticking is so bad as to make quite impossible the removal of the glass without mechanical deformation of the metal inside; this is particularly serious when the diameter of the castings is small, as here. Formerly, the tube had been washed with a dilute solution of Nujol in petroleum ether, and then dried by heating in vacuum; this left enough deposit to accomplish the purpose. It was, however, difficult to control the operations exactly, and failures were numerous, either because the deposit had been entirely oxidized away, or because so thick a deposit was left that gas bubbles were generated in sufficient number to break the continuity of the castings. The new material used for coating the molds was an enamel made by the General Electric Co. for insulating copper wire. This was made very dilute with $CHCl_3$, the interior of the mold was washed with it, allowed to drain in a vertical position for several hours, and then the mold was heated progressively with a Bunsen burner, first driving off the volatile matter, and then carbonizing the residue, which was left on the inner walls as a black deposit of carbon. The thickness of this deposit was now appropriately reduced by heating the entire mold in an electric

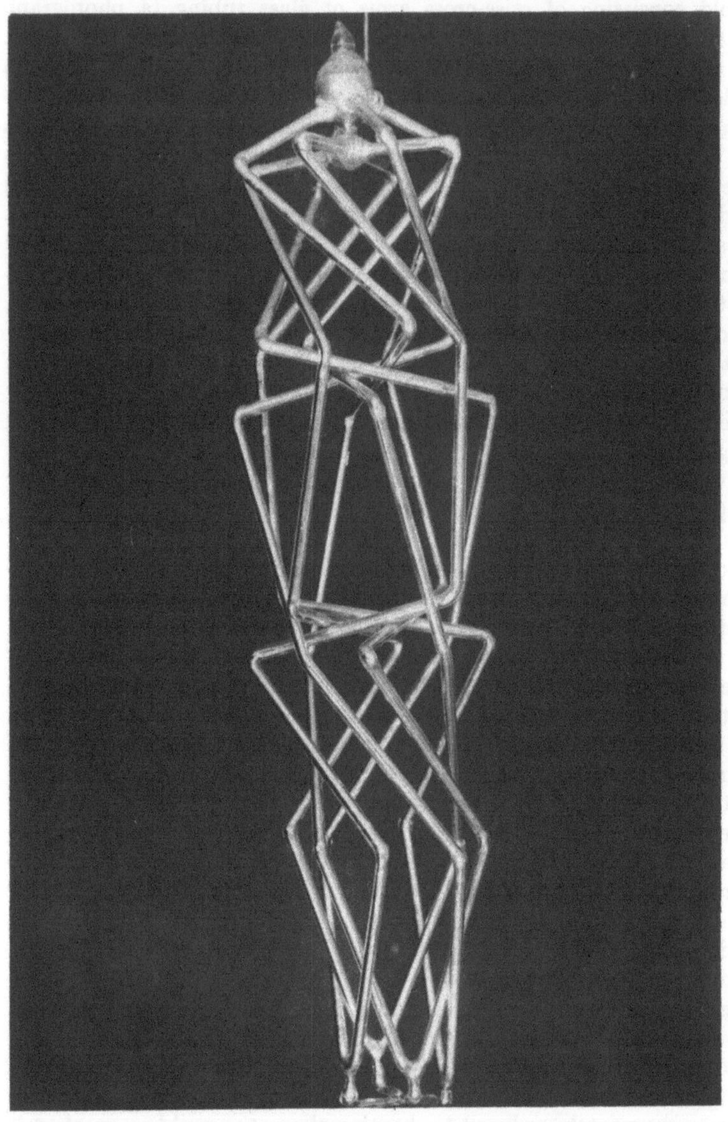

Figure 1. Photograph of the glass mold in which many single crystal rods with a wide range of orientation may be cast in one operation.

furnace to a temperature of 400° or 500°. If the mold is sealed to prevent oxidation, the deposit is unaffected by this treatment, but if air is allowed to enter, the deposit is oxidized and completely disappears in the course of 10 minutes or so. By controlling the amount of air and the time of heating, the deposit may be reduced to such a thickness that there is a just perceptible blackening of the walls, which is the ideal to strive for.

Filling the mold with molten metal could be accomplished in various ways; the diameter of the glass tubes used was so small that the metal mensicus would not break, even in an inverted position. It was therefore possible, when convenient, to fill the mold from an upper chamber, allowing the molten metal to run down one arm and up the other with no danger of gas bubbles breaking the continuity. In some cases I found it convenient to fill by sucking the molten metal up into the mold from an open lower reservoir. I had used this method some time ago in preliminary work, and it has been recently described in a publication by Sachs.[5] I found it necessary to suck the metal up very slowly, otherwise gas bubbles were apt to appear. The rate of rise may be made as slow as desired by connecting the upper end of the mold to a large vessel, which is exhausted very slowly through a fine capillary attached to the vacuum pump. To prevent the metal in the open reservoir from oxidizing during the filling, the furnace may be filled with an atmosphere of CO_2.

The Apparatus for Measuring Thermal e.m.f.

The previous apparatus consisted of two temperature baths at different temperatures, with the single crystal rods leading from one bath to the other, and with copper leads soldered to the ends of the crystals. The e.m.f. was measured as a function of the temperature difference. The apparatus could be used with only one crystal rod at a time, so that besides the disadvantage of slowness, slight temperature irregularities in the different runs might superpose vagaries on the differences due to differences of orientation in the different specimens. The new apparatus was so designed that 16 different samples could be measured with a single set-up, thus very much increasing the speed, and decreasing any error due to temperature irregularities. The 16 crystal rods, each about 3 mm. in diameter and about 7.5 cm. long, were passed through small stuffing boxes, arranged in a row in a block of bakelite 2.5 cm. thick. This block of bakelite was clamped, between asbestos packing, between two heavy cast brass cylinders about 7 cm. inside diameter, provided with openings with machined faces to press against the bakelite. A section is shown in Figure 2 and an elevation in Figure 3. The

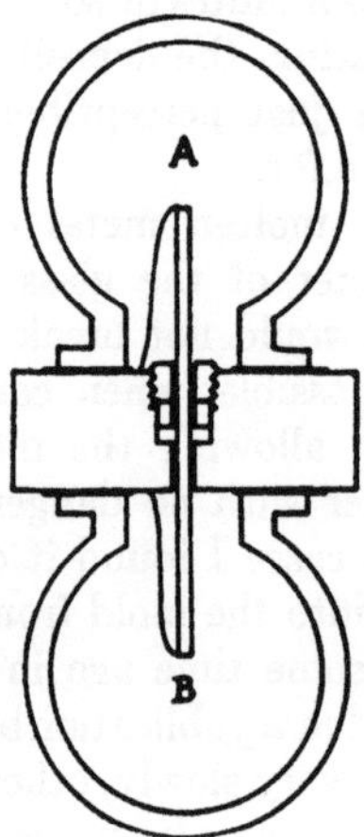

FIGURE 2. Section of part of the apparatus for measuring thermal e.m.f. of single crystal rods. The rod reaches from A to B. The two ends A and B are maintained at different temperatures by two streams of oil, circulated through two regulated sources.

dimensions are such that the ends of the crystal rods project to about the middle of the brass cylinders. The cylindrical brass castings are connected at top and bottom to two separate reservoirs, each filled with kerosene. A rapid circulation is maintained through the cylindrical castings, past the projecting ends of the crystals, into the reservoirs at the top and into the castings at the bottom. The reservoirs are maintained at different constant temperatures by thermostatically controlled electric heating arrangements. In addition to the stirrers at the upper ends of the castings, the reservoirs also have independent stirrers, in order to keep the temperature as uniform as possible. One of the baths was maintained at a constant temperature of about 20°. Since sometimes the room rose above this temperature, artificial cooling was maintained by a current of tap water circulating through a coil of copper tube placed in the reservoir. The thermostatic control with electric heating was superposed on this constant cooling. The other bath was varied in temperature, the temperatures chosen being 37°, 54°,

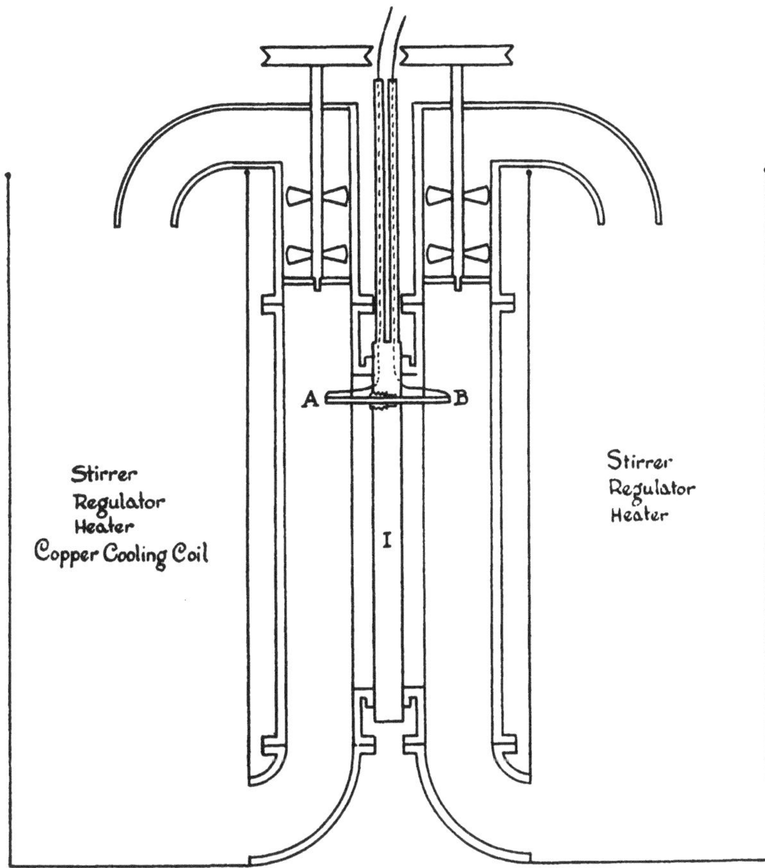

FIGURE 3. Section in elevation of part of the apparatus for measuring thermal e.m.f. of single crystal rods. There is provision for measuring 16 rods at the same time. A single sample rod is shown at AB. The others are mounted in a similar way in the rest of the bakelite block I.

70°, and 88°. Temperatures were read with calibrated mercury-in-glass thermometers. The temperature regulator was a conventional mercury-in-glass affair; four of these regulators were used, each set permanently to one of the four temperatures. After readings at one temperature had been completed, the regulator for the next higher temperature was substituted for the former one, and temperature was run up rapidly to the desired point with a heavy heating current. The operation of changing temperature and coming to equilibrium again at the new point occupied from 10 to 15 minutes.

Copper leads, all from the same spool of wire, were soldered to the two ends of the crystal rods, brought out through a stand-pipe arrangement at the upper end of the bakelite block, and connected to the arrangement for measuring e.m.f. Any one of the 16 rods could be connected to the measuring arrangement through a bank of 32 all-copper switches, mounted in a closed box, and operated by long handles from outside to eliminate parasitic e.m.f.'s due to slight temperature inequalities. The circuit by which e.m.f. was measured is shown in Figure 4. The method is essentially a deflection method, this being much more rapid than the potentiometer method used before, and sufficiently accurate. A Pye galvanometer, G, which has a low period and a steady zero, was used, with mirror and scale. The circuit contained various resistances in series or in parallel, R_1 and R_2, with the galvanometer, by which its sensitivity could be appropriately varied, so that for each different sort of metal the throw for the maximum temperature difference was made about 40 cm. There were also arrangements by which an acid Weston cell, E_1, which permits a fairly large current to be drawn from it without deterioration, could be connected into the circuit in order to calibrate the galvanometer. This calibration was made at the beginning and the end of every run and varied only slightly from day to day. The e.m.f. of the acid cell was checked by comparison with a standard Weston cell, E_2, which could also be thrown into the same circuit in such a way that the comparison was made by a null method, thus drawing no current from the standard cell. R_3, which is a decade box reaching to 10,000 ohms in steps of 1 ohm, and R_4, which reaches to 100,000 in steps of 10,000, were used in making the various calibrations. Error from galvanometer drift and parasitic e.m.f.'s in parts of the apparatus beyond the thermocouple switch were limited by using the double throw of the galvanometer, instead of its deflection from rest. The period of the galvanometer was of the order of 5 seconds, so that it was possible to complete the measurements on a single rod in 20 seconds or less, and the entire series in 5 minutes. Constancy of the bath temperature was checked by repeating at the end of the series the measurement on the first couple; this also checked whether temperature equilibrium had been attained before beginning the run.

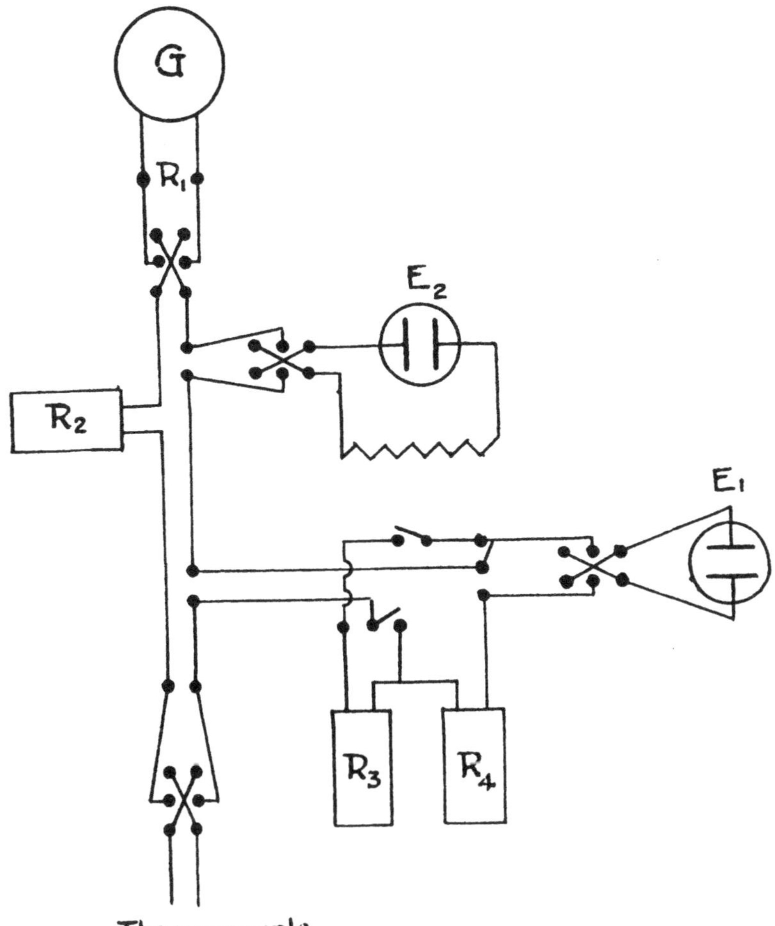

FIGURE 4. The circuits for measuring thermal e.m.f.

Before the adoption of the method just described, another piece of apparatus was built and tested, in which 16 crystal rods were led between two heavy copper bars, 5 cm. on a side, maintained at different temperatures. The crystal rods dipped into wells drilled in the copper bars, filled with stagnant oil, and of a diameter only slightly greater than that of the rods. The difficulty with this apparatus was that temperature equality between the crystal bars and the copper bars was not attained; this was tested by thermocouples directly soldered to the ends of the rods. Apparently the only way by which a positive temperature control can be attained is to immerse the crystal rods in a rapidly circulating stream of liquid.

Methods of Calculation.

The first step in the computation was to reduce the double throws of the galvanometer to volts by means of the calibration. Next a small correction, which could be determined by linear interpolation, was applied to these voltages to bring the temperature of the cold junction to exactly 20°; the actual temperature was usually a fraction of a degree below this. Then, by linear interpolation between 20° and the maximum temperature, the voltages at the intermediate temperatures were calculated which would have been found if the relation between temperature and voltage had been linear. The difference between the linear and the actual voltage was then plotted against temperature. The difference curve was, in almost every case, parabolic within the error of measurement; the condition for this is that the curve should be symmetrical about the mid point and that the ordinates at the one-quarter and the three-quarter points should be three-quarters of the maximum ordinate. A parabolic difference curve means that within experimental error the relation between thermal e.m.f. and temperature is of the second degree. The constants in this second degree relation could then be found at once in terms of the total voltage, e_0, between 20° and the maximum temperature, and the deviation from linearity at the mid point, Δ. In fact:

$$(\text{thermal e.m.f.}) = \left(\frac{e_0 + 4\Delta}{t_0}\right)(\tau - 293.1) - \frac{4\Delta}{t_0^2}(\tau - 293.1)^2,$$

where t_0 is the temperature range from 20° to the maximum, and τ is Abs. temperature.

If Kelvin's relation holds, that is, if at every temperature the thermal e.m.f. of a crystal rod against any homogeneous substance is a linear function of $\cos^2\theta$, θ being the angle between the length of the rod and the crystal axis, it is evident that e_0 and Δ must also be linear functions of $\cos^2\theta$. e_0 and Δ were accordingly plotted against $\cos^2\theta$, and the best smooth curve drawn through the points. If the relation turned out to be linear, then the values of e_0 and Δ at the extreme points, that is, for $\cos\theta = 1$ and $\cos\theta = 0$, were read from the curve. From the e_0 and Δ found in this way the values of the constants in the two-power series in temperature for thermal e.m.f. against copper could be calculated at the two extreme orientations. From the series formula the Peltier heat against copper and the Thomson heat can be found by differentiation. If e_0 or Δ was not linear against $\cos^2\theta$, as was the case for two metals, then the procedure just described must be repeated at enough orientations between 0° and 90° to define the behavior.

DETAILED DATA.

Zinc. The material was Kahlbaum's best zinc. A considerable number of castings were made: of these there were two sets in two many-arm molds with from 30 to 50 orientations, and at least five castings in the simpler mold with 12 orientations.

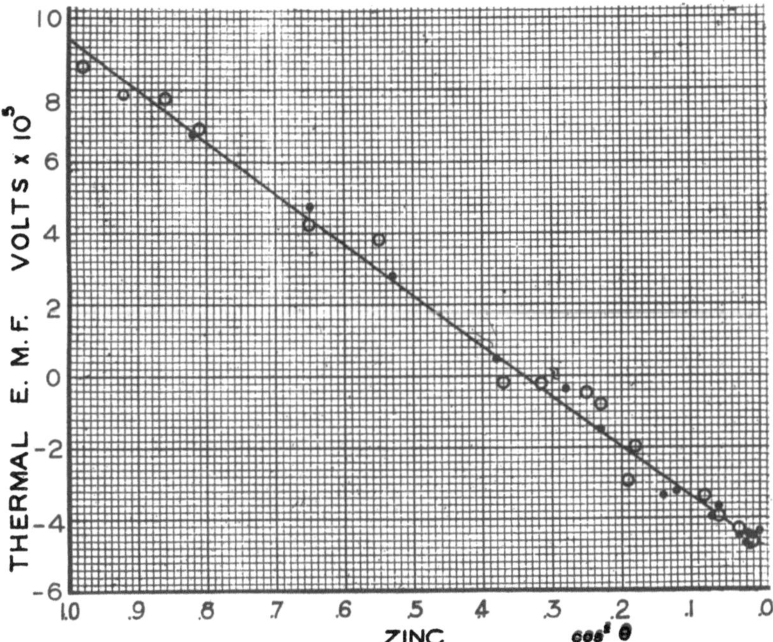

FIGURE 5. Thermal e.m.f. of single crystal zinc against copper between 20° and 88° C. The abscissa is the square of the cosine of the angle between the length of the rod and the principal crystal axis.

Measurements of thermal e.m.f. were made on two sets of 16 rods, one set from each of the larger molds, selected to cover as uniformly as possible the entire orientation range. The angle of the basal plane with the length was determined by cleavage, but instead of cleaving at room temperature, the rod was cooled in liquid air and cleaved while still cold by pressing into it the point of a knife. This trick of cleaving in liquid air is due to Boydston;[6] it is well worth while, the cleavage being much easier to produce and much sharper, and there being less danger of bending the rod.

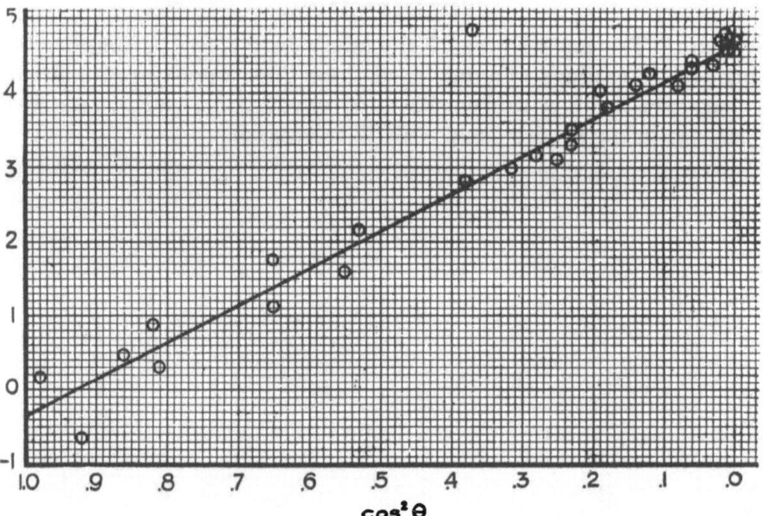

FIGURE 6. Shows the deviation from linearity in the relation between temperature and thermal e.m.f. of single crystal zinc against copper. The deviation is shown in 10^{-6} volts at the mean temperature 54°, the temperature range being from 20° to 88° C. The abscissa is the square of the cosine of the angle between the length of the rod and the crystal axis.

The results of the measurements of thermal e.m.f. are plotted in Figure 5 and Figure 6; in Figure 5 the total e.m.f. of the couple Zn-Cu between 20° and 88° is plotted against $\cos^2\theta$, and in Figure 6 the deviations from linearity at the mean temperature also against $\cos^2\theta$. All the e_0's of one set of rods have been corrected by the constant amount, 1×10^{-6} volts, to bring the absolute values into agreement with those of the other set. The difference in absolute values is doubtless due to a parasitic e.m.f. in parts of the circuit beyond the reversing switch, due to different temperature conditions in the room during the two sets of readings, and is evidently without significance, since the quantity in which we are interested here is the variation of e.m.f. with direction in the crystal. The deviations from linearity of the two sets of 16 lay on the same straight line without correction, as should be the case if the difference in e_0 is due to parasitic effects as suggested.

It will be seen from the figures that e_0 and Δ are both linear functions of $\cos^2\theta$; that is, the relation of Kelvin is satisfied for zinc. The results contained in the figures may be used to give the following formulas:

$$(\text{t.e.m.f.})_{Zn_{\parallel}-Cu} = 1.372\times 10^{-6}\times(\tau-293.1)+3.03\times 10^{-10}\times(\tau-293.1)^2 \text{ Volts};$$

$$(\text{t.e.m.f.})_{Zn_{\perp}-Cu} = -0.428\times 10^{-6}\times(\tau-293.1)-3.97\times 10^{-9}\times(\tau-293.1)^2.$$

Whence:

$$(\text{t.e.m.f.})_{Zn_{\parallel}-Zn_{\perp}} = 1.800\times 10^{-6}\times(\tau-293.1)+4.27\times 10^{-9}\times(\tau-293.1)^2,$$

and since

$$P=\tau\frac{dE}{d\tau} \quad \text{and} \quad \sigma=\tau\frac{d^2E}{d\tau^2},$$

we have for the Peltier and Thomson heats:

$$P_{Zn_{\parallel}-Zn_{\perp}} = \tau\times[1.800\times 10^{-6}+8.54\times 10^{-9}\times(\tau-293.1)],$$

$$\sigma_{Zn_{\parallel}-Zn_{\perp}} = \tau\times 8.54\times 10^{-9}.$$

The designation $Zn_{\parallel}$ means that the crystal axis (of six-fold symmetry) is parallel to the length of the rod. The sign convention is the usual one. If $(\text{t.e.m.f.})_{Zn_{\parallel}-Zn_{\perp}}$ is positive, current flows from $Zn_{\parallel}$ to $Zn_{\perp}$ at the hot junction.

It is probable that within the limits of error the relation between temperature and e.m.f. is not quite represented by a second degree expression in the temperature, but the curves of deviation from linearity are systematically unsymmetrical, in that the deviation at

¾ of the temperature range (71°) is slightly greater than at the ¼ temperature (37°). The difference is of the order of 4×10^{-7} volts. It is in such a direction as to decrease the curvature at the lower temperature and increase it at the higher temperature end of the range; this means that σ increases with rising temperature somewhat more rapidly than proportionally to the absolute temperature, as given by the formula. In order to make partial allowance for this slight departure from symmetry, the deviation, Δ, at the mean temperature was calculated by taking 0.4 of the sum of the deviations at the ¼, ½ and ¾ points, this being the fraction to be used if the deviation is accurately parabolic. An idea of the smoothness of the readings, and so of the accuracy, is given by the fact that the actual Δ at the ½ point usually differed from the calculated Δ by only 0.1 or 0.2 mm. deflection of the galvanometer, corresponding to from 4 to 8×10^{-8} Volts.

Compared with my previous measurements,[7] these new results show very important differences. The previous results were obtained with 8 rods, with a maximum angle between basal plane and length of 55°. The results indicated by the previous measurements were that P is not linear against $\cos^2 \theta$ (or against the specific resistance, which is the same thing), but that σ (or Δ) is linear. These new and greatly improved results do not substantiate this conclusion, but both P and σ are linear within satisfactory limits of error. The values given in the previous paper for Peltier heat and Thomson heat of Zn_{11} against Cu were obtained by an extrapolation to nearly double the range of measurement, and cannot, therefore, be expected to be accurate. It follows that the difference between the P and the σ for parallel and perpendicular directions in the former paper (p. 132) cannot be accurate. In fact, the former paper gives for $P_{11-\perp}$ at 0°, $1.001 \times 10^{-6} \times \tau$ against $1.629 \times 10^{-6} \times \tau$ found here; and for $\sigma_{11-\perp}$, $6.2 \times 10^{-9} \times \tau$ against $8.5 \times 10^{-9} \times \tau$ found now. The previous value for the basal plane parallel to the length avoids error from the long range extrapolation, but here there is another source of error in that the copper was not from the same source in the two sets of experiments. In fact the previous measurements give for $P_{Zn_{\perp}-Cu}$ at 0° the value $-.147 \times 10^{-6} \times \tau$, whereas the present value is $-.269 \times 10^{-6} \times \tau$. The difference, which amounts to about a tenth of a micro-volt per degree, does not seem too large for the thermal e.m.f. between two different varieties of commercial copper.

The conclusion that thermal e.m.f. satisfies Kelvin's law agrees with the conclusion reached by Linder[2] from his measurements up to 200°. Linder was of the opinion, however, that at 300° and 400° his departures from the relation were greater than experimental error. It seems to me, however, that this effect of Linder's might well be due to the very steep temperature gradient at the junction, which would make difficult the exact determination of the temperature, and which would have an increasing effect at high temperatures.

After the completion of the thermo-electric measurements, I made new measurements of specific resistance. There were several reasons for doing this. I desired to submit the $\cos^2$ law of Kelvin for resistance to a more accurate check than had yet been made; my previous measurements had been made on only eight samples scattered over a range of orientation of 60° instead of the whole 90°, and there were rather serious differences between the values of Grüneisen and Goens[8] and of myself for the ratio of the resistance parallel and perpendicular to the axis, Grüneisen and Goens finding the ratio 1.080 against my 1.036. At first, on repeating the measurements, disturbingly irregular results were obtained. After considerable work, the source of the irregularities was found in the very great sensitiveness of resistance to slight mechanical deformations. If the rod is slightly bent in removing from the glass mold, or if it sticks to the glass so that deformations are produced by unequal thermal expansion between the mold and metal on cooling, the resistance is materially increased. The reason that this source of error had not been more disturbing in the previous measurements was that the larger diameter of the rods, 6 mm. against 3mm. used here, made deformation incidental to cracking away the glass mold much less likely. It is perhaps surprising that thermal e.m.f. should be much less sensitive to deformation than resistance; this was proved not only by the much greater regularity of the results for thermal e.m.f., which were obtained on rods which must have suffered some deformation, but also by special tests, in which the rods were deliberately bent and straightened, with no measurable change in thermal e.m.f.

It finally proved impossible to obtain measurements of the resistance which were sufficiently regular on rods which had been removed from the molds, and accordingly the resistance was measured before removing the glass. This was done as follows. The potentiometer method of measuring resistance was used. Current terminals were attached by soldering copper leads to the two ends of the rod, the glass being broken away a sufficient distance to make this possible. The potential terminals were two needles, mounted on a frame at a fixed distance apart, and pressed by springs into contact with the metal through two V-shaped notches ground through the glass wall of the mold with a fine glass-cutting wheel dressed to a sharp edge with a diamond. The notches were ground to such a depth as to just break through the glass. In this way, the only appreciable deformation was confined to the ends near the current terminals, and so was without effect in the region between the potential terminals, which determines the resistance. The angle of the basal plane with the length was determined after the measurements by cooling in liquid air, and by cleavage, as before. The diameter was measured with a micrometer after the glass mold was broken away; the fluctuations from uniformity of diameter in the space between the potential terminals was of the order of 1%.

The specific resistances at 20° of two sets of rods, obtained with all these precautions, are shown in Figure 7 plotted against $\cos^2\theta$. Except for one bad point, the results are fairly regular. In selecting from these data the most probable values for the specific resistance, I assumed that almost all sources of error would give too high a resistance, and that therefore the lowest of the values found are most probably correct. Figure 7 shows that within experimental error specific resistance is linear against $\cos^2\theta$, so that the relation of Kelvin is satisfied. In fact there has never, as far as I know, been any reason to think that Kelvin's relation for resistance might not be correct.

In Figure 7 are also shown my previous values for resistance parallel and perpendicular to the axis. These both differ by a small and nearly constant amount from the value now found. The new value for the ratio of the two resistances is 1.039 (6.06/5.83) against 1.036 found before by extrapolation over a range of 30°. The absolute difference between my former and present results is perhaps to be explained by greater purity in my new supply of zinc; the fact that the difference is constant over the entire range of orientation is against the supposition that the previous high values were due to deformation. Plotted in the same diagram are also the values of Grüneisen and Goens. Their value for the resistance when the axis is perpendicular to the length is almost exactly the same as my new value, but their value for the parallel orientation is much higher.

It seems to be unquestionable that their large value is an effect of distortion, for it is of course true that a crystal is very deformable when the axis is parallel to the length, but comparatively rigid when perpendicular.

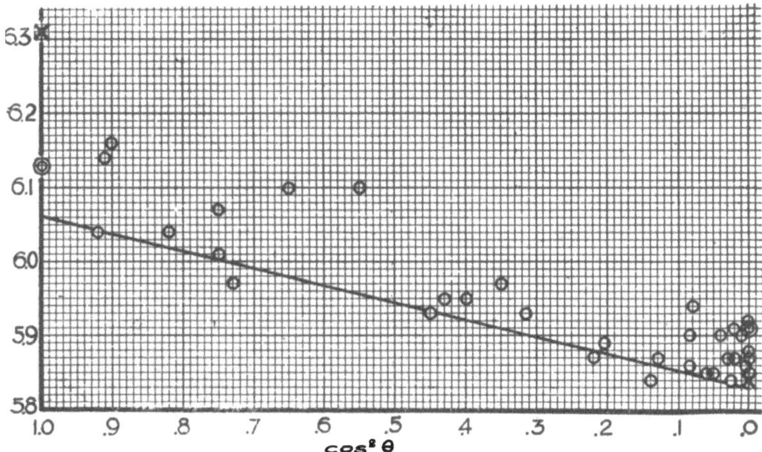

FIGURE 7. Specific resistance at 20° C in 10^{-6} ohms per cm. cube of single crystal zinc plotted against the square of the cosine of the angle between the crystal axis and the length of the rod. The double circles are my previous values and the crosses those of Grüneisen and Goens.

Cadmium. The material was Kahlbaum's best. At first the attempt was made to fill the molds by suction, drawing the melted Cd up from an open crucible, but this failed because of the unexpectedly rapid oxidation of the molten Cd, even at temperatures close to the melting point. The oxidation was sometimes so rapid that the coating of brown oxide spontaneously became red hot. The difficulty was not avoided by filling the furnace as completely as possible with a stream of CO_2, led in at the bottom and flowing out at the open upper end. The rapidity of oxidation seems to increase very markedly as the last traces of impurity are removed from the metal; the purest Cd of American make which I could obtain oxidized very much more slowly. The filling of the mold was finally accomplished by running the molten metal down from a sealed reservoir attached to the upper end of one of the branches. The two-branch mold with 12 sections was used. It was exhausted and flushed several times with CO_2 before filling.

Cd is perhaps the most difficult of the metals worked with here to compel to crystallize as one grain, the tendency to switch from one grain to another in the middle of a straight arm being particularly strong. The general behavior is consistent with the explanation that fresh nuclei of the solid are particularly likely to form from the liquid if it is subcooled by a small amount. The remedy is to make the temperature gradient unusually steep in the region where crystallization takes place, in this way reducing to a minimum the extent of the region in which there is any possible subcooling. A steep gradient in the desired locality was produced by lowering the mold into oil, and by various manipulations of the position of the surface of the oil and the temperature of the furnace, which can be learned only by trial. In general, the temperature of the furnace should be further above the melting point than is necessary for other metals.

The determination of the angle between the crystal axis and the length offered some difficulty. Cadmium does not cleave, even at liquid air temperature, and the small size of the rods made difficult the method previously used of marking the reflection pattern on a wooden ball slipped over the rod. At first I tried to determine the orientation by means of the specific resistance, but these results were very irregular, cadmium being even more deformable than zinc, and particularly difficult to prevent from sticking to the glass mold. After various unsuccessful attempts, including an attempt by locating the slip planes when stressed beyond the elastic limit, I returned to the optical method, not attempting, however, to locate the entire reflection pattern, but only the basal plane. A simple arrangement was made for this purpose, in which the crystal rod was mounted in a goniometer holder on an optical bench in a beam of parallel light from an electric arc, and the angle of reflection determined by sighting through two small peep holes. The error may easily be made less than 1° with very simply constructed apparatus. The difficulty with the method is in being sure that the plane of reflection is actually the basal plane. If the basal plane is approximately parallel to the length of the rod, the reflection from it is so much stronger than the reflection from any other plane that there is seldom any trouble. But if the basal plane is inclined at a high angle, then the reflection from it may be weaker than from one of the secondary planes more nearly parallel to the length, and in such a case, decision must be made with the help of other data. There are never more than a few discrete possibilities to choose between, and a knowledge of the specific resistance, or thermal e.m.f., or perhaps the slip pattern makes possible an unambiguous choice.

Measurements of thermal e.m.f. were made on 64 different samples, four sets of 16 each. This comparatively large number was used because the first measurements, in which the angle was determined from the specific resistance, had indicated a systematic departure from Kelvin's $\cos^2\theta$ law, and if this were really the case, I was anxious to settle it as certainly as possible. The suspected departure did not substantiate itself, however, on increasing the number of readings and determining the orientation in the better way.

The method by which the results were calculated for cadmium differed somewhat from that outlined above. The temperatures of the four sets of runs were so nearly alike that a very small correction sufficed to reduce the readings of any one set to a mean and standard set of temperatures: 20°, 37.20°, 54.03°, 72.00° and 88.14°. Four different plots of the thermal e.m.f.'s of all of the 64 rods between 20° and each of the other temperatures were then made against $\cos^2\theta$. In each case the points so obtained lay on a straight line, confirming Kelvin's law. The observed points for the interval 20°–88.14° are reproduced in Figure 8. Kelvin's law having thus been established, it remained only to get the best e.m.f.'s for the parallel and the perpendicular orientations; these were found from the intercepts of the best straight lines drawn through the experimental points. Next, the best second degree curve in the temperature was passed through these intercepts. In this way the following results were found:

$$(\text{t.e.m.f.})_{Cd_{\parallel}-Cu} = (1.357 \times 10^{-6})(\tau - 293.1) - 0.790 \times 10^{-8}(\tau - 293.1)^2 \text{ Volts},$$

$$(\text{t.e.m.f.})_{Cd_{\perp}-Cu} = -(1.616 \times 10^{-6})(\tau - 293.1) - 1.872 \times 10^{-8}(\tau - 293.1)^2.$$

Whence:

$$(\text{t.e.m.f.})_{Cd_{\parallel}-Cd_{\perp}} = (2.973 \times 10^{-6})(\tau - 293.1) + 1.082 \times 10^{-8}(\tau - 293.1)^2,$$

and for the Peltier and Thomson heats:

$$P_{Cd_{\parallel}-Cd_{\perp}} = \tau \times [2.973 \times 10^{-6} + 2.164 \times 10^{-8}(\tau - 293.1)],$$

$$\sigma_{Cd_{\parallel}-Cd_{\perp}} = \tau \times 2.164 \times 10^{-8}.$$

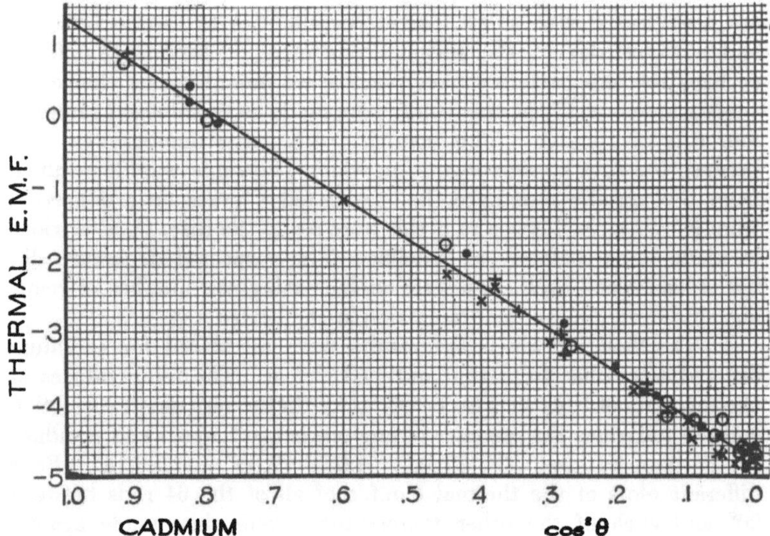

Figure 8. Thermal e.m.f. of single crystal cadmium against copper between 20° and 88.14° C. The ordinates are on an arbitrary scale, one large division corresponding to 40.15×10^{-6} volts. The abscissas are the squares of the cosine of the angle between the crystal axis and the length. The different designations of the observed points refer to different set-ups of the apparatus.

As in the case of zinc, the second degree curve does not quite reproduce the measurements, but there are small differences which are probably larger than experimental error. At the three temperatures 37.2°, 54.0°, and 72.0° the actual $(\text{t.e.m.f.})_{Cd_{\parallel}-Cu}$ is to be found by adding to the values given by the second degree formula -0.2×10^{-6}, $+0.2 \times 10^{-6}$, and $+0.1 \times 10^{-6}$ Volts, respectively, and similarly, to obtain the actual $(\text{t.e.m.f.})_{Cd_{\perp}-Cu}$ the above formula is to be corrected by -0.4×10^{-6}, -0.6×10^{-6}, and $+0.9 \times 10^{-6}$ Volts, respectively. The correction term has the effect of decreasing the curvature, and therefore σ, at the low temperature end, and increasing it at the high temperature end.

We now compare these new results with those previously found. The previous values were obtained with 14 rods, distributed over practically the entire orientation range. Because the previous results were obtained without extrapolation, better agreement is to be expected than in the case of zinc. There is, however, the uncertainty in the previous results that the orientation was determined in terms of the specific resistance, which we have just seen does not give regular results.

The previous results agree with the new ones in that Kelvin's relation was found to hold. At 0° C the previous values for $P_{Cd_{\parallel}-Cd_{\perp}}$ was $+ 2.95 \times 10^{-6} \times \tau$ against $+ 2.54 \times 10^{-6} \times \tau$ now found, and for $\sigma_{Cd_{\parallel}-Cd_{\perp}}$, $+ 1.7 \times 10^{-8} \times \tau$ against $+ 2.2 \times 10^{-8} \times \tau$ now found.

Of the many sets of castings, only one was perfectly satisfactory for the determination of specific resistance, all the others sticking slightly to the glass or else having been partially deformed. Readings were made on this set before removing the glass mold, connections to the potential terminals being through notches ground in the glass, as in the case of zinc. The results are shown in Figure 9. It is evident that the cos² law is satisfied within experimental error. My previous values are shown as double circles in the figure; these agree rather closely with the new values. I believe that a study of the data will suggest that the new value for the resistance when the basal plane is perpendicular to the length, 8.24, is somewhat preferable to the former value 8.30, in spite of the fact that the distribution of orientations in the previous work was somewhat wider. The values of Grüneisen and Goens[8] are also shown in the figure. These are higher than my values at both orientations by such an amount that their ratio of the resistances (8.45/7.11 = 1.188) does not differ by much from mine (8.24/6.82 = 1.207). The difference between my results and those of Grüneisen and Goens is in the direction to be explained by impurity or deformation in their material.

Antimony. The material was Kahlbaum's; I have not been able to obtain antimony from any other source of sufficient purity. The molds in which the single crystal rods were cast were made of quartz tubing; it would have been possible to use glass combustion tubing, but the softening point of this glass is so near the melting temperature, and other mechanical difficulties in handling the glass are so great, that the much greater convenience of the quartz much outweighed the greater expense. Fortunately antimony does not stick to quartz (or to glass either) so that it was not necessary to coat the mold, which would have been difficult at temperatures above a red heat. The first batch of castings was made in one of the complicated many-arm molds. The greatest difficulty with antimony

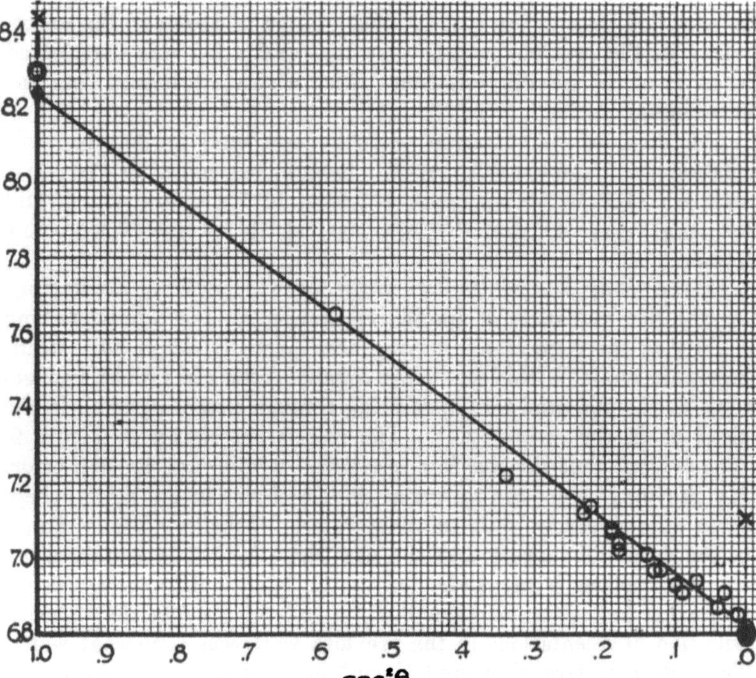

FIGURE 9. The specific resistance in 10^{-6} ohms per cm. cube at 20° of single crystal cadmium against the square of the cosine of the angle between the crystal axis and the length. The double circles show my previous values for the extreme orientations, and the crosses the values of Grüneisen and Goens.

is in the appearance of bubbles during filling; many of these appeared in the filling of the many-arm mold, and they could not be eliminated by any means that suggested themselves, so that many of the arms were of no use. I therefore returned to the much simpler mold with only six arms, made by bending a straight piece of quartz tubing at the desired angles. This was filled by slow suction from above, and the tube sealed off. The melted antimony was contained in an open porcelain crucible in the bottom of the furnace, and oxidation was sufficiently prevented by filling the furnace with a stream of CO_2. The advantage of this simple method is that if a bubble of gas appears during the filling, the mold can be allowed to empty itself and the process repeated until successful. Two of these six-arm molds could be filled at the same time from the same crucible, and thus 12 castings obtained by the same lowering operation. The higher orientations (cleavage perpendicular to the length) are more reluctant to appear with antimony than with most of the other metals, and even when once obtained, manipulation is difficult because of extreme brittleness. It is almost impossible to crack the quartz mold away from a rod with cleavage plane perpendicular to the length without breaking the rod. The method which I finally adopted was to grind the mold away from two sides, holding the quartz tube pressed lengthwise against a flat grinding surface, until the quartz was entirely ground through, as shown in section in Figure 10, when the remains of the mold will drop off.

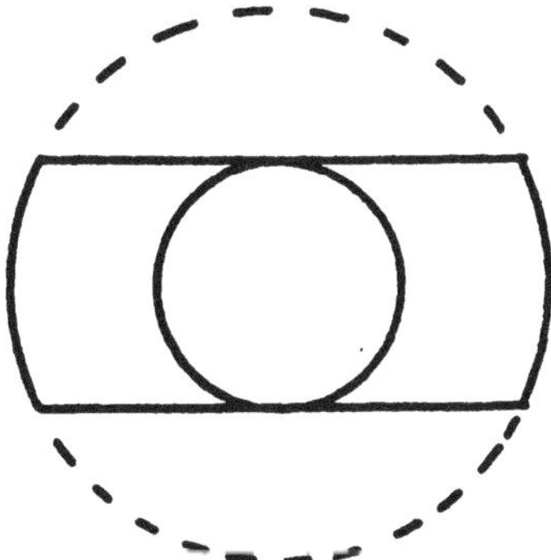

FIGURE 10. Shows the method of grinding away the quartz mold from the single crystals of antimony in order to avoid breaking the crystal when the cleavage plane is very nearly perpendicular to the length.

The orientation was determined with a simple goniometer from the cleavage, made with a knife at room temperature.

Three sets of runs were made to get the thermal e.m.f. The first set was on 12 rods obtained from the complicated many-arm mold. The second set was on 16 rods selected so as to give the widest range of orientation from four batches of castings made in the simple 6-arm molds. None of these rods was of very high orientation. The third set was on two pieces, both from the same original rod, with the cleavage plane at 83° to the length. These pieces were so short that it was not possible to use the regular arrangements; a new bakelite block was therefore made, machined so as to be only 3 mm. thick at the part where the rods were inserted, instead of 2.5 cm., and with special stuffing boxes of minimum depth. These two pieces from the same rod gave identical results, and they are shown in the following figures slightly separated.

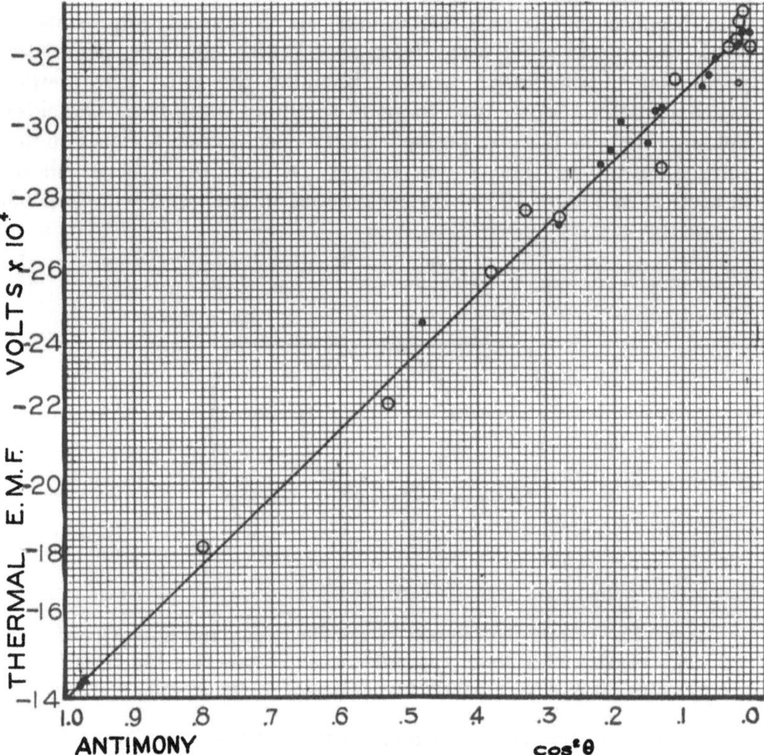

FIGURE 11. Thermal e.m.f. between 20° and 88° C in 10^{-4} volts of single crystal antimony against copper as a function of the square of the cosine of the angle between the crystal axis and the length.

The results for antimony were not as regular as for the other metals; doubtless the reason is connected with the short length of the specimens, so that irregular temperature fluctuations, due to greater proximity of the walls, were more important. The results for thermal e.m.f. between 20° and 88° are shown in Figure 11 for all specimens, and in Figure 12 for the deviation from linearity at the mean temperature. It is seen that within the limits of experimental error there is no reason to suspect a departure from Kelvin's $\cos^2$ law. The great scattering of the points for the deviation from linearity shows that there is considerable error in the curvature deduced from these results.

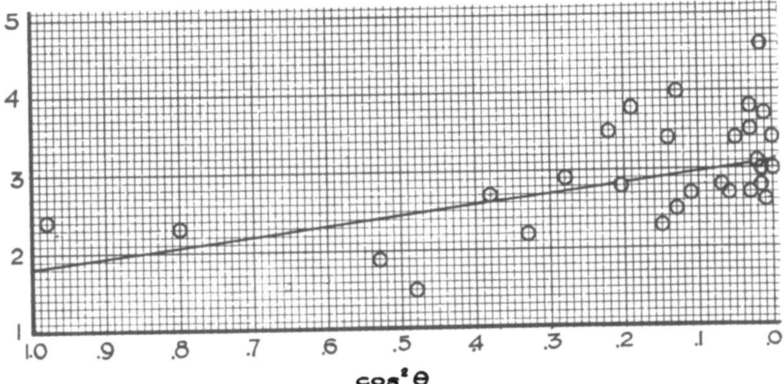

FIGURE 12. The deviation from linearity in the relation between temperature and thermal e.m.f. of single crystal antimony against copper. The deviation is shown in 10^{-5} volts at the mean temperature 54°, the temperature range being from 20° to 88° C. The abscissa is the square of the angle between the crystal axis and the length of the rod.

In spite of this comparatively large uncertainty, it is again probable that a second degree series in the temperature does not sufficiently reproduce the results. The departure from linearity was not quite symmetrical about the mean, the deviation being greater on the higher temperature side, so that the actual curvature is less at the lower temperature end and greater at the high temperature end than given by the second degree expression. The dissymmetry is of the order of 10^{-5} Volts, or ½% of the maximum t.e.m.f. I do not believe that the data justify an attempt to make a more precise statement about a possible departure from the second degree relation.

Expressing the results in the form of the best second degree series, the following values were found:

$$(\text{t.e.m.f.})_{Sb_{11}-Cu} = -19.5 \times 10^{-6} \times (\tau - 293.1) - 1.55 \times 10^{-8} \times (\tau - 293.1)^2 \text{ Volts,}$$

$$(\text{t.e.m.f.})_{Sb_\perp - Cu}$$
$$= -46.2 \times 10^{-6} \times (\tau - 293.1) - 2.66 \times 10^{-8} \times (\tau - 293.1)^2.$$

Whence:

$$(\text{t.e.m.f.})_{Sb_{\parallel} - Sb_\perp}$$
$$= + 26.7 \times 10^{-6} \times (\tau - 293.1) + 1.11 \times 10^{-8} \times (\tau - 293.1)^2,$$

and

$$P_{Sb_{\parallel} - Sb_\perp} = \tau \times [26.7 \times 10^{-6} + 2.22 \times 10^{-8} \times (\tau - 293.1)],$$
$$\sigma_{Sb_{\parallel} - Sb_\perp} = \tau \times 2.22 \times 10^{-8}.$$

In the previous paper I was able to measure the t.e.m.f. of only one sample of antimony, with the cleavage plane parallel to the length. For this I found:

$$(\text{t.e.m.f.})_{Sb_\perp - Cu}$$
$$= - 61.6 \times 10^{-6} \times (\tau - 293.1) - 8.5 \times 10^{-8} \times (\tau - 293.1)^2.$$

This former value is without doubt much too high; the reason for it is not evident.

The value of the Peltier heat between $Sb_{\parallel}$ and $Sb_\perp$ is so large as to have a special significance that must be emphasized. At 20° C the formula above gives for $P_{Sb_{\parallel} - Sb_\perp}$ the value $293 \times 26.7 \times 10^{-6} = 0.0078$ Volts. This means that 1 coulomb absorbs 0.0078 joules ($= 7.8 \times 10^4$ ergs) in flowing from the parallel to the perpendicular direction. The amount absorbed by one electron is therefore $- 1.59 \times 10^{-19} \times 7.8 \times 10^4 = - 1.24 \times 10^{-14}$ ergs. Now the energy of a gas molecule at 20° C is 6.05×10^{-14} ergs, so that in antimony one electron, in flowing from a direction of motion perpendicular to one parallel to the axis of the crystal, absorbs *as heat* 0.2 of the classical equipartition energy. It will be seen later that the corresponding quantity for bismuth is about twice as great. This is evidently a most important point for any theory of conduction. It would seem to indicate that the thermal energy of the conduction electrons must be of the same order of magnitude as given by classical equipartition. It is not obvious what account would be given of this situation by such a theory as that of Sommerfeld,[8] in which the part of the energy of the electrons which varies with temperature is supposed to be much less than the classical amount.

The determination of the specific resistance of this new material was a comparatively straightforward matter, the danger of deformation being very small because of its brittleness. The results are shown in Figure 13. Again the $\cos^2$ law is satisfied within experimental error. My previous values are also shown in the figure; these were obtained from only four specimens, with the cleavage plane at 0°, 0°, 47°, and 66° to the length respectively. It is not surprising that the agreement is not better. It is to be noted that the abnormal result previously found is sustained, namely that the

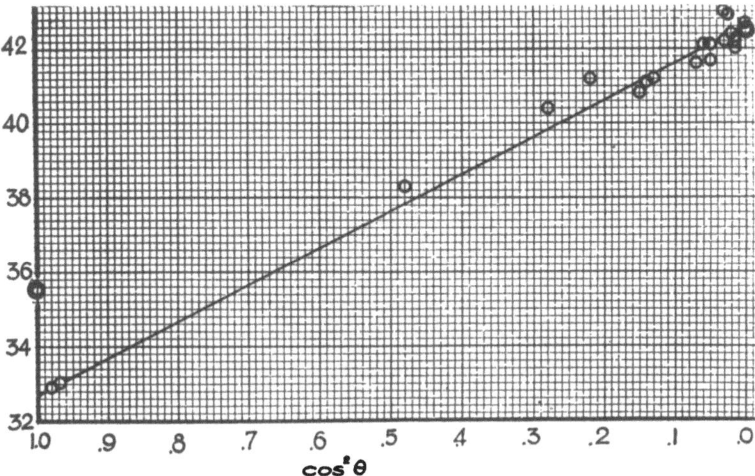

FIGURE 13. The specific resistance in 10^{-6} ohms per cm. cube at 20° C of single crystal antimony as a function of the square of the cosine of the angle between the crystal axis and the length. My previous values for the extreme orientations are shown by the double circles. The best values seem to be: 32.7×10^{-6} for the cleavage plane perpendicular to the length, and 42.6 for the parallel orientation.

resistance across the cleavage plane is less than parallel to the plane; in all other metals the behavior is the reverse. The discrepancy between my previous results and the new one for cleavage plane perpendicular to the length is in the direction to be accounted for by minute fissures on the cleavage plane in the former sample.

The Pressure Coefficient of Resistance of Single Crystal Antimony. Advantage was taken of the greater range of orientation of the specimens now available to extend my previous measurements[9] of the effect of pressure on resistance. The previous measurements were made on two samples with the cleavage plane at 0° and 47.5° to the length. The new measurements were made on three specimens, with the cleavage plane at 7°, 42.5°, and 83° to the length. Measurements were made in the routine manner with the potentiometer. Current and potential leads were both soldered to the specimen, using very fine leads and a minimum amount of solder for the potential leads. Measurements were made at 30° and 75°, thus incidentally giving values for the temperature coefficient of resistance. The transmitting medium was petroleum-ether. In this way the irregular effects with the samples with high orientations previously found were completely avoided; these effects were ascribed to minute fissures produced by the viscosity under pressure of the transmitting medium. The regularity of the new results is shown by the following values for the average departure of a single reading from a smooth curve, in terms of percentage of the maximum pressure effect, at 30° and 75° respectively for the three orientations in the order given above: 0.50%, 4.0%; 0.18%, 0.22%; 0.08%, 0.17%. The larger error for the cleavage plane parallel to the length is explained by the much smaller value of the pressure coefficient in this direction.

The change of resistance produced by pressure cannot be represented within experimental error by any simple expression. For the 83° and the 42.5° orientations, the deviations from linearity are unsymmetrical about the mean pressure, the greatest deviations occurring at pressures less than the mean. The asymmetry is more pronounced for the 42.5° orientation. For the 7° orientation, the deviation from linearity reverses sign, being negative for low pressures and positive for high pressures; this effect is the more marked at the low temperature. In view of all these complications, the results are not indicated by a formula, but by a table (Table I), giving the resistance at 1000 kg intervals of pressure at 30° and 75°. In making the calculations, the relative resistance of each sample was first expressed in terms of its resistance at atmospheric pressure and 30° C as unity. These numbers were then multiplied by the specific resistance at 30° and atmospheric pressure to give the figures in the table. The specific resistance at 30° and atmospheric pressure was obtained from Figure 13, using in passing from 20° to 30° the values of the mean temperature coefficient shown in Table I. This mean temperature coefficient was obtained by linear extrapolation of the resistance measured at 30° and 75° during the pressure runs. The resistance of antimony is not linear with temperature, so that extrapolation is not quite correct, but it can only be a very little in error, because the range of temperature of the measurement, 30°–75°, is so nearly in the middle of the range 0°–100°.

The resistances listed in Table I are thus nearly specific resistances; they are, however, not quite this, because the resistances under pressure were obtained with terminals fixed to the specimen. To get specific resistances from these measurements, a correction must be applied for the distortion under pressure. From the elastic constants which I have previously determined it may be easily computed that when the basal plane is parallel to the length, the measured resistance at 10,000 kg must be corrected by subtracting 1.69% to convert to specific resistance, and when the basal plane is perpendicular to the length, the correction is an addition of 0.68%. These figures are also approximately the figures for the 7° and 83° directions respectively given in the table.

TABLE I.

Resistance of Antimony.

Pressure kg./cm.²	Resistance × 10^6					
	Basal Plane at 7° to length		Basal Plane at 42°.5 to length		Basal Plane at 83° to length	
	30° C	75° C	30° C	75° C	30° C	75° C
0	44.41	53.37	40.01	48.81	34.60	42.47
1000	44.58	53.38	40.30	49.09	35.22	43.21
2000	44.76	53.40	40.66	49.43	35.95	43.94
3000	44.93	53.43	41.06	49.72	37.05	44.89
4000	45.11	53.51	41.52	50.09	37.78	45.94
5000	45.28	53.56	42.00	50.48	38.61	46.83
6000	45.45	53.65	42.49	50.89	39.60	47.86
7000	45.63	53.74	43.01	51.30	40.63	48.92
8000	45.80	53.78	43.53	51.73	41.70	50.02
9000	45.98	53.81	44.05	52.16	42.80	51.16
10000	46.16	53.82	44.58	52.59	43.93	52.32
11000	46.33	53.82	45.13	53.03	45.07	53.48
12000	46.50	53.78	45.66	53.38	46.20	54.60
Mean Temp. Coefficient, 0°–100°	.00495		.00551		.00572	

Comparing these new results with the previous values, the most important qualitative features found before are verified, but the numerical agreement is not very close. For the 7° direction, the minimum of resistance with increasing pressure is found again. As before, this minimum of resistance occurs first at higher temperatures. The pressure coefficient perpendicular to the basal plane is, as before, found to be very much greater than parallel to the plane. The difference is so great that at some pressure not very far beyond the maximum of 12,000 the initial abnormality of a resistance greater parallel than across the cleavage plane will be wiped out. Previously the disappearance of the abnormality was found actually to take place within the experimental range of pressure.

Tin. Kahlbaum's tin was used. Two sets of measurements of thermal e.m.f. were made, on two sets of 16 rods, selected from two different castings in the complicated mold with many branches. The orientation of these small rods was not easy to determine. Tin has no cleavage and the basal plane is absent from the reflection pattern, so that to determine the orientation from the reflection pattern would have demanded the location of several of the other planes, and not enough of these were usually present to make this possible. Accordingly I had to be satisfied with a measurement of specific resistance to give the orientation. Fortunately, tin is not as deformable as zinc or cadmium, and furthermore its resistance is comparatively insensitive to deformation, as was shown by direct trial, and also by the fact that the measurements of thermal e.m.f. were very much more regular when plotted against resistance than had been the case with zinc or cadmium.

In Figure 14 is plotted the total e.m.f. against copper between 20° and 87.8° as a function of the specific resistance at 20°. The readings of one set have been corrected by a small constant amount to make their mean agree with the other set; the difference between the two sets is doubtless due to different parasitic e.m.f.'s during the two runs, as has already been explained in connection with zinc. Within experimental error, thermal e.m.f. could be represented by a second degree series in the temperature. The deviations of the observed points from linearity at the mean temperature are shown in Figure 15. It is evident that within error the deviation is a linear function of the specific resistance, as is demanded by Kelvin's law. The total e.m.f., on the other hand, seems to be very definitely not a linear function. To a sufficient degree of approximation, the

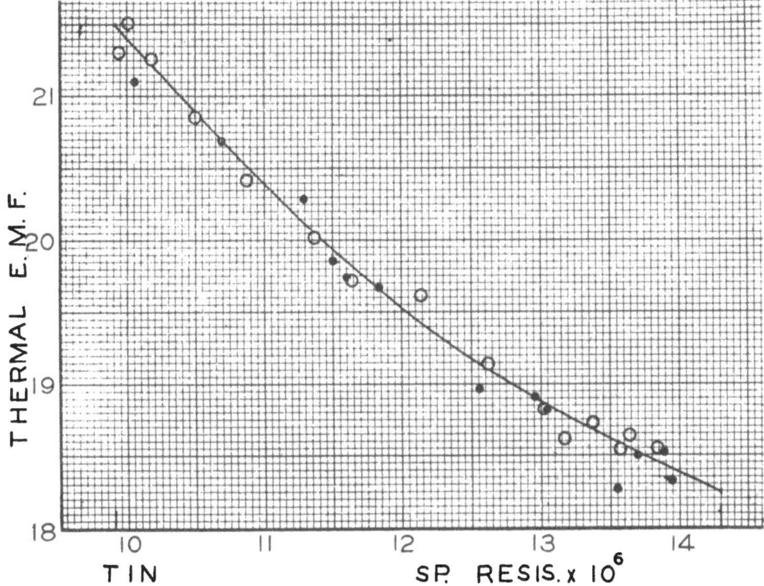

FIGURE 14. Thermal e.m.f. between 20° and 87.8° of single crystal tin against copper as a function of the specific resistance of the tin at 20°. The ordinates are centimeters deflection of the galvanometer; one centimeter is the equivalent of 1.096×10^{-5} volts.

e.m.f. is a second degree function in $\cos^2\theta$, so that the results can be reproduced by describing the behavior at the 0°, 45°, and 90° orientations. The results are as follows:

$$(\text{t.e.m.f.})_{Sn_{\parallel}-Cu} = (2.650 \times 10^{-6}) \times (\tau - 293.1) + 4.29 \times 10^{-9} \times (\tau - 293.1)^2 \text{ Volts},$$

$$(\text{t.e.m.f.})_{Sn_{45}-Cu} = (2.909 \times 10^{-6}) \times (\tau - 293.1) + 3.44 \times 10^{-9} \times (\tau - 293.1)^2,$$

$$(\text{t.e.m.f.})_{Sn_{\perp}-Cu} = (3.307 \times 10^{-6}) \times (\tau - 293.1) + 2.47 \times 10^{-9} \times (\tau - 293.1)^2,$$

$$(\text{t.e.m.f.})_{Sn_{\parallel}-Sn_{45}} = (-.259 \times 10^{-6}) \times (\tau - 293.1) + 0.85 \times 10^{-9} \times (\tau - 293.1)^2,$$

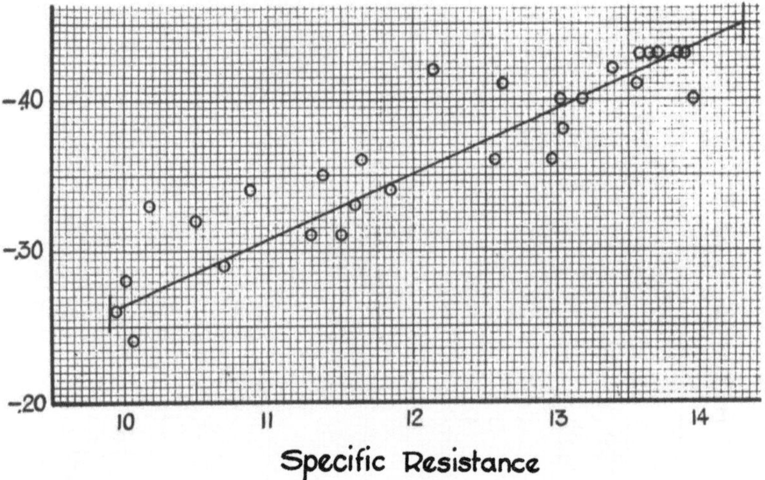

FIGURE 15. The deviation from linearity in the relation between temperature and thermal e.m.f. of single crystal tin against copper. The deviation is shown in terms of centimeters deflection of the galvanometer (1 cm. = 1.096×10^{-5} volts) at the mean temperature 53.9°, the temperature range being from 20° to 87.8° C. The abscissa is the specific resistance at 20° $\times 10^6$.

$$(\text{t.e.m.f.})_{Sn_{\parallel} - Sn_{\perp}}$$

$$= (-.657 \times 10^{-6}) \times (\tau - 293.1) + 1.82 \times 10^{-9} \times (\tau - 293.1)^2,$$

$$P_{Sn_{\parallel} - Sn_{45}} = \tau \times [-.259 \times 10^{-6} + 1.70 \times 10^{-9} \times (\tau - 293.1)],$$

$$\sigma_{Sn_{\parallel} - Sn_{45}} = \tau \times 1.70 \times 10^{-9},$$

$$P_{Sn_{\parallel} - Sn_{\perp}} = \tau \times [-.657 \times 10^{-6} + 3.64 \times 10^{-9} \times (\tau - 293.1)],$$

$$\sigma_{Sn_{\parallel} - Sn_{\perp}} = \tau \times 3.64 \times 10^{-9}.$$

The previous results* for tin were obtained from 14 specimens, of which only one had the basal plane approximately perpendicular to the length, the 13 others having the angle less than 40°; the results, furthermore, were much more scattering than the present ones, so that there is no doubt that the new results are much to be preferred. The value formerly found for the Peltier heat at 0° C

* In the previous paper,[1] the signs of the Peltier heat for tin given in Table II on page 132 should be negative instead of positive.

from the parallel to the perpendicular direction was $\tau \times (-.75 \times 10^{-6})$ against $\tau \times (-.73 \times 10^{-6})$ found here. The goodness of the agreement must be in part fortuitous. The previous measurements showed no difference in the Thomson heat between the perpendicular and the parallel directions; the difference found here is small. The e.m.f.'s against copper previously found were about 0.20×10^{-6} Volts lower than the new values; the difference is doubtless due to difference in the quality of the copper.

There are, of course, no measurements of the resistance as a function of the orientation, as there were for the other metals, since the method of determining the orientation did not permit this.

Bismuth. The material was Kahlbaum's. A large number of castings were made in the complicated many-arm molds. The orientation was determined from the cleavage plane; in a few cases the cleavage was made at liquid air temperatures, but cleavage at room temperature of rods as small as these was so easy that this refinement was not necessary. Four sets of measurements of thermal e.m.f. were made from four sets of 16 rods, selected to cover the orientation range.

Within experimental error the relation between e.m.f. and temperature is of the second degree in the temperature. In Figure 16 is shown the total e.m.f. between 20° and 88° against copper of the 64 specimens, plotted against $\cos^2\theta$, and in Figure 17 the deviations from linearity at the mid temperature. As in the case of tin, the total e.m.f. seems to definitely depart from Kelvin's $\cos^2$ law. The departure is not large, and the results can be sufficiently reproduced by giving the formulas for the 0°, 45°, and 90° orientations. The results follow:

$$(\text{t.e.m.f.})_{\text{Bi}_{\parallel}-\text{Cu}} = 1.084 \times 10^{-4} \times (\tau - 293.1) - 7.27 \times 10^{-8} \times (\tau - 293.1)^2 \text{ Volts},$$

$$(\text{t.e.m.f.})_{\text{Bi}_{45}-\text{Cu}} = .868 \times 10^{-4} \times (\tau - 293.1) - 3.46 \times 10^{-8} \times (\tau - 293.1)^2,$$

$$(\text{t.e.m.f.})_{\text{Bi}_{\perp}-\text{Cu}} = .572 \times 10^{-4} \times (\tau - 293.1) + 0.26 \times 10^{-8} \times (\tau - 293.1)^2.$$

Whence follows:

$$(\text{t.e.m.f.})_{\text{Bi}_{\parallel}-\text{Bi}_{45}} = .216 \times 10^{-4} \times (\tau - 293.1) - 3.81 \times 10^{-8} \times (\tau - 293.1)^2,$$

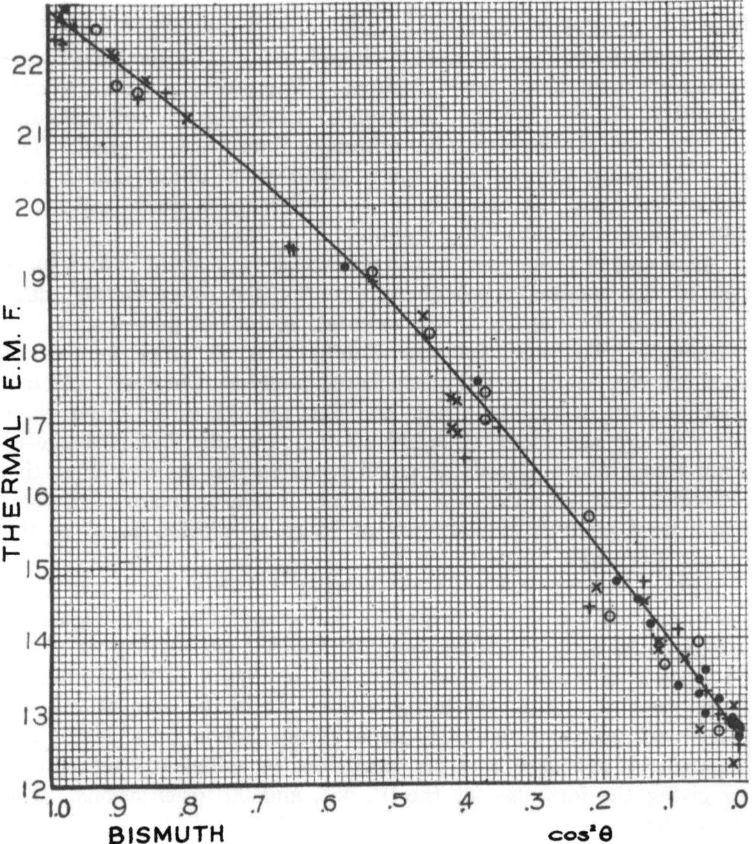

FIGURE 16. Thermal e.m.f. between 20° and 88° of single crystal bismuth against copper as a function of the square of the cosine of the angle between the crystal axis and the length. The ordinates are centimeters deflection of the galvanometer; one centimeter is the equivalent of 3.10 × 10⁻⁴ volts. The different symbols on the observed points refer to runs with different set-ups.

$(\text{t.e.m.f.})_{Bi_{11}-Bi_{\perp}}$

$$= .512 \times 10^{-4} \times (\tau - 293.1) - 7.53 \times 10^{-8} \times (\tau - 293.1)^2,$$

and:

$$P_{Bi_{11}-Bi_{45}} = \tau \times [.216 \times 10^{-4} - 7.62 \times 10^{-8} \times (\tau - 293.1)],$$

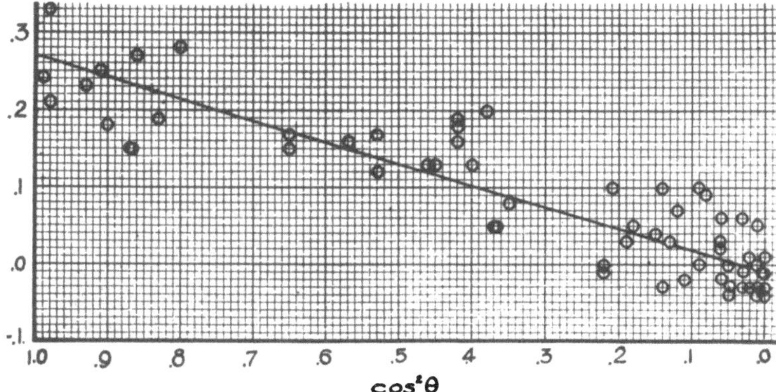

FIGURE 17. The deviation from linearity in the relation between temperature and thermal e.m.f. of single crystal bismuth against copper. The deviation is shown in terms of centimeters deflection of the galvanometer (1 cm. = 3.10 × 10⁻⁴ volts) at the mean temperature 54°, the temperature range being from 20° to 88°. The abscissa is the square of the cosine of the angle between the crystal axis and the length.

$$\sigma_{Bi_{\parallel}-Bi_{45}} = \tau \times [-\ 7.62 \times 10^{-8}],$$

$$P_{Bi_{\parallel}-Bi_{\perp}} = \tau \times [.512 \times 10^{-4} - 15.06 \times 10^{-8} \times (\tau - 293.1)],$$

$$\sigma_{Bi_{\parallel}-Bi_{\perp}} = \tau \times [-\ 15.06 \times 10^{-8}],$$

My previous results for bismuth covered a range of orientation of only 21°, so that a comparison with previous results is profitable only for the perpendicular orientation. The former value for the Peltier heat against copper was $0.578 \times 10^{-6} \times \tau$ at 0° C, against $0.571 \times 10^{-6} \times \tau$ found now. The agreement is good; better agreement is to be expected here than for the other metals because the absolute magnitude of the effects are so much larger for bismuth. Since my previous results were published, Boydston[10] has measured the thermal e.m.f. of single crystal bismuth. He used constantan as the metal of comparison, so that his results are not directly comparable with mine. It may be deduced from his Figure 1 on page 913 of his paper that the thermal e.m.f. between 0° and 100° C of a couple composed of $Bi_{\perp} - Bi_{\parallel}$ is about 5.4×10^{-3} Volts, against 4.7×10^{-3} to be deduced from my formulas above. Boydston's material covered a range of orientation up to within 23° of perpendicularity of basal plane to length, so that no large error is to be expected in his extrapolation to 90°. A more probable source of error I believe to be impurity in Boydston's bismuth. His metal was obtained from Mallinckrodt, and contained 0.04% of silver. This is not so stated in his paper, but I have learned from correspondence that the omission was due to an oversight. Now I have already found that silver[11] in amount even smaller than this may be responsible for important changes in the properties; I shall return to this question in discussing the resistance.*

In my previous work[9] the specific resistance had been found from a series of measurements on specimens covering an orientation range of only 45°. The points were furthermore rather scattering (Figure 6 on page 531 of my previous paper). In deducing from those measurements the most probable value of resistance, the guiding idea was that the lowest values were most likely to be correct, since high values can easily be produced by slight fissures on the cleavage plane due to deformation. The values deduced for specific resistance at 20° were: $\rho_{\perp} = 109 \times 10^{-6}$ and $\rho_{\parallel} = 138 \times 10^{-6}$. The ratio of the two resistances was thus only 1.27, against nearly 1.6 reported by practically all early observers.† Adopting a suggestion endorsed

* The thermo-electric properties of bismuth as a function of orientation have also been measured by Borelius and Lindh,[12] but it is not possible to compare their results with mine because of the very great variations which they find are produced by the temperature history of the specimen. About all that can be said is that their thermal e.m.f. between the two directions is of the same order of magnitude as that found here.

† Grüneisen (Handbuch der Physik, vol. XIII, p. 12) quotes Borelius and Lindh as giving the ratio 1.34, but it is not obvious from what part of their work this figure was obtained. Probably it was deduced from Figure 2 on page 613 of their paper in Ann. Phys. 51, 1916. But Borelius and Lindh were apparently not willing to commit themselves as to just what the ratio of the specific resistances really is, in view of the large effect which they attributed to fissures. Thus on page 614 of the paper just quoted they say: "This survey makes it very probable that the excess resistance (parallel to the axis over that perpendicular to the axis) becomes very small or entirely vanishes under sufficient pressure." In a later paper (Ann. Phys. 53, p. 124, 1917), in describing the symmetry relations of the resistance, they give for the resistance at 20° C, perpendicular, at 45°, and parallel to the axis the figures 1.24, 1.41, and 1.92 $\times 10^{-6}$, which makes the ratio 1.55, but they remark that the figure 1.92 can be brought down to 1.51 by pressure. The general impression which the work of Borelius and Lindh gives is that in their opinion most, if not all, of the differences observed between different directions in bismuth are due to fissures or internal stresses.

by Borelius and Lindh, I attributed the difference to the effect of fissures. Since that work, a new determination has been made by Schneider,[6] who finds 1.62 for the ratio; Schneider is apparently inclined to accept my explanation of the discrepancy as due to the effect of fissures. It seemed therefore important to settle this question as definitely as possible. I was in a position to much improve on my former results because I could now cover the entire range

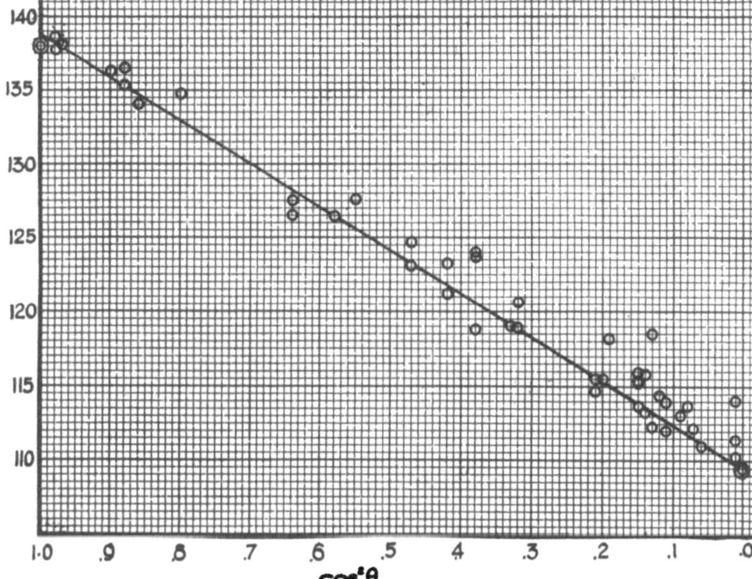

FIGURE 18. Specific resistance in 10^{-6} ohms per cm. cube at 20° of single crystal bismuth as a function of the square of the cosine of the angle between the crystal axis and the length.

of orientation. I did not make measurements on the crystals before they had been removed from the glass mold, as had been done for zinc and cadmium, but I selected from the large number of specimens those which had received little or no bending, and which sufficiently covered the entire range of orientation. All the results for the specific resistance of these selected specimens, *making no discards*, are shown in Figure 18 plotted against $\cos^2\theta$. It is evident in the first place, that Kelvin's $\cos^2$ law for specific resistance is satisfied.

In the figure are also shown as double circles my previous values for the parallel and perpendicular orientations; it is evident that these values were substantially correct.

Since the experimental work of this paper was done, a paper has appeared by Kapitza (Proc. Roy. Soc. 119, 358, 1928), in which a very careful investigation was made of the resistance as a function of direction. Kapitza grew his crystals by a modified method in which the crystal was not constrained by a mold. He found specimens with abnormally high resistance, but a study of the effect of compression on these led him to accept the explanation that the effect is due to minute cracks. He gives as the best values for resistance: 139×10^{-6} perpendicular to the cleavage plane, and 107 parallel to it. The ratio is 1.30, not far from my value 1.27. I believe that my value is somewhat to be preferred, because Kapitza's values for the parallel orientation showed considerable scattering, from 107 to 114, but in any event there can now be no question whatever that the true ratio cannot possibly be as high as found by the early observers.

I now believe that there are two possible explanations for the large value found for this ratio by other observers; one is the effect of fissures, as already suggested. This is substantiated by my own observations made on specimens in which the bending was known to be severe, which gave points lying very far above the curve of Figure 18 (these points are not shown in the diagram), by the observation of Borelius and Lindh[12] that the resistance of a rod with cleavage plane perpendicular to the length receives a large discontinuous decrease on applying a small longitudinal pressure in such a direction as to close the hypothetical fissures, and now by the recent work of Kapitza. But that this is not the only explanation became obvious on examining some of the specimens of Schneider, which had the high ratio, and which he was so kind as to send. On applying hydrostatic pressure, these showed no discontinuity whatever of resistance, as must have been the case if there had been cracks. I found that the temperature coefficient of resistance of these samples of Schneider was only 0.0032 instead of 0.0044 to be expected for pure bismuth. I have already found that a minute trace of silver has a large effect on temperature coefficient, and Schneider's bismuth is known to contain 0.04% silver. It seems possible that during the slow cooling incident to casting in unicrystalline form the silver may segregate on the basal planes in such a way as to increase the resistance to flow perpendicular to this plane, but not to affect it very much for parallel flow. Silver is one of the commonest impurities in bismuth, and it is also very difficult to remove, so that the same source of error may have been present in the material of some of the other observers.

On the Rotational Symmetry of Thermal E.M.F.

Altogether, I spent much more time in the investigation of bismuth than of the other metals, since the chance of finding new effects in bismuth is higher because of the larger numerical values to be expected. One of the questions investigated was that of the rotational symmetry of the thermo-electric properties. The symmetry is assumed to be of this sort in the argument of Kelvin, but if Kelvin's $\cos^2$ relation does not hold, as apparently it does not for bismuth, it becomes important to examine this assumption. I briefly touched on this question in my previous paper; I had not realized the importance of the question on gathering the data for that paper, and my argument rested on the general smoothness of the results and two fortuitous measurements for bismuth. I have now made a systematic examination of this question. In one set of 16 rods, not only was the orientation of the cleavage plane with respect to the length determined, but also the orientation of the secondary cleavage planes; these two together are sufficient to completely define the orientation of the rod with respect to the crystal. The secondary cleavage planes were determined from the triangular pattern which close examination will almost always disclose in the basal cleavage plane. Secondary cleavages may be started from the sides of this triangular pattern. The set of rods chosen all had the basal plane nearly parallel to the length, since for this orientation the effect would be expected to be a maximum. It is obvious that if the basal plane is perpendicular to the length, the orientation of the triangular pattern must be without effect on the longitudinal e.m.f. No correlation whatever was found between the thermal e.m.f. and the orientation of the secondary cleavage plane, thus establishing that within the limits of error indicated by the points in Figure 16 the thermo-electric properties have rotational symmetry about the trigonal axis, as assumed by Kelvin. Of course this proof applies only to bismuth, but it is probable that the other metals have rotational symmetry also. Direct proof would have been much more difficult for any of the others, except for antimony, but the regularity of the results when plotted against only a single parameter, the angle between axis and length, makes it probable that no other parameter enters.

Pressure Coefficient of Resistance. Previously I had measured the effect of pressure on the resistance of two samples of single crystal bismuth,[13] with the basal plane parallel and at 33° to the length. These measurements are now extended to three samples, with the basal plane as 4°, 41°, and 82° to the length. Measurements were made on each of these specimens at 30° and 75° over a pressure range of 12000 kg. The potentiometer method was used as usual. Terminals were soldered to the specimens, so that the following results must be corrected by the change of dimensions to obtain the true relative specific resistances. The accuracy of the measurements was that usual in this work; the average departure of a single reading from a smooth curve for all six runs with three samples was 0.10% of the maximum pressure effect, the extreme variations in accuracy being from 0.02% to 0.15%. Within these limits of error the results can be reproduced by a two-power series in the pressure; this was not the case with the previous measurements. The results follow, pressure being in kg/cm^2:

TABLE II.

Angle between Basal Plane and Length	$\frac{\Delta R}{R_{(0.30°)}}$	$\frac{\Delta R}{R_{(0.75°)}}$
4°	$+1.047\times10^{-5}p+2.96\times10^{-10}p^2$	$+1.055\times10^{-5}p+2.67\times10^{-10}p^2$
41°	+1.564 +4.93	+1.513 +3.92
82°	+2.011 +8.10	+1.975 +6.14

The average temperature coefficients between 0° and 100° of these three samples were: 0.00441, 0.00431, and 0.00437 respectively. These values are nearly as high as the best which I have ever obtained for the purest electrolytic bismuth (00441), and are evidence of the high purity of this material. Before the war, the best bismuth which I could get from Kahlbaum had a coefficient of only 0.00332; it is evident that the quality of Kahlbaum's bismuth has improved, or possibly the process of very slow solidification to produce the single crystal itself brings about further purification.

In general, the new results under pressure bear out the results found previously in a narrower range. The pressure coefficient nearly doubles as the cleavage plane changes from parallel to perpendicular to the direction of flow, and except for orientations very close to parallel, the pressure coefficient is less at the higher temperatures. The effect of pressure is therefore to accentuate the excess of resistance across the cleavage planes over that parallel to the planes; one would perhaps be inclined to expect the opposite behavior.

Discussion of Various Questions Connected with the Thermo-Electric Behavior of Crystals.

On the Connection between the Symmetry of the Peltier and Thomson Heats. The conclusion reached in my former paper was that the *total* thermal e.m.f. of a crystal couple need not satisfy Kelvin's law. The total e.m.f. may be analyzed into contributions made by the Peltier heat and the Thomson heat. The conclusion reached in that paper was that the Thomson heat does satisfy Kelvin's law, but that the Peltier heat does not. This conclusion is also substantiated by the new data of this paper, since in the two cases which did not satisfy Kelvin's law, bismuth and tin, the deviation from linearity, which determines the second temperature derivative and so the Thomson heat, was nevertheless linear against $\cos^2\theta$ (or against the specific resistance). We now have to inquire whether this difference between the symmetry of P and σ is a real difference, or whether it is merely apparent, due to the unavoidably greater error in the determination of σ.* There is, of course, a thermodynamic connection between P and σ, namely:

$$\frac{d}{d\tau}\left(\frac{P}{\tau}\right) = \frac{\sigma}{\tau}.$$

Let us now assume that σ satisfies Kelvin's relation, so that we may put $\sigma = f_1(\tau) + f_2(\tau)\cos^2\theta$, and ask what is thereby involved about the symmetry of P. The equation may be integrated at once, giving:

$$\frac{P}{\tau} = \int_0^\tau \frac{f_1(\tau)}{\tau}\,d\tau + \cos^2\theta \int_0^\tau \frac{f_2(\tau)}{\tau}\,d\tau + \psi(\theta).$$

* I had gone through considerations of this sort a number of years ago, but had not published them. The importance of the matter was recalled to my attention by a conversation with Dr. Goens in 1926.

$\psi(\theta)$ is an arbitrary function of integration, so that as far as the differential relation between P and σ is concerned, there is no reason why P should be linear in $\cos^2\theta$ at constant temperature merely because σ is.*

But now the third law of thermodynamics suggests that

$$\lim_{\tau=0}\left(\frac{P}{\tau}\right) = 0.$$

Call

$$\lim_{\tau=0}\int_0^{\tau}\frac{f_1(\tau)}{\tau}\,d\tau = A_1, \quad \text{and} \quad \lim_{\tau=0}\int_0^{\tau}\frac{f_2(\tau)}{\tau}\,d\tau = A_2,$$

and we have

$$\psi(\theta) = -A_1 - A_2\cos^2\theta,$$

so that in general:

$$P = \tau\left\{\left(\int_0^{\tau}\frac{f_1(\tau)}{\tau}\,d\tau - A_1\right) + \cos^2\theta\left(\int_0^{\tau}\frac{f_2(\tau)}{\tau}\,d\tau - A_2\right)\right\},$$

and at constant temperature this is linear in $\cos^2\theta$. It seems, then, that the third law demands that the Peltier heat satisfy Kelvin's law if the Thomson heat does, and the explanation of the apparent difference in symmetry found experimentally is the greater error of the Thomson heat.

Although the conclusion reached by the argument in the last paragraph appears to me very probable, I do not believe that it can be regarded as entirely rigorous. The question comes in inferring from the third law that P/τ goes to 0 at 0° Abs. Strictly, I suppose that the third law means only that the total entropy change accompanying any finite change in a condensed system at 0° Abs. should vanish. Now consider a couple at 0° Abs. composed of two crystal rods of different orientations. A current may be made to flow in such a system, but the total heat absorbed by the current is necessarily zero because a Peltier heat at one junction is exactly neutralized by a Peltier cold at the other. If the system is so arranged that the Peltier heat arises from the transfer of static charge so that this compensation does not occur, then the system is left charged after the passage of current, and the surface heats of charg-

* The integration as given is not rigorously exact, because in a non-cubic crystal in which the thermal expansion is different along the different axes $\cos\theta$ is a function of temperature, but the effect is very small.

ing make contributions to the total heat absorbed, so that again the total heat absorbed would be zero.

On the Transverse Thermal Effect, and on Kelvin's Axiom of the Superposition of Heating Effects. Kelvin based his proof of the symmetry of the Peltier heat when a current leaves a crystal surface, and also his proof of the existence and the symmetry of the transverse heating effect,[14] on his so-called "Principle of the Superposition of Thermo-Electric Action," which is: "Each of any number of co-existing systems of electric currents produces the same reversible thermal effect in any locality as if it existed alone." In the paper[15] in which I announced my experimental discovery of the transverse effect,* I stated that Kelvin's principle was internally inconsistent, and that therefore his prediction of the existence of the transverse effect could not be justified. I now find that this statement of mine was not correct, but that if due account is taken of the transverse effect itself, the principle may be maintained and the symmetry relations given by Kelvin are a consequence of the principle.

To straighten the matter out, it will be necessary to go back a considerable distance and correct another erroneous deduction which I made before I knew of the existence of the transverse effect, and which is essentially modified by it.[16] Consider the junction between two metals, one of them a single crystal and the other isotropic, shown in Figure 19. It is evident that thermodynamics requires that the total heat absorbed when a current flows from the single crystal, C, to the isotropic metal, A, in the region at constant temperature, t, be independent of the precise details of the transition

* When I wrote that paper I supposed that I was the first to find the transverse effect experimentally. Dr. Borelius has, however, recently kindly called my attention to the fact that he and A. E. Lindh described the experimental investigation of this effect in 1917 (Ann. Phys. 53, 97, 1917), in a paper entitled "Resistance, Peltier Heat and Electrical After-Effects, outside and within a Magnetic Field, particularly in Crystalline Bismuth." They made quantitative measurements. I have compared their numerical results with those to be expected from my own measurements of difference of thermo-electric properties of bismuth in different directions, and find agreement within about 30%. The agreement is as good as could be expected in view of the unfavorable dimensions of their specimen, and there can be no doubt that they actually observed the effect. I am sorry that I did not know of their work when I wrote my paper. It is interesting that since I wrote my paper the experimental detection of the effect has also been announced by Terada and Tsutsiu in Japan.

from one metal to the other in the region, so that the total heat absorbed by the current is the same for the two arrangements of the figure. I drew from this the conclusion that the Peltier heat at the surface depends only on the direction of flow of the current within the body of the crystal with respect to the crystal axis, and is independent of the orientation of the surface by which it leaves the crystal. It is now obvious that this conclusion is erroneous, because it neglects the transverse effect at the free surfaces of the crystal. Heat is absorbed when the current passes, not only at

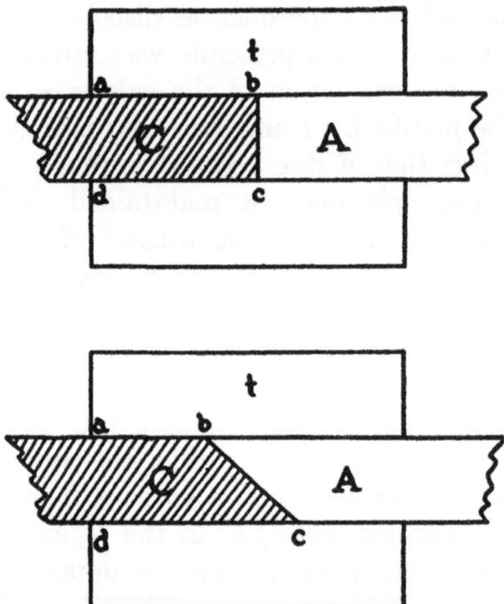

FIGURE 19. Two different methods of making the junction between an isotropic metal and a single crystal in a region at constant temperature. The total heat absorbed by unit quantity of electricity in passing from crystal to isotropic metal is the same in the two arrangements of the figure.

the surface *bc* which separates the two metals, but also at the transverse surfaces *ab* and *dc*. In the upper arrangement of Figure 19, the surfaces *ab* and *dc* are equal in extent, and the heat absorbed at *ab* is exactly neutralized by the heat given out at *dc*. But in the lower arrangement, this neutralization does not take place. It is evident that the correct theorem is that the total heats absorbed at the surface AB and BC, Figure 20, is equal to that which would be absorbed if the crystal were cut on AC. The heat absorbed at

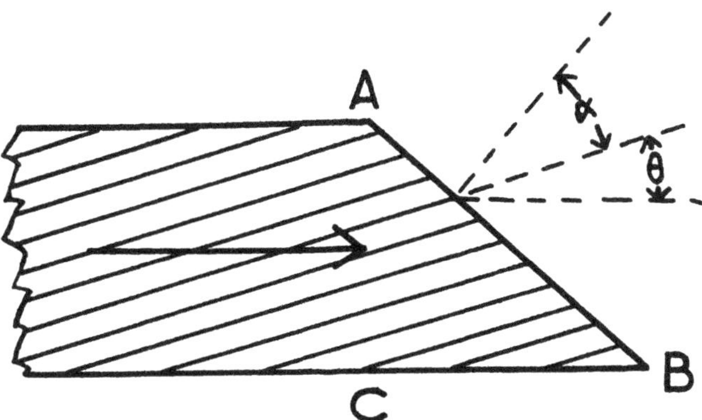

FIGURE 20. Diagram for the calculation of the relation between the reversible heats at various surfaces.

AB is proportional to the total quantity of electricity crossing the surface; hence if we take AC = 1, the cross section of the crystal equal to unity, and denote the current density by i, the heat will be merely proportional to i. This heat evidently involves two parameters: θ, the angle which the current within the crystal makes with the crystal axis, and α, the angle between the crystal axis and the normal to the surface. We may, therefore, put for the heat absorbed in unit time at AB, i P (θ, α). The transverse heat absorbed at BC by current parallel to the surface is evidently proportional to current density and the extent of the surface. Call the heat per unit current density per unit area T. T is evidently a function of the angle between the crystal axis and the surface. There is a sign convention that must be observed here. T reverses sign accordingly as the rotation through an acute angle which carries the current (or the surface CB) into the axis carries one inside or outside the crystal. There must also be a sign convention for α. The relation of heat equality is now evidently (since BC = tan $(\alpha + \theta)$):

$$P(\theta, \alpha) + T(\theta) \tan(\alpha + \theta) \text{ is independent of } \alpha.$$

The sum of these two expressions, or the value of P when $\alpha = -\theta$ $[= P(\theta, -\theta)]$, is what is given by the usual experiments with thermo-couples. In order to find P and T separately, it would be necessary to experiment with surfaces so large that the heat could be localized.

The conclusion of this analysis is, therefore, that the local heat absorbed when a current leaves a crystal and enters an isotropic conductor is a function both of the direction of flow with respect to the crystal axis in the crystal and the orientation of the surface. Now apply this to Kelvin's principle of the superposition of heating effects. My former proof that this cannot be correct rested on the assumption that the heat absorbed on crossing the surface AB by any of the components into which the current i was resolved was a function only of the direction of the component with respect to the crystal axis, and did not involve the orientation of AB with respect to the axis. But this assumption is erroneous. On using the correct and more general expression above, and also using Kelvin's explicit expressions for P and T

$$\left[\begin{array}{ll} P \equiv P(\theta, -\theta) & = P_{\perp} \cos^2 \theta + P_{11} \sin^2 \theta \\ T(\theta) & = (P_{\perp} - P_{11}) \cos \theta \sin \theta \end{array}\right],$$

everything is found to carry through consistently, the total heat is independent of the way in which the current is resolved into components, and the principle of superposition is logically possible.

The only question remaining is, therefore, whether as an experimental fact the symmetry relations of Kelvin are satisfied? We have seen that they are approximately satisfied, but in the case of bismuth and tin, the departures are apparently beyond experimental error. As far as I can see, there is no reason for thinking that the principle must necessarily be satisfied. It is without doubt true, however, that the principle holds to a first approximation in crystals hitherto examined, and in constructing a theory of these effects, it would doubtless be legitimate to begin with this assumption as a first approximation.

The fact that Kelvin's principle can be logically maintained involves a serious modification of a conclusion which I had drawn from my previous experimental results.[17] Granting my previous argument, neglecting the transverse effects, then it was easy to show that Kelvin's principle could be maintained only for one special method of resolving the current, namely parallel and perpendicular to the crystal axis. From the fact that experimentally Kelvin's symmetry relations were approximately satisfied I drew the conclusion that the electrons which constitute the current are actually compelled to travel either perpendicular or parallel to the crystal axis, by some mechanism more or less of the nature of channels. This conclusion must now be abandoned; from the approximate validity of Kelvin's relation no conclusion whatever can be drawn as to the paths of the electrons.

Summary.

An improved method of making single metal crystals has been developed by which it is possible to cast from the same melt a number of single crystal rods of a wide range of orientation. An apparatus has been developed by which the thermal e.m.f. of 16 rods may be measured simultaneously, so that the only variable factor in the results is the crystal orientation. The thermal e.m.f. between 20° and 88° C of single crystal Zn, Cd, Sb, Sn, and Bi against Cu has been measured, and from the results the thermal e.m.f., Peltier heat, and difference of Thomson heats between rods of the same metal of different orientations is calculated. It is emphasized that in the case of Sb and Bi the heat absorbed by an electron when its direction of motion changes from perpendicular to the axis to parallel to the axis is very large, being respectively 0.2 and 0.4 of the energy of a gas molecule at the same temperature. This is a difficult point for any theory like the recent one of Sommerfeld, in which the part of the energy of an electron which varies with the temperature is supposed small compared with the classical amount. The Kelvin-Voigt law that thermal e.m.f. is a linear function of $\cos^2\theta$, θ being the angle between crystal axis and length of the rod, is verified for Zn, Cd, and Sb, but there are deviations for Sn and Bi which seem distinctly greater than possible experimental error, so that it would appear that in general the Kelvin-Voigt law is only an approximation. Within experimental error the Thomson heat does satisfy the law in all cases; it is only the contribution to total e.m.f. made by the Peltier heat that fails to satisfy the law within experimental error. In the case of Bi a special examination was made of the Kelvin-Voigt assumption that thermal e.m.f. has rotational symmetry about the crystal axis, and no appreciable deviation was found.

The specific resistance of these metals has also been studied as a function of orientation. The possible error from distortion is so great in the case of Cd that a special method of measurement had to be used. The Kelvin-Voigt symmetry relation for resistance is satisfied within experimental error. Especial attention was given to the resistance of Bi, and the previous low value perpendicular to the cleavage plane verified. The pressure coefficient of resistance of Sb and Bi was redetermined, so that now this quantity is known over the entire range of orientation.

In the theoretical discussion it is shown that the third law of thermodynamics gives considerable plausibility, although not complete certainty, to the thesis that the symmetry of the Peltier and the Thomson heat must be the same; this means that the apparent experimental difference in the symmetry of the two heats is merely an effect of the much greater error in the Thomson heat. It is shown that, contrary to the statements previously made by me, Kelvin's axiom of the superposition of thermal effects of currents is not internally inconsistent, so that his proofs of the symmetry relations and of the existence of a transverse temperature effect are logically possible. Experiment, however, seems opposed to the truth of the axiom. The revised discussion further shows that the surface heat where a current leaves a metal crystal is a function both of the direction of flow with respect to the crystal axis within the crystal and of the orientation of the surface, contrary to my former statement. It is also shown that these crystal phenomena no longer offer a basis for a proof that the electrons must move in the crystal along something analogous to fixed channels.

THE JEFFERSON PHYSICAL LABORATORY,
Harvard University, Cambridge, Mass.

REFERENCES.

[1] P. W. Bridgman, Proc. Amer. Acad. 61, 101, 1926.

[2] E. G. Linder, Phys. Rev. 29, 554, 1927.

[3] G. W. Schneider, Phys. Rev. 31, 251, 1928.

[4] P. W. Bridgman, Proc. Amer. Acad. 60, 305, 1925.

[5] G. Sachs, Mitteilungen aus dem Materialprüfungsamt und dem Kaiser Wilhelm-Institut für Metallforschung, 1927, Neue Folge, Heft 5.

[6] R. W. Boydston, Phys. Rev. 30, 911, 1927.

[7] Reference 1, page 114.

[8] E. Grüneisen und E. Goens, ZS. f. Phys. 26, 223, 1924.

[8a] A. Sommerfeld, ZS. f. Phys. 47, 1, 1928.

[9] P. W. Bridgman, Proc. Amer. Acad. 60, 361, 1925.

[10] Reference 6.

[11] P. W. Bridgman, Proc. Amer. Acad. 57, 114, 1922.

[12] G. Borelius und A. E. Lindh, Ann. Phys. 51, 607, 1916.
[13] Reference 9, page 353.
[14] Kelvin, Mathematical and Physical Papers, Vol. I, page 267.
[15] P. W. Bridgman, Proc. Nat. Acad. Sci. 13, 46, 1927.
[15a] T. Terada and T. Tsutsui, Proc. Imp. Acad. (Tok.) 3, 132, 1927.
[16] Reference 1, page 124.
[17] Reference 15, page 50.

THE EFFECT OF PRESSURE ON THE RIGIDITY OF STEEL AND SEVERAL VARIETIES OF GLASS.

By P. W. Bridgman.

Presented Oct. 10, 1928. Received Oct. 15, 1928.

CONTENTS.

Introduction.

The change of rigidity or shearing modulus under pressure seems to have never been determined. The measurement of this effect is not easy, and for a number of years I have been searching for a suitable method. With the development of the sliding contact potentiometer method of measuring small displacements,[1] which I have already applied to the measurement of various small motions under pressure, it was evident that a means was at hand, and in this paper the application of such a method is described. When I first approached the problem, I felt that a method which would give merely the order of magnitude of the effect would be worth while, and my attack was conducted in this spirit. It seemed probable that the effect of pressure on the shearing modulus of a very compressible substance such as glass would be much higher than on such an incompressible substance as steel, and that to a first approximation the effect on steel could be neglected in comparison with that on glass. The first method adopted was a differential method, therefore, which gave the difference between the effects on steel and glass; by setting the effect on steel equal to zero, I hoped to get a first approximation to the effect on glass. The numerical magnitudes obtained were not, however, in the expected range, so that it appeared very questionable whether the assumption of a vanishingly small effect on steel was justifiable. It became necessary, therefore, to devise a second method allowing an absolute determination on the steel which had been used as the standard of comparison. It turned out that the magnitude of the effect on glass does not differ by a very large amount from that on steel, so that there was no real necessity in having made the first measurements on a substance of so comparatively little physical interest as glass. I plan now to extend the measurements to other metals besides steel.

Experimental Methods.

In view of the smallness of the effect, it was necessary that the displacement, the change of which under pressure was to be measured, should be as large as possible. The helical spring satisfies this requirement, and also the necessary limitations imposed by the small size of the pressure apparatus, better than any other arrangement, such, for example, as the twist in a long slender rod. Furthermore, the extension of a helical spring of ordinary dimensions under load involves to a sufficient approximation only the geometrical dimensions and the shearing modulus, so that the required

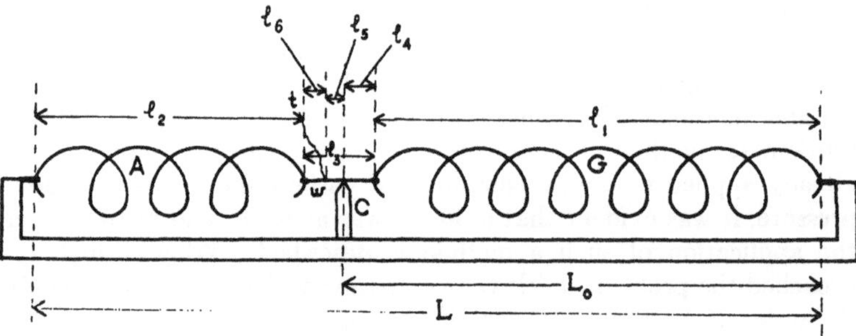

Figure 1. Scheme of the apparatus for determining the difference of the effect of pressure on the stiffness of two springs. If there is a change of relative stiffness under pressure, there is a shift of the point of coupling.

information could be calculated simply from the change of extension of the spring. The apparatus adopted for measuring the differential effect is shown in Figure 1. The glass spring G and the steel spring A are coupled together by a wire of manganin w. The springs are stretched to perhaps twice their normal length and attached at either end to a steel frame. The manganin wire slides over a contact C fixed to the steel frame, and there is a potentiometer terminal t attached to the wire. The entire assembly is exposed to hydrostatic pressure; if there is a change under pressure of the relative stiffness

80 — 2296

of the springs, there will be a motion of the point of coupling. If A becomes relatively stiffer than G, A shortens while G is lengthened, and conversely. The amount of this motion is measured in the conventional way on a potentiometer, current passing lengthwise in w, and the difference of potential between C and t being determined.

The method for measuring the absolute effect, shown in Figure 2, is even simpler. The spring of steel is suspended vertically, carrying at the bottom end a weight P, separated from the spring by a length of manganin wire w, which carries the lead wire t and, as in the first method, slides on the contact C rigidly connected to the upper support of the spring. The whole arrangement is immersed in the liquid by which hydrostatic pressure is transmitted. If the shearing modulus increases or decreases under pressure the spring shortens or lengthens accordingly, and the amount of motion is obtained electrically from the difference of potential between C and t.

It is obvious that a number of experimental precautions must be observed to make such apparatus function properly, and there are a number of corrections to be applied in making the calculations. Perhaps the most obvious and important difficulty is friction at the sliding contact C. The changes are so slight, the whole motion under 12000 kg. being of the order of 0.5 mm., that very slight friction here would entirely mask the effect. To avoid error from friction, the differential apparatus was mounted horizontally in a pressure cylinder which could be rotated through 180° about the horizontal axis of the springs. The rotatable pressure apparatus which this demands had been already constructed, and was the same as that which had been used in measuring the viscosity of liquids; it could be used again for this purpose with only minor alterations. While the pressure was being changed, the cylinder was rotated through 180°, so that the contact C, Figure 1, was over the wire. In this position the weight of the springs and the wire produce a slight amount of sag, so that the contact was broken, and there is therefore no frictional resistance to the springs taking the exact position of equilibrium. To make the reading, the cylinder was rotated back through 180°, the wire w dropping back into contact with C, without encountering any friction. The contact so made was often so light that there were various coherer effects from outside disturbances, which made readings difficult. The resistance of the contact was decreased by passing through it current from a small bell-ringing magneto, in the same way that I have previously done in making resistance measurements.

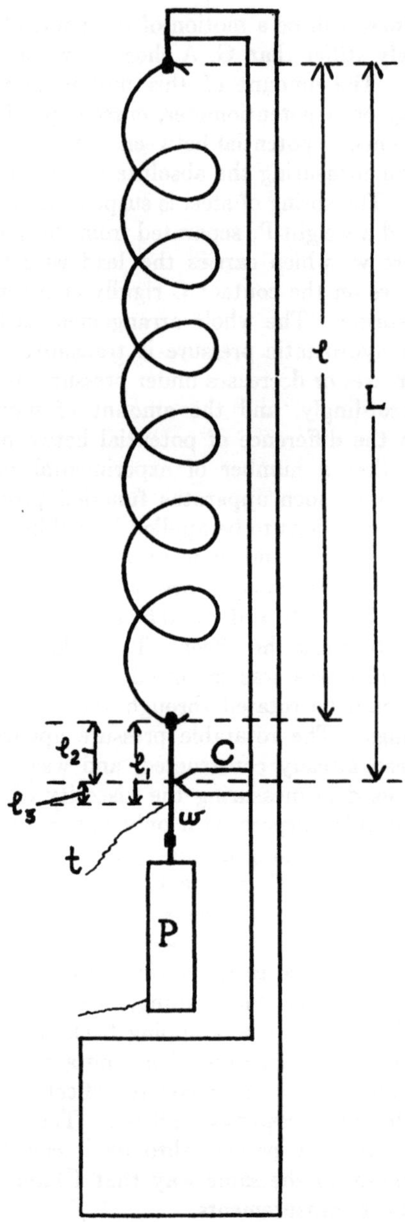

Figure 2. Scheme of the apparatus for determining the effect of pressure on the stiffness of a spring. The spring changes length when its stiffness changes.

In the case of the absolute apparatus (Figure 2), friction at the contact C during change of pressure was avoided by so mounting the apparatus that it could be very slightly tilted from the vertical, thus breaking contact.

It is evident that the various leads must be so flexible as to have negligible stiffness compared with the spring. The leads, t, were made of copper wire .004 cm. in diameter, wound into a long helix. The current connection at the glass spring end of the manganin wire in Figure 1 was a helix of this same wire, lying inside and concentric with the glass spring. The other current connection was made through the steel spring, which was grounded to the apparatus. In the absolute apparatus (Figure 2), the steel spring, grounded to the apparatus, was used for one current lead, and the other was a helix of the same fine wire grounded to the weight.

The measurements were made in a straight forward manner, and do not require detailed description. The pressure range was 12000 kg./cm.[2] The temperature of most of the measurements was 30°, which was maintained with a thermostat. Rough measurements were made on two of the specimens to find whether there was any large temperature effect. Measurements were usually made at 2000 kg. intervals in the order: 0, 4000, 8000, 12000, 10000, 6000, 2000, and 0. The accuracy was not sufficient to justify an attempt to determine departures of the effect from linearity.

Corrections and Method of Calculation.

The corrections play an important part, so that it will be necessary to describe them in considerable detail. Consider first the differential method.

The force exerted by each spring is proportional to its extension from its unstressed length, and the condition of equilibrium when the two springs are coupled together is that the force exerted by the two springs should be equal. These conditions give

$$F_1 = k_1(l_1 - l_1'), \quad F_2 = k_2(l_2 - l_2'),$$

with the equilibrium condition

$$k_1(l_1 - l_1') = k_2(l_2 - l_2').$$

Besides the quantities which are sufficiently explained in Figure 1, l_1' and l_2' are the lengths of the two springs under no extensive force. The condition of equilibrium holds both before and after the application of pressure, so that we also have

$$(l_1 - l_1')dk_1 + k_1(dl_1 - dl_1') = (l_2 - l_2')dk_2 + k_2(dl_2 - dl_2'),$$

where the differentials are the increments produced by a given increment of pressure. Call now the linear compressibility of the two springs χ_1 and χ_2, that of the manganin χ_3, and that of the steel frame χ_4 (we choose the subscript 2 to denote the steel spring, so that for this particular apparatus χ_2 and χ_4 are to a sufficient approximation equal to each other). Using the condition that $l_1 + l_2 + l_3 = L$, and also that $dl_1' = -l_1'\chi_1 dp_1$, etc., a first equation is obtained:

$$(l_1 - l_1')dk_1 + k_1(dl_1 + l_1'\chi_1 dp)$$
$$= (l_2 - l_2')dk_2 + k_2\{(l_2'\chi_2 + l_3\chi_3 - L\chi_4)dp - dl_1\}. \quad (1)$$

In this equation, l_1, l_1', l_2, and l_2' are known from geometrical measurements of the apparatus; χ_2, χ_3, and χ_4 are known by independent experiment; k_1 and k_2 must be found by measurements of the extension of the spring under known weights; dk_2 is supposed known from the auxiliary experiments on the absolute pressure effect by the second method, and dp is the arbitrary increment of pressure, so that the only unknowns are dl_1 and dk_1. It is our next task to calculate dl_1 in terms of the measured change of resistance between the points C and t, thus permitting a determination of dk_1, and thus eventually of $d\mu_1$, the change of shearing modulus.

To connect dl_1 with the measured change of resistance, we have the relations $l_1 + l_4 = L_0$, and the similar equation obtained by differentiating this with respect to pressure; $l_4 + l_5 + l_6 = l_3$, and the corresponding differentiated equation; and $l_5 = R_0/\rho_0$ and the corresponding equation at pressure dp, which is

$$l_5 + dl_5 = \frac{R_0 + \Delta R_0}{\rho_0(1 + \alpha dp)}.$$

Combining these equations gives finally:

$$dl_1 = \frac{\Delta R_0}{\rho_0} + dp\left\{\chi_3(l_3 - l_6) - \chi_4 L_0 - \alpha\frac{R_0 + \Delta R_0}{\rho_0}\right\}. \quad (2)$$

In this equation, α is the pressure coefficient of the resistance of the wire w per unit length. It is the pressure coefficient of resistance measured in the usual way with terminals rigidly attached to the wire, corrected by the linear compressibility.

Let us next examine how to calculate dk_2 from the measurements with the absolute apparatus. By writing the equations $l + l_2 = L$, $l_2 = l_1 - l_3$, $l_3 = R_0 / \rho_0$, and the corresponding equations after pressure has been applied (see Figure 2 for notation), we find for the change of length of the spring in terms of the change of resistance, etc.:

$$dl = -\chi pL + \chi_m l_1 dp + \Delta R_0/\rho_0 - \{(R_0 + \Delta R_0)\alpha dp\}/\rho_0.$$

χ is the linear compressibility of the frame, and χ_m of the manganin wire.

To connect with the change in shearing modulus, we have the formula for a helical spring given by Miller:[2]

$$\mu = \frac{2Ps^3\cos^2\alpha}{\pi a^4\phi_0^2(l - l_1)}.$$

Here

μ = shearing modulus of spring.
a = radius of the wire of the spring.
P = total longitudinal pull.
l' = initial length along axis of helix.
l = length under load along axis of helix.
α = angle between spires of spring and horizontal.
s = total length of spring wire ($l = s\sin\alpha$).
ϕ_0 = total angular twist of unstretched spring, in radians.

Differentiate this equation logarithmically, obtaining:

$$\frac{d\mu}{\mu} = \frac{dP}{P} + 3\frac{ds}{s} - 4\frac{da}{a} + 2\frac{d(\cos\alpha)}{\cos\alpha} - 2\frac{d\phi_0}{\phi_0} - \frac{d(l - l')}{l - l'}.$$

We have the relation

$$\frac{ds}{s} = \frac{da}{a} = -\chi dp.$$

If the material of which the helix is composed is in a state of ease, as we assume is approximately the case, then $d\phi_0 = 0$, because the effect of a change of pressure is merely a change of linear dimensions without a change of angle.

Using the connection between l and s, gives

$$\frac{dl}{l} = \frac{ds}{s} + \frac{\cos\alpha}{\sin\alpha}d\alpha.$$

This enables us finally to write:

$$\frac{d\mu}{\mu} = \frac{dP}{P} + \chi dp - 2\tan^2\alpha\left(\frac{dl}{l} + \chi dp\right) - \frac{dl}{l - l'} - \frac{l'\chi dp}{l - l'}.$$

On the right hand side everything may be found. dP is the change in the stretching force on the spring exerted by the weight. To calculate it, the density of the weight must be known and its compressibility, the density of the transmitting liquid, and its change under pressure.

Next to connect the change of shearing modulus, $d\mu$, with the change of stiffness of the spring, we find from the formula for μ that

$$k = \frac{\pi}{2} \frac{a^4 \mu \phi_0^2}{s^3 \cos^2 \alpha}.$$

Differentiate this logarithmically, using the various relations already employed, and we get

$$\frac{dk}{k} = -\chi dp(1 - 2\tan^2 \alpha) + 2 \tan^2 \alpha \frac{dl}{l'} + \frac{d\mu}{\mu}. \qquad (3)$$

There are two relations of this kind, with the appropriate subscripts, one for the glass spring and one for the steel spring. For a given increment of pressure, $d\mu/\mu$ is, of course, determined entirely by the material, and does not depend on the particular geometry of the individual. For the steel spring, $d\mu/\mu$ is supposed known from the measurements with the absolute apparatus, so that for any particular experiment on the differential effect between glass and steel, dk/k for the steel spring (that is, dk_2/k_2) may be found. We are now in a position to return to equation 1, in which everything is now known except dk_1 for the glass spring. Solving this equation for dk_1, we now go back to equation (3) and solve for $d\mu/\mu$ for the glass, the quantity finally desired.

The magnitude of some of the correction terms is as great as that of the uncorrected effect. Thus in formula (2) for dl_1, if we call the uncorrected dl, $\Delta R_0/\rho_0$, the corrected dl will in some cases, where the pressure effect is comparatively small, be found to be of the opposite sign and greater numerically than the uncorrected dl. The largest part of the correction in this case arises from the term $\chi_4 L_0 dp$, that is, the term for the change of dimensions under pressure of the frame which holds the springs. More usually, however, the corrected dl differed from the uncorrected dl by something of the order of 20%. In finding $d\mu/\mu$ for steel by the direct method, we may call the term $dl/(l - l')$ the uncorrected effect. The actual effect was a little more than one half the uncorrected effect. By far the largest part of the correction here arises from the change with pressure of the buoyancy of the weight. The difference between $d\mu/\mu$ and

dk/k for the steel spring was of the order of 10%. In most cases $d\mu/\mu$ for the glass spring differed from its dk/k by something of the order of 20%.

Detailed Description of Experiments.

The Steel. Different steel springs were made for the absolute measurements and for each of the differential measurements. Greatest sensitiveness demands that the steel spring have such a stiffness that its total extension when coupled against the glass is the same as that of the glass, and since the stiffness of the glass springs varies greatly, it was necessary to vary the steel springs also. The general order of magnitude of the dimensions of the steel springs was: outside diameter 0.75 cm., length 1.5 cm., with 30 turns. The springs were wound in a lathe over mandrels of varying diameters; they were all made from the same coil of hard drawn piano wire, 0.025 cm. in diameter. Presumably the steel had a carbon content of about 1.25%.

The Glass. I am much indebted to Dr. Littleton, of the Research Laboratory of the Corning Glass Works, for providing six different varieties of glass. The compositions, supplied to me by W. C. Taylor, the chief chemist of the Corning Glass Works, were approximately as follows:

A is a potash lead silicate of very high lead content.

B is the same as Pyrex. A typical composition for this is:
SiO_2 81.4; B_2O_2 11.5; Na_2O 4.0; Al_2O_3 2.1; CaO 0.2; MgO 0.3.

C is a soda potash lime silicate.

D is a soda zinc borosilicate.

E is a soda lead borosilicate, opacified with calcium and aluminum fluorides.

F is a soda lime silicate containing a small percentage of boric oxide.

The glass was furnished in the form of solid rods of circular section, from 6 to 8 mm. in diameter. For converting these rods into helical springs I am very much indeed indebted to Professor Harold Pender, of the University of Pennsylvania, who developed the apparatus by which this was done, and to Dr. Charles Weyl, also of the University of Pennsylvania, who kindly supervised the actual work during the absence of Professor Pender in Europe. The glass rods were first drawn down to a diameter of the general order of 0.025 cm. The machine by which this is done consists essentially of a device by which the rod is fed through a brass casting maintained at the softening temperature of the glass, and is pulled out on the further side and wound up on a wheel (the rod is so flexible in small diameter that it can be wound on a wheel of large diameter without breaking), rotating at a definite speed with respect to the feeding speed. In this way a slender rod is produced whose section is a controllable fraction of the section of the original rod. The slender rod is now wound into a helix in another specially constructed machine, which consists of a core of carbon on which the rod is wound, the core being rotated inside a brass casting maintained at the proper temperature, and fed transversely as it rotates in order to give the helix the proper pitch. As furnished me, the helixes were open wound, with about 15 turns per cm. For my purpose it was better to convert these into closely wound helixes, which I did by slipping inside the helix a closely fitting core of aluminum, compressing the helix with a weight sliding on the core, and then slowly warming in an electric furnace to the softening temperature. By watching the heating, it is easy to stop the operation at such a point that the turns are closely in contact, but without sticking to each other. Finally, the ends of the helix were bent so as to give the proper means of attachment at the ends by very circumspect manipulation with a microscopic gas flame, issuing from a steel capillary such as is used in hypodermic needles.

Compressibility of the Glass. One of the corrections involves the linear compressibility of the glass. Since this correction may be important, and since the compressibility of glass varies greatly with the composition, it was necessary to make a direct determination of the compressibility of each variety of glass. This was done with the apparatus which I have described as "the lever apparatus for short specimens," and which has been used in measuring many other compressibilities.[3] The glass was cut from the original rod to a length of 2.7 cm., and the ends ground flat. In most cases the rod was of such a diameter that it could be used without further modification, but in one or two cases where the original diameter was too high, the diameter of these rods was reduced by grinding by the proper amount; I did this in order not to introduce the internal strains which might have been the result of drawing down to the proper size by heating. Although not immediately needed, the compressibility was determined at two temperatures, 30° and 75°; these measurements of the compressibility of glass have a certain interest for their own sake, and supplement determinations which I have already published for glass of other compositions.[4] The results are shown in Table I.

TABLE I.

COMPRESSIBILITY OF GLASS.

Designation of Glass	$\Delta V/V_0 = -(ap + bp^2)$, pressure in kg/cm²					
	30°			75°		
	a	b	Deviation*	a	b	Deviation*
A	30.54×10^{-7}	-24.3×10^{-12}	.09%	30.60×10^{-7}	-22.6×10^{-12}	.07%
B	30.12	+ 6.1	.10	29.72	+ 6.7	.11
C	24.69	−22.0	.44	25.44	−21.2	.42
D	25.97	+ 4.0	.10	26.30	+ .2	.08
E	27.78	+ 2.5	.10	27.71	+ 3.8	.10
F	23.29	− 6.1	.07	23.92	−10.1	.18

* This column shows the average deviations of a single reading from the smooth curve in terms of percentage of the maximum effect (at 12000 kg), and is an indication of accuracy.

Direct Measurement of the Shearing Modulus of Steel under Pressure. The steel spring for this determination had 48 turns, an outside diameter of 7.5 mm., and a length, when extended by the weight in a position to begin measurements, of 6.35 cm. The constant k was determined by hanging varying weights on the spring and observing the extension with a cathetometer; its value was 2.3, force being measured in gms., and extension in cm.

The liquid by which pressure is transmitted must be a perfect insulator, and should be relatively incompressible; furthermore its compressibility must be already known, or it must be specially determined. Kerosene was chosen as most nearly satisfying the several conditions. There is, however, a disadvantage in the use of kerosene, in that it becomes so viscous at high pressures that readings could not be made at pressures higher than 6000 kg., the contact refusing to open and shut when the apparatus was tilted at higher pressures. This defect could have been avoided by the use of petroleum ether, but the compressibility of this is much higher, and therefore the correction for the changing buoyancy of the weight is larger. Furthermore, the compressibility was not known with sufficient accuracy, so that, all things considered, the advantage seemed to lie with kerosene. The compressibility of kerosene was taken from a previous determination.[5]

The weight was composed of several metals, but was mostly of gold (7.073 gm. of gold, with 1.617 gm. of brass in the form of a holder for the gold, and 0.440 gm. of steel, the half weight of the spring). Gold was used in order to reduce the correction for the change of buoyancy with pressure to a low value. The correction was about 1% at the maximum pressure of 6000 kg. It is evident that with so small a correction the demands on the accuracy of the compressibility of the kerosene are not high. In addition to the correction arising from the compressibility of the kerosene, there is a correction arising from the compressibility of the metals of the weight; this was so small as to be just on the verge of the perceptible.

In Figure 3 all the observations are shown, the readings of the potentiometer slide wire from which the extension of the spring was obtained being plotted against the setting of the slider on the bridge by which pressure was measured in terms of the change of resistance of the manganin gauge. It will be seen that the relation between pressure and change of deformation of the spring is linear within a small margin of error. The change of length of the spring calculated from these readings was a shortening of 0.77 mm. at the

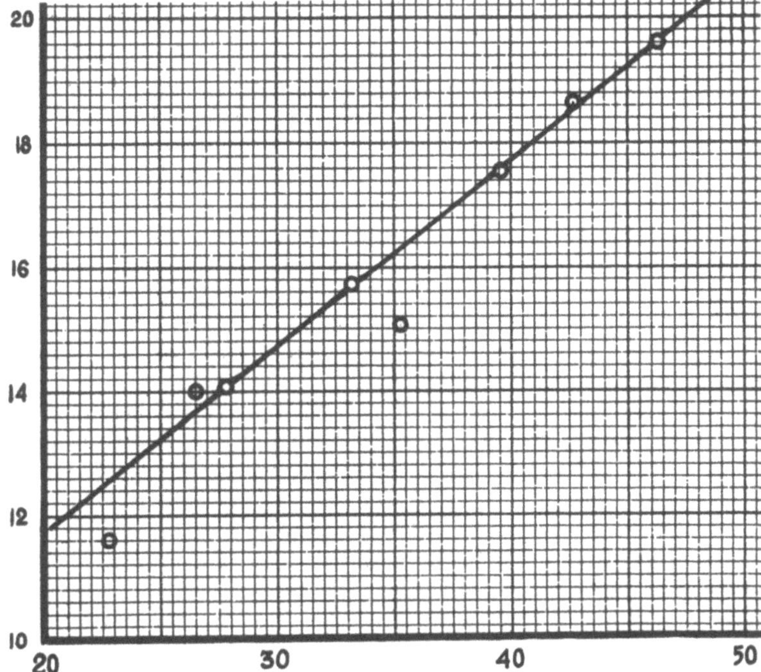

FIGURE 3. Reproduction of readings for the change of stiffness of a steel spring obtained with the apparatus of Figure 2. The ordinates are the settings in cm. on the potentiometer, and the abscissas the pressure in arbitrary units, the range of pressure between the extreme points being about 6000 kg. The extreme effect corresponds to an increase of shearing modulus of steel of about 1 per cent.

maximum pressure of 6000. From this, by means of the formulas already given, the change of shearing modulus is found to have the value:

$$\frac{1}{\mu}\left(\frac{\partial\mu}{\partial p}\right)_\tau = +\,2.16 \times 10^{-6},$$

pressure being expressed in kg./cm.² This means an increase of a little over 2% under a pressure of 10000 kg./cm.²

An attempt was made to obtain readings at 75° as well as at 30°. It was possible to run to higher pressures at 75°, because the viscosity of the kerosene is so much lower that there was no trouble from sticking contacts. Readings were made at 12000, 10000, and 8000, but at lower pressures nothing could be obtained because of electrical disturbances due to the chattering contacts of the thermal regulator. At higher pressures the viscosity of the kerosene prevented this difficulty. The difficulty was not serious, but to have remedied it, troublesome changes would have been necessary, and it did not seem worth while, especially since the subject is to be taken up again, and the value of the temperature coefficient, if it had been obtained, could not have been used in connection with the measurements on glass. The three points obtained at 75° lay on a straight line of 40% smaller slope than the points at 30°, so that it is probable that at 75° the increase of shearing modulus of steel under pressure is materially less than at 30°, but one cannot be certain of this until all the corrections at the higher temperature have been more carefully determined.

Differential Measurements on Glass. The differential measurements were made, as has been already explained, with the axis of the springs in a horizontal position, so that the correction for the changing buoyancy of the transmitting liquid could be neglected. This made it possible to transmit pressure with petroleum ether, which offers the advantage that the increase of viscosity under pressure is so small that the measurements could be pushed to 12000. Readings were usually made only at 30°, but for two of the varieties of glass, readings were also made at 75°. The results for the various kinds of glass differed greatly in regularity; this is partly to be explained by differences in the absolute magnitude of the effect, and partly by the dimensions of the springs, the springs which were wound out of smaller diameter rods being obviously more sensitive to disturbing effects. A set of observations is reproduced in Figure 4; this is one of the better ones, although not the best.

In order to get as much idea as possible of the elastic properties of the various kinds of glass, the shearing modulus was calculated from the constants of the springs. The accuracy of this determination is low, and the results must be used only for orienting purposes. The chief source of error was in the diameter of the glass. This enters as the fourth power into the formula for the modulus, and could be determined only with low accuracy. The diameter of the actual glass of the spring could not be conveniently measured; instead I determined the mean diameter of several of the straight lengths left over from the winding operation, and these might vary by as much as 10% in one or two cases. As an additional check, the absolute shearing modulus of the steel was determined from the constants of the steel springs. There were six of these springs,

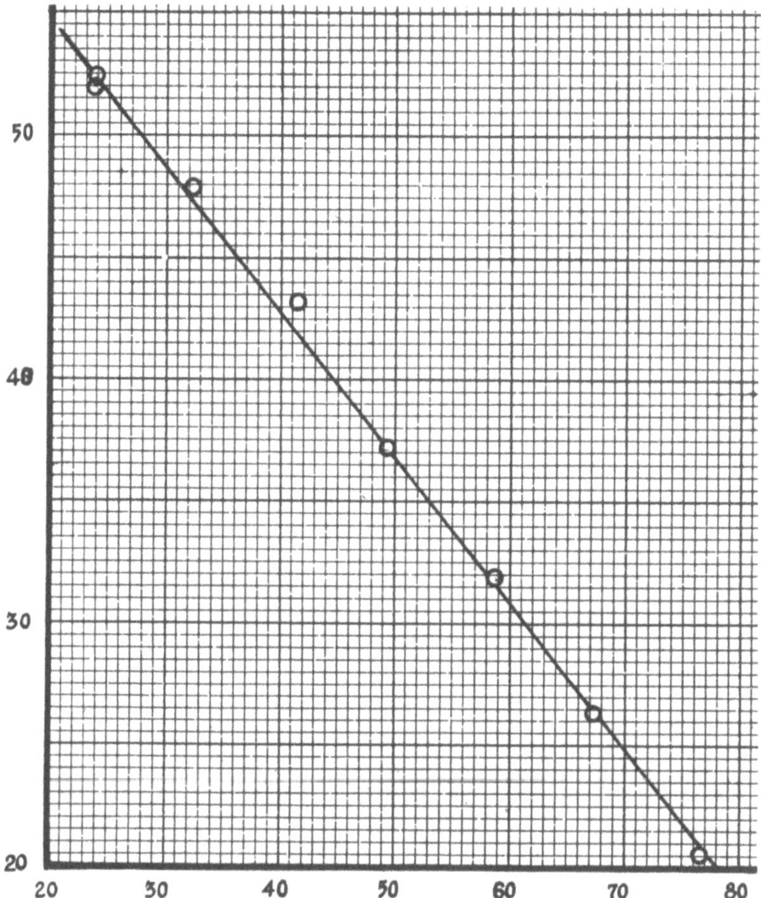

FIGURE 4. Reproduction of readings for the change of relative stiffness of steel and glass E, obtained with the apparatus of Figure 1. The ordinates are potentiometer settings in cm,, and the abscissas are pressures in arbitrary units, the range of pressure between the extreme points being about 12000 kg. The extreme effect corresponds to a decrease of rigidity of the glass with respect to the steel of about 10 per cent.

all of different dimensions, which were used to give the constant. The diameter of the steel wire may be safely assumed uniform since the wire was all from the same spool, but the chief source of error with the steel was in the outside diameter of the helix, which enters as the third power into the modulus. The outside diameter was measured with a micrometer, but since the spring is very flexible, it was difficult to be sure that it was correctly obtained. The following values were found for the shearing modulus of the steel springs used with glass springs A, B, C, D, E, and F respectively: 6.77×10^{11}, 6.94, 7.99, 8.02, 7.30, and 7.45, average 7.4×10^{11}. Kaye and Laby's Tables gives for the shearing modulus of steel of 1% carbon content 8.12×10^{11}.

The final results obtained with the different varieties of glass are shown in Table II. The negative sign of the effect was a surprise to me; by very crude analogy with the action of pressure in enormously increasing the viscosity of liquids, I had expected a rather large increase of shearing modulus. On reflection, however, the negative sign does not seem so strange in view of the fact that the compressibility of a number of different kinds of glass has been shown to increase with increasing pressure.[4] In fact, turning to the table of compressibilities, it will be seen that just those glasses, B, D, and E, which have the abnormal increase of compressibility with pressure also have the largest decrease of shearing modulus, and the two glasses, A and C, which are most normal in their decrease of compressibility, also have the smallest numerical change of shearing modulus under pressure.

The temperature effect was measured on samples B and E. The only conclusion that can be drawn is that the effect is not large. B at 75° showed a displacement of the contact point 8% less than at 30°, and the displacement of the contact of E at 75° was 6% greater than at 30°. The effect of temperature on the various corrections was not determined, so that the statement above, that the temperature coefficient of the pressure coefficient is small, seems to be all that is justified.

Effect of Pressure on Other Elastic Constants.

Since an isotropic substance has only two independent elastic constants, we are now in a position to find the effect of pressure on Young's modulus, E, and Poisson's ratio, σ. We have the relations:

$$\sigma = \frac{3 - 2\mu c}{6 + 2\mu c},$$

$$E = \frac{9\mu}{3 + \mu c}.$$

TABLE II.

Designation of Glass	Shearing Modulus of Glass, in Abs. C. G. S. units	Pressure Coefficient of Shearing Modulus in kg. units, $\frac{1}{\mu}\left(\frac{\partial\mu}{\partial p}\right)_\tau$.	Probable Accuracy as Shown by Deviations of a Single Reading from a Smooth Curve	
			(a) in terms of percentage of maximum measured effect	(b) in terms of cms. displacement of contact point
A	2.60×10^{11}	-0.62×10^{-6}	15.2%	.0012 cm.
B	2.31	−8.45	.8	.0009
C	2.33	−2.15	4.8	.0017
D	3.14	−8.02	5.4	.0032
E	2.54	−8.80	1.4	.0010
F	(6.64)	−3.86	16.	.0021

Here μ is the shearing modulus, as before, and c is the volume compressibility, defined by the relation

$$c = -\frac{1}{v}\left(\frac{\partial v}{\partial p}\right)_\tau.$$

Differentiation of these equations gives:

$$\frac{1}{\sigma}\frac{d\sigma}{dp} = \frac{-9\mu c}{(3-2\mu c)(3+\mu c)}\left\{\frac{1}{c}\frac{dc}{dp} + \frac{1}{\mu}\frac{d\mu}{dp}\right\},$$

$$\frac{1}{E}\frac{dE}{dp} = \frac{1}{3+\mu c}\left\{3\frac{1}{\mu}\frac{d\mu}{dp} - \mu c\frac{1}{c}\frac{dc}{dp}\right\}.$$

$1/c\ dc/dp$ may be found from the formulas for compressibility, as in Table I, to have the value

$$a + \frac{2b}{a}.$$

$\frac{1}{\mu}\frac{d\mu}{dp}$ is known, so that all the quantities are known which are required for the calculation of the pressure derivatives.

The velocity of a wave of shear is also of interest. This is given by

$$w = \sqrt{\frac{\mu}{\rho}},$$

whence:

$$\frac{1}{w}\frac{dw}{dp} = \frac{1}{2}\left[\frac{1}{\mu}\frac{d\mu}{dp} - c\right].$$

The results are contained in Table III. The calculation was not carried through for glass F because of the great uncertainty in the absolute value of its shearing modulus.

In general it appears that the changes in the various elastic constants are of the same order of magnitude as the change of compressibility already found. There is no general rule about the sign. The fact that the velocity of a wave of shear in glass decreases with increasing pressure may be of some geological interest.

I am indebted to my assistant, Mr. Stephen Stark, for most of the readings of this paper.

The Jefferson Physical Laboratory,
Harvard University, Cambridge, Mass.

TABLE III.

EFFECT OF PRESSURE ON VARIOUS ELASTIC CONSTANTS.

Substance	E Young's Modulus Abs. C. G. S.	σ Poisson's Ratio	$\frac{1}{E}\frac{dE}{dp}$ (p is in kg. /cm.^{2})	$\frac{1}{\sigma}\frac{d\sigma}{dp}$	$\frac{1}{w}\frac{dw}{dp}$ w is velocity of a wave of shear
Steel	20.8×10^{11}	.30	$+2.8 \times 10^{-6}$	$+2.6 \times 10^{-6}$	$+.78 \times 10^{-6}$
Glass A	6.2	.19	+2.2	+18.1	−1.8
" B	5.6	.22	−8.2	+1.5	−5.7
" C	5.9	.26	+0.7	+13.7	−2.3
" D	7.4	.18	−7.5	+3.3	−5.3
" E	6.2	.21	−8.0	+4.6	−5.8

80 — 2313

References.

[1] P. W. Bridgman, Proc. Amer. Acad. 58, 170, 175, 177, 1923.
Amer. Jour. Sci. 10, 483, 1925.
[2] J. W. Miller, Jr. Phys. Rev. 14, 146, 1902.
[3] Reference 1, page 175.
[4] P. W. Bridgman, Amer. Jour. Sci. 10, 359, 1925.
[5] P. W. Bridgman, Proc. Amer. Acad. 48, 357, 1912.

GENERAL SURVEY OF THE EFFECTS OF PRESSURE ON THE PROPERTIES OF MATTER

By Prof. P. W. BRIDGMAN, Hollis Professor of Mathematics and Natural Philosophy, Harvard University

The Fourteenth Guthrie Lecture, delivered April 26, 1929

PROF. RICHARDSON has intimated that an acceptable subject for this lecture would be a general survey of my experiments on the properties of matter under high pressure, and I have accordingly made this choice. I feel, however, that this choice demands some apology on two counts: first, because little of what I have to say is new; and second, because the subject, concerned as it is with the properties of matter in bulk, is not to-day one of the most lively interest. There are, nevertheless, certain points of view from which such a topic may be regarded, on the contrary, as especially timely. If the Bohr-Heisenberg principle, with the radical change in our physical point of view which it involves, turns out to be correct, it will not be possible to make indefinite further progress in the direction of the analysis of the very small, and physics must soon return to the task of explaining the properties of matter in bulk, a task which has been temporarily laid aside. Furthermore, the Bohr-Heisenberg principle suggests that matter in bulk may have properties not deducible from the measurable properties of its smallest parts; if this is the case, it becomes doubly important to investigate the large scale properties of matter under those especially simple changes of condition produced by hydrostatic pressure. But in order that hydrostatic pressure may produce significant changes, it is necessary that it be of the same order of intensity as the atomic or molecular forces themselves, and these, for ordinary solids or liquids, are of the order of tens of thousands of kilograms per square centimetre. The point of the experiments upon which I am to report is that the pressures are of this order of magnitude.

This is not the place to go into questions of technique. It will be enough to say that by the use of a packing of special design, which automatically becomes tighter at higher pressures, the problem of leak disappears, so that the only limit to the pressure attainable is set by the cohesive strength of the walls of the containing vessels.* By the use of alloy steels, and of vessels of small size, so that the beneficial effects of heat treatment may be extended throughout the entire wall, pressures up to more than 20,000 kg./cm.^{2} can be handled and measured with an accuracy of 0·1 per cent. Most of my experiments, however, are made to only

* The way in which the automatic tightening of the packing is produced in the case of the plunger by which pressure is generated is shown in Fig. 1. This represents a section of a cylinder

12,000 kg., in the interests of economical life of the apparatus. There was, of course, a great deal of preliminary development work to be done in finding the best designs of the containing vessels, in developing methods of measuring pressure, and in devising means of measuring various properties of small amounts of substances enclosed in heavy steel vessels, but I leave this aspect of the subject to discuss the effects of pressure on various properties of matter.

Perhaps the most significant and certainly the simplest of the effects of pressure is diminution of volume. The change of volume of any truly homogeneous substance, free from internal strains, is entirely reversible with pressure; no permanent change of volume has ever been observed in such materials up to pressures at least as high as 25,000 kg. Contrary statements sometimes found in the literature are to be explained either by flaws in the material or else by failure of the pressure to be truly hydrostatic, as when pressure is transmitted by an oil which freezes under pressure, thus producing permanent changes of figure in the material.

in which pressure is generated in a liquid, *L*, by the advance of a piston of hardened steel, *P*, driven by a hydraulic press. The mushroom-shaped plug, *A*, carries a collar, *C*, of soft rubber packing. The plug with the packing is pushed forward by the piston *P* acting on the intermediary ring of hardened steel *D*. The space, *E*, at the end of the stem of the mushroom is empty, so that the

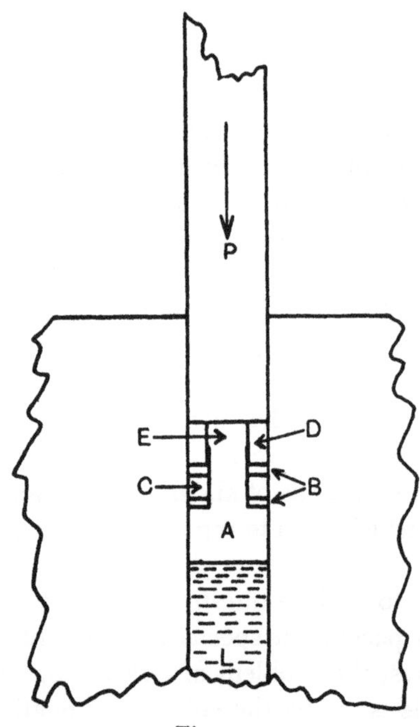

Fig. 1.

total pressure exerted by the ring of rubber must equal the total pressure exerted by the liquid on the head of *A*, and since the area of the ring is less than the area of the head, the pressure in lb./in.^{2} in the rubber is always greater than that in the liquid by the ratio of the areas, and the liquid can never leak. The packing is prevented from escaping by rings, *B*, of mild steel or copper. Further details of the technique are described in *Proc. Amer. Acad. Arts and Sci.*, **49**, 627–643 (1914).

The compressibility of fluids is, of course, in general much greater than that of solids. There is no essential distinction in compressibility between a substance ordinarily liquid and one of the so-called permanent gases, beyond the initial few thousand kilograms of pressure, which is far higher than the critical pressures between liquid and gas. It was shown by Amagat that air, for example, at a pressure of 3000 kg. is as dense as water.

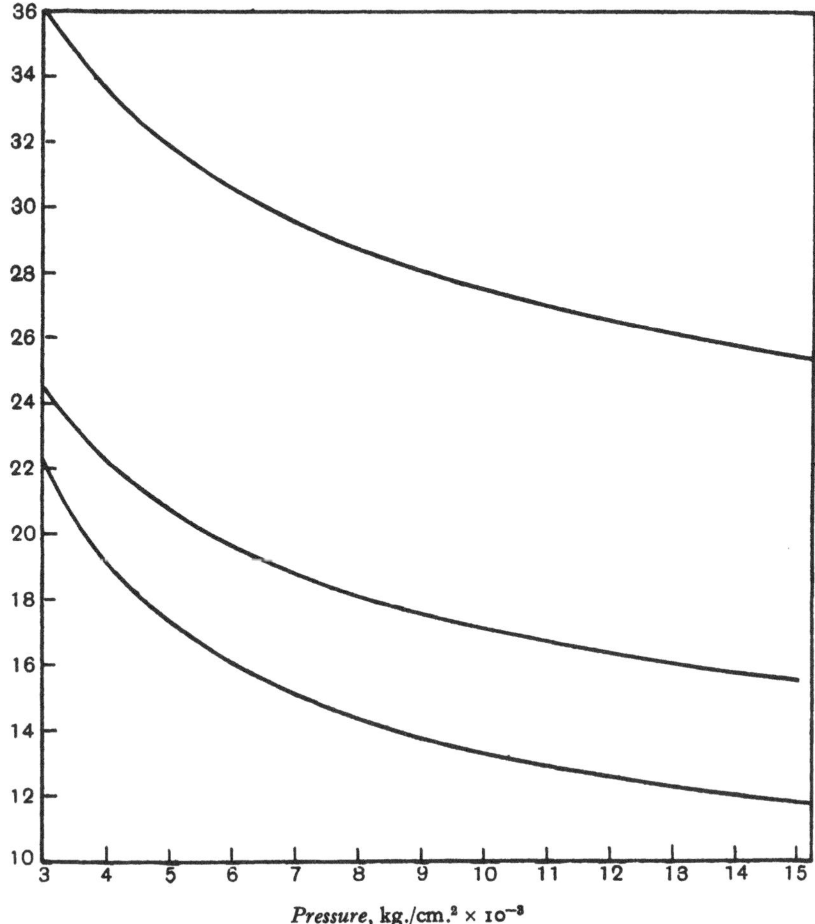

Fig. 2. The volume in c.c. per mol. as a function of pressure at 65° of nitrogen, hydrogen, and helium, reading from the top down.

The chief experimental difficulty in measuring the volume of gases at high pressures is the enormous initial volume, which makes it necessary to introduce the gas into the high pressure apparatus in small bombs in which it is subject to a high preliminary compression. In Fig. 2 is shown the volume in cm.^{3} per gm. molecule of N_2, H_2, and He between 3000 and 15,000 kg./cm.^{2}. The order of volumes is what is to be expected, monatomic He having a smaller volume at all pressures than biatomic H_2, which in turn has a smaller volume than N_2, also biatomic, but with a much more complicated molecule. In Fig. 3 the product pv of these three gases is plotted after the manner of Amagat as a function of pressure. Departures of pv from unity are a rough measure of the departure of the behaviour of the substance

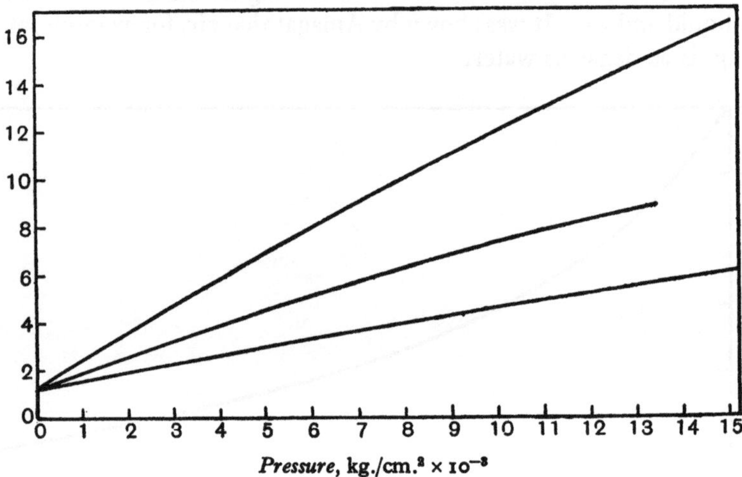

Fig. 3. The product pv as a function of pressure for nitrogen, hydrogen and helium (curves reading from the top down). p is the pressure in kg./cm.² and v is the volume of that amount of gas which under a pressure of 1 kg./cm.² occupies 1 c.c. at 0° C.

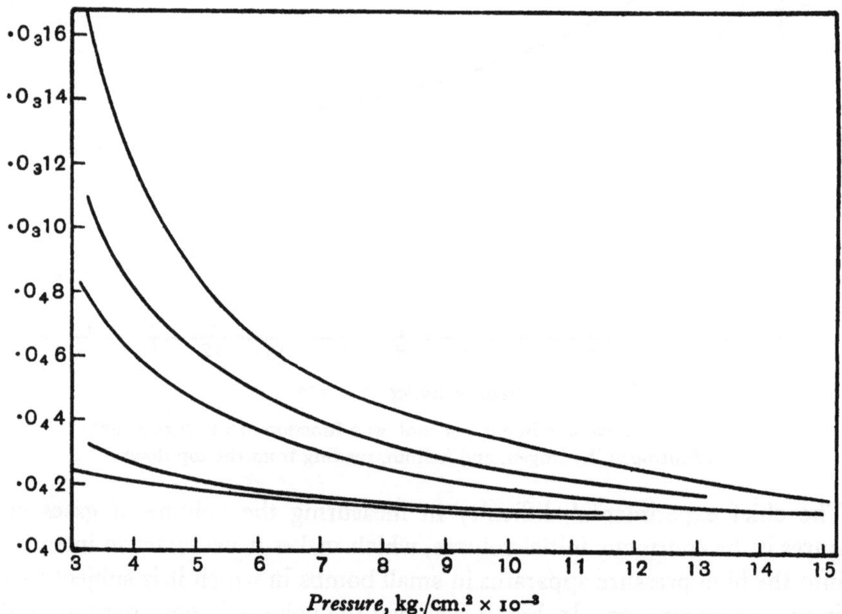

Fig. 4. The instantaneous compressibility, $\frac{1}{v}\left(\frac{\partial v}{\partial p}\right)_T$, as a function of pressure, of helium at 55°, hydrogen at 65°, nitrogen at 68°, CS_2 at 65°, and water at 65°, reading from the top down.

from that of a perfect gas. The departures found by Amagat up to 3000 kg. are seen to continue to become greater at very nearly a constant rate up to the highest pressure reached. Again, the behaviour is as is to be expected; He approaches most nearly to the perfect gas condition, although at 15,000 kg. its volume is six times greater than if it had remained a perfect gas. N_2 departs most widely from

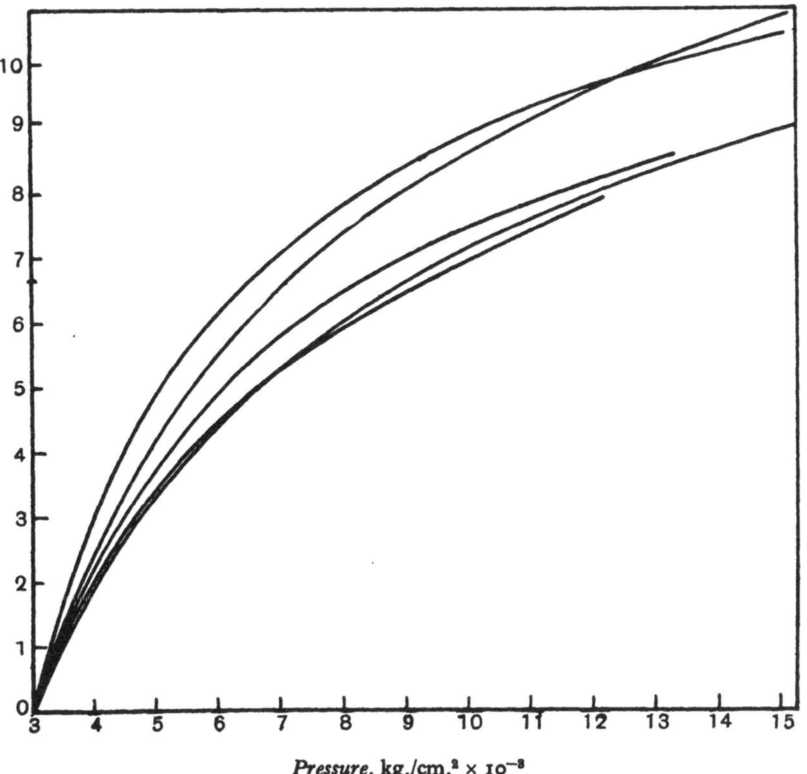

Fig. 5. The change of volume in cc. per mol as a function of pressure, reckoned from 3000 kg./cm.^{2} as the fiducial pressure. At 10,000 kg./cm.^{2} the order of the curves, reading from the top down, is: helium at 55°, nitrogen at 68°, hydrogen at 65°, argon at 55°, and ammonia at 30°.

the perfect gas, while H_2 is intermediate. In Fig. 4 the "instantaneous" compressibility, $\frac{1}{v}\left(\frac{\partial v}{\partial p}\right)_\tau$, of the three gases is shown and also, for purposes of comparison, the instantaneous compressibility of liquid CS_2 and water. At the lower pressures, in the neighbourhood of 3000 kg., the compressibility of the gases is several fold greater than that of the liquids, but at the highest pressure, 15,000 kg., the compressibility of all is nearly the same.

Thus far the behaviour of gases at high pressures has been a natural enough extension of the behaviour at low pressures. There are, however, effects at high pressure which would not naturally be inferred from the behaviour at low pressures. Fig. 5 shows one such effect, namely, the change of volume in cc. per mol, starting from the volume at 3000 kg. as the fiducial volume, for the five gases H_2, He, NH_3,

N_2 and A. The significant feature is the crossing of the curves for N_2 and He; it is evident, furthermore, that the curves for H_2 and A will also cross at a pressure slightly higher than that shown in the figure. The qualitative significance of this is as follows. At comparatively low pressures, the decrease of volume of a gas has its origin in a decrease of the empty space between the atoms or molecules, but as the molecules are pushed into closer contact, this effect becomes exhausted, and at high pressures this contribution to compressibility disappears. But there is another factor in the compression, namely, the actual loss of volume of the molecules themselves, and this evidently may persist at pressures where the initial effect no longer exists. This dual mechanism is doubtless the explanation of the striking difference between He and N_2. The atom of He is much smaller than the molecule of N_2, so that at low pressures the decrease of volume of He is greater than that of N_2. But at high pressures, where the important factor is the loss of volume of the atoms or molecules themselves, the decrease of volume of N_2 becomes greater than that of He, because the structure of the N_2 molecule is so much more complicated than that of the He atom that it has in it the potentiality of much greater loss of volume. The same sort of considerations also explain the relative behaviour of H_2 and A. The molecule of H_2 is normally smaller than the atom of A, so that at low pressures the compressibility of H_2 is greater than that of A. But at high pressures the rôles are reversed, because the electronic structure of atomic A is so much more complicated than the electronic structure of molecular H_2 that the atom of A is capable of much greater loss of volume than the molecule of H_2.

This behaviour, so definitely shown by these gases, is typical of the behaviour of all substances at high pressures. Beyond the first few thousand kilograms, the major part of the loss of volume is provided by the atoms or molecules themselves, and those substances with the most complicated or the most loosely constructed molecules have the greatest compressibility at high pressures, although at low pressures the behaviour may be the reverse. All this does not mean that a molecule is not properly to be regarded as a field of force rather than as a little nugget of matter in the old-fashioned sense, but it does mean that when the molecules are pushed closer together, there are qualitative changes in the interaction of their force fields similar to those which the older picture suggested.

Passing next to the volume behaviour of liquids, the most immediately striking fact is that the volume changes of many liquids tend to approach much more nearly to equality at high pressures than at low. Measurements on fourteen common liquids, including the first five alcohols, ether, CS_2, C_2H_5Cl and water, show an extreme variation in the loss of volume under 12,000 kg. from 21 per cent. for water, the least compressible, to 33 per cent. for ether, the most compressible, which thus under 12,000 kg. loses only 50 per cent. more volume than water, although its initial compressibility is four or five times greater. The compressibility of liquids drops very rapidly with increasing pressure as is shown in Fig. 6. At 12,000 kg. the compressibility of common organic liquids varies from 1/14th to 1/20th part of the initial compressibility; half of the drop to the final value is accomplished in the first thousand kilograms, and at 6000 kg. the drop is 95 per cent. completed.

Qualitatively, there is thus much similarity between liquids and gases. As in the case of gases, the initial high compressibility of liquids arises from the taking out of slack between the molecules, whereas the part of the compressibility which

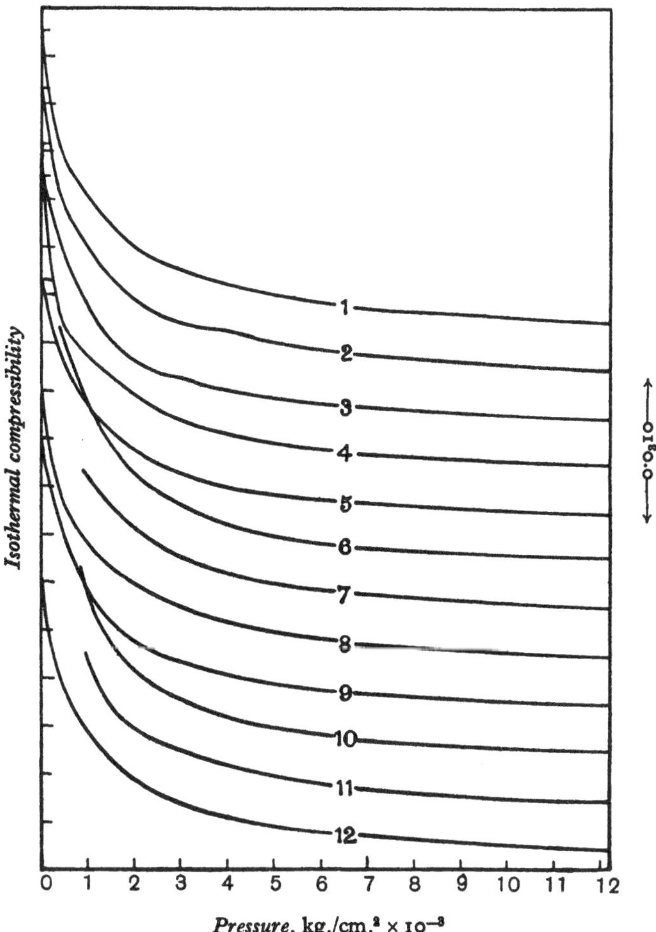

Fig. 6. The average compressibility between 20° and 80° of twelve liquids as a function of pressure. In order to prevent overlapping, the origin of each curve has been displaced one square with respect to the next. The scale of the curves is shown on the right hand side. The origin is so situated that the compressibility of each of the twelve liquids at 12,000 kg./cm.² is between $0{\cdot}0_5 1$ and 0. The numbers on the curves indicate the liquids as follows: 1, methyl alcohol; 2, ethyl alcohol; 3, propyl alcohol; 4, isobutyl alcohol; 5, amyl alcohol; 6, ether; 7, acetone; 8, CS_2; 9, PCl_3; 10, ethyl chloride; 11, ethyl bromide; and 12, ethyl iodide.

persists to high pressures probably arises from the decrease of volume of the molecules themselves. Liquid metals, as is to be expected, have a lower order of compressibility than organic liquids; it is also surprising that glycerine, which has a fairly complicated molecule, is only two-thirds as compressible as water at 12,000 kg.

By measuring the volume as a function of pressure at different temperatures it is possible to find how thermal expansion varies with pressure. The results for twelve different liquids are shown in Fig. 7. As was to be expected, thermal expansion drops with increasing pressure, but it is perhaps surprising that the decrease is much less than the decrease of compressibility; the thermal expansion under 12,000 kg. is on the average between one-fourth and one-fifth of its initial value, against four times as great a decrease in compressibility. Furthermore, the thermal

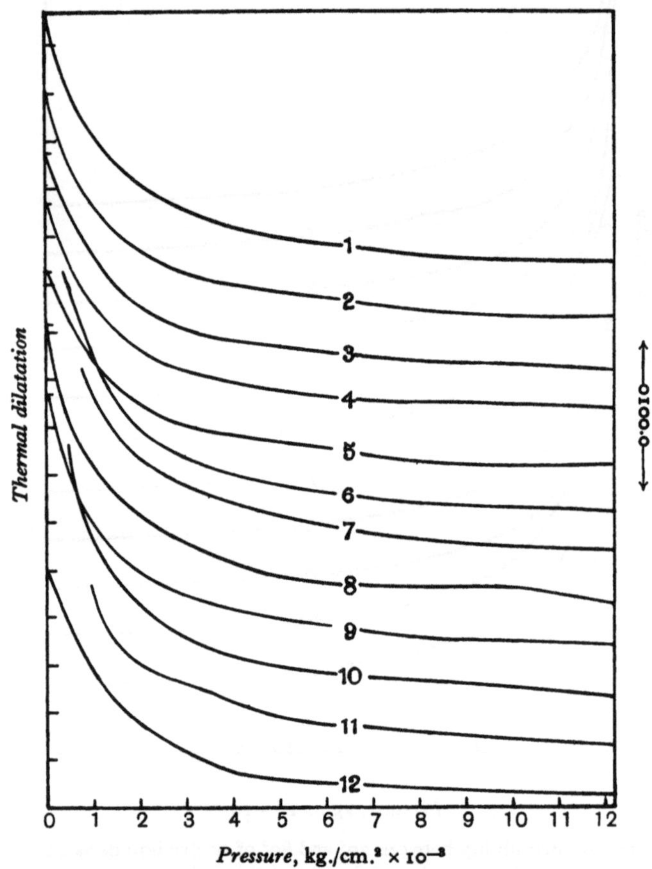

Fig. 7. The average thermal expansion between 20° and 80° of twelve liquids as a function of pressure. The general method of representation is the same as in Fig. 6, except that the origin for each curve is so situated that the expansion at 12,000 kg./cm.² is between 0·0002 and 0·0003.

expansions of fourteen liquids, with the exception of water, approach much more closely to equality under 12,000 kg. than do the compressibilities, the extreme variation being by only 25 per cent., from $2 \cdot 4$ to $3 \cdot 0 \times 10^{-4}$.

It is natural to expect parallelism between the effects of high pressure and low temperature, since both have a tendency to constrain freedom of internal motion, but the parallelism is far from complete, because as temperature approaches 0° abs., the compressibility drops only slightly, while the thermal expansion drops to zero, but at high pressures the compressibility drops much more than the thermal expansion.

At atmospheric pressure the thermal expansion of all liquids increases with increasing temperature, but in the neighbourhood of 3000 kg. there is a reversal, and at higher pressures thermal expansion is greater at lower temperatures than at higher. The explanation is probably connected with the known fact that if the molecular restoring forces in a solid are linear functions of the molecular displacement, thermal expansion vanishes, so that a high thermal expansion means high departure from linearity. The smaller the volume of a substance the greater in general the departure of the forces from linearity, since the repulsive forces which predominate at small volumes vary inversely as some high power of the distance. At constant pressure the volume is less at low temperature, so that the departure from linearity would be expected to be greater, thus accounting for the greater thermal expansion. At low pressures there is a reversal of this behaviour because there is another mechanism active, the liquid tending to approach more nearly to a gas, with its high thermal expansion, the higher the temperature.

Superposed on the broad features just described, common to the behaviour of many liquids, there are small-scale specific differences of an almost indescribable complexity. An inspection of the experimental data would make it evident that an enormous number of parameters would be necessary to describe in full detail the behaviour of even a single liquid, and the small-scale complexities of different liquids are without discoverable relation to each other. This means that no such thing as a general equation of state for liquids can exist, and that the most that can be expected is to find an equation which shall reproduce the broad common features of behaviour described above. Very few attempts have been made in this direction, and as far as I know, none have been successful.

The volume behaviour of solids is qualitatively different from that of liquids or gases. In the first place, the compressibility is, as a general rule, less, as is to be expected. As far as I know, there is no exception to the rule that the compressibility of a substance in the solid phase is less than that of its own melt; this is true for normal substances which contract on freezing, and also true for water and bismuth which expand on freezing, and for which the contrary might be expected. It is thus evident that the lattice structure of itself imparts a certain stiffness foreign to the liquid. In the second place, the initial domain of high compressibility followed by very much lower compressibility, which is characteristic of liquids, is much less prominent in solids, the compressibility of which drops off comparatively little at high pressures. It is therefore probable that by far the larger part of the compression of a solid has its origin in the compression of its atoms. A striking example of this is caesium, the atom of which is highly complicated, and the compressibility of which is greater than that of any other metallic element. In Fig. 8 the volume of Cs is plotted against pressure and for comparison the volume of ether, the most compressible organic liquid. The decrease of volume of ether under the first thousand kilograms is nearly twice that of Cs, but the compressibility of Cs persists at high pressures, so that at 12,000 kg. the total volume decrement of

Cs has become very nearly as large as that of ether, and its actual compressibility is materially larger than that of ether.

As a general rule the compressibility of solids decreases at high pressures by amounts which are greater the greater the compressibility, as might be expected, but there are exceptions. The compressibility of pure quartz glass, SiO_2, and a number of compound glasses in which the content of SiO_2 is high, increases with increasing pressure by an amount far beyond experimental error. It has sometimes been thought that molecular stability demands that the compressibility becomes less as the volume becomes less, but this conclusion evidently rests on an incomplete analysis. I have recently found the same anomaly, that is, compressibility increasing with increasing pressure, in the element cerium. The behaviour of potassium is also highly significant in this respect when compared with that of the other alkali metals. Fig. 9 shows that the compressibility of potassium drops off with increasing pressure much less than that of Rb or Cs, so that although initially the compressibility of Cs, for example, is nearly twice as great as that of K, at 12,000 kg. it has become considerably less. The persistence of compressibility of K is to be connected with the abnormally loose structure of the atom of K, which is shown by the fact that if the atomic volume of the alkali metals is plotted against atomic number, the volume of K is abnormally high. It is interesting, and probably of considerable significance, that at high pressures in the case of K there is a reversal in the direction of a simple function of the compressibility, which we do not need to explicitly define for our immediate purpose, of such a character that this function starts to approach the value which it would have if K were a perfect gas produced by the disintegration of its atoms into their component electrons and protons.

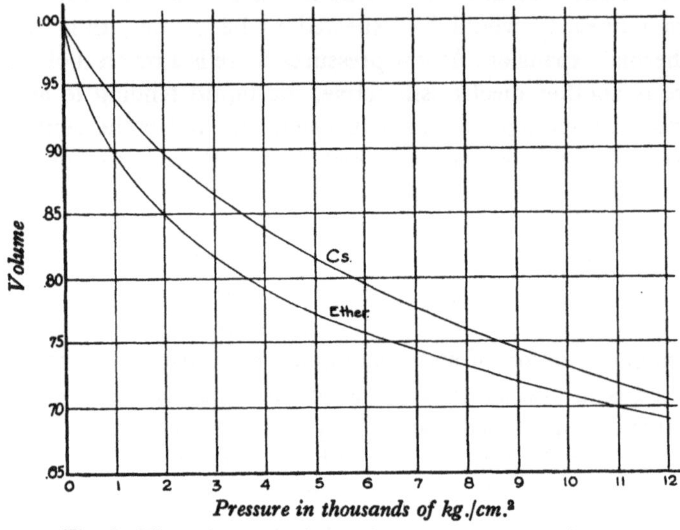

Fig. 8. The volume of ether and CS_2 as a function of pressure.

Of particular significance is the effect of pressure on single crystals. Only in cubic crystals is the change of volume under hydrostatic pressure the same in every direction, so that to determine completely the effect of pressure on the volume of single crystals the linear compressibility must be measured in several directions. A large number of crystals have been studied from this point of view, and also the variation under pressure of electrical properties in different directions has been measured. The results are too numerous and complex to attempt to summarise here, except to state that the linear compressibility may vary much more with direction than might be expected from the difference of atomic spacing in different directions. Thus the linear compressibility of Zn parallel to the hexagonal axis is seven times as great as at right angles, whereas the ratio of the atomic spacing in the two directions is only 7 per cent. different from that for spherical atoms in normal hexagonal piling, and in fact the spacing is compressed along the axis,

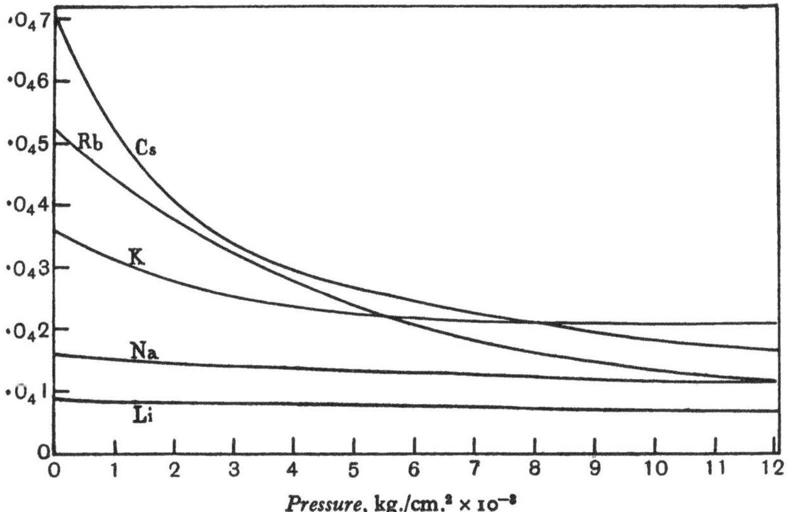

Fig. 9. The instantaneous compressibilities, $\frac{1}{v}\left(\frac{\partial v}{\partial p}\right)_T$, at 0° C. of the five alkali metals as a function of pressure.

while the compressibility is greater along the axis. The behaviour of tellurium is highly unusual, in that there is negative compressibility along the trigonal axis; that is, when exposed to hydrostatic pressure a crystal of Te expands in the direction of the axis. It is difficult to believe that effects of this kind can be explained in terms of atoms with spherical symmetry.

Perhaps the simplest of all crystals are the alkali halides. The compressibility of most of these has been measured by Slater up to pressures so high that good values could be found for the change of compressibility with pressure. This effect is at present beyond the reach of theory; although several theories have been proposed which give fairly good values for the initial compressibility, they all give changes of compressibility with pressure which are wide of the mark.

Next in simplicity after changes of volume come perhaps the changes of state produced by high pressure. We have to consider only the phenomena of change from the liquid to the solid or from one solid to another, for the critical pressures between liquid and gas are of the order of a few hundred kilograms, and therefore

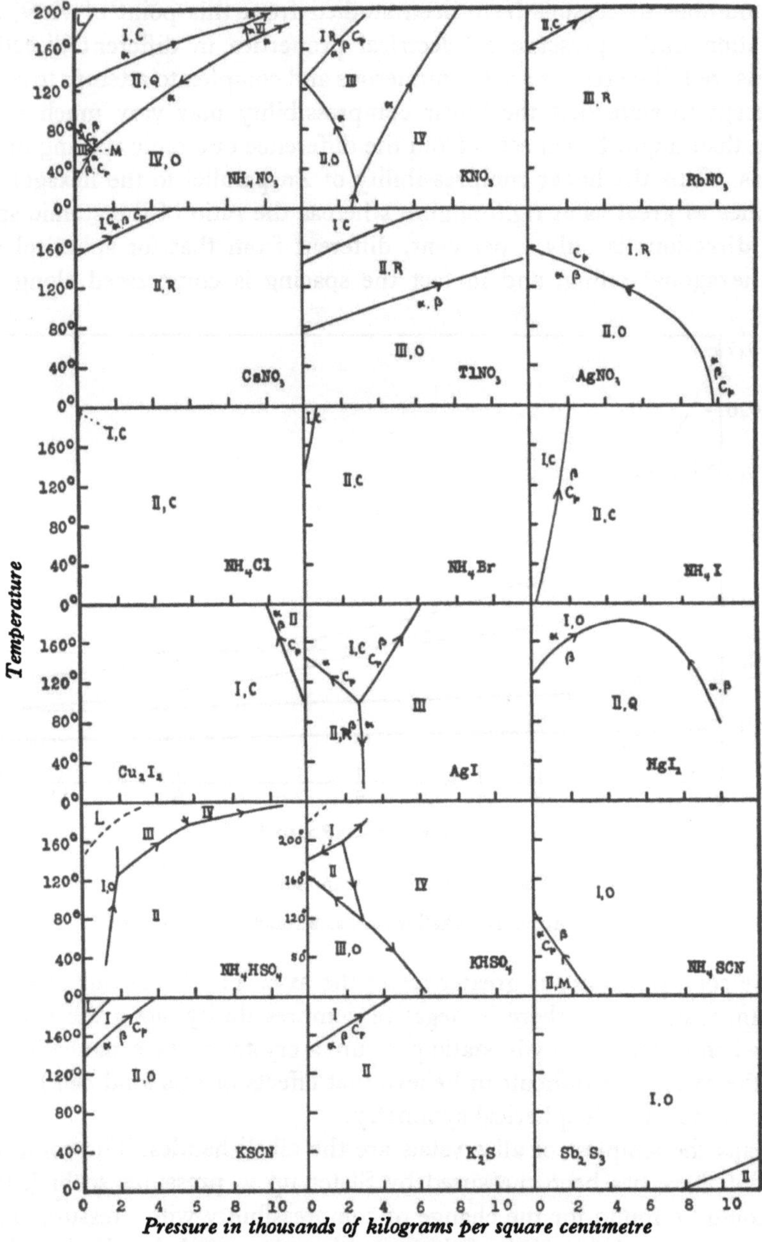

Fig. 10. The phase diagrams of a number of substances, showing the polymorphic transitions under pressure.

much lower than the pressures of interest here. There have been various ideas as to the ultimate relation between liquid and solid at high pressures. The most natural of these, based on pure analogy with the critical phenomena between liquid and gas, was that there is a similar critical point between liquid and solid. Another idea prominently advocated by Tammann was that there is a maximum melting temperature, above which a substance is capable of existence only in the liquid phase, no matter how high the pressure. Experiment, however, seems favourable to neither of these points of view, for measurements on nearly forty substances up to 12,000 kg. and on water up to 21,000 kg. show that if either of these two possibilities ever occur, there must be a complete reversal of the universal trends in the experimental range of pressure. All melting curves, whether rising as is normal, or falling like those of water and bismuth, are concave toward the pressure axis. Furthermore, the difference of volume between liquid and solid plotted against melting pressure gives a curve convex toward the pressure axis, and the slope, dt/dp, of the melting curve plotted against temperature, gives a curve convex toward the temperature axis. A little consideration will show that these universal features demand that there can be neither critical point nor maximum. So far as I can see, there is no reason to think that the melting temperature may not be raised indefinitely by the application of sufficiently great pressure, but the rate at which the temperature is raised by a definite increment of pressure becomes continually less as the pressure becomes higher.

The phenomena of polymorphic transition from one crystalline phase to another show, on the other hand, no such regularities. A number of examples of the phase diagrams of polymorphic substances are shown in Figs. 10 and 11. One of the most interesting of these is that of water which is capable of existence in at least five solid forms under the proper conditions of temperature and pressure. The modifications called II and III in the diagram were first found by Tammann. Of the five modifications, ordinary ice is the only one which is less dense than the liquid, and therefore the only one whose melting temperature is depressed by increasing pressure. Camphor, with six modifications, is the only substance hitherto investigated with more modifications than water. NH_4NO_3 has five modifications, like water, in the ordinary temperature range, but at low temperatures there is another modification which there is some reason to think is merely one of the high temperature modifications reappearing at a lower temperature. Inspection shows that there is not the slightest similarity between the diagrams of the different substances. In general the phase diagrams of substances which are closely related are much less similar than their chemical similarity might lead one to expect. The diagrams of CCl_4 and CBr_4 are a case in point.

The only generalisation that can be made with regard to all these polymorphic changes is that there seems never to be a critical point between two different crystal modifications. This means that continuous transition from one space lattice to another never occurs, as seems natural enough. Apart from this, the most varied sorts of behaviour are possible. There are a comparatively large number of transition curves of the ice type, that is, those in which the modification stable at the higher temperature has the smaller volume, so that the transition temperature decreases with rising pressure. The transition curves may be either concave or convex toward the pressure axis. They may have vertical tangents with either maximum or minimum pressures, and one example is known (HgI_2) of a horizontal tangent with maximum temperature. The mere existence of curves with horizontal or vertical tangents is of significance. At a horizontal tangent, the transition occurs with no volume change but with a finite latent heat; on the high-pressure side of such a horizontal tangent the phase which has the larger volume has the smaller compressibility, a somewhat paradoxical effect. Similarly, at a vertical tangent the transition occurs with no thermal effect, but a finite change of volume; on the low-temperature side of such a vertical tangent the phase stable at the higher temperature has the lower specific heat, again paradoxical.

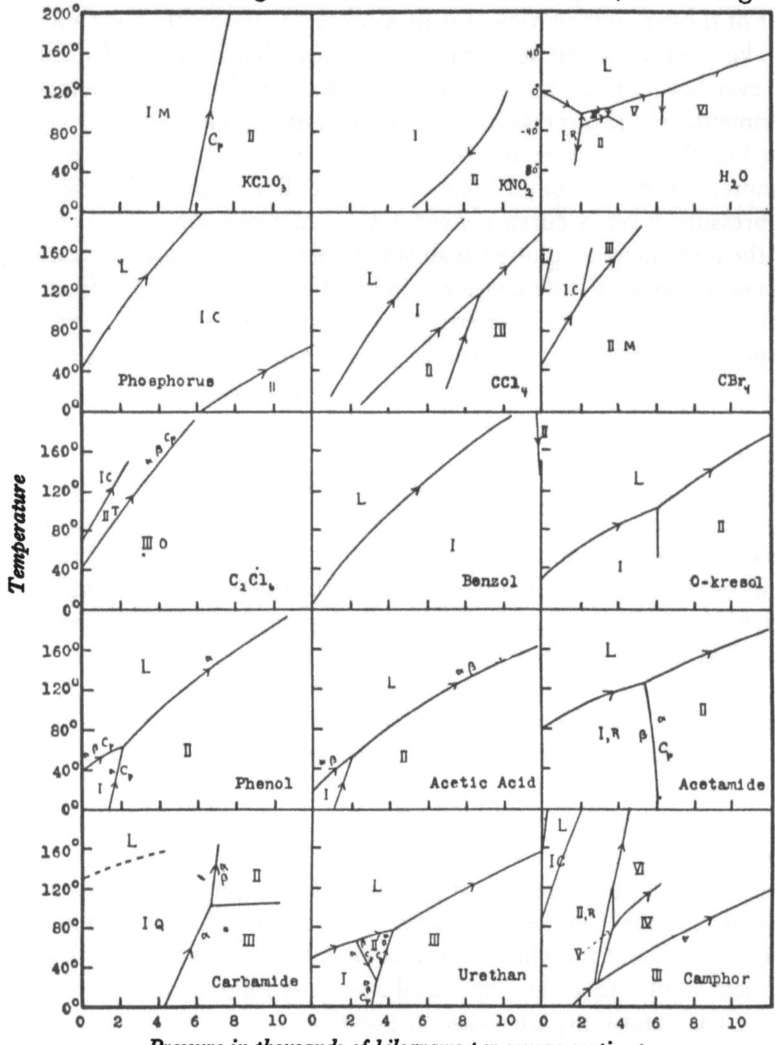

Fig. 11. More phase diagrams, similar to those of Fig. 10.

In addition to the information shown in the diagrams I have in a number of cases been able to measure the difference of compressibility between different polymorphic forms of the same substance. The surprising result is that in the majority of cases that one of two phases which has the larger volume and which therefore is stable at the lower pressure is the less compressible. It is also true in general that the phase of larger volume has the higher crystalline symmetry. It seems as if the molecules or atoms have some internal structure analogous in some respects to a framework of high symmetry and rigidity, such that at low pressures, where the intensity of the external forces is low enough to permit that the natural arrangement be assumed, the molecules take an arrangement of high symmetry and low compressibility, with the projecting parts of the frameworks of different atoms in register with each other, but at high pressures the natural forces are overcome, the frameworks are pushed out of register, and the system collapses to an arrangement of smaller volume, of lower symmetry, and, because the frameworks are the most rigid part of the molecule, of greater compressibility. I doubt whether the complicated facts of polymorphism, or of the volume relations in liquid, can ever be explained in terms of simple central forces between atoms, although such an explanation is all that has hitherto been attempted, and in spite of the partial justification which such an explanation has lately received from the results of wave mechanics applied to simple systems.

It would appear, then, that a study of polymorphic behaviour gives a very delicate method of analysing the properties of matter in bulk, but at present the complications in the way of a theory are rather formidable. A systematic study of this field should start with the simplest systems. Among the series of substances which have been measured under pressure there are two comparatively simple. One of these is the series NH_4Cl, NH_4Br, NH_4I; these experience a polymorphic transition with very large volume change, and the lattice structure of the two modifications is known from X-ray studies at atmospheric pressure. The other of the series is RbCl, RbBr, RbI; the pressure of the transition is in the neighbourhood of 5000 kg., so that a direct X-ray determination of the lattice structure of the high pressure modification has not as yet been possible, but other considerations make it very probable that it is the same body centred cubic type as CsCl. No adequate explanation has yet been given of the occurrence of these two types of lattice in the alkali halide series, but enough has been done to show that the type of lattice which has the minimum free energy may be very sensitive to slight changes in the character of the interatomic forces.

Consider next the effect of pressure on electrical resistance. Out of 48 pure metals which have been measured, the resistance of 39 decreases under hydrostatic pressure by amounts varying with the character of the metal from 1 per cent. under 12,000 kg. for Co to 73 per cent. for K. The relation between change of resistance and pressure is not linear, but always the curve of resistance against pressure is convex toward the pressure axis, which means that the effect of a given increment of pressure becomes less at high pressures, as seems natural. In the case of K the initial rate of change is so great that resistance would entirely vanish at a pressure of 5600 kg. if it continued at the initial linear rate. Six or seven of the 48 metals, Li, Ca, Sr, Sb, Bi, the low-pressure modification of Ce, and perhaps Ti, increase in resistance under pressure, and in all these cases the curve of resistance against pressure is convex toward the pressure axis, just as in the case of the metals with negative coefficient. But now the significance of convexity is that the effect of a given increment of pressure becomes greater at high pressures, which is not to be expected. It is particularly to be noticed that the curvature is such for these two types that the curve for one could be regarded as a prolongation of that of the other, so that it is natural to ask whether the resistance of all metals does not ultimately increase at sufficiently high pressures, the initial difference between different metals being ascribable to different internal pressures. Such an expectation is much strengthened by the discovery of three metals whose resistance does actually pass through a minimum with increasing pressure. The first of these is Cs, which is the most compressible metal, and therefore might be expected to show the effect most easily. The resistance of this at room temperature passes through a minimum of about 0·71 of the initial value between 4000 and 5000 kg. The next is Ba, which stands to the right of Cs in the periodic table, and which has a minimum at 0·97 of the initial resistance at about 8000 kg. Very recently I have found that Rb, which stands above Cs in the periodic table, also has a minimum at 18,000 kg. at 31 per cent. of its initial resistance. Measurements on K show no minimum up to 19,000 kg., but an easy extrapolation by first differences indicates pretty certainly a minimum in the neighbourhood of 24,000 kg. at 18 per cent. of its initial resistance. A similar, but much more uncertain extrapolation of the measurements on Na up to only 12,000 kg. is not unfavourable to the existence of a minimum, but indicates that it occurs at pressures probably considerably higher than 24,000 kg. However, on taking the last step in the series of the alkali metals, all regularity disappears, for the resistance of Li increases from the start with increase of pressure. In Fig. 12 the resistance under pressure of the alkali metals is shown. Furthermore, on passing across to the second column in the periodic table, the irregularities are more striking than the regularities. The resistance of Be decreases under pressure against the increase of Li, Mg decreases as does also Na, but Ca increases against the decrease of K, Sr increases against the decrease of Rb and its increase is greater than that of any other metal, while finally Ba has a minimum as does also Cs, but the whole scale of the pressure effects on Ba is very much smaller than on either Cs or Sr.

It is thus evident that in spite of the extreme simplicity of the change in the lattice structure produced by pressure, the effects of pressure on resistance are complicated, much more complicated than the effects of temperature, for example. I do not believe that any adequate explanation has yet been offered of the effects of pressure on resistance, even by the new wave mechanics. A consideration of the pressure effects makes it fairly certain, however, that the mechanism must be much more complicated than the pure temperature effects would lead us to expect. Particularly, the occurrence of a minimum resistance is suggestive; nearly always a minimum property involves the action of at least two mechanisms, which to a certain extent play against each other.

In addition to the effect of pressure on the electrical conductivity of pure metals, its effect on a number of binary alloys of metals has been measured. Examples are rather frequent here of a positive pressure coefficient of resistance.

One universal generalisation applies to all the measurements yet made on the pressure coefficient of alloys, both by other observers and myself, namely, that the initial effect of adding a small amount of foreign metal to a pure metal is to increase algebraically the pressure coefficient. That is, if the coefficient is initially negative, as it is in the majority of cases, adding a foreign metal makes the coefficient smaller numerically, while if the coefficient is initially positive, it becomes still more positive on addition of another metal. This is capable of a simple geometrical interpretation consistent with the wave mechanics picture of electrical resistance as arising from the scattering of electron waves by irregularities in the atomic structure. For, in addition to the irregularities in the lattice structure of a pure metal

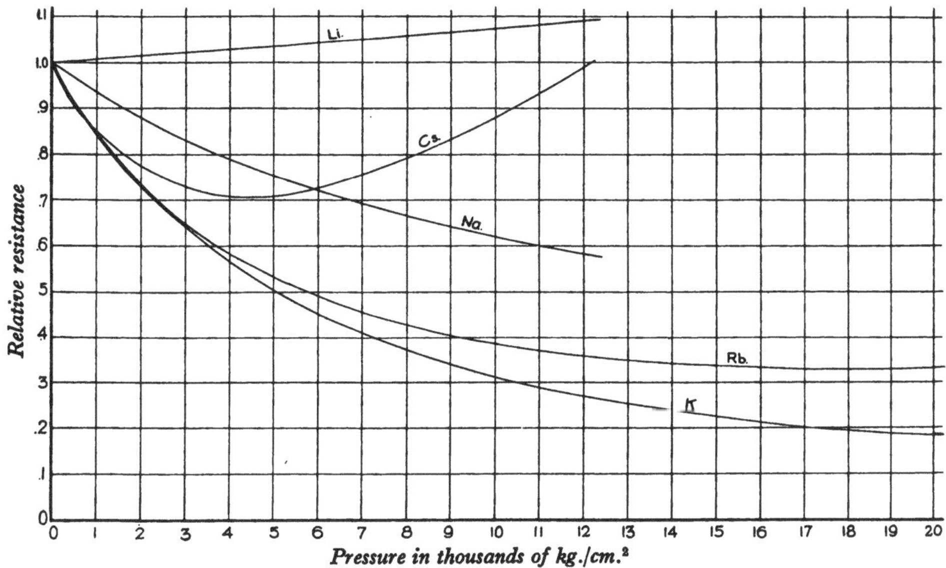

Fig. 12. The relative resistances under pressure of the five alkali metals.

due to temperature agitation, there must in general be irregularity of a purely geometrical character when a foreign metal is introduced, arising from the unequal sizes of the two sorts of atoms. The geometrical discrepancy between the two different sorts of atoms is accentuated as they are pushed into closer contact by pressure, so that the scattering of the electron waves arising from this effect increases, and the pressure coefficient becomes more strongly positive.

The thermal conductivity of metals is also recognised to be an electrical phenomenon, as proved by the universal value of the Wiedemann-Franz ratio. The effect of pressure up to 12,000 kg. has been measured on the thermal conductivity of 11 metals. The measurement is much more difficult to make than of resistance, and the results are not so accurate. Of the 11 metals, the thermal conductivity of 5 increases under pressure and that of 6 decreases. In only two cases does the Wiedemann-Franz ratio increase under pressure, and in the remaining 9 cases it decreases by amounts varying up to 15 per cent. for Ni at

12,000 kg. Since the Wiedemann-Franz ratio varies with pressure, it is obvious that the connection between electrical and thermal conductivity cannot be the universal and simple one of the original theory of Drude. A consideration of the contribution to the thermal conductivity by the atoms as distinguished from the electrons seems to offer a possibility of explaining at least part of the discrepancy, and would suggest that the contribution made by the atoms may be larger than often supposed.

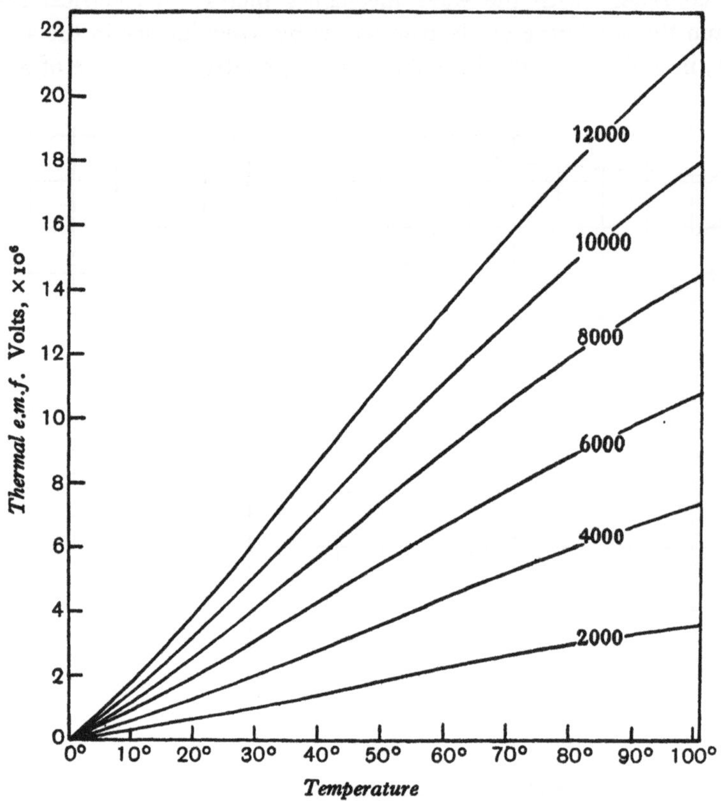

Fig. 13. Thermal e.m.f. of a couple composed of one branch of uncompressed pure platinum, with the other branch composed of the same metal compressed to the pressure in kg./cm.^{2} indicated on the curves, the junctions being at 0° C. and the temperature plotted as abscissa.

Thermal e.m.f. is another electrical property that can be measured under pressure without too great difficulty. The effects are not small, for the thermal e.m.f. of a couple composed of a metal in the uncompressed state and the same metal compressed approaches the order of magnitude of the thermal e.m.f. of couples composed of ordinary dissimilar metals. For example, the thermal e.m.f. of a couple composed of Zn uncompressed and compressed to 12,000 kg., with its two junctions at 0° and 100°, is one-third as great as that of a couple of uncompressed Zn and Pb between the same temperature limits.

The effects of pressure on thermal e.m.f. may be exceedingly complicated and vary greatly from one metal to another, as Figs. 13 and 14 show. In general the effect of pressure on Peltier and Thomson heats is positive, that is, the positive current absorbs heat in flowing from uncompressed to compressed metal, and the Thomson heat absorbed in passing from a low to a high temperature is greater in the compressed than in the uncompressed metal, but the irregularities are much greater than the irregularities in the resistance effects and occur in unexpected places, the irregularity of thermal e.m.f. being greatest for Sn, Fe, and Al, metals without special distinction in other respects under pressure. Just as in the case of resistance phenomena, these complicated effects indicate a complicated mechanism, certainly more complicated than contemplated in any theory yet proposed. There is no discoverable parallelism between the effects of pressure on resistance and on

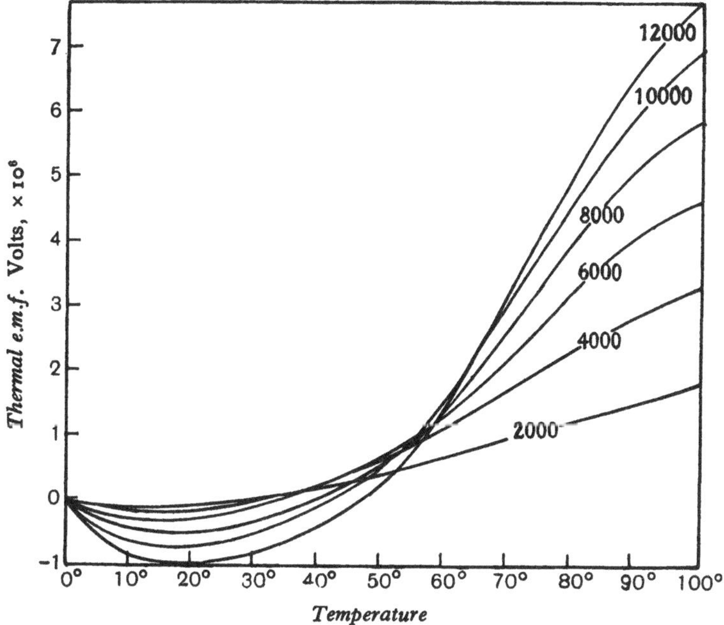

Fig. 14. Thermal e.m.f. of a couple composed of one branch of uncompressed pure iron, with the other branch composed of the same metal compressed to the pressure in kg./cm.² indicated on the curves, the junctions being at 0° C. and the temperature plotted as abscissae.

thermal e.m.f., from which one may perhaps draw the conclusion that essentially different aspects of the electron mechanism are responsible for thermoelectric phenomena, and the phenomena of resistance.

Passing now to another group of phenomena, the thermal conductivity of 15 liquids has been measured to 12,000 kg. The effect is to increase thermal conductivity by fairly large amounts, the factor of increase varying from 1·50 for water, the least compressible of the substances measured, to 2·74 for normal pentane, approximately the most compressible. It is very suggestive that this increase runs roughly parallel with the increase in the velocity of sound under pressure, as calculated from the change of density and the compressibility. I have found a very simple relation, suggested by this fact, for the thermal conductivity of liquids, namely, $k = 2\alpha v\delta^{-2}$. Here k is thermal conductivity in absolute units, α is the molecular gas constant $2{\cdot}02 \times 10^{-16}$, v the velocity of sound, and δ the mean distance of separation of the molecules in the liquid, assuming them piled in simple cubic array. The formula represents with considerable success the thermal conductivity of normal liquids, of water, which is abnormal in so many respects, and also of non-crystalline solids, such as glass or hard rubber.

In gases the phenomena of viscosity are closely connected with those of thermal conductivity, but in liquids the difference in behaviour of the temperature coefficients of thermal conductivity and viscosity makes it highly probable that the mechanisms of the two effects are different. This view receives strong support from the pressure effect on viscosity, which has been measured for 43 liquids. In all cases, except the abnormal one of water and then only over a limited range of pressure and temperature, viscosity increases under pressure, and by amounts varying enormously from substance to substance. The factor of variation under 12,000 kg. is 1·33 for mercury, and 10^7 for eugenol ($C_3H_5.C_6H_3.OH.OCH_3$). In fact, the pressure effect on viscosity and its variation from substance to substance is much greater than any other known pressure effect. Viscosity increases geometrically with pressure, that is, the logarithm of viscosity is approximately a linear function of pressure. There is a very close correlation between the magnitude of the pressure effect and the complexity of the molecular structure, the pressure coefficient being least for monatomic Hg and greatest for complicated organic substances like eugenol. It is evident that no purely kinetic mechanism of viscosity, such as we have in gases, is competent to explain such enormously large effects. There would seem to be little question that a large part of the viscosity of a liquid is purely mechanical in origin, arising from the jambing together or interlocking of the molecules. Such a mechanism is consistent with the very large pressure effects, and the enormous variations of this effect with molecular complexity.

Finally, I may mention an investigation not yet finished on the effect of pressure on the shearing modulus or rigidity of solids. Hitherto the only elastic constant which has been measured under pressure is compressibility, but there would be considerable interest in determining how pressure affects all the constants. Unfortunately this is a matter of great experimental difficulty, and up to the present I have been able to measure the effect of pressure only on the shearing modulus of several isotropic substances. The rigidity of Fe increases under pressure, as would be expected, the increase being about 2·5 per cent. for 12,000 kg. The rigidity of glass, on the other hand, decreases by amounts varying with the composition from 0·7 per cent. to 11 per cent. for 12,000 kg. The effect is largest for those glasses which have the abnormal increase of compressibility with pressure, as might be expected. By combining this change in the shearing modulus with the known changes in density, it may be calculated that the velocity of a wave of shear decreases under 12,000 kg. by amounts varying from 2 to 6 per cent., a fact of some possible geological interest. By combining the pressure coefficient of rigidity with the pressure coefficient of compressibility, the effect of pressure on Young's modulus may be found; this may increase or decrease under pressure depending on the composition of the glass, by amounts varying from + 2·5 per cent. to − 10 per cent. under 12,000 kg.

IRREVERSIBLE TRANSFORMATIONS OF ORGANIC COMPOUNDS

IRREVERSIBLE TRANSFORMATIONS OF ORGANIC COMPOUNDS UNDER HIGH PRESSURES

(Preliminary Paper)

BY P. W. BRIDGMAN AND J. B. CONANT

JEFFERSON PHYSICAL LABORATORY AND THE CHEMICAL LABORATORY OF HARVARD UNIVERSITY

Communicated April 8, 1929

The observation was made some years ago that egg white was coagulated by the application of a pressure of 5000–7000 atm. for 30 minutes.[1] The present investigation was undertaken in order to discover what other types of essentially irreversible organic transformations could be brought about by the application of pressures greater than 3000 atmospheres, and to study in more detail the effect of pressure on protein solutions. We have as yet tried but relatively few of the many substances and mixtures which might be expected to undergo changes when highly compressed and which we hope eventually to study.

The experiments were performed by filling small glass tubes of about 2 cc. capacity with the organic material. The pressure was transmitted to the organic material by means of mercury. The tubes were inverted in a steel container with their drawn-out ends open under a few centimeters of mercury. The steel container was then placed in the usual high-pressure apparatus employed in the Jefferson Physical Laboratory[2] and the pressure transmitted to the mercury by means of kerosene. The substances and mixtures which were not appreciably affected by subjection to a pressure of 10,000 atmospheres for 24 hours were as follows: amylene, pinacone, tert. amyl alcohol, diacetone alcohol, aniline acetate dissolved in aniline, solid maleic acid, benzoquinone dissolved in isopropyl alcohol, phenol dissolved in 20 per cent aqueous formaldehyde.

Isoprene, 2,3-dimethyl-1,3-butadiene, styrene and indene were partially polymerized by being subjected to a pressure of 9000 atmospheres for 24 hours. In the case of the styrene and indene the material was still liquid but on evaporation the polymer was obtained as an amorphous glassy material. The di-enes (isoprene and dimethyl butadiene) under the same conditions were converted into a jelly-like solid. On standing in the air this diminished considerably in bulk and weight by evaporation of the liquid hydrocarbon, leaving a rubber-like solid, slightly yellow but transparent. Both isobutyraldehyde and *n*-butyraldehyde were converted to a soft, waxy solid by the action of 12,000 atmospheres for 40 hours. The reaction was evidently not complete since the material smelled strongly of the original aldehyde. The solid product was not one of the known polymers, since on standing 24 hours at room pressure and temperature it slowly changed to a liquid. This liquid was largely the unchanged aldehyde but contained also a very small amount of some higher boiling material which was insoluble in water and non-acidic. It will be necessary to prepare considerable quantities of the solid products from these aldehydes in order to discover the nature of the transformation which is involved in this case.

A few experiments were carried out on the rate of polymerization of isoprene at different pressures. A sample of commercial isoprene was employed. It has a very disagreeable odor and left a very slight greasy residue on evaporation. Polymerization by the action of 12,000 atmospheres pressure for 50 hours yielded a tough transparent rubber-like solid with only a slight odor. On standing in the air there was practically no shrinkage of the solid product and it is evident that the polymerization was practically complete. The transparent solid obtained by the action of 9000 atmospheres for 24 hours was initially quite soft, and on standing 24 hours in the open lost 60 per cent of its weight and shrank to a denser rubber-like solid. On the assumption that the loss in weight by evaporation on standing represents the amount of unpolymerized isoprene, the results of our preliminary experiments may be expressed as follows:

PRESSURE (ATMOSPHERES)	TIME (HOURS)	PER CENT POLYMERIZED
12,000	50	Practically complete
9,000	24	40–45
6,000	48	10
3,000	68	Only a trace

In order to determine whether or not the mercury which was used to transmit the pressure had a catalytic influence, a glycerine-water mixture was used in one experiment at 6000 atmospheres. The extent of the polymerization was the same as in the experiments in which mercury was employed.*

For the study of the action of pressure on protein solutions, carboxyhemoglobin was employed since we were particularly interested in obtaining an insoluble form of hemoglobin. Several years ago Dr. Alice R. Davis observed the coagulation of blood under high pressures using the same apparatus as employed in our experiments (experiments unpublished). The carboxyhemoglobin was prepared from crystallized oxyhemoglobin (horse) by pumping off the oxygen and saturating with carbon monoxide. The solutions used were about 10 per cent in hemoglobin and contained potassium phosphates in a concentration of 0.14 mole per liter. About 10 cc. of the hemoglobin was subjected to pressure in each experiment. After removing from the pressure apparatus, the tube appeared to be filled with a reddish precipitate. This precipitate was separated from the solution by centrifuging and the carbon monoxide content of the solution determined by the usual Van Slyke method. The precipitate resembles in appearance and behavior the "denatured" carboxyhemoglobin obtained by the action of alcohol on carboxyhemoglobin. It is insoluble in water but soluble in dilute alkalies.

On the assumption that the loss in carbon monoxide content of the solution represents the amount of denatured carboxyhemoglobin, the following figures on the rate of the process were obtained:

INITIAL CONCENTRATION G. PER 100 CC.		TIME (HOURS)	FRACTION OF PROTEIN DENATURED	REACTION VELOCITY CONSTANT (FIRST ORDER REACTION)
Series (A)	pH = 7.1		Pressure 9000 atm.	Temp. 20–25°
11.0		15	0.07	0.0048
		39	0.19	0.0054
		113	0.37	0.0041
Series (B)	pH = 6.4		Pressure 9000 atm.	Temp. 20–25°
5.7		18	0.63	0.055
3.0		18	0.64	0.057
1.5		18	0.60	0.051

The data given above indicate that the denaturation of carboxyhemoglobin by pressure is essentially a first-order reaction and that the rate is a function of the acidity of the solution. A more detailed study of the factors effecting the denaturation is now being made and we hope to prepare enough of the denatured protein to examine it thoroughly.

We are indebted to Mr. W. A. Zisman for assistance in manipulating the pressure apparatus and to Dr. R. V. McGrew for preparing and analyzing the hemoglobin solutions.

* Experiments with different samples of isoprene have shown that impurities have a marked catalytic influence. Thus a sample of material prepared from limonene and freshly distilled was only 30% polymerized in 20 hours at 12,000 atmospheres. The same material after standing several weeks was 80% or 90% polymerized under the same conditions. Similarly two different samples of commercial isoprene showed even greater differences in their rate of polymerization. We are inclined to attribute the catalytic effects to the presence of peroxides. (Footnote added to proof July 19, 1929. J. B. C.)

[1] P. W. Bridgman, *J. Biol. Chem.*, **19**, 511, 1914.

[2] P. W. Bridgman, *Proc. Amer. Acad.*, **49**, 11, 1914.

DIE EIGENSCHAFTEN VON METALLEN UNTER HOHEN HYDROSTATISCHEN DRUCKEN

P. W. Bridgman, Cambridge (Mass.)

Die Drucke, die bei verschiedenen technischen Prozessen erreicht werden, betragen fast nie mehr als 100 kg/cm²; die Substanzen, die hierbei diesen Drucken unterworfen werden, sind meistens Gase, wie z. B. bei Dampfmaschinen oder bei Gasreaktionen unter hohem Druck. Gewöhnliche Flüssigkeiten, die solchen Drucken unterworfen werden, zeigen keinerlei besonders bemerkenswertes Verhalten, es kann z. B. auch die Volumenänderung von Wasser in einer hydraulischen Presse in den meisten Fällen vernachlässigt werden. Wenn man die Eigenschaften von Flüssigkeiten in dieser Weise beeinflussen will, so muß man zur nächsten Größenordnung, d. i. 1000 kg/cm², übergehen. Mit solchen Drucken wird im Laboratorium schon seit einer Reihe von Jahren gearbeitet. Diesbezügliche Apparate wurden von verschiedenen feinmechanischen Firmen gebaut, die Eigenschaften einer Reihe von Flüssigkeiten bei diesen Drucken wurden von verschiedenen Seiten ausführlich untersucht. Eine Grenze wurde diesen Versuchen durch beginnende Undichtigkeit gesetzt. Der maximale Druck mit Apparaten dieses Typs betrug ungefähr 3 000 kg/cm². Amagat in Paris und Tammann in Göttingen müssen hier durch ihre außerordentlich wichtigen Arbeiten auf diesem Gebiet genannt werden. Ebenso Cohen in Utrecht, der interessante Ergebnisse bei Drucken erhielt, die allerdings selten 2 000 kg/cm² erreichten. Viele Forscher begnügten sich mit Drucken, die 1 000 oder 1 500 kg/cm² nicht überschritten.

Wenn man die Eigenschaften von festen Körpern, insbesondere von Metallen wesentlich beeinflussen will, so sind Drucke von der nächsten Größenordnung, d. i. 10 000 kg/cm², nötig. In diesem Aufsatz sollen einige Versuche beschrieben werden, die Verfasser über die Eigenschaften von Metallen bei Drucken dieser Größenordnung ausgeführt hat. Diese Drucke sind hoch genug, um nicht nur das lineare, sondern auch das quadratische Glied der Druckfunktion verschiedener Eigenschaften erfassen zu können. Für jeden theoretischen Schluß aus den Versuchsresultaten ist dies sehr wichtig. Die meisten meiner Versuche wurden bei Drucken von ungefähr 12 000 kg/cm² ausgeführt, in einigen Fällen wurde bei 20 000 kg/cm², zwei- oder dreimal sogar bei 30 000 kg/cm² gearbeitet. Um eine Vorstellung von einem Druck von 30 000 kg/cm² zu erhalten, sei gesagt, daß derselbe Druck am Grunde eines Ozeans von 300 km Tiefe herrschen müßte (wobei die Kompressibilität des Wassers vernachlässigt ist), oder in einer Tiefe von 120 km unter der Erdrinde, vorausgesetzt, daß die mittlere Gesteinsdichte 2,5 beträgt.

Diese hohen Drucke konnten auf Grund einer einfachen Anordnung, die jede Undichtigkeit absolut verhindert, erreicht werden. Das Prinzip ist aus der schematischen Abb. 1 zu ersehen: Der Kolben P drückt den Zapfen A durch den gehärteten Ring R, die Unterlage C, die aus weichem Stahl besteht, und die Kautschukdichtung B. Die zu komprimierende Flüssigkeit befindet sich unter A bei

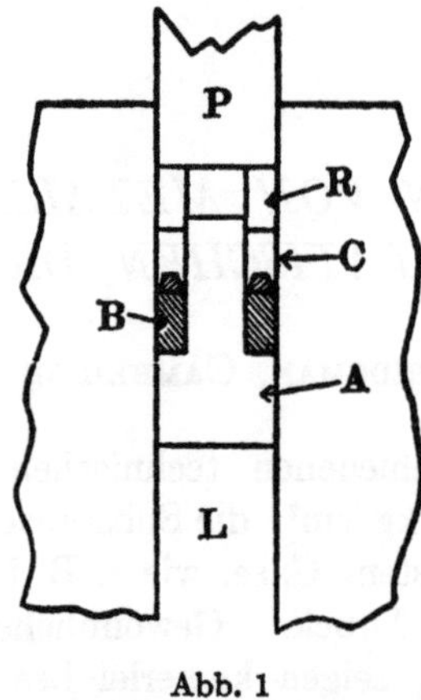

Abb. 1

L. Der Zapfen A ist mit einem Stiel versehen, der lang genug ist, um bis zum Ring R, aber nicht lang genug, um bis zum Kolben P zu reichen. Wenn wir nun das Gleichgewicht bei A betrachten, so sehen wir, daß der Flüssigkeitsdruck über dem unteren Ende von A ausgeglichen werden muß durch den Druck, der durch die Dichtung B auf eine Fläche ausgeübt wird, die kleiner ist als die Fläche A, und zwar um die Fläche des nicht gestützten Stieles. Daraus folgt, daß der hydrostatische Druck pro Flächeneinheit innerhalb der Dichtung B immer um einen gewissen Prozentsatz höher ist als der in der Flüssigkeit, so daß die Flüssigkeit nie bis zum Kolben vordringen kann. Die einzige Grenze, die dem erreichbaren Druck gesetzt ist, besteht in der Festigkeit der Gefäße. Oftmals wurde der Apparat einige Tage lang den höchsten Drucken ausgesetzt, ohne irgendwelche bemerkenswerte Undichtigkeit zu zeigen.

Als ich meine Versuche begann, war noch wenig über die maximale Festigkeit von Gefäßen, wenn sie innerem Druck ausgesetzt werden, bekannt. Die Theorie verlangt, daß der maximale Druck derselbe sei wie die Zugfestigkeit, so daß ein Stahl beispielsweise mit einer Zugfestigkeit von 10 000 kg/cm² selbst dann keinem höheren Druck standhalten könnte, wenn die Wände des Gefäßes unendlich stark werden. Glücklicherweise trifft diese theoretische Erwartung nicht zu; tatsächlich können Drucke erreicht werden, die viel höher sind als die obengenannte Grenze. Dies kommt in folgender Weise zustande: Wenn ein Druck zuerst auf das Innere eines Zylinders ausgeübt wird, so wird die Elastizitätsgrenze an den inneren Wandungen eher erreicht, als an den äußeren, d. h. die äußeren Teile des Gefäßes werden weniger beansprucht. Wenn die Wandung des Zylinders dünn ist, tritt Bruch ein, ist sie dagegen stark, so ist kein Bruch zu befürchten, die inneren Teile des Zylinders geraten jedoch in einen mehr oder weniger plastischen Zustand, durch den die Beanspruchung auf die äußeren Teile übertragen und dadurch mehr ausgeglichen wird. Wenn jetzt der Druck nachläßt, tritt eine leichte Schwindung der äußeren Teile ein, während die inneren Teile bei vollkommenem Nachlassen des Drucks in einem mit Druckspannungen behafteten Zustand verbleiben, gleich der inneren Rohrwandung eines bandagierten Gewehrs. Bei der nächsten Druckeinwirkung befinden sich die inneren Teile bereits in einem komprimierten Zustand und erreichen ihre Streckgrenze erst, wenn die äußeren Teile einen größeren Anteil der

Beanspruchung aufgenommen haben, wie bei der ersten Belastung. Auf diesem Wege ist es möglich, innerhalb eines starken Zylinders (unter starkem Zylinder ist hier ein Zylinder gemeint, dessen äußerer Durchmesser 8 oder 10 mal so groß ist als der innere Durchmesser), innere Drucke zu erreichen, die viermal so hoch sind, als die Zugfestigkeit des Materials. Unter diesen extremen Drucken kann allerdings Bruch eintreten, aber ein starker Zylinder hält ohne weiteres eine unbegrenzte Zahl von Druckbeanspruchungen aus, die das Doppelte der Zugfestigkeit betragen.

Es ist klar, daß dieses Verfahren der Erzeugung von Druckspannungen im Inneren von Rohren durch Drucke, welche die Elastizitätsgrenze überschreiten, uns eine bessere und billigere Methode der Herstellung von Geschützrohren gibt, als die übliche durch Aufziehen eines Rohres auf ein zweites. In der Tat hat diese Methode auch in den Vereinigten Staaten ihre praktische Anwendung gefunden.

Es ist aber auch noch ein anderes Problem zu lösen, bevor genaue Untersuchungen angestellt werden konnten und das war, eine geeignete Methode ausfindig zu machen, um diese hohen Drucke messen zu können. Natürlich konnte keines der sonst bei elastischen Deformationen üblichen Meßinstrumente benutzt werden, da das Metall Drucken ausgesetzt ist, die oberhalb der Elastizitätsgrenze liegen. Nach vielen Versuchen ergab sich als geeignetste Methode der Gebrauch einer einfachen Abänderung der von Amagat angegebenen Methode. Das Eichverfahren von Amagat konnte oberhalb 3 000 kg wegen der rasch auftretenden Undichtigkeiten nicht verwendet werden. Dies konnte jedoch unter Beachtung dreier Punkte vermieden werden: Zuerst wurde ein Kolben von nur geringem Durchmesser (1,6 mm) verwendet; es wurde ferner der Zylinder, in dem sich der Kolben bewegt, einem äußeren Druck ausgesetzt, wodurch die Bildung eines Zwischenraumes zwischen Kolben und Zylinder vermieden wurde; schließlich wurde der Druck bei dem Kolben durch eine Mischung von Glukose und Glyzerin, die unter hohen Drucken nicht einfriert, aber trotzdem sehr viskos ist, ausgeübt. Auf diese Weise war es möglich, absolute Messungen des Druckes bis zu 13 000 kg/cm^2 mit einer Genauigkeit von 0,1 Prozent anzustellen. Nachdem eine absolute Eichung des Druckes gefunden war, konnte man eine bequemere, hiervon abgeleitete Standardmessung einführen. Zu diesem Zweck wurde die Änderung des Widerstandes von Manganin verwendet, die zuerst von Lissel als Ergebnis von Versuchsreihen bis 3 000 kg vorgeschlagen wurde. Die Vorteile dieser Eichmethode sind in den Eigenschaften des Materials begründet, Temperatur-Korrekturen sind dabei nicht notwendig. Die Kalibrierung ist sehr einfach, da das Anwachsen des Widerstandes eine lineare Funktion des Druckes ist, so daß also eine Kalibrierung bei nur einem Druck notwendig ist. Eine Reihe meiner Messungen wurde mit einem Mangalininstrument ausgeführt, dessen Eichung auf den Erstarrungsdruck des Quecksilbers bei 0° C bezogen war. Mittels der absoluten Eichung wurde gefunden, daß Quecksilber bei einem Druck von 7 640 kg/cm^2 erstarrt, dieser feststehende Punkt kann bei der Eichung in derselben Weise wie die Schmelzpunkte verschiedener Substanzen bei Atmosphärendruck als Fixpunkte in der Thermometrie, benutzt werden. Der numerische Wert des Druckkoeffizienten c von Manganin bedingt bei einem Druck von 10 000 kg Anwachsen des Widerstandes um 2,3 Prozent. Es ist klar, daß eine solche Eichung für die Messung niederer Punkte bei gewöhnlichen Versuchen nicht sehr geeignet ist.

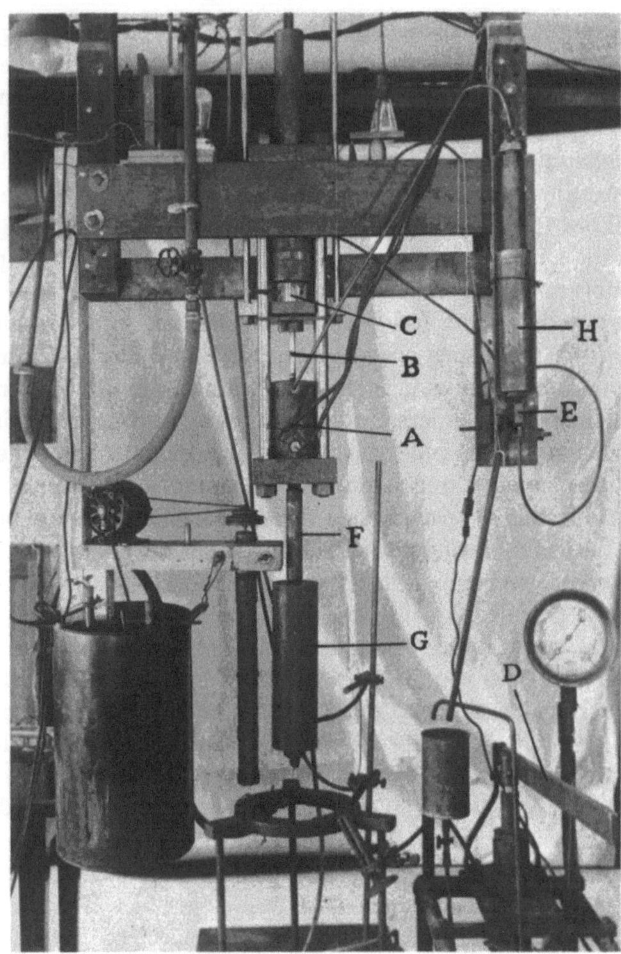

Abb. 2

Abb. 2 zeigt den Apparat, mit dem die meisten Messungen ausgeführt wurden. Der Druck wird in dem oberen Zylinder A durch den schmalen Kolben B erzeugt, dessen Durchmesser 1,27 cm beträgt und der durch den breiteren Kolben C (6,35 cm Durchmesser) einer hydraulischen Presse betrieben wird, die durch die Handpumpe D betätigt wird. Mit Hilfe der Druckventile bei E kann die Pumpe D mit dem Kolben C oder dem hydraulischen Verstärker verbunden werden. Der Verstärker dient dazu, um in dem Hochdruckteil der Apparatur einen Anfangsdruck von 2 000 kg/cm² zu erzeugen. Dieser Anfangsdruck komprimiert zufällig im Apparat befindliche Luft auf ein zu vernachlässigendes Volumen und verursacht bereits eine gewisse Kompression der Flüssigkeit. So ist es möglich, 12 000 kg mit einem einzigen Kolbenhub des Hochdruckkolbens zu erreichen, was anders, da Druckventile im Hochdruckteil der Apparatur fehlen müssen, nicht erreicht werden kann. Der obere Zylinder A ist durch das Rohr F mit dem unteren Zylinder G verbunden, der je nach dem Versuch bewegt und ausgewechselt werden kann.

Die geringe Größe der Apparatur hat verschiedene Vorteile. Erstens ist es, da die Volumina sehr klein sind, möglich, die Höchstdrucke mit Hilfe einer Handpumpe in einigen Minuten zu erreichen. Die geringe Größe vermindert ferner die Gefahr von Explosionen, die manchmal eintreten. Schließlich ist es bei Verwendung kleiner Stahlgefäße möglich, die mechanischen Eigenschaften des Stahles durch Wärmebehandlung gleichmäßig zu verbessern. Ein solches Gefäß ist in weit höherem Maße den Anforderungen gewachsen als ein größeres Gefäß, bei dem zu befürchten ist, daß die Wärmebehandlung nicht die gesamte Masse des Stahles gleichmäßig verbessert hat.

Nachdem nun die Methoden der Herstellung und Messung hoher Drucke geschildert wurden, soll im folgenden auf die Wirkung derselben auf Metall eingegangen werden.

Die wichtigste Änderung, die durch Druck erzielt werden kann, ist die polymorphe Umwandlung von Stoffen, von der eine große Zahl von Beispielen bekannt ist. Am bemerkenswertesten ist wohl die Polymorphie des Wassers. T a m m a n n fand, daß bei einem Druck von 3 000 kg gewöhnliches Eis sich in eine von zwei neuen Modifikationen umwandelt, die beide nicht wie gewöhnliches Eis eine geringere, sondern eine höhere Dichte haben als Wasser. Ich habe gefunden, daß bei höheren Drucken zwei andere Formen auftreten, die um verschiedene Beträge dichter sind als flüssiges Wasser. Bei 75° C und einem Druck von ungefähr 21 000 kg/cm^2 erstarrt Wasser. Ähnliche Effekte sind bei vielen organischen und anorganischen Substanzen bekannt, treten aber verhältnismäßig selten bei Metallen auf. Ich habe nur zwei Beispiele gefunden; das eine betrifft Kadmium, das sich bei 0° C und einem Druck von 3 000 kg in eine andere kristalline Modifikation umwandelt, die ihrerseits bei 5 500 kg abermals eine Umwandlung zeigt. Die Volumenänderungen, die diesen Wechsel des Kristallsystems begleiten, sind sehr gering. — Der Effekt wurde von mir nur an Einkristallen von Kadmium gefunden, es wurde nur die Längenänderung und nicht die Volumenänderung gemessen. Die linearen Veränderungen sind nach verschiedenen Richtungen verschieden, sie mögen in manchen Richtungen größer sein als die Volumenänderungen. Das zweite Beispiel einer polymorphen Umwandlung wurde am Cer festgestellt, das sich bei 30° C und 7 600 kg bzw. 75° C und 9 400 kg in eine neue Modifikation verwandelt. Es sollte erwartet werden, daß Wismut, das beim Erstarren sich ausdehnt, in Analogie zu gewöhnlichem Eis ebenfalls eine neue Hochdruckform bildet, eine Umwandlung konnte jedoch unterhalb 20 000 kg nicht festgestellt werden. Die Angaben von L u d w i g, daß eine Umwandlung eintritt, ist zweifellos auf irgendeinen apparativen Fehler zurückzuführen.

Eng verbunden mit den polymorphen Umwandlungen, die durch Druck erzeugt werden, sind die Umwandlungen vom flüssigen in den festen Zustand. Ein Ergebnis der elementarsten Thermodynamik sagt aus, daß die Schmelztemperatur einer Substanz durch Druckerhöhung steigt, wenn das spezifische Volumen der flüssigen Phase größer ist als das der festen. Da die meisten Metalle sich beim Schmelzen ausdehnen, so folgt daraus eine Erhöhung des Schmelzpunktes. Die numerischen Werte der Änderung der Schmelztemperatur durch Druck sind von Wichtigkeit. Es wurde bereits erwähnt, daß der Gefrierdruck von Quecksilber bei 0° C 7 640 kg beträgt: der Druck bewirkt also eine Erhöhung des Schmelzpunktes von ungefähr

39°. Ein Druck von 12 000 kg bewirkt eine Erhöhung der Schmelztemperatur des Quecksilbers um ungefähr 60°. Die Alkalimetalle, die weich und in hohem Maße komprimierbar sind, zeigen ähnlich große Effekte. Unter ihnen ist die Druckabhängigkeit am geringsten bei Lithium, am größten bei Caesium; um einen Überblick über die Größenordnung der Schmelzpunkterhöhung zu geben, sei mitgeteilt, daß der Schmelzpunkt von Kalium durch einen Druck von 19 000 kg von 62° C auf 180° C erhöht wird. Da der Druckeffekt bei den Metallen mit dem niedrigsten Schmelzpunkt am größten ist, so überkreuzen sich bei hohen Drucken die Schmelzkurven; so wurde in der Tat gefunden, daß bei Drucken oberhalb 9 000 kg die Schmelztemperatur von Kalium höher ist als die von Natrium, obwohl sie bei Atmosphärendruck 36° C niedriger ist. Es ist sehr wahrscheinlich, daß bei genügend hohen Drucken eine vollkommene Umkehr eintritt, so daß von den Alkalimetallen Caesium den höchsten und Lithium den niedrigsten Schmelzpunkt besitzen.

Die Schmelzkurve ist gegen die Druckkoordinate konkav, d. h. die Schmelztemperatur steigt mit steigendem Druck immer schwächer an. Es liegt aber kein Grund vor, anzunehmen, daß die Werte für die Schmelztemperatur nicht jeden beliebig hohen Wert annehmen können, Voraussetzung dafür ist lediglich die Anwendung von genügend hohen Drucken, wenn nicht vorher zuerst Atomzerfall eintritt.

Wismut ist eins der Metalle, die sich beim Erstarren zusammenziehen, sein Schmelzpunkt wird daher durch Druck auch erniedrigt. Ein Druck von 12 000 kg erniedrigt den Schmelzpunkt von Wismut von 271° C auf 218° C. Im Gegensatz zu den übrigen Metallen nimmt die Schmelzpunkterniedrigung bei hohen Drucken in steigendem Maße zu.

Von diesen Zustandsänderungen abgesehen dürfte am wesentlichsten die Änderung des Volumens unter Druckeinfluß sein. Die Kompressibilität von gewöhnlichen Metallen ist gering; die Volumenverminderung von Eisen beträgt beispielsweise bei 10 000 kg/cm^2 nur 0,6 Prozent. Obwohl die Volumenänderungen, wenn sie im abs. Maß ausgedrückt werden, nur gering sind, so werden sie doch nicht so klein, wenn sie mit den Werten der thermischen Ausdehnung verglichen werden. In vielen Fällen ist die Volumenverminderung eines Metalles bei 10 000 kg und Zimmertemperatur größer als die Volumenverminderung bei Atmosphärendruck und Abkühlung von Zimmertemperatur zu 0° abs. Die Kenntnis dieser Volumenänderungen läßt also auch wertvolle theoretische Folgerungen erwarten. Es ist klar, daß zur Messung dieser Änderungen besonders verfeinerte Methoden herangezogen werden müssen. Eine von mir ausgearbeitete elektrische Methode gestattet durch Anwendung eines auf den Versuchskörper befestigten Schleifkontaktes Änderungen zu messen, die weniger als eine Wellenlänge optischen Lichts betragen. Diese Methode wurde bei Kompressibilitäts-Messungen an 40 oder 50 Metallen verwendet. Die Kompressibilität dieser Elemente schwankt innerhalb weiter Grenzen; am wenigsten komprimierbar ist Iridium, dessen Volumenverminderung bei 10 000 kg nur 0,25 Prozent beträgt, am stärksten Caesium, das bei 10 000 kg eine Volumenverminderung von 27 Prozent zeigt und bei noch höheren Drucken stärker komprimierbar ist als selbst die am meisten komprimierbaren organischen Flüssigkeiten, wie Äther. Die Genauigkeit der

Untersuchungen war groß genug, um die Veränderung der Kompressibilität mit dem Druck zu zeigen; naturgemäß können diese Änderungen nur in ziemlich großem Abstand mit Genauigkeit ausgeführt werden. Praktisch in allen Fällen sinkt die Kompressibilität, wie erwartet, mit steigendem Druck. Dieses Absinken ist jedoch nicht einmal annähernd so groß bei den Metallen wie bei organischen Flüssigkeiten. Der Effekt beruht zweifellos auf der Kompression der Atome selbst. Das Kaliumatom ist besonders komprimierbar, es sind Anhaltspunkte dafür vorhanden, daß bei Drucken von 15 000 kg bereits eine Zerstörung der Atome eintritt.

Es ist mithin nicht notwendig, daß eine Substanz bei höheren Drucken weniger komprimierbar wird wie einige Ausnahmen zeigen. Es gibt verschiedene Glasarten, deren Kompressibilität mit dem Druck wächst. Unter den metallischen Elementen bildet Cer ein Beispiel des Anwachsens der Kompressibilität bei hohem Druck.

Von besonderem Interesse sind Einkristalle der nicht kubischen Metalle wegen ihres anisotropen Verhaltens bei Druckbeanspruchung; die Verschiedenheit der Kompressibilität nach verschiedenen Richtungen nimmt manches Mal beträchtliche Werte an. So ist die lineare Kompressibilität von Zink nahezu siebenmal größer parallel zur hexagonalen Achse als senkrecht zu ihr. Tellur zeigt das überraschende und paradoxe Verhalten einer negativen Kompressibilität längs der trigonalen Achse; wenn ein Tellureinkristall gleichmäßigem hydrostatischen Druck ausgesetzt wird, so tritt Verlängerung in Richtung der Achse ein. Natürlich verkürzt sich der Kristall in den senkrecht dazu liegenden Richtungen, und zwar um einen mehr als ausreichenden Betrag, um die longitudinale Ausdehnung aufzuheben, denn das Volumen muß unter Einwirkung hydrostatischen Druckes sich natürlich vermindern.

Die Kompressibilität ist nur eine der elastischen Konstanten eines Metalles. Eine Kenntnis der Druckfunktion aller elastischen Konstanten ist von großer Wichtigkeit, unglückseligerweise stellen sich jedoch der Erforschung dieses Gebietes große experimentelle Schwierigkeiten in den Weg, außerdem liegen noch keinerlei Versuchsergebnisse darüber vor. Ich habe jedoch unlängst eine Methode ausgearbeitet, durch die der Schermodul eines isotropen Metalles unter Druck gemessen werden kann und ich habe gefunden, daß er bei Stahl unter 10 000 kg um 1,6 Prozent steigt. Es war dieses Ansteigen wohl zu erwarten, wenn auch keine Notwendigkeit dafür vorliegt, tatsächlich habe ich auch bei einigen Gläsern gefunden, daß der Schermodul unter Druck abnimmt. Diese Tatsache scheint von einiger Bedeutung für die Geologie zu sein.

Die Veränderung des elektrischen Widerstandes der Metalle unter Druck ist eine wichtige Eigenschaft, die verhältnismäßig leicht gemessen werden kann. Ich habe die Druckabhängigkeit des Widerstandes bei einer großen Zahl von Elementen und einigen Legierungen untersucht. Im allgemeinen sinkt der Widerstand mit steigendem Druck. Größenordnungsmäßig variieren die Werte von etwas unterhalb 1 Prozent bei 10 000 kg für Kobalt bis zu 0,3 des Anfangswiderstandes bei Kalium. Gewöhnlich ist der Effekt um so größer, je mehr das Metall komprimierbar ist, ein Parallelismus, der nicht immer genau zutrifft; numerisch ist die Abnahme des Widerstandes ungefähr zehnmal größer als die Volumenabnahme. Die Abnahme des Widerstandes geht nicht linear mit dem Druck, bei hohen Drucken fällt sie, wie erwartet, langsamer, so daß die Kurve des elektrischen Widerstandes in Abhängigkeit vom Druck konvex zur Druckachse liegt.

Nach der klassischen Theorie der elektrischen Leitung von D r u d e ist eine Abnahme des Widerstandes unter Druck nicht zu erwarten, eher sogar ein Anwachsen desselben, da durch die dichtere Packung der Atome die freie Weglänge der Elektronen verkürzt wird. Die Tatsache, daß der Druckkoeffizient entgegengesetzt ist, als man es nach der Theorie erwartet, beweist, daß der Mechanismus ganz anderer Natur sein muß, als von der klassischen Theorie angenommen wird. Die Erklärung des Druckkoeffizienten des Widerstandes ist für jede Theorie bedeutungsvoll. Die meisten Theorien sind in ihrer Erklärung des Druckkoeffizienten des elektrischen Widerstandes nicht sehr erfolgreich gewesen, erst in der letzten Zeit sind durch die Anwendung der F e r m i s t a t i s t i k durch S o m m e r f e l d un ihre Erweiterung durch H o u s t o n unter Benutzung einiger Gesichtspunkte der neuen Wellenmechanik vielversprechende Ausblicke zur erfolgreichen Behandlung dieses Problems gezeitigt worden. Nicht alle Metalle zeigen die Abnahme des Widerstandes unter Druck, es sind vielmehr einige, deren Widerstand zunimmt, unter ihnen Wismut, das sich ja in manchen Beziehungen abnorm verhält und Caesium und Barium, deren Widerstand zuerst abnimmt, ein Minimum durchläuft und bei den höchsten Drucken zunimmt. In allen diesen Fällen ist der Charakter der Kurve derselbe, konvex gegen die Druckachse, wie für jene Metalle, deren Widerstand abnimmt. Es ist daher zu vermuten, daß bei Anwendung genügend hoher Drucke der elektrische Widerstand aller Metalle mit zunehmendem Druck ansteigt. Dies scheint jedenfalls von großem theoretischen Interesse.

Der elektrische Widerstand von nicht kubischen Einkristallen zeigt unter Druck dasselbe anisotrope Verhalten wie bei der Kompressibilität. Der Widerstand ist natürlich nach verschiedenen Richtungen hin verschieden, ebenso der Druckkoeffizient des Widerstandes. Jedoch sind im allgemeinen die Unterschiede des elektrischen Widerstandes nicht so groß wie bei der Kompressibilität. So zeigt Zink, das längs der hexagonalen Achse siebenmal stärker komprimierbar ist, als senkrecht dazu, einen Druckkoeffizienten des Widerstandes der parallel zur Achse nur zweimal so groß ist als senkrecht dazu. Die Unterschiede des Widerstandes eines nicht kubischen Kristalles in verschiedenen Richtungen waren ebenfalls durch die klassische Theorie nicht zu erklären, auch hier scheint die Wellenmechanik mehr Aussichten zur Lösung dieser Frage zu zeigen.

Der Druckeffekt des elektrischen Widerstandes ist bei Legierungen viel komplizierter als bei einfachen Metallen. Hier ist z. B. keine allgemeine Regel über den Verlauf der Kurve. Wie schon anfangs bemerkt, steigt der Widerstand von Manganin l i n e a r mit dem Druck, weshalb diese Legierung auch zu Eichzwecken besonders geeignet ist. Eine Reihe von Fällen ist bekannt, bei denen eine Legierung positiven Druckkoeffizienten des Widerstandes besitzt, während ihre beiden Komponenten in reinem Zustand negative Koeffizienten aufweisen. Es gibt z. B. Eisen-Nickel-Legierungen, die einen positiven Koeffizienten haben, der numerisch dreimal größer ist als die negativen Koeffizienten der reinen Metalle. Hier ist noch viel theoretische Arbeit zu leisten, um die Effekte zu erklären.

Die thermische Leitfähigkeit der Metalle ist ebenfalls von großem theoretischem

Interesse. Die experimentellen Schwierigkeiten bei einer Messung sind jedoch viel größer als beim elektrischen Widerstand, der Druckeffekt auf die thermische Leitfähigkeit der Metalle ist deshalb nicht mit derselben Genauigkeit bekannt als der Effekt auf den elektrischen Widerstand. Es scheint sich jedoch zu ergeben, daß die thermische Leitfähigkeit unter Druck zunimmt, und zwar manchmal mehr und manchmal weniger als die elektrische Leitfähigkeit. Die thermische Leitfähigkeit unter Druck wurde noch nicht bei den Metallen gemessen, deren elektrische Leitfähigkeit abnimmt. Die klassische Elektronentheorie läßt erwarten, daß der Druckkoeffizient der elektrischen und thermischen Leitfähigkeit der gleiche ist, da das Verhältnis der beiden Leitfähigkeiten (Wiedemann-Franz-Gesetz) nach der Theorie eine universale Konstante gibt, unabhängig vom Metall und daher unabhängig vom Druck. In der neuen Wellenmechanik liegen bis jetzt noch keine Ansätze über die Gültigkeit des Wiedemann-Franz-Gesetzes unter Druck vor.

Die hier beschriebenen Eigenschaften gehören zu den einfachsten Materialeigenschaften, die bei hydrostatischen Drucken untersucht worden sind. Eine Reihe anderer interessanter Effekte bei hohen Drucken ergeben sich dadurch, daß der Druck in den festen Teilen der Apparatur nicht hydrostatisch ist. Es ist so möglich, Bruch herbeizuführen unter Bedingungen, die von den gewöhnlichen sehr verschieden sind, und dadurch wichtige Schlüsse zu ziehen im Hinblick auf technologische Fragen. Es würde zu weit führen, auf die Details dieser Fragen einzugehen, es sei daher lediglich das Ergebnis mitgeteilt. Es ist möglich, zu zeigen, daß durch sorgfältig gewählte Bedingungen, unter denen Bruch bei hohem Druck hervorgerufen werden kann, keine der gewöhnlichen Bruchbedingungen im technologischem Sinne gultig sind, daß wahrscheinlich eine Charakterisierung des Bruches überhaupt nicht existiert. Nähere Einzelheiten sind in einem Vortrage des Verfassers vor dem 2. Kongreß für angewandte Mechanik in Zürich 1926 zu ersehen.

Andere interessante Phänomene, die zu ihrer Untersuchung sehr intensive und hohe nichthydrostatische Kräfte verlangen, sind die Permeabilität oder Porosität. Man findet häufig die Meinung, daß feste Metalle in Flüssigkeiten unter sehr hohen Drucken porös werden. Diese Meinung geht wahrscheinlich zurück auf die Versuche von Amagat, der flüssiges Quecksilber in feinen Tröpfchen durch Stahl mit einem Druck von 3 000 kg durchpreßte. Es ist durchaus wahrscheinlich, daß derartige Effekte zurückzuführen sind auf feine Sprünge und Risse im Metall; mir ist es nie geglückt, diese Erscheinung bei gesundem Metall zu reproduzieren. Bei höheren Drucken jedoch tritt ein ähnlicher Effekt ein. Es ist nicht möglich, Drucke über 6 000 kg mit Quecksilber in Stahlgefäßen hoher Festigkeit zu erzeugen, da das Gefäß, einerlei wie stark seine Wände sind, zerspringt. Dieser Vorgang ist halb chemischer, halb mechanischer Natur. Quecksilber amalgamiert Eisen, jedoch nicht unter gewöhnlichen Bedingungen, wo eine Schutzhaut von Oxyd die Oberfläche des Eisens schützt. Wenn auf die Wände des Stahlgefäßes durch Quecksilber ein Druck ausgeübt wird, so öffnen sich die Zwischenräume zwischen den Atomen, und wenn der Druck hoch genug ist, so werden die Quecksilberatome, die verhältnismäßig klein sind, in diese Räume, in denen kein Oxyd vorhanden ist, hineingedrängt, so daß sofort Amalgamierung eintritt. Die Reaktion geht immer rascher vor sich, amalgamierter Stahl ist viel weniger widerstandsfähig als gesunder Stahl, so daß die durch den Druck hervorgerufenen Kräfte in den amalgamierten Teilen stärker wirken. Die Reaktion wird außerordentlich beschleunigt, das Quecksilver dringt mit explosiver Geschwindigkeit in den Stahl ein und führt zum Zerreißen des Stahles. Die Bruchstücke solcher Gefäße zeigen an den Bruchflächen starke Amalgamierung. Weicher Stahl zeigt diesen Effekt nicht in dem Maße wie ein Stahl von hoher Festigkeit, da die inneren Teile eines starken Zylinders aus weichem Stahl die Neigungen besitzen, plastisch zu werden. Dadurch schließen sich die Zwischenräume, bevor der Druck hoch genug wird, um das Quecksilber in das Metall eindrigen zu lassen.

Ähnliche Effekte können erwartet werden bei niedrigeren Drucken als 6 000 kg und hohen Temperaturen; es ist dies zweifellos ein Gefahrenmoment bei Hochdruck-Quecksilber-Turbinen.

Andere Substanzen vermögen ebenso wie das Quecksilber den festen Stahl zu durchdringen. Wasserstoff entweicht bei Drucken oberhalb 9 000 kg mit explosiver Heftigkeit durch die Wände von Stahlgefäßen, Sauerstoff zeigt bei Drucken oberhalb 15 000 kg ein ähnliches, jedoch nicht so heftiges Verhalten. Keine dieser Effekte treten jedoch ein, wenn das Metall einem gleichmäßigen hohen hydrostatischen Druck ausgesetzt ist, so daß die interatomaren Zwischenräume kleiner anstatt größer werden.

Eingegangen am 11. Oktober 1928.

TRANSVERSE THERMO-MAGNETIC AND THERMO-ELECTRIC EFFECTS IN CRYSTALS

ON THE NATURE OF THE TRANSVERSE THERMO-MAGNETIC EFFECT AND THE TRANSVERSE THERMO-ELECTRIC EFFECT IN CRYSTALS

By P. W. Bridgman

Jefferson Physical Laboratory, Harvard University

Communicated August 23, 1929

It is known that there is a close formal parallelism between the Ettingshausen temperature gradient set up in an isotropic conductor at right angles to an electric current and a magnetic field, and Kelvin's temperature difference between two opposite sides of a crystal rod, the length of which is oblique to the crystal axis, and in which a longitudinal electric current flows. Not only are these effects geometrically similar, but both reverse sign when the direction of current flow changes, and are, as far as known, proportional to current strength. The parallelism may be carried through for the other effects. The analogue of the Nernst transverse e. m. f. under a longitudinal heat current is the longitudinal e. m. f. in a crystal produced by a transverse temperature difference; the analogue of the Hall effect is the transverse potential gradient in a crystal carrying a longitudinal electric current arising from the fact that the lines of current flow are not perpendicular to the equipotential surfaces, and similarly the analogue of the Righi-LeDuc transverse temperature gradient is the transverse temperature gradient in a crystal carrying a longitudinal heat current arising from the fact that the lines of heat flow are not perpendicular to the surfaces of constant temperature. One might therefore be tempted to look for an underlying similarity of mechanism in a crystal and an isotropic substance made non-isotropic by a magnetic field; it is the purpose of this note to show that in spite of the close parallelism there is a fundamental difference of sign in the relations connecting the Ettingshausen with the Nernst coefficient and their corresponding crystal analogues, which indicates that the effects must be essentially different in character.

There are two diametrically opposite ways of regarding the Ettingshausen effect. The first of these I presented to the Solvay Congress in 1924, and published later in the *Physical Review*.[1] According to this point of view the Ettingshausen temperature difference between two opposite sides of a plate carrying a longitudinal electric current in a magnetic field is not accompanied by a transverse flow of heat from the hot to the cold side of the plate, but the temperature difference is maintained without heat flow, by what I called a "thermomotive" force, precisely as the two terminals of a battery on open circuit are maintained at a difference of electric potential by the "electro-motive" force of the battery. The difference of temperature between the two sides of the plate may be utilized to operate a thermodynamic engine, allowing heat to flow from the hot side through the engine to the cold side, the return flow being from *cold to hot* in the plate. When such a heat current flows between the sides of the plate, a longitudinal e. m. f. is set up by the Nernst effect (the Nernst effect and the Ettingshausen effect thus appear as one the inverse of the other), and the longitudinal electric current flowing against this Nernst e. m. f. puts into the system the energy extracted by the thermodynamic engine. On equating the energy put in to the energy extracted, the relation $Q = kP/\tau$ is found between the Nernst coefficient Q, the Ettingshausen coefficient P, thermal conductivity k, and absolute temperature τ. This relation is checked by experiment within the accuracy of the measurements.

The second point of view with regard to the Ettingshausen effect is that of Lorentz,[2] also presented to the Solvay Congress. According to this, the Ettingshausen temperature difference is accompanied by a transverse current of heat in the plate by conduction from the hot to the cold side. This point of view may also be made to give a quantitative relation. The argument which I now give is not that employed by Lorentz, but is patterned after one given by me in the preceding paper on the application of thermodynamics to the thermo-electric circuit. The continual conduction of heat in the plate means a continual dissipation of available energy, which means an increase in the entropy of the entire universe, which under these conditions demands a continual rise of the average temperature of the plate. This continual rise of the temperature of the plate must be at the expense of energy put into the system by the current, and the source of this is obviously the Nernst e. m. f. associated with the transverse heat flow. Working out the quantitative relations gives exactly the same result as before except for a difference of sign, the result now being $Q = -kP/\tau$. The reason for the difference of sign is obvious, for according to the second point of view the current puts energy into the system when the transverse heat current in the plate is from hot to cold, whereas according to the first point of view, energy is put in when the heat current is from cold to hot. Since the negative sign is directly contrary to experiment, the first point of view must be the correct one, and the necessity for the concept of thermo-motive force is established.

Lorentz, in the discussion which he added in the printed Solvay Report several years after the meeting of the Congress, clearly formulated the essential difference between these two points of view, and recognized the necessity of the concept of thermo-motive force. He also commented on the care which must be used in applying the le Chatelier principle to cases like this. Suppose that we apply a transverse temperature difference to opposite sides of a plate in a magnetic field. This temperature difference is accompanied by a transverse heat flow, which gives rise to a longitudinal e. m. f. by the Nernst effect. Now complete the external circuit, allowing a longitudinal electric current to be driven by the Nernst e. m. f. This electric current will produce a transverse temperature difference by the Ettingshausen effect. In what direction is it? It is natural to expect that it must be in such a direction as to diminish the applied temperature difference. But detailed analysis will show at once that such a diminution of the temperature difference would demand the minus sign in the relation between Q and P which is given by the second point of view, and which is contrary to experiment. The Ettingshausen effect is in such a direction that the transverse temperature difference becomes greater. The le Chatelier principle can be maintained only by noting that, in spite of the greater transverse temperature difference, the transverse heat flow has become less, because of the thermomotive force tending to drive heat up the temperature gradient. In other words, in order to maintain the le Chatelier principle, heat flow must be regarded as the fundamental thing and not temperature difference. It is obvious that the le Chatelier principle might easily prove a false guide in fresh fields.

The transverse temperature difference in crystals predicted by Kelvin may now be subjected to an analysis exactly like that above for the Ettingshausen effect. The transverse temperature difference is an experimental fact; the question is whether this temperature difference is maintained without heat flow by a thermo-motive force in the crystal, or whether there is a continual flow of heat and so continual dissipation of energy when a steady current flows obliquely in a crystal. The answer to this can be given by examining the sign of the inverse effects. If opposite sides of a long crystal bar oblique to the axis are maintained at a difference of temperature, a longitudinal e. m. f. arises, which will drive a current through the bar on completing the external circuit. This current produces in turn a transverse temperature difference—does this oppose or aid the original temperature difference?

The inverse effect, that is, the longitudinal e. m. f. generated by a transverse temperature difference, is very easy to observe, but as far as I am aware, little attention has been paid to the sign of the effect. I have just found with a single crystal of bismuth that the current generated by the inverse effect is in such a direction as to decrease the original temperature difference. That is, the direct and the inverse effect are connected by a relation with the opposite sign from that of the Nernst-Ettingshausen effects. A crystal is not the seat of a thermo-motive force, but if the transverse difference of temperature is allowed to establish itself, there is a continual heat current, and continual dissipation of energy.

The precise connection between the direct and the inverse effect in a crystal may now be found by the counterpart of the argument already indicated. Consider a crystal plate of breadth b and unit depth carrying a longitudinal current of density i (total current ib), and assume for simplicity that the crystal has rotational symmetry about the crystal axis (as do all metals yet investigated) and that this axis lies in the plane of the face b, and makes an angle θ with the length. Furthermore, suppose the bar so long in comparison with the breadth that the Peltier heat at the ends may be neglected in comparison with the transverse heating effects. Consider now a piece of this plate of unit length. Let T denote the transverse generation of heat per unit area of transverse face per unit current density, and let k denote the transverse thermal conductivity. T and k are in general functions of the angle θ, but the precise nature of the functional relation does not concern us here. The difference of temperature between the two sides of the plate is evidently biT/k and the increase of entropy in unit time due to irreversible heat flow transversely is $(biT/k)(iT/\tau^2) = (bi^2T^2)/k\tau^2$. Denote the longitudinal e. m. f, per unit length per unit transverse heat current by e_t. Then the e. m. f. in this case is e_tiT, the energy input of the current against this e. m. f. is e_ti^2Tb, the entropy rise per unit time due to this energy input is e_ti^2Tb/τ, and equating this to the rise of entropy due to irreversible heat conduction gives

$$e_t = \frac{T}{k\tau},$$

the connection between the direct and the inverse effects.

The sign convention with regard to e_t has already been suggested and the positive sign in the relation just deduced has been found to agree with experiment. A direct (quantitative) experimental verification of the relation is not superfluous, however, as it is conceivable that there might be still some transverse thermomotive action, of less than the full amount. It would be interesting also to find whether the transverse heat current or the transverse temperature gradient is active in the production of the longitudinal e. m. f. This might be tested by finding whether a crystal bar, in which a transverse thermo-motive force is developed by a longitudinal current in a magnetic field, but in which there is no transverse heat flow, is the seat of a longitudinal e. m. f.

It is interesting to carry the analysis a little further. If there is no development of heat within the body of the crystal, the net rise of temperature accounted for by the energy input of the current must be described in thermal terms as due to a difference of the transverse heat at the two faces which are at different temperatures. This gives: $\Delta\tau \frac{d}{d\tau}(iT) = e_t . Ti . ib$, and substituting the values already found for $\Delta\tau$ and e_t gives $\frac{dT}{T} = \frac{d\tau}{\tau}$, or $T = c\tau$, exactly as in the analysis for the ordinary Peltier heat, when the Thomson heat is neglected. We may suspect that this relation will be found not to agree with experiment, and that, therefore, there must be a generation of heat in the body of the crystal when a longitudinal electric current flows across a transverse temperature gradient, that is, a transverse Thomson heat. Denote by σ_t the heat so absorbed per unit time per unit depth per unit temperature difference per unit current density. The equation of energy balance is now

$$\Delta\tau\left[i\sigma_t + \frac{d}{d\tau}(iT)\right] = e_t . Ti . ib,$$

whence

$$\sigma_\tau + \frac{dT}{d\tau} = \frac{T}{\tau}.$$

Expressing T in terms of e_t gives at once

$$\sigma_\tau = -\tau\frac{d(e_t k)}{d\tau}$$

for the transverse Thomson heat.

The precise analogy between these formulas and those for the ordinary thermo-electric circuit is at once obvious, T taking the place of the Peltier heat, σ_t the ordinary Thomson heat, and $e_t k$ the thermoelectric power (that is, the e. m. f. per unit temperature difference) of the couple. The method of argument used here is the same as that in my previous paper on the application of thermodynamics to the thermo-electric circuit, so that the uncertainty in the result of Kelvin arising from neglect of the irreversible effects is avoided, and we may be sure that these relations are the rigorous result of thermodynamic principles, with no assumptions involving the neglect or irreversible aspects of the phenomena.

The formal thermo-magnetic analogy of the Thomson transverse effect in crystals is an absorption of heat by a heat current flowing transversely in a bar carrying a longitudinal electric current of density i in a perpendicular magnetic field in amount equal to the fraction PHi/T of itself per unit length measured transversely. This may be proved at once from the equation of energy balance.

[1] P. W. Bridgman, *Phys. Rev.*, Dec., 1924, 644–651. Report of the Fourth Solvay Congress, "Conductibilité Électrique des Métaux," 352–354.

[2] H. A. Lorentz, *Fourth Solvay Congress*, 354–360.

THE ELASTIC MODULI OF FIVE ALKALI HALIDES.

By P. W. Bridgman.

Presented Oct. 9, 1929. Received Nov. 7, 1929.

CONTENTS.

Introduction.

A knowledge of the elastic moduli of crystals is of much theoretical importance, but comparatively few have been determined. Perhaps the simplest of all crystals from the point of view of lattice structure are the alkali halides, but of these the elastic moduli only of NaCl have been determined at all satisfactorily, and of the others, determinations have been made on only two samples of KCl, of which one was obviously imperfect. The difficulty is in obtaining single crystals free from flaws in large enough pieces; the conditions are very much more exacting than in determining the cubic compressibility, for example.

By a development of the method which I have used for the production of single crystals of the metals,[1] it has been possible to produce single crystals of five alkali halides, namely NaCl, NaBr, KCl, KBr, and KI, large enough to permit cutting from them rods on which measurements of the elastic moduli could be made.

METHODS.

Method of Making the Crystals. The method is that of slow cooling from the melt, the material to be crystallized being placed in a suitable mold, and slowly moved from the interior of a furnace maintained above the melting temperature to an external colder region. The greatest difficulty in extending this method from metals, for which it works easily, to salts, is the great propensity of salts to crack into small pieces on cooling to room temperature. This is an effect of temperature contraction combined with the great brittleness of the salts. Temperature contraction may result in cracking either when the salt sticks to the walls of the mold, or when there is too steep a temperature gradient while withdrawing from the furnace. The difficulty with steepness of gradient can to a large extent be avoided by suitable design, but it is not possible to reduce the gradient indefinitely without danger that the isothermal surface at the solidifying temperature have more than one sheet, when crystallization may start from more than one nucleus. The difficulty from sticking to the walls of the mold was in large part avoided by making the mold of platinum foil 0.001 inch thick, as was first done by Ramsperger and Melvin.[2]

In order to determine the elastic constants it was necessary to get the specimen into the form of a slender rod. In order to avoid casting large blocks in single crystal form from which the rods could be cut, the casting was made at once in the form of a rod. The natural method would be to make the mold in the form of a thin tube, like the glass capillaries in which the metals were cast, but this did not prove successful. There were difficulties in making the mold out of the thin foil, but the greatest difficulty was in getting the casting out of the mold without cracking it. Finally the molds were made in the form of open boats, used in a horizontal position. This demanded, of course, that the axis of the furnace be placed horizontally rather than vertically. The boats were bent from pieces of foil 8.3 cm long and 1.9 cm wide; the ends were welded together, and the edges bent double to increase the transverse stiffness. At first the joints in the boats were made with pure gold solder, but this was given up because gold is soluble in the molten salts, and it was very difficult to use such a small quantity of gold as not to appreciably discolor them. One very important point in making the boats is that the platinum must be slightly crinkled, instead of being left perfectly smooth; the crinkling provides enough elastic yield to prevent the cracking which otherwise takes place. The cold casting could be removed from the boat without great difficulty by working the platinum foil away from the casting with the thumbs of the two hands simultaneously, as will readily suggest itself on trial. It was never possible to use the foil more than once. No substitute was found for platinum, although several were tried. The most nearly successful was silver, but this slowly dissolves in the molten salt.

In the furnace the boats with their contents were protected from drafts and any furnace fumes by enclosing them in quartz test tubes, with stoppers ground so as not to be quite a perfect fit, thus allowing thermal expansion of the air within. Several of the quartz test tubes with their contents were made together into a bundle with a suitable holder, and withdrawn simultaneously from the furnace, so that several castings could be made at the same time.

The furnace was made in two parts with independent windings. In one half the temperature was maintained by a suitable current 10° or 20° above the melting temperature, and in the second part at a temperature somewhat below the melting temperature; the proper relation between the two temperatures had to be determined by trial. The temperatures of the two parts of the furnace were read on thermocouples, so connected with potentiometer circuits as to give a sensitivity of considerably better than a degree. Temperature was maintained constant within a degree or so by manual regulation during the process of withdrawal. Current was supplied by an independent generator of much constancy, so that the manual regulation was not difficult. The molds were drawn from the hot part to the colder part at a rate of the order of 10 cm per hour. When the mold was entirely in the colder part, the current was gradually cut down, and finally the furnace allowed to cool over night. Too rapid cooling after solidification was responsible for the loss of a number of crystals.

After the castings are removed from the mold, they must be checked to make sure that they are one grain. A simple way of doing this is to rub the surface of the casting lightly with a moist cloth, etching it slightly. When the moisture has dried off, it is very easy to tell by the appearance of the reflection pattern whether there is more than one grain. If the crystal passed this test, it was then formed into a slender cylinder. This was done in a jeweller's lathe, and of course required some care, but after some practice it was not difficult to get rods sometimes 5 cm long and 0.28 cm in diameter, round and true over the entire length to 0.0002 cm.

After the measurement of the elastic constants, to be described later, the orientation of the crystal axes was determined by making the principal cleavage planes with a knife blade, and measuring the angles with a protractor. The angles of all three planes were measured; the consistency of these three measurements thus furnished an internal check on the correctness of the orientation. The measured angles were adjusted so as to make the sum of the squares of the three direction cosines equal to unity, dividing the necessary adjustment among the angles proportionally roughly to the squares of the sines of the angles. The adjustment was practically never larger than a degree or two. The cleavage planes were determined at both ends of the rods, thus giving an additional check that the rods were truly one grain, and on the orientations.

The nature of the material will be described in detail in the presentation of data for the individual substances. One general remark is to be made, namely that slight and obscure impurities play a much more important part than I appreciated at first. In several cases a long series of unsuccessful results with salt from one source was followed by immediate success when the material was replaced by salt from another source, of no greater presumptive purity. In some cases the original material contained enough water so that there was violent snapping and jumping about of the salt on heating to the melting temperature. These materials were dried by a preliminary fusion, either in quartz or in a platinum crucible. An especially effective method of preliminary treatment is to cast the salt by the regular procedure for making single crystals, melting it in a large quartz test tube and slowly lowering in a vertical position through the furnace. The lower part of the resulting block was usually as clear as glass, except for the numerous cracks, and this clear material was used in making the final castings in the platinum boat.

Measurements of the Moduli. A cubic crystal has three independent elastic moduli, so that at least three different measurements must be made. There is a great deal of latitude as to what these may be; if perfect accuracy were attainable, either three independent different sorts of elastic deformation might be measured on a single specimen, or measurements of the same sort of deformation, as a bending, for example, might be made on three different samples of different orientations. The experimental accuracy is far from perfect. however, so that many observations, of different kinds of deformation on different samples, were desirable. The sort of deformation that is easiest to measure is the bending and twisting of rods, and the measurements made in connection with this paper were of the bending and twisting of rods of a variety of orientations. In addition, the linear compressibility may be measured. This has already been determined by Slater[3] for all the salts of this paper; it can be measured with high accuracy, is not much affected by imperfections in the specimen, and there are values by other observers against which the values of Slater may be checked. In calculating the elastic moduli from the data of this paper, I have assumed that Slater's values for the compressibility are correct.

The experimental determinations of bending and twisting were made by methods conventional enough not to need extended description. In determining the bending, the round rod was supported at each end on V-shaped knife edges, the distance between the end supports being about 2.5 cm. The load was applied to a scale pan slung from the center with a double loop of cotton thread. Pressed up by a spring against the under side of the rod at the center between the loops of thread was a link arrangement, by which any vertical displacement of the center was transmitted to a conventional rocking mirror. The two ends of the rod were pressed against the V-shaped notches by springs with greater force than the upward force on the link. The scale distance was about 4.5 m, and the width of the mirror rocker 0.11 cm, so that the magnification was about 8000. The load was applied to and removed from the pan with a spring arrangement to avoid shock. In most cases the load was 20 gm, but for two of the less deformable salts was 40 gm. The displacements of the scale observed in the telescope for this load were of the order of 1.5 cm, and this could easily be read to 0.1 mm. At first, readings were made at both half and full load. The reading with half load came out so consistently equal to half the reading with full load that continued check of this point seemed superfluous, and the later readings were made only with full load. The readings were made immediately after the application and removal of load, so that strictly these are measurements of the adiabatic moduli; the difference between adiabatic and isothermal modulus is disregarded in the following. The effective bending modulus of a round rod cut from a cubic crystal should be the same for all orientations of the rod about the axis; this was always checked by making readings on each rod for four orientations spaced 90° apart. At each orientation, five readings of the displacement under full load were made. The readings in different orientations did not always agree, but differed by only a few per cent, and in no consistent manner; the average of all the readings was taken in calculating the effective bending modulus.

The twist was determined with apparatus that had already been used in determining the twist of slender single crystal rods of the metals. The specimen is held rigidly at one end, and at the other end is attached to the arm through which the twisting couple is applied through an arrangement of knife edges so that the couple acts practically without friction. The maximum twisting moment applied to these rods was about 170 gm cm. The twist is observed in telescope and scale in two mirrors, attached by a spring arrangement of obvious enough design to the central part of the rod. The mirrors must be far enough from the ends of the rods to avoid end effects; the usual distance between the mirrors was of the order of 2 cm, and the clear length of the rod about 4 cm. The twist was determined from the difference of reading of the two mirrors. The proportionality between twist and moment was tested in every case, and no deviation found within experimental error, which might be perhaps as much as 1%. The scale was 4.5 m distant, and the difference of the two deflections of the order of 1 cm, and the optical system was good enough to permit the certain estimation of 0.1 mm. Ten or a dozen readings of the twist of each rod were usually made, and the average taken.

The rods were attached to the two end pieces with cement. At first, elaborate precautions were taken not to heat the rods in applying the cement more than a few degrees, and paraffine and other low melting substances were used as cement. These were all unsatisfactory, and many times the specimen slipped. Finally it was found that these slender rods can stand much apparent abuse, and they were cemented into the holders with deKhotinski cement, melted into position with a small gas flame, which many times fell directly on the specimen without cracking it. It is surprising how difficult it is to crack one of these rods by unequal heating. I hoped to be able to get the cleavage planes by a local heating, instead of fracturing with a knife blade, but without success. Cooling in liquid air was also ineffective in producing the desired fracture. There must, however, judging by the behavior of the specimens as they come from the furnace, be a region near the melting point where the salt is very brittle and very sensitive to temperature inequalities.

The fundamental elastic moduli are connected with the bending and twisting moduli by formulas which may be found in Voigt.[4] For the bending of a round bar, we have the formulas:

$$u = \frac{WL^3 s_{33}'}{12\,\pi\,a^4}, \tag{1}$$

$$s_{33}' = s_{11} - 2\,\{s_{11} - s_{12} - \tfrac{1}{2}\,s_{44}\}\,\{\gamma_2^2\,\gamma_3^2 + \gamma_3^2\,\gamma_1^2 + \gamma_1^2\,\gamma_2^2\}. \tag{2}$$

Here u is the displacement under the weight W of the center of the rod, the length of which is L and the radius a. γ_1, γ_2, and γ_3 are the direction cosines of the three crystal axes referred to the length of the rod. s_{11}, s_{12}, and s_{44} are the three elastic moduli to be determined, referred in the conventional way to the principal crystal axes. That is, stress and strain are connected by the relations:

$$\left.\begin{aligned} e_{xx} &= s_{11}\,X_x + s_{12}\,Y_y + s_{12}\,Z_z \\ e_{yy} &= s_{12}\,X_x + s_{11}\,Y_y + s_{12}\,Z_z \\ e_{zz} &= s_{12}\,X_x + s_{12}\,Y_y + s_{11}\,Z_z \end{aligned}\right\} \qquad \left.\begin{aligned} e_{yz} &= s_{44}\,Y_z \\ e_{zx} &= s_{44}\,Z_x \\ e_{xy} &= s_{44}\,X_y \end{aligned}\right\}.$$

For the twist we have:

$$n = \frac{N}{\pi a^4}\,(s_{44}' + s_{55}'), \tag{3}$$

$$(s_{44}' + s_{55}') = 4\,(s_{11} - s_{12}) - 4\,\{s_{11} - s_{12} - \tfrac{1}{2}\,s_{44}\}\,\{\gamma_1^4 + \gamma_2^4 + \gamma_3^4\}. \tag{4}$$

Here n is the twist per unit length, and N is the twisting moment.

Also, the cubic compressibility χ is given by:

$$\chi = 3\,(s_{11} + 2\,s_{12}). \tag{5}$$

We now have the problem of determining the best values for the three moduli s_{11}, s_{12}, and s_{44} from measurements on a large number of specimens. Equation (1) may be solved for s_{33}' in terms of the dimensions and the measured displacement u, and (3) may be solved for $s_{44}' + s_{55}'$ in terms of twist and dimensions, and these values substituted in (2) and (4). The result of this is two linear equations in the three unknown moduli, the coefficients being known functions of the orientations. Next, taking the compressibility as known from equation (5) express s_{12} in terms of the known compressibility and s_{11}. Substituting this value for s_{12} back into (2) and (3) gives two linear equations in the two moduli s_{11} and s_{44}, the coefficients being function of the orientations, displacements, and compressibility. Finally, by dividing these equations through by the constant term, they may be written in the form:

$a_{11}\,s_{11} + a_{12}\,s_{44} = 1$, (6), from bending measurements.

$a_{21}\,s_{11} + a_{22}\,s_{44} = 1$, (7), from twisting measurements.

For any particular rod, after the measurements have been completed, the values of a_{11} and a_{12} are known from the measurement of bending, or a_{21} and a_{22} from the measurements of the twist. But whatever the values of a_{11} etc., they must be consistent with a single definite pair of values of s_{11} and s_{44} when substituted into (6) or (7). But now if we regard a_{11} etc. as variables, then (6) or (7) is the equation of straight line with intercepts $1/s_{11}$ and $1/s_{44}$, the intercepts, and hence the line represented, being the same for either (6) or (7). Writing (6) or (7) in the common form

$$a_1 s_{11} + a_2 s_{44} = 1, \tag{8}$$

the following procedure is at once suggested. From the measurements of elastic distortion, dimensions, and orientation, obtain for each specimen an equation in the form (8). Plot the coefficients a_1 and a_2 as points in an $a_1 - a_2$ plane. There will be as many of these points as there are independent measurements of elastic distortion. These points must all lie on the same straight line, the intercepts of which on the axes are $1/s_{11}$ and $1/s_{44}$. The best values for the moduli from a number of independent measurements I assume to be given by the intercepts of the best line that can be drawn through all the points, and the internal consistency of the measurements on different samples is shown by the accuracy with which the points lie on a single line.

It is obvious that the greatest accuracy would be obtained by so locating the points as to lie as near the axes as possible. An inspection of the equations shows at once what this involves. If one of the principal axes lies along the length of the rod, only s_{11} occurs in (2), and only s_{44} in (4), so that bending and twisting measurements on rods of these orientations would give directly s_{11} and s_{44}. Unfortunately, this orientation is one that does not often spontaneously occur, but the rods naturally grow from the melt in intermediate directions, as will be apparent from the detailed presentation of the data. Only in the case in which I repeated Voigt's measurements on NaCl cut from large natural blocks was the most favorable orientation exactly attained.

Detailed Data.

KI. This was the first substance attempted, and much more time was spent on it than on any of the others. A considerable part of the time was spent in finding the various best details of the method. All the early work was done with material from Kahlbaum, of the highest purity which they manufacture, but not their "K" grade. I anticipated that KI would be the easiest of all the materials to crystallize, because Slater had found it so, but although a great many attempts were made, only five specimens in all were obtained on which it was possible to measure the elastic distortion. Even these specimens were not all satisfactory, as it was not always certain that the specimens were truly one grain, and the results were very scattering. After no further improvement in the technique seemed easily possible, this material was given up, somewhat in despair, and KBr tried, which was obtainable in Kahlbaum's "K" grade. The first attempt was brilliantly successful. The obvious explanation of the difference was in the difference of purity of the two materials, and it therefore became important to obtain KI of greater purity. Finally I was fortunate to obtain some of the old stock of Powers, Weightman, and Rosengarten, now no longer manufactured. This pròved adequate, giving clear castings with sharp cleavage planes from which the rods were readily machined in the lathe. To a cutting tool the consistency of KI is much like that of horn or celluloid; in progressing through the series of halides toward smaller atomic weights the hardness and accordingly the difficulty in turning increases.

Five specimens were made with the pure material, which gave good single crystals and satisfactory determinations of the elastic moduli. The results obtained with these five specimens, and also with the best of the Kahlbaum material, are shown in Figure 1, the coefficients of s_{11} and s_{44} being plotted in the way already described. The points all lie on a straight line within the limits of error. In determining these coefficients Slater's value used for the compressibility was 8.54×10^{-12} Abs. C. G. S. It is unfortunate that the orientations were not more favorably distributed. It will be seen that only one of the specimens approximately satisfied the condition of most favorable orientation with one cleavage plane perpendicular to the length. The method used for producing the crystals gives no direct control of the orientation, although it might be modified to do so, and one has to trust more or less to luck to get what is needed. A number of other perfect single grain specimens were made in addition to those measured, in an endeavor to get some with more favorable orientations, but without success.

An attempt was made to determine the effective Young's modulus, that is, the modulus s_{33}', directly by measuring the extension when stretched by a weight, but the shortness of the specimen made it difficult to eliminate various irregularities arising from the end

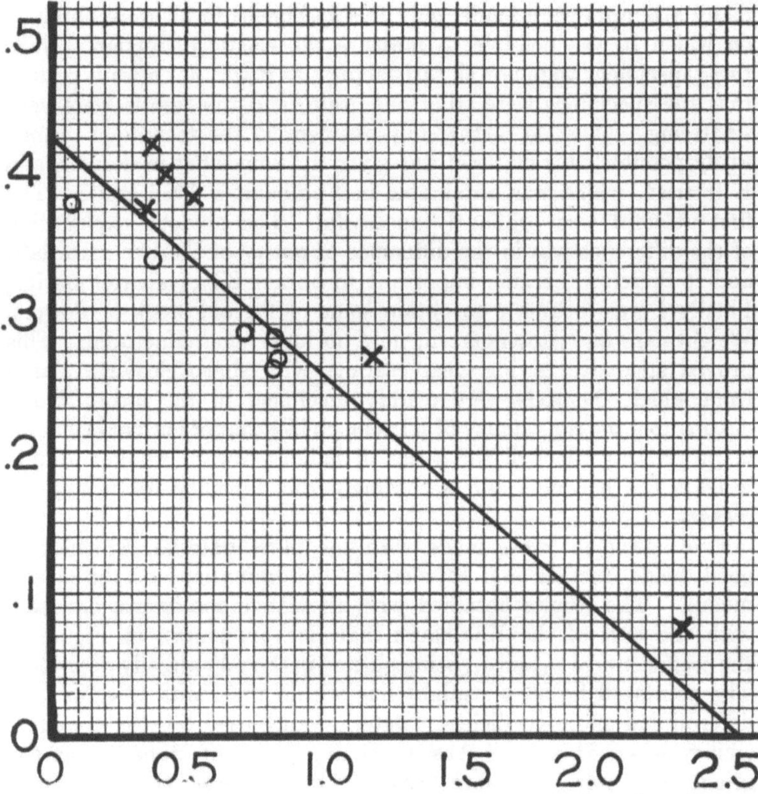

Figure 1. The coefficients of $s_{11} \times 10^{-11}$ in equation 8 plotted as abscissas against the coefficients of $s_{44} \times 10^{-11}$, for potassium iodide. The points obtained from bending measurements are shown as crosses, and those from twisting measurements as circles. If there were no experimental error the points should all lie on a straight line.

effects, and the attempt was abandoned in favor of the more convenient method by bending.

KBr. The material used was Kahlbaum's, grade "K," as has already been explained. Measurements were made on eight specimens, of which only three were measured for torsion. It is more difficult to get a good specimen for the torsion measurements than for bending, because the length must be greater.

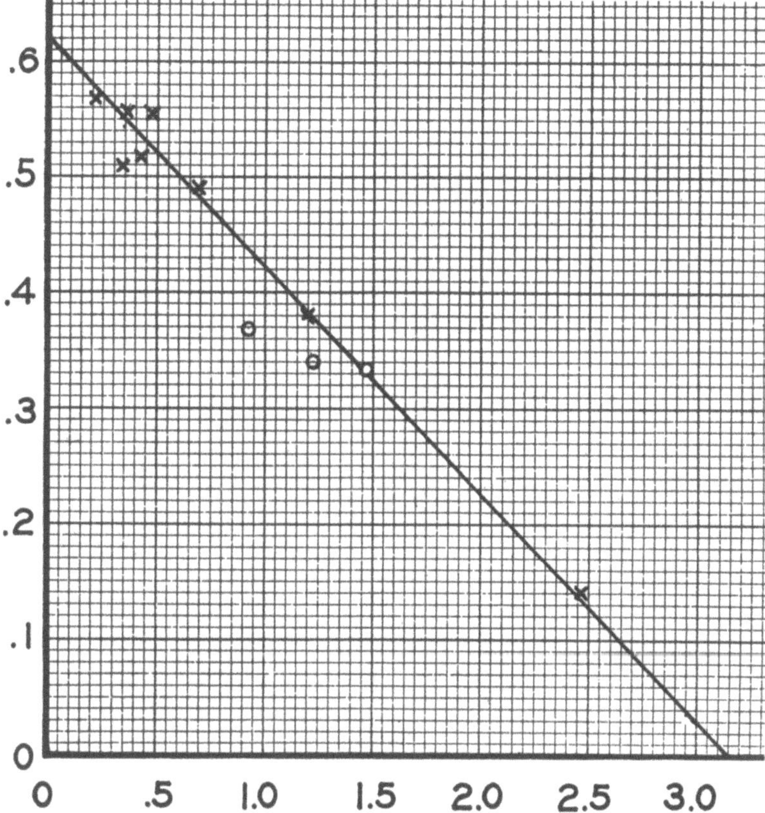

Figure 2. The coefficients of $s_{11} \times 10^{-11}$ in equation 8 plotted as abscissas against the coefficients of $s_{44} \times 10^{-11}$, for potassium bromide. The points obtained from bending measurements are shown as crosses, and those from twisting measurements as circles.

The coefficients of s_{11} and s_{44} are plotted in Figure 2, and again it will be seen that the points lie on a straight line, as is demanded by the crystal symmetry. Slater's value for the compressibility of KBr is 6.69×10^{-12}.

KCl. This was from Kahlbaum, grade "K." In spite of the high melting point, this was distinctly the easiest to work with of the materials tried. The rods were beautifully clear, always one grain, long enough for the torsion measurements, and of considerable

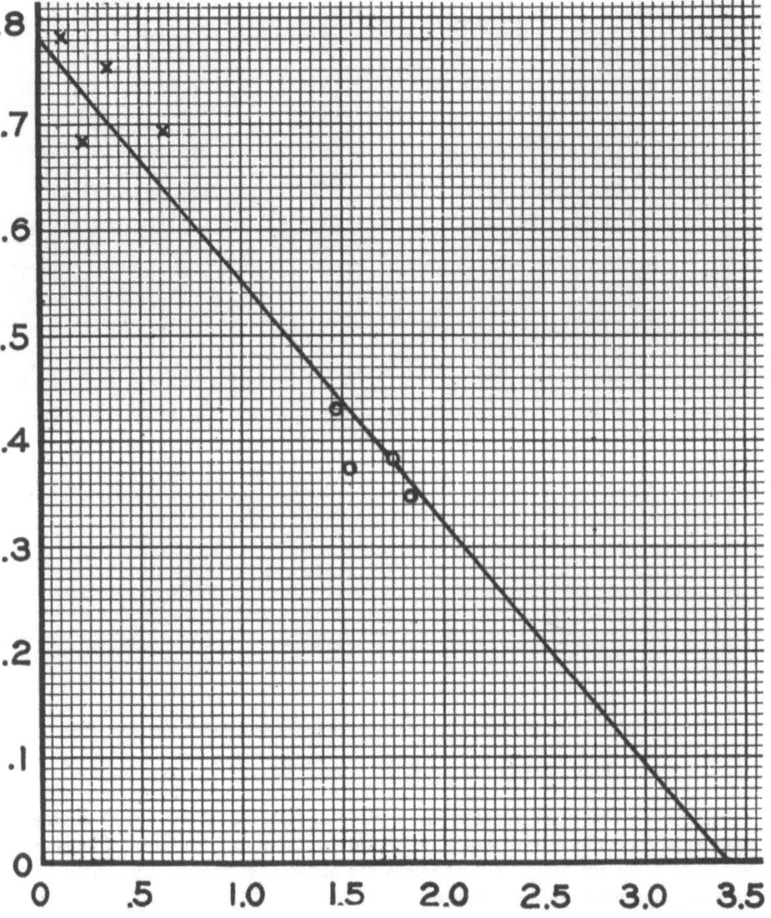

Figure 3. The coefficients of $s_{11} \times 10^{-11}$ in equation 8 plotted as abscissas against the coefficients of $s_{44} \times 10^{-11}$, for potassium chloride. The points obtained from bending measurements are shown as crosses, and those from twisting measurements as circles.

85 — 2370

mechanical strength, so that turning was an easy process. Bending and torsion measurements were made on four specimens; the coefficients of s_{11} and s_{44} are plotted in Figure 3. The points lie on a straight line within the limits of error. Slater's compressibility is 5.64×10^{-12}.

NaBr. Attempts to prepare single crystals of this from the purest material that I could purchase were not successful. Professor Baxter was, however, so kind as to have prepared from his stock of atomic weight material some of this salt of unusual purity. Enough of the pure material was available for three crystals; all of this was cast successfully as single grains. Unfortunately the one of these which would have given the most accurate information, with a cleavage plane nearly perpendicular to the length, broke in turning. This, of course, is much more likely to happen when the cleavage plane is perpendicular, an exasperating situation. Satisfactory measurements were made of the bending and twist of the two remaining specimens, and the results are shown in Figure 4, plotted in the regular way. The 40 gm. weight, instead of that of 20 gm., was used in making the bending measurements. Slater's compressibility for this is 5.08×10^{-12}.

NaCl. These measurements were made mostly by way of check, there being no a priori reason to think that the material on which Voigt worked was not satisfactory. The material used was natural rock salt, which can be obtained in large clear blocks. In addition to the natural material one single grain specimen was prepared by crystallization from the melt from salt of Powers, Weightman and Rosengarten. The bending and twisting of this specimen was measured, and also its compressibility, as will be described later.

The natural crystal blocks were of German origin, obtained through Ward's Natural History Establishment. Three rods were cut from one block, and two from another. They were cut perpendicular to the cleavage planes, so that bending and twisting measurements gave directly s_{11} and s_{44}. Three rods from the first block gave for s_{11} .216, .209, and $.212 \times 10^{-11}$ respectively, and the two from the second: .230 and $.232 \times 10^{-11}$. It would thus appear that there are differences between the elastic moduli of rods cut from different blocks greater than the experimental error. This result emphasizes a conclusion that has been becoming increasingly evident lately, namely that it is very seldom that natural crystals are the unique things that have been supposed. Twisting measurements were made on only one rod from each block; these gave for s_{44} .782 and

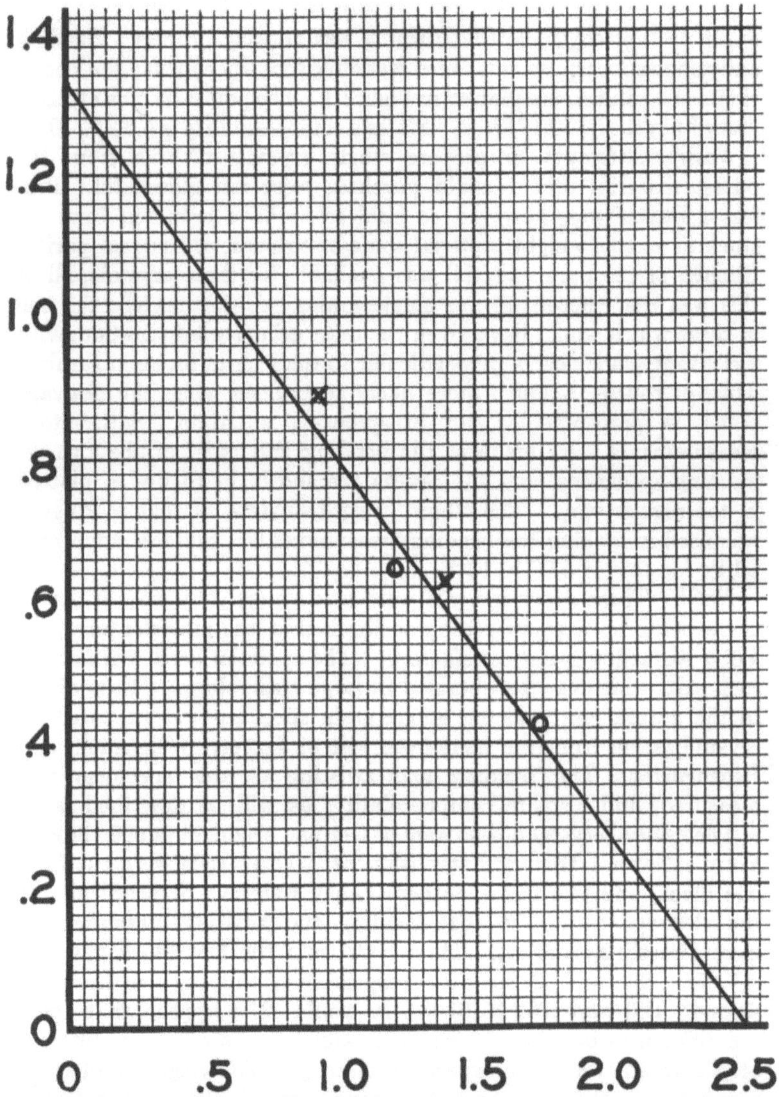

Figure 4. The coefficients of $s_{11} \times 10^{-11}$ in equation 8 plotted as abscissas against the coefficients of $s_{44} \times 10^{-11}$, for sodium bromide. The points obtained from bending measurements are shown as crosses, and those from twisting measurements as circles.

.742 $\times 10^{-11}$ respectively. The values which Voigt gives for natural NaCl are: $s_{11} = .243 \times 10^{-11}$, $s_{12} = -.0527 \times 10^{-11}$, $s_{44} = .788 \times 10^{-11}$. In the following I have selected as the most probable values: $s_{11} = .23 \times 10^{-11}$ $s_{12} = -.05 \times 10^{-11}$, $s_{44} = .78 \times 10^{-11}$. The values of bending and twist of the sample crystallized from the melt were not inconsistent with these values. The orientation was not favorable, however, so that no weight can be attached to the values of the moduli which would be given by a rigorous calculation from the two measurements.

The compressibility of the sample crystallized from the melt was measured in the regular way to a maximum pressure of 12000 kg/cm² at 30° and 75° C, and the following results found:

$$\left.\begin{array}{l} \text{At } 30°, -\Delta V/V_0 = 41.82 \times 10^{-7}\, p - 50.4 \times 10^{-12}\, p^2 \\ \text{At } 75°, -\Delta V/V_0 = 43.44 \times 10^{-7}\, p - 51.9 \times 10^{-12}\, p^2 \end{array}\right\}.$$

Here pressure is expressed in kg/cm² units, as in all my compressibility measurements. The average deviations of single readings from a smooth curve at these two temperatures were 0.13 and 0.21 % respectively, which is about the average accuracy for material of this nature.

There is some interest in this compressibility measurement since nearly all previous measurements of the compressibility of NaCl, of which there have been a number, have been made on the natural crystal. One would expect the artificial material, which starts with a chemical of known high purity, to be more pure than the natural material, and hence the compressibility measured with it to be more significant.

The compressibility values just given yield, when converted into Abs. C. G. S. units:

$$\left.\begin{array}{l} \text{At } 30°, -\Delta V/V_0 = 4.263 \times 10^{-12}\, P - 5.23 \times 10^{-23}\, P^2 \\ \text{At } 75°, -\Delta V/V_0 = 4.428 \times 10^{-12}\, P - 5.38 \times 10^{-23}\, P^2 \end{array}\right\}.$$

Slater's value for the natural crystal at 30° was $4.20 \times 10^{-12}\, P - 4.60 \times 10^{-23}\, P^2$, and at 75° $4.33 \times 10^{-12}\, P - 4.74 \times 10^{-23}\, P^2$. The second degree formulas just given were calculated by me from Slater's results, which were tabulated in somewhat different form. The difference between the results of Slater and me is not large, but the sign of the difference is unexpected, the natural crystal being initially less compressible. The difference between the two results is not as great as would appear at first, there being a compensation between the first and second degree terms; the total volume change at the maximum pressure of 12000 kg/cm^2 given by Slater's formula is not quite ½ % *greater* than given by mine.

The initial compressibility at 20° of the artificial crystal is 4.23×10^{-12} by extrapolation from my formula above, and 4.17×10^{-12} by extrapolation from Slater's for the natural crystal. The value of Adams, Johnston, and Williamson[5] for the natural crystal at the same temperature was 4.12×10^{-12}, and that of Madelung and Fuchs[6] 4.14×10^{-12}. Richards[7] is the only observer who has previously measured the compressibility of both the natural and the artificial crystal, and he finds the natural crystal somewhat *more* compressible, but since he gives only two significant figures for the natural crystal and three for the artificial, it is probable that too great significance should not be attached to this difference. His value for the artificial material was 4.30×10^{-12} and 4.5 and 4.6×10^{-12} for two samples of the natural crystal. Both his values are distinctly larger than those of other observers.

The maximum variation in compressibility between any of the observed values is not great enough to introduce appreciable changes in the values given for the moduli as calculated from observations on bending, twist, and compressibility.

Other Salts. Unsuccessful attempts were made to crystallize some of the other alkali halides, but no great effort was expended on these, and others should not be deterred by my failure. The lithium salts are difficult to manage either because they are very hygroscopic, or because they cannot be obtained in sufficient purity. NaI could not be obtained in sufficient purity. Professor Baxter was so kind as to partially purify my commercial material by recrystallization from aqueous solution. This treatment would be expected to remove any organic impurities, but this did not prove to be sufficient. RbCl was also attempted; I had a very small supply left from the material prepared by Professor Richards, the polymorphic transitions of which under pressure I had measured. Dr. Paul Anderson was so kind as to remove by electrolysis the iron which had got into it during the pressure manipulations, but this did not prove to be sufficient.

Summary and Discussion.

In the following table are given the values of the elastic moduli obtained from the compressibilities and the intercepts on the axes of the lines already shown in the figures for the various salts. The number of significant figures given for NaCl is less than for the others because of the uncertainty arising from the variation in the different samples of the natural crystal.

TABLE.

ELASTIC MODULI IN ABS. C. G. S. UNITS.

Salt	s_{11}	s_{12}	s_{44}
NaCl	$.23 \times 10^{-11}$	$-.05 \times 10^{-11}$	$.78 \times 10^{-11}$
NaBr	.400	−.115	.754
KCl	.294	−.053	1.27
KBr	.317	−.047	1.61
KI.	.392	−.054	2.38

The most important question in connection with these elastic moduli, and which was decisive in my decision to attempt their measurement, is whether Cauchy's relation holds. Expressed in terms of the elastic constants this is $c_{12} = c_{44}$; in terms of the moduli it is more complicated, being

$$(s_{11} - s_{12})\,(s_{11} + 2\,s_{12}) = -\,s_{12}\,s_{44}.$$

In the case of the general crystal the relations of Cauchy are more complicated, and reduce the 21 constants required by general considerations of thermodynamics and crystal symmetry to 15. These relations were deduced by Cauchy on the assumption that the forces between the elements of the crystal structure have spherical symmetry. It is known that in general the relations cannot be correct, as shown by the experimental measurement of the elastic constants of a large number of different crystals by Voigt. An adequate theoretical account is given of the situation by Born's theory of crystal lattices;[8] the failure of the Cauchy relation is shown to be connected with the possibility of a rigid relative displacement with respect to each other of the elementary lattices of which the crystal is composed. The fact that Cauchy's relation does not apply to cubic crystals of the metals must be taken, according to Born's theory, to mean that the displacements of the atomic nuclei relative to the surrounding cloud of electrons play an essential part in the elastic phenomena of metals. The situation is simpler with regard to the alkali halides than with regard to the metals. If the alkali halide structure can be adequately represented by interpenetrating cubic lattices of positive and negative ions, the attractive forces being the electrostatic forces and the repulsive forces some high inverse power of the distance, then Born's theory reduces to Cauchy's, so that the Cauchy relation should hold. Now there is considerable physical evidence, apart from the elastic behavior, to suggest that this is an adequate picture of the interatomic forces, and consequently the expectation is that the Cauchy relation will hold. The expectation that the relation will hold is much strengthened by the new wave mechanics,[9] because according to this a complete electron shell must have perfect spherical symmetry, and not the sort of cubic symmetry that Born at first supposed, and furthermore the ions of the lattice are composed of complete shells.

The experimental check of the Cauchy relation has, however, been meagre. The only satisfactory material hitherto available has been NaCl, for which the measurements of Voigt[10] showed the relation satisfied. Voigt[10] also measured KCl, but his material was not satisfactory in that only two specimens were available, not perfectly free from flaws. The values of the elastic constants obtained from the two specimens of KCl were quite different. Voigt was inclined to take the average of the constants from the two specimens as giving the most probable value, in which case the Cauchy relation was far from satisfied. Försterling[11] remarked, however, that one of Voigt's specimens was evidently much poorer than the other, because the cubic compressibility calculated from its elastic constants failed by a large amount to agree with the direct experimental value of other observers, whereas the other specimen gave agreement. Discarding the poorer sample, the other sample of KCl approximately satisfied Cauchy's relation. In general, therefore, the experimental check could not be called satisfactory. It was still less satisfactory when it is considered that NaCl and KCl might give a fortuitous check because of their position somewhere near the middle of the alkali halides series, whereas the more extreme members of the series might not give a check. A very similar situation arose with respect to the inverse ninth power of repulsive force demanded by Born's original theory of compressibility; the ninth power was found for NaCl, but not for the extreme members of the alkali halide series.

The following procedure was adopted in examining experimentally whether the Cauchy relation holds for the salts measured above It will be remembered that, assuming the experimental determination of compressibility to have higher accuracy than the other elastic moduli, the results of the bending and twisting measurements on rods of different orientations could be plotted in such a way that all points must lie on a single straight line. The intercepts of this line determined the moduli s_{11} and s_{44}. In general, there is no necessary connection between s_{11} and s_{44}, so that a two-parameter family of lines is available to represent the results. But if the Cauchy relation holds, there is an additional relation between the moduli, so that if s_{11} is given, s_{44} is determined, and the possible straight lines by which the experimental results can be represented become a one-parameter family. In detail, $s_{12} = \frac{1}{2}(k - s_{11})$ where k is the linear compressibility, and substituted into the Cauchy relation this gives

$$s_{44} = \frac{k(3s_{11} - k)}{s_{11} - k}.$$

This determines the s_{44} intercept when the s_{11} intercept and the compressibility are fixed.

The lines given in all figures have been drawn subject to this restriction. Although in each of the figures it is probable that exactly the line shown would not have been drawn if a least squares determination had been made of the best line for each figure taken by itself, nevertheless in each case the line shown comes very close to the line that would have been drawn with no ulterior consideration. Hence the conclusion may be drawn that the Cauchy relation is satisfied at least very closely by all the salts, and there is no reason to think that it may not be satisfied within a considerably smaller margin than the error of these experiments. As a first approximation, the assumption of the Cauchy relation seems to be justified.

Although the theory is thus justified as far as the Cauchy relation goes, it is still very far from complete. A complete theory would allow a detailed calculation of the numerical values of all three moduli. This it was not able to do in the case of NaCl, and the situation is not altered now by the determination of these new constants. In fact, the situation has become somewhat more complex, because although there is a certain regularity in the progression of the compressibility in passing from one alkali halide to another, and there is also a regularity in s_{11}, there is no such regularity in s_{12} and s_{44}, whereas some sort of regularity would doubtless be given by any first simple theory.

Summarizing, the three elastic moduli of KI, KBr, KCl, NaBr, and NaCl have been determined, and the compressibility of an artificial crystal of NaCl measured to 12000 kg/cm². The Cauchy relation between the elastic moduli is found to be satisfied, thus checking the expectation from Born's theory of crystal lattices, now put on a firmer basis by the new wave mechanics. There is, however, no theory which gives the exact numerical values of the moduli; there are irregularities in the sequence of the moduli which will probably be difficult for a simple theory.

It is a pleasure to acknowledge the skillful assistance of Mr. George Langreth in the manipulations of producing the single crystals. I am also much indebted for financial assistance to the Milton Fund of Harvard University.

The Jefferson Physical Laboratory,
Harvard University, Cambridge, Mass.

References.

[1] P. W. Bridgman, Proc. Amer. Acad. 60, 307, 1925.

[2] H. C. Ramsperger and E. H. Melvin, Jour. Opt. Soc. Amer. & Rev. Sci. Inst. 15, 359, 1927.

[3] J. C. Slater, Proc. Amer. Acad. 61, 135, 1926. Phys. Rev. 23, 488, 1924.

[4] W. Voigt, Kristallphysik, B. G. Teubner, 1910, pp. 637 and 739.

[5] L. H. Adams, E. D. Williamson and John Johnston, Jour. Amer. Chem. Soc. 41, 1, 1919.

[6] E. Madelung und R. Fuchs, Ann. Phys. 65, 289, 1921.

[7] T. W. Richards and G. Jones, Jour. Amer. Chem. Soc. 31, 176, 1909.

[8] M. Born, Atomtheorie des Festen Zustandes, B. G. Teubner, 1923.

[9] A. Sommerfeld, Wellenmechanischer Ergänzungsband zu Atombau und Spektrallinien, Vieweg, 1929, p. 100.

[10] W. Voigt, reference 4, p. 741.

[11] K. Försterling, ZS. f. Phys. 2, 172, 1920.

THE EFFECT OF PRESSURE ON THE RIGIDITY OF SEVERAL METALS.

By P. W. Bridgman.

Presented Oct. 9, 1929. Received Nov. 7, 1929.

CONTENTS.

Introduction.

Determinations have already been published of the effect of pressure on the rigidity of several varieties of glass.[1] The experimental method which was finally developed to make those measurements was just as capable of dealing with metals as with glass, and since the measurements on metals have a greater intrinsic interest than those on a substance of variable composition like glass, the method would have been applied in the first instance to metals, had not the complete preparations already been made for the measurements on glass, due to an erroneous preconception as to the relative magnitudes of the effect in metal and glass. In this paper the method is applied to eight metals up to 12000 kg/cm^2.

Method.

The method is essentially the same as that of the previous work. It is a differential method; two helical springs are stretched against each other, one spring consisting of the metal on which the effect is to be determined, and the other spring being of some standard metal on which the effect has already been determined by some absolute method. The standard metal was steel cut from the same spool of wire as the steel of the previous paper, and the absolute effect on it was determined by the method described in that paper, and in fact the actual value there found was used, making no new experimental determination. The two helical springs are coupled through a short length of manganin wire, sliding over an insulated contact attached to the same frame which maintains the springs stretched. If there is a change in the relative stiffness of the two springs when hydrostatic pressure is applied, there will be a motion of the point of coupling which can be measured by determining the change in the potential drop between the contact fixed to the frame and another contact attached to the wire, a current flowing lengthwise of the manganin wire.

The experimental arrangements were practically the same as in the previous paper. The apparatus was essentially like that shown in Figure 1 of that paper, except that the sliding contact was now located in the middle of the frame, so that the two springs were of equal length, this being the disposition to give maximum sensitiveness. The dimensions of the springs were so adjusted that the stiffness of the two springs should be the same, this condition being demanded by maximum sensitiveness. The pressure part of the apparatus was the same as before; the springs were mounted in the same cylinder as before, and were rotated as before to secure freedom from friction.

The method of applying the corrections and calculating the change in rigidity from the measured changes on the potentiometer was improved and simplified as compared with that used in the previous paper. The fundamental formula for the spring was taken as before to be:

$$P = \tfrac{1}{2}\pi \frac{a^4 \mu \varphi_0^2}{s^3 \cos^2 \alpha}(l - l'),$$

where P is the stretching load, a the radius of the wire of which the spring is wound, μ the rigidity of the material of the wire, l the stretched length and l' the unstretched length of the *helical* part of the spring, s the total length of wire in the helical part, α the angle between the turns of the helix and a plane perpendicular to the axis, and φ_0 the total angle through which one end of the helix is rotated compared with the other. When there are two springs stretched against each other, P for the two springs is the same, and may be eliminated from the two equations for the two springs. Also when pressure is applied, and there is a shift of the position of the coupling point, the same equality continues to hold, the values of a, μ, l, etc. on the right hand side being now the values under pressure. The effect of pressure on all the terms on the right hand side can be written out in terms of the geometry, known compressibilities, and the change under pressure of l given by the potentiometer measurements, except for the changes under pressure of the rigidity, which remains in the equation as unknown. If the pressure coefficient of one of the rigidities is known, as that of the steel spring by independent experiment, the other may be calculated from the equation in terms of known quantities. In the equation above, φ_0 is to be taken as not changing under pressure, the original spring being in a state of ease and so stretching without twist.

The equations actually used in the computations are:

$$(1-\chi_1 p)\left(1+\frac{d\mu_1}{\mu_1}\right)\left[1+\frac{\cos 2\alpha_1}{2\cos^2\alpha_1}\left(\frac{dl_1}{l_1}+\chi_1 p\right)\right]\left[1+\frac{dl_1+l_1'\chi_1 p}{l_1-l_1'}\right]$$

$$=(1-\chi_2 p)\left(1+\frac{d\mu_2}{\mu_2}\right)\left[1+\frac{\cos 2\alpha_2}{2\cos^2\alpha_2}\left(\frac{dl_2}{l_2}+\chi_2 p\right)\right]\left[1+\frac{dl_2+l_2'\chi_2 p}{l_2-l_2'}\right],$$

$$dl_1 = -\frac{\Delta R_0}{\rho_0}+\frac{\alpha p}{\rho_0}(R_0+\Delta R_0)+p\,[\lambda_1\chi_1+l_4\chi_3-L_0\chi_4],$$

$$dl_2 = -\,dl_1+p\,[\lambda_1\chi_1+\lambda_2\chi_2+l_3\chi_3-L\chi_4].$$

The meaning of most of the letters used in the formulas is obvious from figure 1. λ_1 and λ_2 are the lengths of the non-helical parts

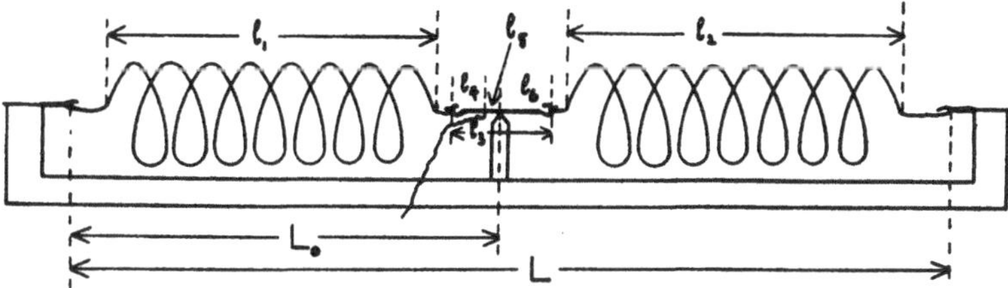

FIGURE 1. Scheme of the apparatus for measuring the effect of pressure on the relative rigidity of two different metals.

of the springs, that is, the lengths along the axis of the hooks on the two ends by which attachments were made. The various χ's denote the linear compressibilities of the various parts of the apparatus, taken as intrinsically positive. χ_1 is the linear compressibility of the metal to be measured, the length of the helical part of the spring of which is l_1, χ_2 is the linear compressibility of the comparison spring of piano wire, χ_3 that of the manganin connecting wire, and χ_4 that of the supporting frame of mild steel. α is the pressure coefficient of resistance of the manganin per unit length. α is obtained from a measurement of the pressure coefficient with the terminals rigidly attached combined with the linear compressibility. In the following, $\Delta R_o/\rho_o$ will be referred to as the uncorrected value of the elongation.

The advantage of the formulas in the form given is that the stiffness of the individual springs does not enter, and no experimental determination of the amount of stretch of each spring under known weights at atmospheric pressure is necessary. The formulas of the previous paper did involve this stiffness, which was determined by direct experiment. The formulas used there were more complicated, and the stiffness appeared in two places. It now appears that the stiffness cancelled from the final result, a fact which evidently would have much simplified the previous calculations if I had noticed it in time. A second advantage of the formulas in the new form is that the measurements obviously give differential results. The difference between the pressure coefficients of the rigidities of tungsten and platinum, for example, is correct whether the value used for steel is in error or not, and if at any future time it should appear that the value used for steel was in error, the values given here can be corrected by adding the same correction to them all.

There follows now the detailed presentation of data.

Detailed Data.

In general comment on the substances chosen for investigation in the following it would appear that a number of rather uncommon metals have been included. The reason for this is in the limitations of the method. The metal has to be one which when drawn into wire and wound into a helical spring will permit a fair amount of stretch without assuming permanent set. There are not a great many metals which satisfy this requirement, and in the following I have measured all such that were readily available. The requirement of high elastic limit rules out a number of the common metals such as gold, silver, and copper.

Tantalum. This is the only metal for which more than one set of measurements was made, so that from this some idea may be obtained of the probable accuracy and significance of the results. The wire I owe to the kindness of the Research Laboratory of the General

Electric Co. at Schenectady, who supplied it to me from their regular stock. It was not very uniform in geometrical shape, the external diameter varying from 0.0223 to 0.0239 cm. Three sets of measurements were made; two of these were with the identical steel and tantalum springs, one measurement being made near the beginning of the series, and the other near the end. The purpose of the repetition was to determine the probable error due to the apparatus in measurements on the same metal. The third measurement was made on steel and tantalum springs of different dimensions from the first, and the agreement of this result with the two others will give some idea of the significance of these measurements as determining absolute constants of the metals, unaffected by accidental differences in different samples.

All the measurements of this paper on the different metals are much more irregular than measurements of most other pressure effects; in particular there is much greater tendency for a few points in each run to lie very far off the curve. There is sometimes a distinct tendency for the points to lie on one or the other of two distinct lines, the appearance being as if the springs were capable of assuming one or the other of two positions of equilibrium. The explanation of this may be found in the slight rotary motion sometimes possible to the springs due to their manner of suspension. It must be remembered that the conditions here are very much more severe than in the usual pressure experiment, the apparatus being rotated and the contact purposely broken and made several times before each reading. Perfect readings demand that the contact assume a definite unique position at each pressure, irrespective of the number of times it has been made or broken. Slight particles of dirt in the transmitting medium, which are almost impossible to avoid, produce an effect, and this is doubtless the explanation of the few points with exceptionally large discrepancies.

Of the three runs with tantalum, the first was the best. Eleven readings were made to 12000 kg/cm^2. Of these two were discarded; the remaining nine lay on a straight line with an average departure from it by a single point of 6%. The corrections were of such a magnitude as to reduce the "uncorrected" value for the shortening of the tantalum spring to a final value about one-half as large. The actual shortening under 10000 kg was 0.0034 cm. On the repetition of the experiment several months afterwards, thirteen measurements were made. Two of these had to be discarded; the remainder lay on two straight lines of the same slope, the distance between the lines being one-third of the maximum effect. The average departure of a single reading from one or the other of these two lines was 3.3% of the maximum effect. The first set-up gave, after all corrections, an increase of rigidity of 1.49% for 10000 kg, and tne repetition an increase of 1.65%. The first result is to be preferred. The run with the second set of springs gave an opposite sign for the uncorrected effect. Thirteen readings were made to 12000 kg, of which three had to be discarded, the best of them lying off the curve by 30%. The average departure from a straight line of a single one of the remaining readings was 5.4% of the maximum effect. The finally corrected result was an increase of rigidity under 10000 kg of 0.31%, about one-fifth of that found with the first springs. The determinations with the second springs are not so good as with the first, as shown by the greater irregularity of the experimental points, and the fact that the relative stiffness of the tantalum and the steel spring was not so well adjusted to give the maximum sensitiveness. Nevertheless there can be no question, I think, that all the difference between the results with the two samples cannot be ascribed to experimental error, but the two springs were doubtless actually different in their behavior under pressure. This is not so surprising when it is considered that in a drawn wire there are many internal strains, which may have been particularly great in this wire, as suggested by the lack of uniformity in the diameter.

The conclusion to be drawn from these measurements is that the rigidity of homogeneous tantalum increases under pressure by something of the order of 1% for 10000 kg.

Molybdenum. This wire I also owe to the stock of the General Electric Co. It was not perfectly round, but varied in diameter from 0.0261 to 0.0270 cm. One successful run was made to 12000 kg; several unsuccessful attempts were first made, in which there was one trouble or another with the connections and contacts. Ten readings were made in the successful run; these were best represented by a straight line, but the departures of some of the points were rather large. The average departure from a line of a single reading (no discards) was 19% of the maximum effect. The independent correction terms were nearly as large as the uncorrected elongation, but there are both positive and negative corrections, so that the final corrected elongation differs by only 2% from the uncorrected value.

The final result is an increase of rigidity of .15% under 10000 kg.

Tungsten. This was also obtained from the General Electric Co. in the form of wire 0.025 cm in diameter. Readings were made on two days with the same set-up, and three excursions were made to 12000; a number of the single pressure settings gave no readings at all, doubtless because of dirt preventing contact. Twelve readings with positive contacts were obtained; two of these were discarded, the better of them lying off a line by 40% of the maximum effect. The average departure from a straight line of the remaining 10 points was 7% of the maximum effect. The corrections, of which the largest is more than three times the uncorrected effect, combine in such a way as to make the final corrected elongation somewhat more than twice the uncorrected elongation.

The final result was decrease of rigidity of 0.29% under 10000 kg. This unexpected negative result is probably connected with the lack of perfect homogeneity of wire of this metal; when an ordinary tungsten wire is broken by repeated bending it frequently splits, in the neighborhood of the bend, into several long fibres.

Platinum. This was Baker's purest platinum, annealed, 0.0294 cm in diameter. One set-up of the apparatus was used, and readings made to 12000, back to 3000, and up to 9000 again to check at doubtful points. Eight readings in all were obtained with steady contacts. Of these two were discarded, the better lying off the line by 50%. The average departure from a straight line of a single one of the remaining 6 points was 2.6% of the maximum effect. One of the correction terms is 25% larger than the uncorrected displacement; the final corrected displacement is 11% less than the uncorrected displacement.

The final result is an increase of rigidity of 2.4% for 10000 kg.

Zirconium. This material I owe to the kindness of Dr. G. Holst of the Philips Lamp Works at Eindhoven, Holland. It is from the same length of wire as that on which I have already determined the pressure coefficient of resistance.[2] The wire was very uniform and 0.050 cm in diameter. The spring also wound much more uniformly than many of the other metals, indicating probable absence of internal strains. One set-up was used, with two excursions to 12000, making eight readings. Discarding none of these, the average departure of a single reading from a straight line was 5.5% of the maximum effect. Of these eight points, the three worst were at pressures below 4000 kg; the average departure of a single one of the five readings above 4000 was 1.9% of the maximum effect. One of the correction terms was 66% larger than the uncorrected effect; all of the corrections combined in such a way as to make the final corrected displacement 10% less than the uncorrected displacement.

The final result was a decrease of rigidity of 0.17% for 10000 kg.

Palladium. The material was from Baker and Co., annealed wire, 0.029 cm in diameter. The results on this metal are by far the most unsatisfactory of all, and must be taken as merely of orienting value. The reason is the low elastic limit. The original spring, when stretched against the steel spring, was not far from its elastic limit, and under pressure there was an effect which amounted to an increase of set or lowering of the elastic limit. Two excursions to 12000 were made; the up and down points of the two runs lay on four straight lines like a distorted Greek sigma, the change in slope corresponding to a decrease in stiffness of the spring. After the termination of the experiment, the palladium spring was found to have received a considerable permanent elongation.

The effect of pressure on the elastic limit has apparently never been determined; what we have here is virtually a qualitative observation that the limit is lowered by pressure, an unexpected result.

Making due allowance for the effect of permanent set, the individual points of this set of measurements with palladium were perhaps better than those with any other metal, the individual points, except for one discard, lying off the straight lines by only 1 or 2%. In making the calculations, the mean of the first up and down excursion was used, as representing best the virgin wire. One of the correction terms is nearly twice the uncorrected term. The corrections combine in such a way that the final corrected displacement is of the opposite sign from and only 3% of the measured effect. That is, the corrections together almost exactly balance the measured effect.

The final result is an increase of rigidity of 1.08 % for 10000 kg.

Nickel. This was "pure" nickel from Baker and Co., 0.025 cm in diameter. A single run was made to 12000 and back, making seven readings. No points were discarded; the average departure of a single reading from a straight line was 5.7% of the maximum effect. The largest of the correction terms was 50% larger than the measured effect; the final corrected displacement was 30% less than the uncorrected displacement.

The final result is an increase of rigidity of 1.84% for 10000 kg.

Thorium. This was from the same piece of wire whose pressure coefficient of resistance I have measured,[3] and which I owe to the kindness of Dr. Rentschler of the Westinghouse Lamp Works. The section of the wire was far from regular, it was more nearly rectangular in section than either circular or elliptical, and the maximum and minimum diameters were 0.033 and 0.039 cm. Two excursions were made to 12000, obtaining 24 readings; of these 6 had to be discarded, the best departing from a straight line by 20% of the maximum effect. Of the 18 remaining readings, 12 lay on one line and 6 on another of the same slope, the average departure of a single reading from one or the other of these lines being 4.4%. The effect is unusually large, the largest correction term is only about one-half the uncorrected term, and the final corrected displacement is 54% of the uncorrected displacement and of the same sign.

The final result is an increase of rigidity of 5.73 % for 10000 kg.

Discussion.

Not a great deal of significance can be attached to the exact values found above for the pressure coefficient of the shearing modulus, as shown by the failure to obtain the same results on two springs of tantalum. One reason for this has already been suggested in the internal strains always found in any drawn wire; the importance of such strains was shown strikingly enough in the historic struggle between the multi- and the rari- constant theories of elasticity. Furthermore, the pressure coefficient of rigidity is not so definite a thing physically as the pressure coefficient of the compression modulus, for example. For a cubic crystal is equally compressed in all directions by hydrostatic pressure, and the compressibility of an aggregate is determined merely by the compressibility of the individual grains, and is independent of their relative orientations, whereas the rigidity of a microscopic aggregate of cubic crystals is connected in a more complicated way with the three elastic constants of the individual crystal grains, and will vary accordingly as the arrangement of the individual crystals is completely random or has traces of order. All of this makes it evident that the results above have a statistical rather than an individual signifflcance.

In the following table are collected the values found above for the effect of 10000 kg on the rigidity, together with the value previously found for the percentage effect of the same pressure on the incompressibility.[4] This latter is defined as $(2b/a)10^4$, where the *a* and the *b* are the constants in the formula for the change of volume under pressure: $\Delta V/V_o = -ap + bp^2$, p being in kg/cm².

TABLE I.

SUMMARY OF THE EFFECT OF PRESSURE ON ELASTIC BEHAVIOR.

Metal.	Percentage Change of Rigidity under 10,000 kg/cm².	Percentage Change of Incompressibility under 10,000 kg/cm².
W	−.3	+ 8.8
Ta	+ 0.3 to + 1.5	+ 1.1
Mo	+ .15	+ 6.9
Zr	− .17	+ 13.6
Pt	+ 2.4	+ 10.
Th	+ 5.7	+ 28.4
Pd	+ 1.1	+ 8.1
Ni	+ 1.8	+ 7.9
Fe* (steel)	+ 2.2	+ 7.2

* Previous Result

With the exception of tungsten and zirconium, for which the effect is small, the rigidity increases under pressure. An increase is what would be expected from general considerations, and is to be contrasted with the decrease shown by several varieties of glass. Further, the increase of rigidity under pressure is notably less than the increase of incompressibility. This again is what might be expected, and is doubtless connected with the very rapid increase in the force of repulsion between atoms when they are brought closer together than their normal distance of separation. The forces resisting volume compression are for the most part contributed by the pairs of atoms in closest contact. The repulsion between these increases rapidly when the distance between them is made less, so that the compressibility decreases by a comparatively large amount as the volume decreases under increased pressure. The forces resisting shear, on the other hand, contain a larger contribution from the more distant atoms, the component of the force effective in resisting shear contributed by the atoms directly in contact being of the second order. But the forces between distant pairs of atoms do not increase with decreasing distance as rapidly as those between adjacent pairs, and hence the effect of decrease of volume (or of increase of pressure) must be less on rigidity than on incompressibility.

The formulas have been given in the previous paper on the rigidity of glass for the effect of pressure on Young's modulus in terms of the effect on compressibility and rigidity. It will be found in general that for the metals studied above Young's modulus increases under pressure by an amount intermediate between that of the rigidity and the incompressibility. This seems natural in the light of considerations on the nature of the atomic forces like those of the last paragraph.

I am much indebted to my assistant Mr. W. A. Zisman for making the readings and setting up the apparatus. I am also indebted to the Milton Fund of Harvard University for financial assistance.

The Jefferson Physical Laboratory,
Harvard University, Cambridge, Mass.

References.

[1] P. W. Bridgman, Proc. Amer. Acad. 63, 401, 1929.

[2] P. W. Bridgman, Proc. Amer. Acad. 63, 347, 1928.

[3] P. W. Bridgman, Proc. Amer. Acad. 62, 217, 1927.

[4] P. W. Bridgman, Proc. Amer. Acad. 58, 166, 1923, in addition to references 1, 2, and 3.

THE COMPRESSIBILITY AND PRESSURE COEFFICIENT OF RESISTANCE OF SEVERAL ELEMENTS AND SINGLE CRYSTALS.

By P. W. Bridgman.

Presented Oct. 9, 1929. Received Nov. 7, 1929.

CONTENTS.

Introduction.

This paper is in continuation of my program of determining the quantities indicated in the title for all the well defined materials possible, particularly elements and single crystals.[1] The novel feature in this paper is that there are included measurements of the linear compressibility in different directions of single crystals of several organic substances, a class of substance for which these measurements have not hitherto been made. This has demanded the construction of a new type of piezometer, the former type not being applicable because in it pressure is transmitted directly to the specimen by kerosene, in which most organic substances dissolve.

Apparatus and Method.

The design of the new piezometer is indicated in Figure 1. The specimen is submerged in mercury, by which pressure is transmitted to it. The specimen under pressure is connected to a manganin

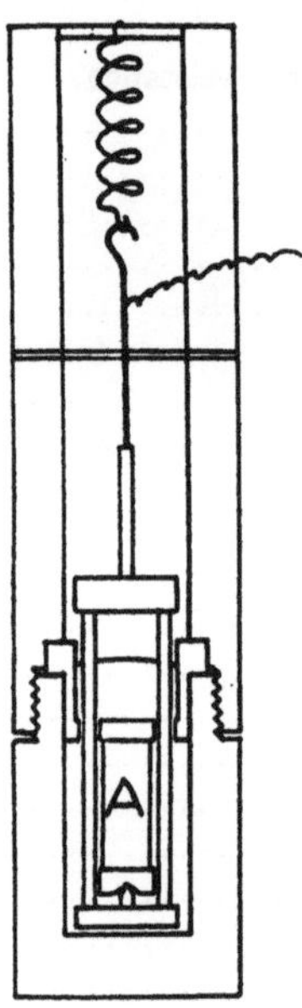

Figure 1. Diagram of the new form of piezometer, in which the specimen, A, is submerged under mercury to prevent action between the kerosene by which pressure is transmitted and the specimen.

wire, held pressed tightly against it by a spring, the wire sliding over a fixed contact, so that the change of dimensions of the specimen under pressure can be determined in terms of resistance measurements on a potientometer, exactly as in the previous types of piezometer. Corrections must be applied to the readings, similar in character to the corrections demanded by the other forms of piezometer, for the linear compressibility of the steel parts, the linear compressibility of the manganin wire, and the change of specific resistance of the manganin under pressure. These corrections have been determined by direct and indirect measurement. In addition to these corrections which can be calculated, there are other corrections arising from small effects which cannot be determined so easily, such for example, as that arising from the compressibility of the insulating washers of mica, or from internal strains in the metal parts at the various screw connections. These corrections were determined together by measuring with the apparatus the compressibility of pure iron, the compressibility of which is already known, and finding the corrections which must be added to the calculated corrections to give the known compressibility. These corrections altogether amounted to a change of length of about 0.002 cm at 10000 kg/cm^2.

It will be noticed that in this apparatus the changes of dimensions are measured directly, without multiplication as in the previous piezometer. The necessity for this arose from the fact that many organic substances are too weak mechanically to withstand the stress to which they must be subjected in the multiplying apparatus. In using the piezometer, small specimens had to be measured in a number of cases, because it is often difficult to obtain crystals of organic substances of any size. In consequence of these two facts, the measured effects were often small, and the corrections were sometimes as much as 50% of the measured effect. The accuracy with this type of apparatus is not as high as that obtainable with the previous types, which still are to be used whenever the substance will allow it.

There follows now the detailed description of the various measurements.

Effect of Pressure on Electrical Resistance.

Indium. This material was of exceptionally high purity, having been made by Professor T. W. Richards in connection with determinations of its atomic weight. I owe my supply to the kindness of Dr. L. H. Hall, to whom was intrusted the task of winding up many of the scientific affairs of Professor Richards after his death.

The potentiometer method of measurement was used. The wire, of about 0.035 cm diameter, was made by cold extrusion through steel dies. The piece measured was about 10 cm. long, bent into a hair pin shape. The terminals were fine wires of pure silver, fused directly to the indium by a simply constructed arrangement utilizing the Joulean heating of a current through the silver. The use of solder in attaching the terminals was undesirable because indium forms with tin and lead an alloy of melting point lower than 75°, at which it was desired to make measurements.

Measurements of the effect of pressure on resistance were made in the regular way up to 12000 kg at 30° and 75°. The pressure measurements went very smoothly; the mean departure of a single point from a smooth curve at 30° being 0.09% of the maximum effect, and at 75° 0.03%. The relation between change of resistance and pressure is very nearly of the second degree, but departs from it

TABLE I.

THE RESISTANCE OF INDIUM UNDER PRESSURE.

Pressure Kg/cm²	$\frac{-\Delta R}{R(0, 30°)}$	$\frac{-\Delta R}{R(0.75°)}$
2000	.0288	.0302
4000	.0549	.0575
6000	.0795	.0831
8000	.1026	.1075
10000	.1245	.1295
12000	.1450	.1504

by more than the experimental error, the pressure of maximum departure from linearity being somewhat less than the mean pressure. The accompanying table gives the effect at pressure intervals of 2000 kg. The initial resistances at atmospheric pressure at 30° and 75° were 1.1472 and 1.3678 respectively, in terms of the resistance at 0°.

The temperature coefficient of resistance at atmospheric pressure between 0° and 100°, obtained by linear extrapolation of actual readings at 25° and 75°, was 0.004904. By combining with the values in the table for the effect of pressure, the corresponding temperature coefficient at 12000 is found to be 0.004876, thus agreeing with the result found for many other metals that the temperature coefficient is little affected by pressure.

I have previously measured the effect of pressure on the resistance of a sample of indium of inferior purity.[2] The mean coefficient to 12000 of the pure metal is about 15% greater than that of the former sample, agreeing with the universal rule that the coefficient of the purer metal is less algebraically than that of its alloys. The temperature coefficient of the former sample was 0.00404, against the value 0.00490 for this new pure material. The new value is higher than any listed value of the temperature coefficient.

Manganese. This material I owe to the kindness of Professor M. A. Hunter of Rensselaer Polytechnic Institute, who prepared it in massive form by electrolysis of the fused salts. The specimen on which the resistance measurements were made was in the form of wire several cm long and of the order of 0.5 mm in diameter. It was most kindly prepared from the massive material by Mr. G. F. Taylor of the General Electric Co. at Schenectady by his method of drawing out a tube of glass or quartz containing the metal heated above its melting point.

Manganese is a metal which previous investigators have found very difficult to get into massive form without the formation of numerous cracks which destroy the continuity of the metal. These cracks have been attributed to a polymorphic transition which manganese passes through on cooling from the melting point to room temperature. The transition is accompanied by a large change of volume. It would be expected that this difficulty would make impossible the formation of wire by the Taylor process. Doubtless the effect was present, for although the quartz capillaries in which the magnanese was drawn out were 10 or 15 cm long, and they were apparently completely filled with a thread of manganese, the thread always broke into lengths varying from a few millimeters to one or two centimeters. These short lengths were very strong mechanically, being very difficult to break with the fingers, and not showing any of the brittleness usually associated with metallic manganese. The places in which the thread broke doubtless corresponded to the cracks usually found in the massive metal. It is possible that the coherent pieces may have been single crystals of the low temperature modification.

Measurements of resistance were made by the potentiometer method. Four terminals were attached to the specimen by springs. It was not until after the completion of the measurements that I was informed by Mr. Taylor, and verified for myself, that it is very easy to solder manganese with soft solder; if I had known this in time, I would have soldered the connections, instead of using springs, because of the greater convenience.

Before making the pressure measurements, the temperature coefficient of resistance was determined at atmospheric pressure. The mean coefficient between 0° and 100°, obtained by linear extrapolation of measurements at 25° and 75°, was 0.00078. There seem to be no published values of this coefficient, but 0.00079 is so much lower than the value customary for a pure metal that there can be little doubt that the metal contains several per cent of impurity. It is perhaps more probable that the impurity was in the original material than that it was given to it during the process of formation into wire.

In view of the probable impurity, the effect of pressure was measured only at 30°, up to 12000. The points lay smoothly, the average deviation of a single reading from a smooth curve being 0.1%. The relation between change of resistance and pressure can be represented within experimental error by a second degree formula:

$$\Delta R/R_0 = -7.012 \times 10^{-6}p + 5.63 \times 10^{-11}p^2.$$

It is thus seen that the resistance decreases with increasing pressure, as is normal, and the curvature is also in the normal direction. Perfectly pure manganese, in accordance with the universal rule, may be expected to have a negative coefficient somewhat larger numerically than the value above.

Chromium. The material was obtained from Eimer and Amend. It had been made by the Goldschmidt process, was quite brittle, and probably had several per cent of impurity. It was very kindly made into wire for the pressure measurements by Mr. Taylor by his method. More difficulty was experienced with this than with manganese, the particular difficulty being sticking of the metal to the quartz tube in which it was drawn out. Removal of the quartz with hydro-fluoric acid was not as successful as with some other metals, and the method finally used was mechanical, breaking away as much of the quartz as possible, and removing the remainder by rubbing with emery paper. The dimensions were about the same as for manganese. There is not the difficulty with chromium from a polymorphic transition that there is with manganese, but its greater brittleness makes the manipulation more difficult.

The potentiometer method of measurement was used. It is not possible to solder chromium with either soft solder or silver or gold. The four terminals were fine platinum wires attached by spot welding; I am much indebted to Dr. H. G. deLászló of the Mass. Inst. of Technology for very skillfully doing this.

At atmospheric pressure the temperature coefficient of resistance between 0° and 100° was found to be 0.000033. This is very low indeed, and indicates high probable impurity. The specific resistance at 30° was 160×10^{-6}, very high, and also suggestive of impurity.

At 30°, resistance decreases linearly with pressure up to 12000, the pressure coefficient being -5.8×10^{-7}. This is very small, but of the normal sign. The pure metal may also be expected to have the same negative sign, but to be considerably larger numerically. Considering the smallness of the effect, the readings were very regular, the average deviation of a single reading from a straight line corresponding to 0.12 mm on the slide wire of the potentiometer; this is equivalent to 1.7% of the maximum change of resistance.

Arsenic. See the next section for a description of the effect of pressure on the electrical resistance.

Compressibility.

Arsenic. Since the measurements of resistance under pressure and of compressibility of this element were closely related, it is best to describe them together. These results are to be regarded only as preliminary; it is evident that arsenic is highly crystalline in its properties, and the final study must be made on single crystals.

The material was from Kahlbaum, grade "K." The compressibility was determined first. I tried to make a single crystal by sealing the arsenic in a heavy quartz capillary of about 2 mm inside diameter, and slowly lowering from an electric furnace maintained above the temperature of the triple point. This attempt to make a single crystal was not successful, but nevertheless the resulting material was highly crystalline in appearance, and there was a strongly marked preferred direction of orientation. The appearance was much like that which I have often found in trying to make single crystals of tellurium by slow cooling; the cleavage planes are usually parallel to the axis of the rod, but the orientation about the axis is at random. In the case of tellurium, more rapid cooling and higher purity in the original material brings the desired result. It is possible that the arsenic may have been contaminated by the quartz during this attempt, since the inside of the capillary was somewhat pitted after the run, which lasted for 15 hours.

The compressibility of this approximately single grain was measured in the regular way with the lever piezometer for short specimens. Two runs were made to 12000 at 30°. The results of the first run were much more irregular than usual, and showed the highly unusual feature of an abnormal curvature, that is, a compressibility increasing with increasing pressure. The second run, made more carefully, showed that the abnormal curvature was illusory; the curve really consists of two parts with a discontinuity between 6000 and 8000 kg, the slope of the low pressure branch being less than that of the high pressure branch. The explanation clearly is a polymorphic transition. It is an unusual feature of this transition that the points corresponding to the low pressure modification could be repeated after the excursion into the domain of the high pressure modification, indicating that the low pressure modification recovers its original orientation. The high pressure modification is therefore probably not formed by the chance occurrence of a nucleus and growth from it at random, but there is some definite relation between the lattices of the two modifications such as is shown, for example, by quartz.

The initial linear compressibility of the low pressure modification was 2.2×10^{-7}, giving for the volume compressibility 6.6×10^{-7} on the assumption of equal compressibility in all directions. The volume compressibility of arsenic at low pressures has been directly determined by Richards[3] to be 4.4×10^{-6}. This great discrepancy cannot possibly be due to experimental error, but must mean that the compressibility of a single arsenic crystal is very different in different directions.

The linear compressibility of the high pressure modification was 2.3×10^{-7}, slightly higher than that of the low pressure modification, but the exact significance of this result cannot be clear until the relative orientation of the two phases is known.

In order to investigate more accurately the transition phenomena, the resistance under pressure of arsenic was measured. For the resistance measurements the arsenic was cast in quartz in the form of a rod several centimeters long and about 1 mm in diameter. The procedure in making the casting was so much like that described by Kapitza[4] in a paper published after this work was done that it is not necessary for me to add anything in the way of description. I also observed the fluorescence of the arsenic vapor in the tube, but I would describe the color as greenish yellow rather than as reddish yellow. My rod was not a single crystal grain like that of Kapitza. The production of the casting required only a few minutes, instead of many hours as before, and there was no pitting of the inside of the quartz capillary. The high purity of this arsenic was verified by measuring its temperature coefficient of resistance at atmospheric pressure. The mean coefficient between 0° and 100°, obtained by linear extrapolation of readings at 30° and 75°, was found to be 0.00474 immediately before the casting had been subjected to pressure, and 0.00472 after several exposures to pressure and the consequent formation of the high pressure modification. This is considerably higher than any listed value for the temperature coefficient.

The pressure measurements of resistance were made in the regular way with the potentiometer; the leads were fine silver wire attached with soft solder. Special pains was taken to get the initial effect on the virgin specimen at low pressures before the transition had occurred. Points were thus obtained on a smooth curve, the resistance *increasing* with increasing pressure. A discontinuity occurred with increase of pressure between 5050 and 6030 kg. The accidental breaking of one of the connections at the maximum pressure prevented measurements of the change of resistance with decreasing pressure. On the next application of pressure the effect on the low pressure modification had changed sign, the resistance now decreasing, with the coefficient -1.52×10^{-6}. A discontinuity was again found with increase of resistance, this time in the neighborhood of 5500 kg. On decreasing pressure an enormous hysteresis was found in the resistance of the high pressure modification, but the discontinuity was at essentially the same place as before, and the pressure coefficient of the low pressure modification was not much altered.

A second piece of arsenic was tried. Trouble with the connections made it impossible to get readings on the virgin specimen. The second application of pressure gave a pressure coefficient of resistance of -1.58×10^{-6} for the low pressure modification, and a discontinuity between 5400 and 5800. The behavior on releasing pressure was approximately the same as before; very great hysteresis in the resistance of the high pressure modification, and a discontinuity, now a little more pronounced than before, at approximately the same pressure, and again approximately the same pressure coefficient for the low pressure modification.

The best value for the transition pressure from the resistance measurements is 5500 kg, somewhat lower than the lower limit found from the compressibility measurements.

These results obtained with arsenic may be summarized as follows. There is a new modification at high pressures, and the transition pressure at 30° is in the neighborhood of 5500 kg. The compressibility of single crystal arsenic must vary greatly with the direction in the crystal, there being a direction in which the compressibility is at least 6.7 times greater than the average. The temperature coefficient of resistance at atmospheric pressure between 0° and 100° is 0.00473. There are several obscure points; why the great hysteresis in the resistance of the high pressure modification, why the difference in the transition pressures from the compressibility and the resistance measurements, and why the positive pressure coefficient of resistance on the virgin specimen, and a nearly constant value of -1.5×10^{-6} on the same material after it has experienced the transition.

The behavior of arsenic has been previously measured under pressure.[5] The previous specimen was very impure, the temperature coefficient of resistance being only 0.00076. The resistance was found to have a pressure coefficient of -3.3×10^{-6}. There was considerable irregularity in the previous results, which was not inconsistent with a transition.

Boron. The specimen was a rod about 1 cm long and 1.5 mm in diameter. It was most kindly loaned to me for these measurements by Mr. Alfred Loomis of Tuxedo Park, who had obtained it from Dr. G. Holst of Eindhoven. It was prepared by the method developed there by van Arkel of thermal decomposition of the vapor of BCl_3 and BBr_3. It is deposited from the vapor on a core of tungsten wire. The diameter of this wire is of the order of 0.05 mm, making the total impurity of tungsten about 0.1%. Except for this, the material may be expected to be very pure.

The measurements were made with the lever piezometer for short specimens. The shape of the piece of boron was not as regular as would have been desirable; it could not be made to fit the piezometer as well as usual, and the results were much more irregular. Another source of irregularity was the smallness of the effect, because the compressibility of boron is very nearly the same as that of iron. The average departure of a single reading from a smooth curve (which within the limits of error was taken as a straight line) was 6.9% of the maximum measured effect, that is, 6.9% of the difference of compressibility between boron and iron. This means an irregularity of only 0.46% on the actual compressibility, but this is considerably greater than the irregularity usual wuth other substances. In view of the irregularity in the measurements, no attempt was made to find the temperature coefficient of compressibility, and measurements were made only at 30°. The results found for the linear compressibility were:

$$-\Delta l/l_0 = 1.837 \times 10^{-7}p - 0.70 \times 10^{-12}p^2.$$

Assuming equal compressibility in all direction, this gives for the volume compressibility:

$$-\Delta V/V_0 = 5.51 \times 10^{-7}p - 2.2 \times 10^{-12}p^2.$$

The second degree terms in these expressions is contributed entirely by the iron. According to this boron is about 6% less compressible than iron.

The crystal system of boron has not been determined, so that this result does not have as much significance as it might, Richards[7] found for the volume compressibility 3×10^{-7} on a specimen known to be of high impurity. If boron turns out to have cubic symmetry, the entire difference between the result of Richards and myself must be ascribed to difference of purity; if not, part of the difference may well be due to the crystal structure. A value somewhat between 5.5 and 3 would fit into the periodic system of the elements better than either. It seems to me probable that boron will be found to be not of the cubic system and that its volume compressibility will be somewhat less than 5.5×10^{-7}.

Titanium. I am indebted for this material to Professor M. A. Hunter, who prepared small coherent slugs of the massive metal by electrolysis of fused salts. The compressibility was measured with the lever piezometer for short specimens. The pieces measured were formed by grinding from the slugs resulting from the electrolysis; two of them were used piled together, of 1.5 and 1.1 cm length. Measurements were made to 12000 kg at 30° and 75°. Within the limits of error, a second degree expression in the pressure reproduces the results. At 30° the average departure of a single reading from a smooth curve was 1.1% of the maximum measured effect, which means 0.3% on the actual compressibility; at 75° the corresponding figures, discarding one reading, were 1.8% and 0.56%.

The crystal system of titanium is known to be hexagonal, with an axial ratio of 1.60. This is very close to the ratio for close packed spheres, so that it is probable that the compressibility of a single crystal of titanium is nearly the same in all directions. Furthermore, the method of preparation of the specimens would not lead to the expectation of any large crystal structure, with any particular direction of orientation preferred. It is therefore probably fairly safe to calculate the volume compressibility from the actual measurements of linear compressibility by assuming equal compressibility in all directions. The results found in this way are:

$$\text{At } 30^\circ \qquad -\Delta V/V_0 = 7.97 \times 10^{-7}p + 0.12 \times 10^{-12}p^2,$$

$$75^\circ \qquad -\Delta V/V_0 = 8.68 \times 10^{-7}p - 4.5 \times 10^{-12}p^2.$$

Two features of these results call for comment; the large increase of compressibility with temperature, and the fact that at 30° the direction of curvature is abnormal.

Apparently the compressibility of titanium has not been previously measured. The value 7.97×10^{-7} fits in well with its position in the periodic table.

Manganese. This material has been already described in connection with measurements of the effect of pressure on resistance. For the compressibility measurements two pieces were ground to the proper shape, of length 0.67 and 0.53 cm; these were then piled together and measured in the lever piezometer for short specimens. Measurements were made at 30° and 75° to 12000 kg. At 30° the average departure of a single reading from a smooth curve was 1.04% of the maximum measured effect, which means 0.27% on the actual compressibility; at 75° the corresponding figures were 1.12% and 0.30%. Both the high and low temperature modifications of manganese belong to the cubic system, so that the volume compressibility may be calculated from the linear compressibility assuming equal compressibility in all directions. The following results were found:

$$\text{At } 30^\circ \qquad -\Delta V/V_0 = 7.91 \times 10^{-7}p - 5.3 \times 10^{-12}p^2,$$

$$75^\circ \qquad -\Delta V/V_0 = 8.08 \times 10^{-7}p - 4.8 \times 10^{-12}p^2.$$

The only previous value is by Richards,[3] who found 8.4×10^{-7} as the average compressibility up to 500 kg. Both my value and that of Richards is very much out of line with what would be indicated by the neighbors of manganese in the periodic table; this would suggest a value not far from 5.6×10^{-7}.

Potassium Alum., *Ammonia Alum.*, *Chrome Alum.* — These three alums were made in large single crystals by slow crystallization lasting several weeks from aqueous solution in a room held at constant temperature near 30°. The material was commercial C. P. The alums are cubic, so that a measurement of the linear compressibility in a single direction is sufficient to determine the cubic compressibility. Potassium alum was the first measured. Under high pressure there was some corrosive action between the alum and the steel parts with which it came in contact. In order to avoid this, the adjacent steel parts were gold plated. The compressibility of gold is so nearly the same as that of steel that no error can be introduced in this way. All measurements were made with the lever piezometer for short specimens. It was not possible to measure the chrome alum at 75° because of decomposition. The following are the results:

		Average Deviation of Single Reading from Smooth Curve.
Potassium Alum.		
At 30°	$-\Delta V/V_0 = 63.03 \times 10^{-7}p - 116.1 \times 10^{-12}p^2,$	.15%
At 75°	$-\Delta V/V_0 = 55.74 \times 10^{-7}p - 93.3 \times 10^{-12}p^2.$	.38%
Ammonia Alum.		
At 30°	$-\Delta V/V_0 = 63.54 \times 10^{-7}p - 102.3 \times 10^{-12}p^2,$	.63%
At 75°	$-\Delta V/V_0 = 61.98 \times 10^{-7}p - 100.8 \times 10^{-12}p^2.$	.45%
Chrome Alum.		
At 30°	$-\Delta V/V_0 = 64.86 \times 10^{-7}p - 112.5 \times 10^{-12}p^2.$	.48%

The compressibilities of these three alums are thus nearly the same. The most striking feature in the results is the very pronounced negative temperature coefficient of compressibility of potassium alum. The temperature coefficient of ammonia alum is also negative, but by a smaller amount. There are no previous values of the compressibilities of the alums for comparison.

Sodium Chlorate. This material crystallizes in the cubic system, so that a measurement of the linear compressibility in a single direction suffices. The crystal was prepared from the aqueous solution. The solution was maintained in a room at 30° for several weeks, and allowed to evaporate slowly. The source of material was commercial C. P. stock, obtained from a large chemical supply house; the slow crystallization should give a product of high purity. The compressibility sample was 0.61 cm. long, cut from a single crystal.

In the measurement of compressibility it was necessary to use the new piezometer for organic substances in which pressure is transmitted to the substance by mercury, not because $NaClO_3$ dissolves in kerosene, but because it combines with it explosively at pressures above 2000 kg, as I had previously found.

Measurements were made to 12000 at 30° and 75°, 14 observations at each temperature. At 30° the average departure of a single reading from a smooth curve was 1.3% of the maximum effect, and at 75° 1.5%. As already explained, the corrections with this method are much larger than with the lever piezometer; in this case the corrections amounted to 50% of the measured effect. The following are the numerical results, obtained immediately from the measurements, without adjustment:

At 30° $-\Delta l/l_0 = 1.765 \times 10^{-6}p - 3.78 \times 10^{-11}p^2$,

At 75° $-\Delta l/l_o = 1.633 \times 10^{-6}p - 1.92 \times 10^{-11}p^2$.

The results as given indicate a smaller initial compressibility at 75° than at 30°, which is abnormal, and also a smaller second degree term at 75°, which again is abnormal. The result of the combination of these two factors is that the average compressibility between 0 and 12000 kg is greater at 75° than at 30°, which is normal. A study of the relative magnitude of the errors of a single reading and of the departures from linearity will show that a great deal of significance cannot be attached to the second degree term. The best way to manipulate the results would seem to be to assume that the second degree term is the same at both temperatures and the average of those found, and then to so adjust the initial compressibility that when combined with the second degree term the average compressibility to 12000 kg found experimentally is reproduced at each temperature. In this way the following results were found, making the additional change from linear to volume compressibility. These are to be accepted as the most probable values on the basis of these measurements:

At 30° $-\Delta V/V_0 = 4.94 \times 10^{-6}p - 9.3 \times 10^{-11}p^2$,

At 75° $-\Delta V/V_0 = 5.28 \times 10^{-6}p - 9.3 \times 10^{-11}p^2$.

There seem to be no previous measurements of the compressibility of this substance for comparison.

Sodium Bromate. This was grown by slow evaporation from aqueous solution in a constant temperature room over a period of several weeks, like sodium chlorate. The original material was commercial stock, C. P.; the slow crystallization should have been effective in further purification. The sample for the compressibility measurement was cut from a single crystal and was 0.67 cm long. The compressibility was measured in the piezometer for organic substances, submerged under mercury. I did not verify by direct experiment that this spontaneously explodes on contact with kerosene at high pressures, but used at once the safe method without further inquiry.

Measurements were made to 12000 kg at 30°; a run was also made to 12000 at 75°, but there was evidently something the matter with the measurements, and on taking the apparatus apart, the sodium bromate was found to have completely decomposed, an effect not shown by the chlorate. At 30°, the two points at the lowest pressures lay off the smooth curves by amounts corresponding to about 1000 kg; the remaining points lay on a smooth curve with an average departure of 0.2% of the maximum effect. The dimensions and the magnitude of the effect were somewhat more favorable than in the case of the chlorate, so that the total correction was only 20% of the measured effect.

Sodium bromate is cubic, so that the cubic compressibility can at once be found from the linear compressibility in a single direction. The results were as follows:

At 30° $-\Delta V/V_0 = 4.320 \times 10^{-6}p - 7.42 \times 10^{-11}p^2.$

The compressibility is this somewhat less than that of the chlorate. This is not the direction of difference which one would at first expect, because the atom of bromine is by itself more compressible than the atom of chlorine.

Sodium Nitrate. This substance has the same crystal symmetry as calcite, so that the compressibility must be measured in two independent directions, parallel and perpendicular to the axis of trigonal symmetry. The original material was Kahlbaum's purest. It was formed into a single crystal block by my method of slow lowering from the furnace in the molten condition. It was melted in a funnel shaped container of pyrex glass; this has the disadvantage that it sticks on solidifying to the sides of the mold and the resultant block contains many cracks, but in spite of this it was possible to get from the block unfractured pieces large enough, the piece parallel to the trigonal axis being 0.97 cm. long, and that perpendicular to it 0.82 cm. Since sodium nitrate is not soluble in kerosene, the measurements were made in the lever apparatus for short specimens. The measurements on the specimen parallel to the trigonal axis were much more regular than on those perpendicular to it. At 30° the average departure of a single reading from a smooth curve for the perpendicular specimen was 0.9% of the maximum measured effect, and at 75° 0.7%. For the parallel specimen the corresponding figures were 0.1% and 0.4%.

The following are the results for the linear compressibilities. Perpendicular to the trigonal axis:

At 30° $-\Delta l/l_0 = 7.093 \times 10^{-7}p - 5.88 \times 10^{-12}p^2$,

At 75° $-\Delta l/l_0 = 7.643 \times 10^{-7}p - 4.41 \times 10^{-12}p^2$.

Parallel to the trigonal axis:

At 30° $-\Delta l/l_0 = 24.36 \times 10^{-7}p - 23.5 \times 10^{-12}p^2$,

At 75° $-\Delta l/l_0 = 23.36 \times 10^{-7}p - 21.4 \times 10^{-12}p^2$.

From these values of linear compressibility the following may be calculated for the cubic compressibility:

At 30° $-\Delta V/V_0 = 38.54 \times 10^{-7}p - 39.2 \times 10^{-12}p^2$,

At 75° $-\Delta V/V_0 = 38.64 \times 10^{-7}p - 35.8 \times 10^{-12}p^2$.

The compressibility at low pressures at 0° has been found by Madelung und Fuchs[8] to be 37.7×10^{-7}, agreeing fairly well with the initial value of the above formulas.

The compressibility of sodium nitrate has several interesting features. The compressibility in the parallel direction is more than three times as great as in the perpendicular direction. In the perpendicular direction the temperature coefficient of compressibility is positive, as is normal, but the numerical value of the coefficient is much larger than normal. In the parallel direction, the temperature coefficient of compressibility is negative, which is abnormal. Calcite is also three times more compressible parallel to the trigonal axis than at right angles, but the absolute magnitudes are only one-third as large.

Rochelle Salt. This belongs to the orthorhombic system, so that observations are necessary in three mutually perpendicular directions. The rods were cut from a single large crystal, which I owe to the kindness of the General Electric Co. It was prepared by the method of slowly lowering the temperature of a saturated aqueous solution. This crystal was prepared at least five years before the measurements were made on it. During these years it has been wrapped in Japanese crepe paper in a tightly closed wooden box; the surface still retained the original glass-like perfection. This material does not dissolve in kerosene, so that measurements were made in the lever apparatus for short specimens. The pressure range was 12000 kg as usual, but it was possible to measure only at 30°, because Rochelle salts loses its water of crystallization at about 37°, making the usual measurement at 75° impossible.

The designation of the directions used in the following is that usual with crystallographers, for example as that used by Mandel[10] in his recent paper on the elastic constants. The *a* and *b* directions are at right angles to each other, and lie in the large basal face, *b* bisecting the angle of approximately 80°, and *c* is perpendicular to both *a* and *b*. The lengths of the specimens measured were as follows: *a*, .73 cm; *b*, .91 cm; and *c*, 1.17 cm. The measurements in all three directions went rather smoothly; the average departure of a single reading from a smooth curve in terms of the maximum effect was 0.9% for the *a* direction, 0.09% for the *b* direction, and 0.29% for the *c* direction. In each direction the departures from a second degree curve were very large, so that the results must be given in a Table, which follows.

TABLE II.

COMPRESSIBILITY OF ROCHELLE SALT AT 30°.

Pressure kg/cm²	$-\Delta l/l_0$			$-\Delta V/V_0$
	a	*b*	*c*	
2000	.00245	.00506	.00333	.01080
4000	.00460	.00969	.00600	.02016
6000	.00651	.01414	.00847	.02885
8000	.00843	.01833	.01085	.03716
10000	.01029	.02217	.01320	.04501
12000	.01209	.02586	.01531	.05237

The cubic compressibility of Rochelle salts has not been previously measured, but all of the elastic moduli have been measured by Mandel, from which the linear compressibilities can at once be calculated. Mandel's[9] values for the moduli in Abs. C. G. S. units are:

$$s_{11} = 46.0 \times 10^{-10},$$
$$s_{22} = 31.4 \times 10^{-10},$$
$$s_{33} = 27.6 \times 10^{-10},$$
$$s_{44} = 59.7 \times 10^{-10},$$
$$s_{55} = 300.4 \times 10^{-10},$$
$$s_{66} = 78.7 \times 10^{-10},$$

$$s_{23} = 16.5 \times 10^{-10},$$
$$s_{31} = -21.4 \times 10^{-10},$$
$$s_{12} = -7.8 \times 10^{-10}.$$

From these the linear compressibilities in kg units may be calculated to be:

Linear compressibility in a direction $(= s_{11} + s_{12} + s_{13}) = 16.8 \times 10^{-7}$,

Linear compressibility in b direction $(= s_{12} + s_{22} + s_{23}) = 40.1 \times 10^{-7}$,

Linear compressibility in c direction $(= s_{13} + s_{23} + s_{33}) = 22.7 \times 10^{-7}$.

By drawing smooth curves through the points given in the table and taking the tangent at the origin, the corresponding initial compressibilities given by my data are: 13.3, 27.1, and 19.5 $\times$ 10^{-7}. These are all considerably lower than the values of Mandel, and the difference is much too great to be explained by mere errors of observation. The difference must be ascribed to difference in the original material; it is well known that the electrical properties of Rochelle salt crystals may be very greatly altered by various processes of seasoning or dessication, and there is no reason not to expect corresponding variability in the elastic properties. The most natural explanation of the difference is difference of water content. It will be seen that the ratio of my compressibilities in different directions is not the same as that of Mandel; this means that different degrees of dessication affect the properties differently in different directions.

Ammonium Tartrate. The C. P. material of commerce occurs in large enough single crystals so that suitable pieces were found by selection. This material is monoclinic, so that, strictly, linear compressibility measurements are necessary in four directions to completely fix the behavior under pressure. However, the a axis, using the usual designation of the axes as given, for example, in Groth, is so nearly at right angles to b and c that the measurement in a fourth direction, exactly perpendicular to b and c, was dispensed with.

This material is not soluble in kerosene, so that the compressibility was measured in the lever piezometer for short specimens. The lengths of the specimens were as follows: a, 0.48 cm; b, 0.87 cm; and c, 1.00 cm. Readings were made to 12000 kg at both 30° and 75°. The departure of a single reading from a smooth curve in terms of the maximum effect was: a direction, 1.2% at 30° and 1.4% at 75°; b direction, 0.3% at 30° and 0.2% at 75°; c direction, 0.16% at 30° and 0.34% at 75°.

Within the limits of error the relation between change of length and pressure can be represented by a second degree curve in the pressure. The results follow:

a direction
At 30° $-\Delta l/l_0 = 8.47 \times 10^{-7}p - 16.8 \times 10^{-12}p^2$,
At 75° $-\Delta l/l_0 = 8.75 \times 10^{-7}p - 19.3 \times 10^{-12}p^2$.

b direction
At 30° $-\Delta l/l_0 = 13.71 \times 10^{-7}p - 9.3 \times 10^{-12}p^2$,
At 75° $-\Delta l/l_0 = 14.51 \times 10^{-7}p - 11.8 \times 10^{-12}p^2$.

c direction
At 30° $-\Delta l/l_0 = 26.85 \times 10^{-7}p - 46.1 \times 10^{-12}p^2$,
At 75° $-\Delta l/l_0 = 27.48 \times 10^{-7}p - 48.2 \times 10^{-12}p^2$.

Approximate volume compressibility:
At 30° $-\Delta V/V_0 = 49.03 \times 10^{-7}p - 79.3 \times 10^{-12}p^2$,
At 75° $-\Delta V/V_0 = 50.84 \times 10^{-7}p - 87.0 \times 10^{-12}p^2$.

Again we find a compressibility in one direction about three times as great as in another. The behavior of the compressibility with temperature and the second degree terms offer no unusual features.

Tartaric Acid. This was prepared by slow evaporation in the constant temperature room from aqueous solution of commercial C. P. material. The crystal is monoclinic; the *a* axis is inclined at 80° to the plane of *b* and *c*, and as in the case of ammonium tartrate the compressibility was determined only along the three crystallographic axes. The lengths of the specimens in the *a*, *b*, and *c* directions were respectively: 0.77, 0.62, and 0.67 cm. This material is also not soluble in kerosene, and the measurements were made in the lever apparatus for short specimens. Measurements were made to 12000 kg at 30° and 75°. In the *c* direction an unusual seasoning effect was found, the points obtained with increasing pressure being irregular and lying far off the descending curve, which by every appearance was the correct curve. Discarding the ascending points of the *c* direction, the average departures of a single reading from a smooth curve at 30° and 75° were respectively: *a* direction, 0.5% and 0.9%; *b* direction, 0.1% and 0.02%; *c* direction, 0.5% and 0.13%.

The results can be reproduced by a second degree expression in the pressure, and are as follows:

a direction
At 30° $-\Delta l/l_0 = 8.19 \times 10^{-7}p - 11.7 \times 10^{-12}p^2$,
At 75° $-\Delta l/l_0 = 8.26 \times 10^{-7}p - 13.8 \times 10^{-12}p^2$.

b direction

At 30° $-\Delta l/l_0 = 50.78 \times 10^{-7}p - 110. \times 10^{-12}p^2,$

At 75° $-\Delta l/l_0 = 53.63 \times 10^{-7}p - 126. \times 10^{-12}p^2.$

c direction

At 30° $-\Delta l/l_0 = 10.89 \times 10^{-7}p - 16.0 \times 10^{-12}p^2,$

At 75° $-\Delta l/l_0 = 13.73 \times 10^{-7}p - 33.1 \times 10^{-12}p^2.$

Approximate volume compressibility, obtained by treating these three directions as if perpendicular:

At 30° $-\Delta V/V_0 = 69.86 \times 10^{-7}p - 148 \times 10^{-12}p^2,$

At 75° $-\Delta V/V_0 = 75.62 \times 10^{-7}p - 186 \times 10^{-12}p^2.$

The temperature coefficient of compressibility in the *b* and *c* directions has the normal sign, but is abnormally large. There is no simple connection between the compressibility of tartaric acid and of ammonium tartrate, the roles of the *b* and the *c* axes being reversed.

Diphenylamine. The origin of this was chemically pure material from Kahlbaum. It was made into large single crystals from the melt in a thermostat. This ordinarily crystallizes in thin plates, but by holding at the melting temperature for several days the plates were induced to grow to about 3.8 mm thickness, sufficient to give the compressibility perpendicular to the face of the plate, the *c* direction, by piling together two pieces. This material is monoclinic but again the three axes are nearly at right angles, and the linear compressibility was determined only in the direction of the axes. The designation of the axes is that of Groth, the *a* and *b* axes being in the flat face, at right angles, and the *c* axis perpendicular to both.

Diphenylamine is soluble in kerosene, so that it was necessary to use the piezometer for organic substances, submerging the crystal beneath mercury. Measurements were made to 12000 kg at 30°; the melting point is at about 54°, so that the customary measurements at 75° could not be made. In the *c* direction the points at 11000 and 12000 lay off the curve, there being evidently some sort of interference with the free motion of the slide wire. With the exception of these two points, the average deviation of a single reading from a smooth curve, in terms of the maximum effect, was 1.1% for the *a* direction, 1.2% for the *b* direction, and 0.4% for the *c* direction. The corrections were not as large a fraction of the total effect as is sometimes the case, being 8% in the *a* direction, 8% in the *b* direction, and 10% in the *c* direction. The lengths of the *a* and the *b* specimens were 0.8 cm.

The relation between change of dimensions and pressure is distinctly not of the second degree, so that the results are best exhibited in the following table:

TABLE III.

COMPRESSIBILITY OF DIPHENYLAMINE AT 30°.

Pressure kg/cm²	$-\Delta l/l_0$ *a*	*b*	*c*	$-\Delta V/V_0$
3000	.0147	.0145	.0164	.0463
6000	.0259	.0256	.0266	.0802
9000	.0358	.0353	.0332	.1079
12000	.0442	.0435	.0383	.1313

The compressibility is very much more nearly equal in the three directions than is usually the case for crystals. This means that the change of volume can be calculated with very little error from the linear changes. The volume compressibility is high for a solid, and is almost exactly the same as for liquid glycerine, the least compressible organic liquid yet measured.

$(NH_3C_2H_5)_2SnCl_6$. This material I owe to the kindness of Dr. R. W. G. Wyckoff of the Rockefeller Institute for Medical Research. Particular interest attaches to it because it is one of the few organic substances which crystallize in the trigonal system, and because its X-ray structure has been completely worked out.[10] The substance crystallizes in the form of hexagonal plates; two plates were available of a thickness of 0.16 and 0.12 cm. Two determinations of compressibility are necessary, one parallel to the hexagonal face and one perpendicular to it. The parallel specimen had a length of 0.65 cm. The perpendicular measurements were made on pieces cut from the two plates, piled together. The measurements had to be made in the piezometer for organic substances, submerged under mercury, for the double reason that it is dissolved by kerosene and it is too soft mechanically to withstand the necessary compression in the lever apparatus. The specimens were prepared by shaving off the excess material with a razor blade. It was not possible in this way to make as perfect specimens as by grinding, and the results are much more irregular than usual, both because of the geometrical imperfection of the specimens and because of their small size.

No attempt was made to measure the compressibility at 75°. The accuracy was not great enough to give departures from linearity. With the parallel specimen the average departure of a single reading from a straight line was 5% of the maximum effect. With the perpendicular specimen, there was some sort of permanent change produced by the pressure, and the points with increasing and decreasing pressure lay on two different lines of approximately the same slope, the average departure of a single reading from one or the other of these lines being about 1.5%. The material is affected in some way by its contact with kerosene under pressure, its mechanical texture becoming more friable, and the appearance being almost as if it had been amalgamated, although retaining its original shape. The corrections were about 5% of the measured effect. The results follow:

30°, average to 12000 kg/cm²,

$$-\Delta l/l_0 = 3.95 \times 10^{-6}p, \text{ parallel to hexagonal face,}$$

$$-\Delta l/l_0 = 5.75 \times 10^{-6}p, \text{ perpendicular to hexagonal face,}$$

$$-\Delta V/V_0 = 12.9 \times 10^{-6}p.$$

The volume compressibility is thus sensibly greater than that of diphenylamine, and is the largest yet measured in a crystalline solid. The difference of compressibility in the two directions is not as great as might be expected.

Bakelite. The new piezometer for organic substances was first checked as to its proper functioning by measuring with it the compressibility of bakelite. Although this is not a chemically well defined substance, it is used to some extent in the construction of instruments, and the results found are therefore tabulated as being of some practical interest. The material investigated was the clear light yellow material, in appearance like amber. The relation between pressure and change of length is not of the second degree. The following are the results, the change of volume being calculated on the assumption of equal compressibility in all directions. Meas-

TABLE IV.

COMPRESSIBILITY OF BAKELITE AT 30°.

Pressure kg/cm²	$-\Delta l/l_0$	$-\Delta V/V_0$
4000	.0250	.073
8000	.0437	.125
12000	.0577	.163

urements were made only at 30°. The compressibility is seen to be high, getting up into the class of organic liquids.

Discussion and Summary.

The effect of pressure on the electrical resistance of four metallic elements is measured. The resistance decreases with rising pressure in all cases, as is normal, with the exception of an initial effect found in arsenic. The compressibility of four elements, seven inorganic single crystals, four organic crystals, and bakelite is measured. A new form of piezometer has been developed for those substances which dissolve in kerosene. Arsenic has a new polymorphic form at high pressures, the transition pressure at 30° being about 5500 kg/cm^2. The compressibility differences of single crystal arsenic in different directions are very large. The compressibility measurements make it probable that boron does not crystallize in the cubic system. The organic crystals in general do not show as large compressibility differences in different directions as inorganic crystals. The absolute compressibility of the organic solids is high, the volume decrease under 12000 kg/cm^2 being of the order of 15%.

I am much indebted to my assistant Mr. W. A. Zisman for setting up the apparatus and making the measurements. I am also indebted for financial assistance to the Rumford Fund of the American Academy of Arts and Sciences and to the Milton Fund of Harvard University.

The Jefferson Physical Laboratory,
Harvard University, Cambridge, Mass.

References.

[1] P. W. Bridgman, Proc. Amer. Acad. 52, 573, 1917; 56, 61, 1921; 58, 151, 166, 1923; 59, 109, 1923; 60, 385, 1925; 63, 207, 347, 1928. Amer. Jour. Sci. 10, 483, 1925; 15, 287, 1928.

[2] P. W. Bridgman, Proc. Amer. Acad. 52, 585, 1917.

[3] T. W. Richards, Publication No. 76 of the Carnegie Institution of Washington, 1907.

[4] P. Kapitza, Proc. Roy. Soc. 123, 292, 1929.

[5] P. W. Bridgman, Proc. Amer. Acad. 56, 113, 1920.

[6] A. E. van Arkel, Chem. Weekbld, 24, No. 8, 1927.

[7] T. W. Richards, Jour. Amer. Chem. Soc. 37, 1643, 1915.

[8] E. Madelung und R. Fuchs, Ann. Phys. 65, 289, 1921.

[9] W. Mandell, Proc. Roy. Soc. 116, 623, 1927.

[10] R. W. G. Wyckoff, ZS. f. Krist. 68, 231, 1928.

THE MINIMUM OF RESISTANCE AT HIGH PRESSURE.

By P. W. Bridgman.

Presented Nov. 13, 1929. Received Nov. 30, 1929.

TABLE OF CONTENTS.

Introduction.

The electrical resistance of most metals decreases with rising pressure, but there are about half a dozen whose resistance increases.[1] Two examples are known, *Cs* and *Ba*, in which the two types of behavior are combined in a single substance, that is, the resistance at first drops, passes through a minimum, and then rises again at higher pressures. In general, if the resistance is plotted against pressure, it will be found that whether resistance increases or decreases with pressure the curvature is always the same, namely convex toward the pressure axis. The possibility of such behavior as that of *Cs* and *Ba* with their minimum resistance is evidently consistent with the common direction of curvature of all the curves. It was therefore natural to ask whether the minimum of resistance would not be found to be characteristic of all metals if pressure could only be pushed high enough. It is natural to look for this effect first in rubidium, the next most compressible metal after caesium; in fact, an extrapolation of the experimental values beyond the pressure range of the previous experiments, which was 12,000 kg., indicated that such a minimum was not impossible within a realizable pressure range. This paper contains an account of the actual attaining of the suspected minimum of rubidium with a new apparatus capable of reaching about 20,000 kg., of very definite indications of such a minimum for potassium at a pressure of 23,500 kg., 3500 kg. beyond that actually reached, and of new measurements on sodium to 18,000 kg., which indicate by extrapolation a minimum near 28,000 kg. In addition, the previous measurements on bismuth are extended to 50% beyond the previous maximum with no change in the trends previously found.

Two new pieces of apparatus were made for the higher pressure range. They were so much like the conventional apparatus of all my high pressure work that no extended description is necessary. Ordinarily there are two pressure cylinders, an upper one in which pressure is produced by the motion of a plunger, and a lower cylinder, connected to the upper one with a heavy pipe, in which is placed the particular substance to be investigated. The connecting pipe and the consequent large volumes are a disadvantage of this type of apparatus when attempting very high pressures. The new apparatus was therefore made in a single piece, being essentially the conventional upper cylinder, made longer than usual so that there was room in the bottom part for the resistance specimen. It is practically the same sort of thing that I used many years ago in measuring the transitions of ice up to 21,000 kg.[2]

The cylinders were of chrome vanadium steel, heat treated and subjected to several seasoning applications of 25,000 to 30,000 kg. in the usual way. The first cylinder was used for all the measurements except those on sodium. It is probable that a flaw was developed in this cylinder by the seasoning process, because a number of times there were leaks for which no adequate explanation appeared. This sort of behavior is the usual preliminary to the ultimate development of a well defined crack. Such a crack did not appear during this work, however, and at the expense of considerable trouble and a number of repetitions, enough runs were obtained without leak to give the information desired.

The second cylinder, that used with sodium, was given a preliminary seasoning like the first, and then, because of press of other work, was laid aside for about six months before making the actual measurements. The maximum pressure reached during the measurements with this was 18,000 kg.; here the steel stretched so much that higher pressures could not be reached, and this in spite of the fact that the initial seasoning had been made to nearly 30,000 kg. That is, the internal stress distribution produced by the original heavy overstrain gradually decays with time, and the elastic limit sinks toward the original value. This is in line with my previous experience, but I was not prepared for so large an effect. It is evident that in order to reach the highest pressures the measurements must be made as soon as possible after the preliminary treatment, and only a short life can be expected for the apparatus. The data obtained for sodium with this cylinder up to 18,000 indicated that the expected minimum would be beyond the range possible with this grade of steel, so that it did not seem worth while to construct a new cylinder for the comparatively small increase of range that would have been possible.

In order to maintain temperature constant, the lower end of the hydraulic press as well as the cylinder itself had to be placed in the temperature bath.

A special three terminal insulating plug was made for this experiment, the depth of the mica washers being 50% greater than normal. The additional friction of the washers against the steel walls due to their additional depth provides an additional element of strength. The manganin gauge with which pressure was measured was wound from the same spool of wire as the standard gauge, and was calibrated by direct comparison with the standard. The resistance was only 25% of the standard, and for that reason the sensitiveness of the pressure readings was not so great as usual, but nevertheless the sensitivity was great enough to correspond to the accuracy of the calibration of the standard. It is to be remembered that the original calibration of the manganin by means of an absolute gauge covered a pressure range of only 13,000 kg.,[3] so that measurements beyond 13,000, which must be made by linear extrapolation, are not so certain as the measurements below 13,000. In view of the smallness of the coefficient of the manganin the uncertainty from this extrapolation is probably low. In any event, a small error in the gauge beyond 13,000 can only affect the pressure of the minimum and not the existence of the minimum itself, which was the point of chief interest in this investigation.

There follows now the detailed presentation of data.

Detailed Data.

Rubidium.—The same metal which had been used in previous measurements and which had been purified by slow fractional distillation,[4] was used again. It had been kept sealed in glass under nujol since the previous measurements. It was extruded to a diameter of about 1.6 mm. and bent into a hairpin 5 or 6 cm. on a side. The potential and current terminals were, as before, fine wires of silver pierced through the rubidium wire and pinched into contact with it. The hairpin was mounted in a glass tube filled with nujol, the terminals were carried over the mouth of the tube and bent back over the outside, and the whole thing attached to a holder carried by the three terminal plug and thrust as a single assembly into the high pressure cylinder. Immediately before assembling in the cylinder the nujol was flushed out and completely replaced with petroleum ether; this was necessary because the viscosity and freezing point of the transmitting liquid must both be low in order to transmit high pressure.

In addition to a number of unsuccessful runs, three runs were made which yielded satisfactory measurements. The first, at 30°, was to a maximum pressure of 17,800 kg., it being impossible to get higher because of leak. This run disclosed a remarkable flattening of the curve of resistance against pressure, so that it seemed highly probable that a maximum was near. The second run reached a maximum of 19,350 kg., where there was a slow leak. This run disclosed a minimum with both increasing and decreasing pressure, the best value for the location of the minimum being 17,900 kg. The discrepancy between readings with increasing and decreasing pressure on this run was so large, however, amounting to about 2.3% of the resistance at the minimum, that improvement seemed desirable. The discrepancy between ascending and descending readings is doubtless due in most part to the chemical action between the transmitting medium and the rubidium; a similar effect has been found in all my previous measurements of the resistance of the alkali metals when the metal has been used in the form of a bare wire. Another explanation of the discrepancy which might seem plausible is hysteresis due to internal strains, but the direction of the hysteresis is in the wrong direction to admit this explanation. In the effort to diminish the discrepancy as far as possible, the petroleum ether with which pressure was transmitted in the third set-up was made as chemically inert as possible by standing in contact with metallic sodium for a number of days before the run. The maximum pressure reached on the third run was 19,800 kg., where there was a slow leak. Again a minimum was found with increasing and decreasing pressure. The discrepancy between the ascending and descending points was now of the order of 1% of the resistance at the minimum; in general the points were much more regular than those of the second run, and were satisfactory enough.

The experimental points of the third run are shown in Figure 1. The points obtained with increasing pressure are doubtless to be preferred to the descending points, since the time during which chemical action may occur is less for these points. The relative resistances between 12,000 and 20,000 kg. at convenient pressure intervals, obtained from the increasing readings, are shown in Table I. A check on these readings is afforded by the readings previously obtained up to 12,000. These previous readings were made at 0° and 35°; by linear interpolation the relative resistance at 30° and

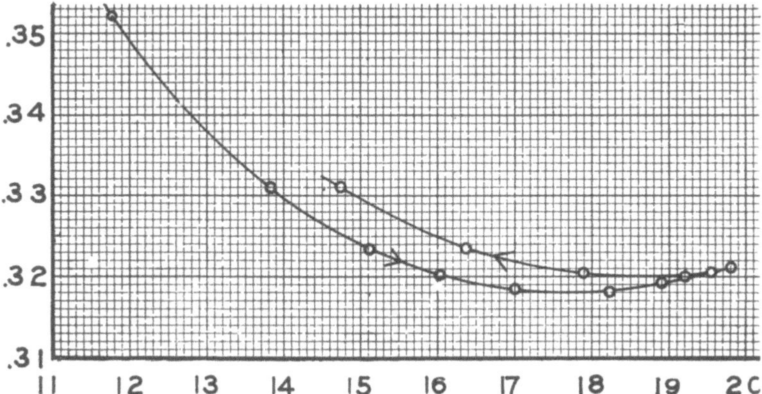

FIGURE 1. The measured relative resistance at 30° of rubidium against pressure in thousands of kg./cm.^{2} as abscissas.

12,000 kg. is found to be 0.340, against 0.349 found now, a difference of 2.6%. The previous value is probably to be preferred, the difference being in the direction to indicate greater chemical action in the present experiment.

TABLE I.

RELATIVE RESISTANCE OF RUBIDIUM AT 30°, FROM THE SMOOTH CURVE THROUGH THE POINTS OBTAINED WTIH INCREASING PRESSURE OF THE PRESENT EXPERIMENT.

Pressure kg./cm.^{2}	R/R_0
12,000	.3490
14,000	.3295
16,000	.3202
17,000	.3184
18,000	.3180
19,000	.3194
20,000	.3220

In Table II is shown the relative resistance at 30° at 1,000 kg. intervals up to 20,000 kg., to three significant figures. Up to 12,000 this table was constructed by linear interpolation between the previous values for 0° and 35°. Above 12,000, the values were obtained by applying a correction of 2.6% to the values of Table I, making a slight readjustment in the neighborhood of 12,000 to secure smooth first differences.

TABLE II.

Most Probable Relative Resistances of Rubidium at 30°, Obtained by Combining Previous and Present Experiments.

Pressure kg./cm.²	R/R_0
0	1.000
1,000	.819
2,000	.702
3,000	.619
4,000	.555
5,000	.504
6,000	.463
7,000	.430
8,000	.403
9,000	.381
10,000	.364
11,000	.350
12,000	.338
13,000	.328
14,000	.321
15,000	.316
16,000	.312
17,000	.310
18,000	.310
19,000	.311
20,000	.314

The most important result is that at 30° the resistance of rubidium passes through a minimum at 17,000 kg., where the resistance is 0.310 of its value at atmospheric pressure at the same temperature. This may be compared with the corresponding result for caesium, which at 0° has a minimum of 4,200 kg. of about 0.71 the initial resistance.

Potassium.—Although in view of the very large increase of the pressure of the minimum on passing from caesium to rubidium it appeared highly improbable that the pressure of the minimum of potassium, granting that it exists, could actually be reached with this apparatus, nevertheless the additional information to be obtained by extending the pressure range by 60% might be expected to give valuable information as to the probable existence and location of such a minimum.

The experimental set-up was practically the same as for rubidium.

The material was from the same original batch as that on which my previous determinations of the effect of pressure on melting point, resistance, and compressibility were made;[5] it has been kept in the meantime sealed in glass under nujol. The readings, which are shown in Figure 2, were very regular, there being very little difference

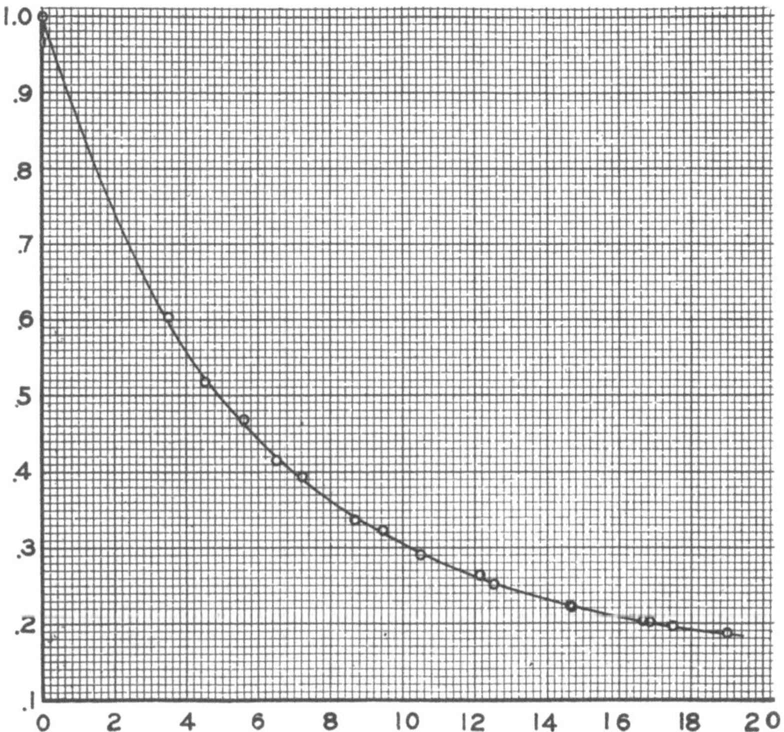

FIGURE 2. The measured relative resistance at 30° of potassium against pressure in thousands of kg./cm.[2] as abscissas.

between the ascending and descending branches. It is to be expected that any discrepancy between ascending and descending points would be much less in this case than for rubidium, both because potassium is less active chemically than rubidium, and because it is not so soft mechanically.

A fairly good check on the new results can be obtained by comparing with the previous measurements on potassium to 12,000 at 0°.

The present value for the relative resistance at 12,000 is 0.261, compared with 0.271 of the previous measurements. The difference, which is small, is in the right direction, and of a not unreasonable magnitude to be accounted for by the temperature difference. The previous measurements were made on potassium at only one temperature, so that a temperature interpolation could not be made such as was possible for rubidium.

The relative resistances in terms of the resistance at atmospheric pressure at 2000 kg. intervals up to 20,000 at 30° are shown in Table III.

TABLE III.

Pressure kg./cm.^{2}	Relative Resistances of Potassium at 30°
0	1.000
2,000	.734
4,000	.557
6,000	.443
8,000	.362
10,000	.305
12,000	.262
14,000	.231
16,000	.208
18,000	.192
20,000	.181

The minimum is seen to lie beyond 20,000. Direct extrapolation of the curve of resistance against pressure in order to find the minimum would not be accurate; this may be better done by plotting the first differences against pressure. In Fig. 3 the differences of relative resistance for 2000 kg. intervals are plotted against the mean pressure of the intervals. This curve is so nearly linear at the high pressure end that it can be extrapolated with little uncertainty. The curve is seen to cross the axis at about 23,500 kg., which may therefore be taken to be the pressure of the minimum. The minimum relative resistance calculated from the extrapolated first differences is 0.175. The pressure of the minimum is thus not so high as might have been anticipated from the relatively large increase in the pressure of the minimum on passing from caesium to rubidium.

Sodium.

The specimen was formed and treated in the same way as *Rb* and *K*. The material was from the same batch as that whose pressure

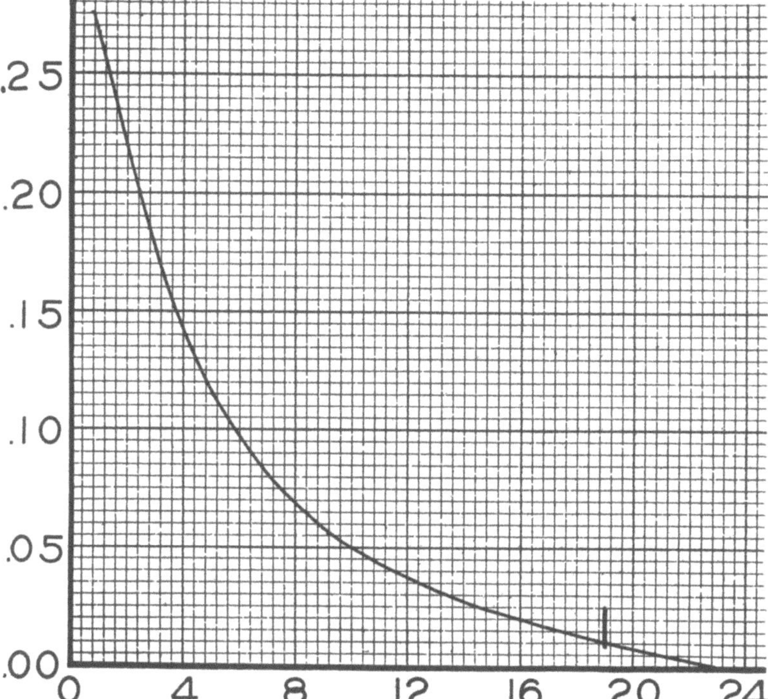

FIGURE 3. The first differences of relative resistance for 2000 kg. intervals of potassium plotted against pressure in thousands of kg.cm.[2] as abscissas.

coefficient of resistance and melting curve under pressure were previously measured,[6] and which gave evidence of high purity by the sharpness of the freezing point. It has been kept since that time in a glass stoppered jar under kerosene. There has been some surface action with the kerosene, but the piece used was cut from the interior of a large block, and should be as pure as the original.

Measurements were made at 30° in the second cylinder; as already explained, the cylinder stretched so much beyond 18,000 kg. that readings could not be obtained at higher pressures. The measured relative resistances are plotted in Fig. 4. There is considerable difference between the ascending and descending readings at the lower pressures, which becomes much smaller beyond 10,000 kg. It is uncertain how much of this is due to chemical action and how much to mechanical distortion. If the mean of the two zero readings is taken as most probably correct, it will be found that the relative resistance at 12,000 kg. is .4210 against .4245 previously found. The difference is only 0.82%. It is probable that the previous values

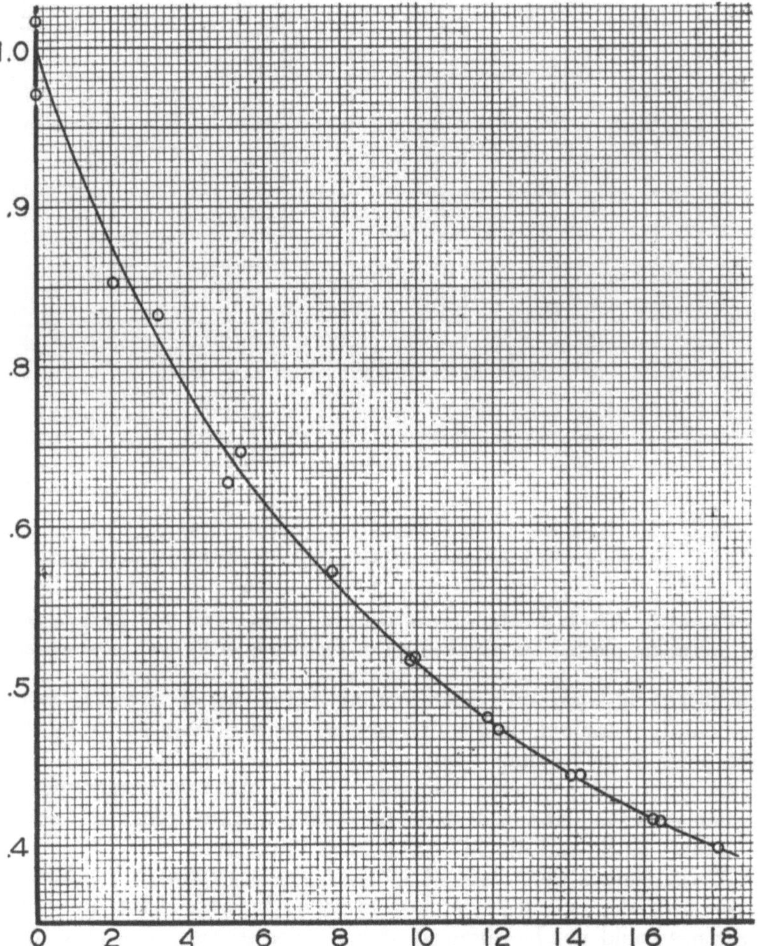

FIGURE 4. The measured relative resistance at 30° of sodium against pressure in thousands of kg./cm.[2] as abscissas.

are to be preferred, since more attention was then paid to the low pressure behavior. Agreement between the two sets of data at

12,000 can be obtained by assuming that the proper zero of these measurements is not the mean of the two observed zeros, but about 0.3% higher. The smooth curve in Fig. 4 has been drawn in accordance with this adjustment.

The relative resistances at 1,000 kg. are shown in Table IV. The four significant figures given are justified by the data, and it is in fact necessary to retain this number of figures to obtain smooth first differences. Compared with the previous results at 12,000 the maximum difference is 3 in the third figure at 4,000 and 5,000, and the average difference 1.

TABLE IV.

Pressure kg./cm.²	Relative Resistances of Sodium at 30°
0	1.0000
1,000	.9338
2,000	.8770
3,000	.8278
4,000	.7850
5,000	.7480
6,000	.7156
7,000	.6866
8,000	.6601
9,000	.6361
10,000	.6141
11,000	.5941
12,000	.5757
13,000	.5590
14,000	.5442
15,000	.5308
16,000	.5185
17,000	.5073
18,000	.4974

In order to estimate whether there is a minimum resistance, the differences for 1,000 kg. intervals are plotted against pressure in Fig. 5. There can be little doubt that the curve will cross the axis, so that it is highly probable that *Na* has a minimum as well as *Cs*, *Rb* and *K*. The dotted extension of the curve in Fig. 5 fits smoothly with the experimentally determined part, shown as a full line, and indicates a minimum near 28,000 kg., with a minimum relative resistance of 0.442.

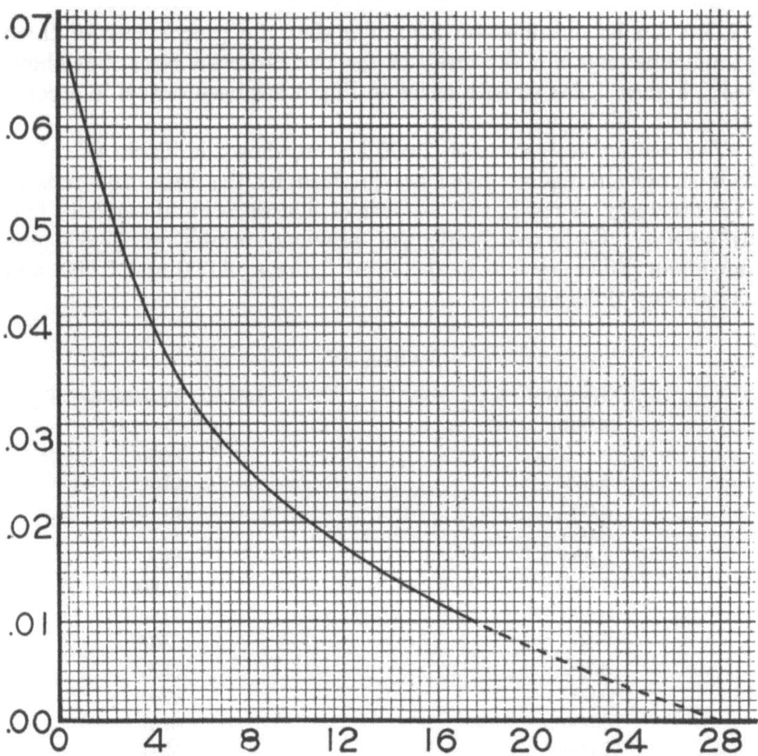

FIGURE 5. The first differences of relative resistance for 1000 kg. intervals of sodium plotted against pressure in thousands of kg./cm.[2] as abscissas.

Bismuth.

My particular reason for trying bismuth to higher pressures than formerly was that because of the abnormal increase of volume of bismuth on freezing it is to be expected that at sufficiently high pressures a new high pressure modification may exist, analogous to the high pressure forms of ice. Any such polymorphic transition should disclose itself by a discontinuity in the resistance.

The material was a single crystal rod of bismuth, with cleavage plane parallel to the length, from the same set of rods whose thermo-electric properties and resistance under pressure have been previously measured.[7] With this rod a single run was made to a maximum pressure of 18,400 kg. at 30°. The points lay perfectly smoothly, with no perceptible difference between the ascending and descending points, and no indication whatever of any sort of discontinuity. The transition, if it exists, must therefore occur at pressure beyond 18,400, or else the transition was suppressed by internal viscosity.

Previously measurements have been made to 12,000 on a crystal with cleavage at 4° to the length, and the results represented by a second degree expression in the pressure. We have to ask in the first place whether the new results agree with the former ones over the common pressure range, and secondly whether the second degree expression may be used to extrapolate beyond 12,000. The formula gives for the relative resistance at 12,000 1.168, against 1.162 found now, and at 18,000 1.284, against 1.273 found now. The differences are thus small, and are in the direction to be accounted for by the slight difference in orientation. The present values may therefore be considered checked, and the use of the previous formulas by extrapolation up to 18,000, and presumably materially higher, is justified.

It is to be remarked that we have here also presumptive evidence of the legitimateness of extrapolating the manganin gauge calibration beyond 13,000. One method of checking such an extrapolation would be to find whether the same pressure is given by extrapolating with the formulas for two different metals; this is essentially what has been done here.

Discussion.

The chief conclusion to be drawn from the results of this paper is that it is almost certain that ultimately, at sufficiently high pressures, the resistance of all the alkali metals increases under pressure. This increase has actually been realized for *Cs* and *Rb*, and *Li* already at atmospheric pressure increases in resistance; the extrapolation for *K* and *Na* is not a large one, and it seems practically certain that they also must ultimately increase in resistance.

It is natural to look for regularities in the minimum data on passing from metal to metal in the alkali series. It is evident in the first place that *Li* is entirely out of the sequence. This perhaps is not surprising in view of the very simple electronic structure of the atom of *Li*, compared with that of the others. Consider now the remaining four metals *Na*, *K*, *Rb*, and *Cs*; the essential data for them are collected in Table V. If the pressure of the minimum is plotted against atomic number, a fairly smooth curve will be obtained, the pressure of the minimum dropping at an accelerated rate in the direction of

Cs. If the relative resistance at the minimum is plotted against atomic number a parabolic curve will be found having a deep minimum at *K*. One is reminded of another respect in which *K* falls out of the sequence, namely in the very loose electronic structure of the atom and the surprisingly persistent compressibility at high pressures emphasized in a previous paper;[8] there does not, however, appear to be any very close correlation between these other effects and the behavior of resistance found here.

TABLE V.

Metal	Atomic Number	Pressure at minimum	$\frac{R(\text{min.})}{R(\text{atoms.})}$
Na	11	28,000*	.442*
K	19	23,500*	.175*
Rb	37	17,800	.310
Cs	55	4,200	.71

* Extrapolated.

The considerations of the introductory paragraph make it natural to suppose that the phenomena shown by the alkalies may be shown by all other metals at high enough pressures; we inquire at what pressures we might possibly find this effect. The alkali earth series is the most natural place to look next. Here barium does have the effect,[9] the pressure of the minimum being about 8600 kg. at 30°, and strontium and calcium[10] have a positive pressure coefficient already at atmospheric pressure like lithium. Beryllium,[11] however, the analog of lithium, has a normal negative coefficient, and magnesium[12] the analog of sodium, also has a negative coefficient. The resistance of *Be* and *Mg*, when plotted against pressure, shows curvature in a direction consistent with a minimum at higher pressures, as do all other metals, and the pressure of the minimum indicated by extrapolation of the second degree formula valid up to 12000 kg. is 46,000 kg. for *Be* and 49,000 for *Mg*. But a second degree formula almost certainly gives by extrapolation too low a pressure for the minimum, as shown by Figures 3 and 5, which would be straight lines if a second degree formula were valid. It is therefore evident that the possible minimum of *Be* or *Mg* is at a pressure so high that it is hopeless to try to reach it with present means, or even to get close enough to allow a secure extrapolation. Searching the rest of the measured elements, one would expect to find the effect most easily with the most compressible metals. Of these, thallium[13] gives by second degree extrapolation a pressure of 46,000 kg. and indium[14] 37,000, which again is certainly too low in view of the detectible departures from the second degree relation below 12,000. It would appear therefore that the only chance of finding the effect in the pressure range at present attainable is with some metal not yet investigated. The largest gap in the investigated metals is in the rare earths: here only a few have been measured, but they are not encouraging. Perhaps the most promising candidates are scandium and yttrium, but these have not yet been produced in sufficiently large quantities in the metalic condition.

The fact that bismuth and antimony have positive pressure coefficients of resistance in the ordinary range requires brief comment. In spite of the fact that the curvature of the plot of resistance against pressure is the same as that of the alkali metals and earths, it is not unlikely that the mechanism of the positive pressure coefficient is different in the two cases. In *Bi* and *Sb* the mechanism is probably special, connected in some way with the crystal structure, as shown by the fact that the coefficient of liquid *Bi* is negative.[15] The positive coefficient of the alkali metals and earths is probably more intimately connected with the structure of the atom, as shown by the fact that the coefficient of liquid as well as solid *Li*[16] is positive. It was shown in a previous paper[17] that it is highly probable that the coefficient of liquid *Cs* will also be positive at pressures high enough, and by analogy, one would expect similar behavior for the other alkali metals.

In speculations as to possible behavior at very high pressures, a significant feature of the behavior of *Cs* should be kept in mind. If the "instantaneous" pressure coefficient $\frac{1}{R}\left(\frac{\partial R}{\partial p}\right)_t$ of *Cs* is plotted against pressure, a curve will be obtained which crosses the axis at 4200 kg., the pressure of the minimum, and which in the low pressure range is rather strongly concave upward, so that a wide extrapolation might be taken to indicate a second crossing of the axis, with a maximum, which would mean that at pressures still higher the pressure coefficient of resistance would again become negative. But this extrapolation is not valid because of a reversal of the direction of curvature of $\frac{1}{R}\left(\frac{\partial R}{\partial p}\right)_t$, that is, a reversal of the third pressure derivative, above 8,000, so that above 8,000 kg. the pressure coefficient becomes increasingly positive, and there is no indicated possibility of a return to normal behavior at excessively high pressures. This has application to the other alkali metals; their third derivatives are such as not to make impossible a second reversal at exceedingly high pressures, but judging by the analogy of *Cs*, this initial possibility will be eliminated at high pressures, and there is no reason to think that the resistance of all these metals will not continue to increase at an ever accelerated pace at pressures far beyond the present experimental range.

The theoretical significance of the reversal of the pressure coefficient at high pressures is at present obscure; there does not seem to be any suggestion of such a phenomenon in the present wave-mechanics pictures of the mechanism of resistance. This must nevertheless be important not only for theories of resistance, but for theories of the structure of the atom, since we have here definite evidence of a reversal in ordinary atomic properties produced by forces which are not high in comparison with the forces between different electrons in the same atom. It may well be that there is here a possibility of obtaining by direct experiment some inkling of atomic behavior under certain stellar conditions.

I am much indebted to Mr. W. A. Zisman for making the measurements of this paper, and to the Milton Fund of Harvard University for financial assistance.

THE JEFFERSON PHYSICAL LABORATORY,
HARVARD UNIVERSITY, CAMBRIDGE, MASS.

REFERENCES.

[1] P. W. Bridgman, Proc. Amer. Acad. 52, 621, 624, 1917; 56, 67, 91, 96, 1921; 60, 404, 1925; 62, 211, 215, 1927.

[2] P. W. Bridgman, Proc. Amer. Acad. 47, 513, 1912.

[3] P. W. Bridgman, Proc. Amer. Acad. 47, 321, 1912.

[4] Third reference under 1, p. 394.

[5] Third reference under 1, p. 390.

[6] Second reference under 1, p. 76, P. W. Bridgman, Phys. Rev. 3, 157, 1914.

[7] P. W. Bridgman, Proc. Amer. Acad. 63, 383, 1929.

[8] P. W. Bridgman, Phys. Rev. 27, 68, 1926.

[9] Fourth reference under 1, p. 215.

[10] Second reference under 1, pp. 91 and 96.

[11] Fourth reference under 1, p. 213.

[12] Second reference under 1, p. 88.

[13] First reference under 1, p. 588.

[14] P. W. Bridgman, Proc. Amer. Acad. 64, 53, 1929.

[15] Second reference under 1, p. 114.

[16] Second reference under 1, p. 67.

[17] Third reference under 1, p. 417.

THE VOLUME OF EIGHTEEN LIQUIDS AS A FUNCTION OF PRESSURE AND TEMPERATURE.

By P. W. Bridgman.

Received Oct. 17, 1930. Presented Oct. 8, 1930.

TABLE OF CONTENTS.

Introduction.

One of the first extended series of measurements of the effects of pressure that I attempted was on the volume of twelve liquids between 20° and 80° C. and up to 12000 kg/cm².[1] Since then I have made many determinations of the compressibility of solids, but have not returned to the question of liquids, in spite of the interest of the subject and the comparatively small number of liquids investigated. My reason for waiting so long before extending the early investigation has been in large part my desire to improve the method there used. That method was extremely simple: it consisted merely in measuring the motion of the piston by which pressure was produced. There were, however, several unsatisfactory features: the correction for the compressibility of the transmitting medium might be as large or larger than the effect being measured; the correction for the distortion of the cylinder, although small, had to be obtained by a calculatoin which is unsatisfactory because the metal is stressed beyond its elastic limit; there was considerable hysteretic difference between the readings with increasing and decreasing pressure with consequent possibility of error in the mean; and the thermal expansions under pressure were likely to be in error because the pressure measuring gauge was subject to the same changes of temperature as the liquid itself. This last point I felt to be of particular importance; for several speculations an exact knowledge of the limiting behavior of the thermal expansion at high pressures is important, so that it was particularly desirable to devise a method capable of giving this with greater accuracy. After a number of trials, extending over several years, and including work with the free piston piezometer,[2] which at first was very promising, but which had to be discarded because it could not give accurate enough values for the thermal expansion, such a method has been devised, so that the way is now open for the routine measurement of a large number of liquids.

In the following are presented the first of the data obtained in this projected campaign on the determination of the volume of a large number of liquids as a function of pressure and temperature. The twelve liquids chosen for the previous investigation were those which had been previously investigated by Amagat; the reason for this choice was to allow comparison with his results in the pressure range up to 3000 kg, and, particularly, to supplement the results at low pressures by Amagat's measurements, the previous method not being well adapted to give accurate values at low pressures. The liquids of Amagat were, however, in many cases of complex structure, and are not those which would naturally be chosen for their significance in theoretical speculations. In the following an effort has been made to select liquids more "normal" in structure, and therefore of greater theoretical interest. In the first place, a number of hydro-carbons are investigated: normal- and iso- pentane, n- hexane and its four isomers, and n- heptane, n- octane, and n- decane. Five liquids are measured which the theoretical investigations of Hildebrand have suggested as of interest, namely, C_6H_6, C_6H_5Cl, C_6H_5Br, CCl_4 and $CHBr_3$. The previous work has been supplemented by measurements of iso-propyl and n-butyl alcohols, the previous measurements having included n-propyl and iso-butyl alcohol. Finally, measurements, merely for check purposes, were made on ether and water and a mixture of glycerine and water.

The liquids chosen for this investigation often suffer from the disadvantage that the range open to measurement, particularly at the lower temperature, is small because of freezing under pressure. However, this limitation is unavoidable; many of the liquids of greatest theoretical interest are easily frozen by pressure.

The previous measurements were made at 20°, 40°, 60°, and 80° C. The temperature behavior proved to be very complicated, there being numerous small scale irregularities superposed on the large scale regularities. These small scale irregularities are, however, so complicated that they are not of much present significance for any theory, and in fact they have been entirely ignored in the theoretical speculations of various writers which have been based on these data. I felt therefore that in this new investigation the emphasis should be placed less on these small scale effects, and have accordingly increased the temperature range from 60° to 95°, but have diminished the number of temperatures from four to three, the new measurements being made at 0°, 50°, and 95°. In this way fewer of the small scale phenomena are disclosed, but greater confidence may be felt in the accuracy of the large scale effects, because of the increase of range.

It has also been possible to cover a greater range of material in the same time by decreasing the number of temperatures. A still wider range of temperature would obviously have been desirable, but this would have demanded essential changes and complications in the whole technique, with a corresponding great decrease in the number of substances that could be examined.

The Method.

The fundamental idea of the method is very simple; the liquid to be measured is sealed into a metal "sylphon" (trade name for a flexible metal bellows), which is then exposed to external hydrostatic pressure. The sylphon shortens until the internal pressure is equal to the external, and from a measurement of the shortening, the change of volume of the liquid inside may be obtained. The advantages of this simple scheme are obvious; the separation of the liquid to be measured from the transmitting liquid is absolute, allowing no contamination; the effect measured is nearly the entire effect, for obviously the corrections are small, and furthermore may be computed with precision, for all the parts involved are subject to hydrostatic pressure and there is no distortion by exceeding the elastic limit; and finally, the pressure gauge may be placed in a vessel separate from that containing the sylphon, so that the temperature of the gauge may be kept independent of the varying temperature of the sylphon, thus eliminating this source of possible error in the previous method of measuring thermal expansion at high pressure.

The method is not entirely smooth sailing, however, but there are several matters, either of construction or function, that require examination. In the first place, there is the question of whether enough distortion can be obtained with the sylphon to permit convenient measurement of the rather large changes of volume involved, amounting to at least 30%. The ordinary sylphon of commerce does not have the requisite flexibility, and furthermore, at the time that this investigation was begun, was not made in sizes small enough to permit its use except by constructing an entire new high pressure plant capable of handling larger volumes. The construction of the necessary tools and jigs for the making of a single small sylphon of the commercial pattern, that is, by forming seamless from a single piece of metal, would have cost $10000, and was therefore prohibitive. The only course open was to build up a sylphon of the desired size by soldering together small sections; this was successfully accomplished, and will be described in detail later. In the second place, there was the fundamental question of whether the interior volume of such a sylphon may not be a complicated function of its length and the hydrostatic pressure. This happily proved not to be the case, but the sylphon has the property that it acts as if it had a fixed effective cross section independent of its extension, so that the change of internal volume may be obtained at once, after the necessary calibration has been made, in terms of its change of length. The check measurements by which this was established will be described in detail later.

The way in which the sylphon is built up in sections is indicated in Figure 1. The individual sections are formed by stamping into a die in a single operation, as indicated in Figure 2. After stamping, the edges of the section are trimmed with a sharp knife. A good many trials were made before the proper dimensions and material were found for the stampings. If the material is too thick, the resulting sylphon is too stiff; if it is too thin, it cannot be formed without tearing. The material finally adopted was "shim" brass, 0.0015 inches (0.0038 cm.) thick. It was cut into true discs in the lathe for forming, and then was carefully annealed to soften it. The forming operation restores it to just the right amount of springiness. Brass 0.002 inches thick was much too thick, and 0.0012 inches was too thin. Materials unsuccessfully tried were: another grade of brass specially manufactured and recommended for deep drawing, as well as grades of phosphor bronze and nickel silver also made for the same purpose, soft iron, silver, platinum, and copper.

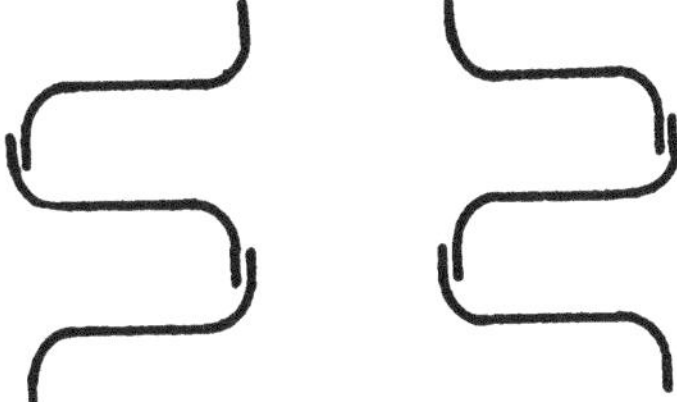

FIGURE 1. Shows the way in which the sylphon is built up in sections.

The sections are held together by first spinning over the edges and then by soldering. A specially built tool of readily suggested design for holding and rotating the sections as they are built up facilitates the assembly. The soldering must be above suspicion. It pays to brighten the edges of the seams with fine emery paper before applying the solder. Minute leaks proved the greatest difficulty in practise; these were fully as likely to be due to flaws in the metal as to imperfections in the soldering, and were often extremely difficult to detect at atmospheric pressure. The best method found for locating a leak at atmospheric pressure was to submerge the sylphon in alcohol while attached to a source of hydrogen at not more than half an atmosphere pressure. The hydrogen gets through fine holes much more easily than air, and alcohol has the advantage of water that it does not deposit on the brass fine bubbles of air from solution. The most satisfactory method of proving that the sylphon is tight is to fill it with a volatile liquid, such as ether or carbon bisulfide, sealing it when the sylphon is stretched so that the liquid will be under a slight excess pressure, and then finding whether the sylphon with its contents changes in weight over an interval of several days. No sylphon which passed this test ever proved defective under pressure.

Considerable skill is necessary in soldering the sections together so as not to put them in such a state of strain that they will snap from one position to another, like the bottom of an oil squirt, at some stage in the stretching. A sylphon which snaps in this way

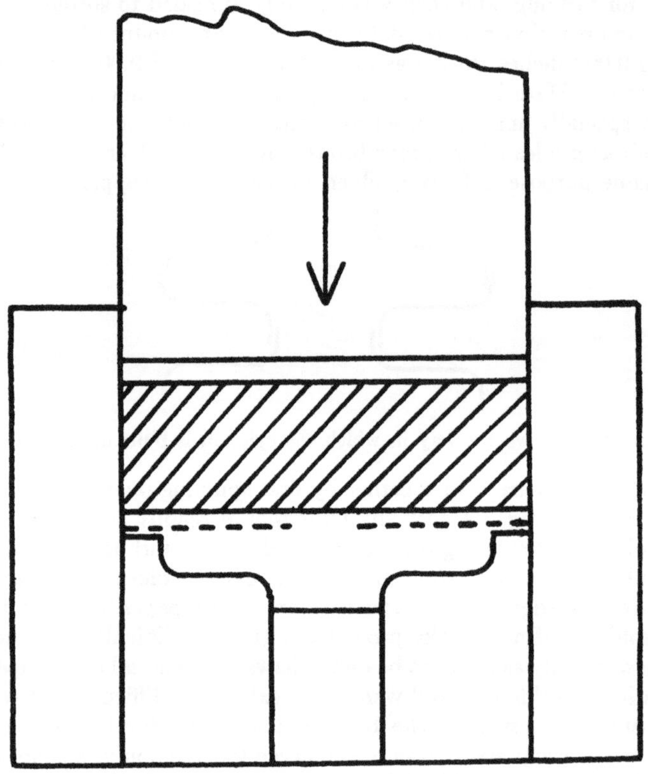

FIGURE 2. Shows the method of forming the sylphon sections by stamping. The thin disc to be formed is indicated by the dotted line. It is forced to the contour of the die by the lead disc, shaded, which is squeezed into the die by the action of the steel piston, indicated by the arrow.

may as well be discarded at once, because the calibration will show jumps in the volume as a function of length, and it is therefore useless for accurate measurements. After some practice my mechanic, Mr. Charles Chase, developed much skill in the production of these sylphons, and is now able to complete one in about half a day; this paper would have been hardly possible without his skillful assistance.

The sylphon as finally constructed consists of nine double sections, of an unextended length of 1 inch (2.5 cm.). The flexibility allows a motion of 0.187 inch (0.475 cm.) in both directions, giving 0.375 inches altogether, or about 32% on the measured volume. The essential parts of the mounting of the sylphon are shown in Figure 3. End pieces are soldered to the sylphon bellows, and the combination is mounted in a brass sleeve which carries the arrangement by which the change of length is measured. This is the conventional sliding contact arrangement, consisting of a manganin wire sliding over an insulated contact. Measurements of the potential difference between the points A and B give the length of the segment AB and so the length of the sylphon to which the wire is attached.

Some sort of a guide has to be provided to constrain the sylphon to shorten and lengthen in the same straight line. This guide is provided by a piston and cylinder integral with the end pieces of the sylphon, playing one within the other, as sufficiently indicated by the drawing. At first the guide was made external to the sylphon, as would naturally suggest itself, but free motion was often interferred with by particles of dirt getting between the bearing surfaces, and the guides were finally put inside with complete success. Complete freedom of motion is obviously a prime requisite, for if the difference of pressure between the inside and the outside of the sylphon reaches much more than half an atmosphere the metal is permanently deformed, and the sylphon must be discarded. In order to avoid seizing of the guides, which often occurs when metal parts in relative motion are made of the same metal, the upper end piece, bearing the piston part of the guide, was made of brass, and the lower part, bearing the cylinder, of copper (not vice-versa because brass is more compressible than copper). To further prevent sticking if any particles of dirt should accidentally be in the sylphon initially, a spiral groove, not shown in the diagram, was cut on the brass piston, in which such particles might lodge. A further advantage gained by placing the guides inside the sylphon is that the internal volume available to the liquid is thereby cut down, thus decreasing the stroke necessary, which otherwise would have been uncomfortably close to the maximum allowable.

The sylphon was filled through the German silver tube at the lower end. During filling the sylphon is first stretched to a definite length and held there with a pin through appropriate holes in the brass sleeve. It is then inverted, a thistle-shape glass containing

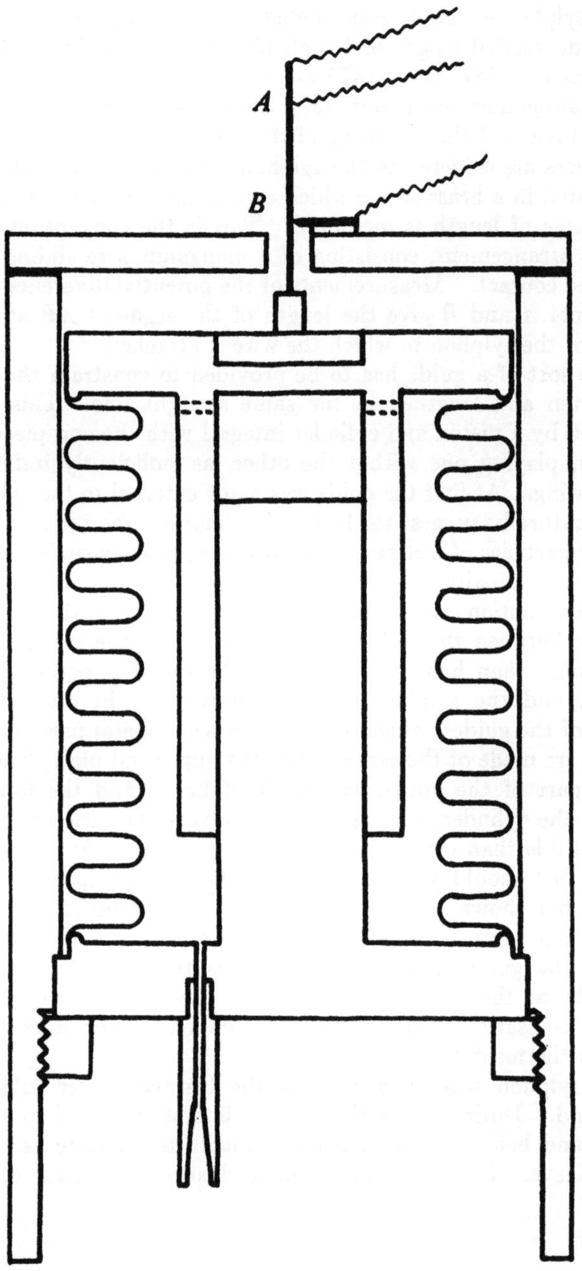

FIGURE 3. The sylphon assembled, with the internal guides and the electrical connections.

the liquid is attached to the German silver tube, and the whole is placed under the receiver of an air pump. The receiver is alternately exhausted and air readmitted until the sylphon is entirely full. The exhaustion also serves the purpose of removing dissolved air from the liquid. During the process of filling, the air must both be removed and readmitted with sufficient slowness to prevent the difference of pressure within and without the sylphon exceeding a safe value. After filling, the German silver tube is filled with a long tapering pin of German silver, and the amount of liquid in the sylphon determined by weighing. The total available volume of the sylphon was about 5 $cm.^3$, and weighings could be made consistent to a fraction of a milligram, so that no appreciable error attaches to the determination of the amount of liquid, After weighing, the sylphon was sealed by soldering into place the German silver pin, the outside of the tube being surrounded with water during the soldering to prevent transmission of heat to the liquid. German silver was used in order to make transmission of heat to the liquid through it as small as possible.

The sylphon thus filled was attached to an insulating three-terminal plug, the various necessary electrical connections were made by soldering, and the whole thing put as one assembly into the high pressure cylinder. This part of the process is so much like that which I have previously employed and described in detail[3] that it is not necessary to go into further detail here.

The first pressure run was almost always made at 0°, starting with the reading at atmospheric pressure, then at a few hundred kilograms, and then at increasing intervals to the maximum pressure which it was judged could be safely reached without freezing, or if there was no danger of freezing, to a maximum of 12000 kg. After reaching the maximum, readings were made with decreasing pressure back to about 500 kg, interspersing the readings with decreasing pressure between those with increasing pressure. The final result was a series of readings separated by intervals of the order of 300 kg at the lower end of the pressure range, and 1000 kg at the upper end. When the apparatus was functioning properly, there was never the slightest hysteresis between the readings with increasing and decreasing pressure, but both lay on the same smooth curve. This constitutes a great advantage over the piston displacement method. It was, however, sometimes necessary to season a new sylphon by a preliminary application of pressure; apparently the soldered seams as left by the soldering iron are capable of slight permanent deformation on the first application of pressure. The transmitting medium was pure petroleum ether; a liquid like kerosene becomes sufficiently viscous at high pressures to permanently deform the sylphon. Obviously excessive viscosity in the liquid within the sylphon may produce irregular results as well as viscosity in the external transmitting liquid. In fact there does seem a distinct tendency in the following for those liquids to give the most irregular results whose increase of viscosity under pressure is likely to be the greatest, that is, those liquids with the most complicated molecules.

In measuring those liquids whose freezing points under pressure were not known a number of sylphons were ruined by carrying the pressure up to the freezing point. The necessity for caution thus having been impressed, a small amount of freezing often being sufficient to do irreparable damage, the tendency in later work was perhaps to stay farther away from the freezing point than was necessary, and it would probably be possible to materially extend the range of data for a number of the following liquids without freezing. Freezing disclosed itself, not in the sudden decrease of volume that might perhaps at first be expected, but usually by a stationary reading, the guides freezing fast to each other, and the decrease of volume being taken up at the expense of permanent distortion of the sylphon. In all, eleven different sylphons were used in the following measurements.

After completion of the measurements at 0°, temperature was raised to 50° while pressure was maintained at about 500 kg. and a series of readings was made similar to those at 0°, with both increasing and decreasing pressure, except that the maximum pressure possible without freezing was now higher. After the 50° run, temperature was raised to 95°, and a similar series of readings made. In many cases the lowest pressure possible at 95° was somewhat higher than 500 kg., since the thermal expansion, if the pressure had been allowed to get lower, would have been sufficient to extend the sylphon beyond its limit. This accounts for the fact that in the following Tables many of the results given for 95° begin with the 1000 kg value instead of with that for 500 kg. After completion of a run at 95°, check measurements were always made again at 0° and 50° at some low pressure in the neighborhood of 500 kg, in order to be sure that the sylphon had experienced no permanent distortion. This test was practically always met within the errors common to all the readings.

In fact, the perfectly single valued character of the readings was a source of great satisfaction; it was possible to make runs at different temperatures on successive days without paying the slightest regard to the intermediate history of the apparatus. On several occasions a much severer test was successfully passed, in that the sylphon, sealed full with liquid, was removed from the holder, and the entire apparatus disassembled and then put together again, with perfect recovery of the original settings. This is all in striking contrast with the piston displacement method, in which a rigorous schedule of manipulation must be followed in order to reduce the hysteresis phenomena to reproducibility.

The completely single valued character of the readings made it possible to obtain accurate values for the thermal expansions from successive pressure runs at different temperatures, instead of being compelled to adopt the more direct procedure of changing temperature at constant pressure, which would have been much more inconvenient, because pressure is always easier to change than temperature. That the two procedures give identical results was established in a preliminary investigation.

In order to convert the measured changes of length of the sylphon into changes of volume, the effective cross section of the sylphon must be found by special calibration. To do this the sylphon was mounted in a special holder in which it could be shortened or lengthened by measured amounts with a micrometer screw, a graduated and calibrated glass capillary was attached to the German silver outlet tube, the sylphon was filled with kerosene, and the height of the kerosene in the capillary determined as a function of the length as given by the micrometer, making about 20 readings in all. The relation between length and change of volume turned out to be linear to better than 0.1 %, a result very gratifying and not expected. Furthermore, it was established by special trial that the volume is not perceptibly affected by pressure differences between the inside and outside of the sylphon up to as much as one-half atmosphere, which is close to the pressure of permanent deformation.

The manganin wire by which the length of the sylphon was determined was also specially calibrated. This again was done by moving it with a micrometer screw over a fixed contact in a specially constructed appliance, which need not be described in detail. Samples of wire were selected which were uniform to considerably better than 0.1%, so that no corrections had to be applied for lack of uniformity.

As indicated in Figure 3, the manganin wire was mounted detachably on the sylphon, so that the same wire could be used with several sylphons, considerably simplifying the manipulations and the calculations.

There are a number of small corrections which must be applied to the readings. The actual formulas for the corrections are somewhat complicated, and will not be written out in detail, since they are simple enough in general idea. The corrections are mostly for the compressibility of the various parts of the apparatus, and include the following.

(1) Correction for the compressibility of the brass sleeve which carries the contacts. Three different sleeves were made and used interchangeably at different times. These were all made from the same bar of brass. Its compressibility was determined by special experiment in the "lever" piezometer, and had the following values

$$\text{At } 30^\circ, \quad -\Delta V/V_0 = 9.207 \times 10^{-7}p - 6.42 \times 10^{-12}p^2.$$

$$\text{At } 75^\circ, \quad -\Delta V/V_0 = 9.309 \times 10^{-7}p - 5.79 \times 10^{-12}p^2.$$

(2) Correction for the compressibility of the brass part of the guide. This was made from the same bar as the sleeve, and therefore the correction has the value just given.

(3) Correction for the compressibility of the copper part of the guide. The compressibility of copper has been previously determined,[4] and the previous value was used.

(4) Correction for the compressibility of the manganin wire. This has also been previously determined, and the previous value was used.

(5) Correction for the compressibility of the sheet brass of which the sylphon is constructed. This was not independently measured, as it would have been difficult. I assumed that its compressibility is the same as that of the brass of the sleeve; the correction is in any event small, and commercial grades of brass do not differ by large amounts, and I believe the error made by the assumption to be inappreciable.

In addition to the various corrections for compressibility, there is a correction for the change of resistance of the manganin wire under pressure. This is the largest of the corrections, and it was determined by direct measurement on the identical piece of wire used at the three temperatures. The following results were found for the pressure coefficient of resistance: At 0°, 2.288×10^{-6}; at 50°, 2.311×10^{-6}; and at 95°, 2.289×10^{-6}.

There are also corrections for the thermal expansion of various parts of the apparatus on changing temperature. For these, values listed in various tables for brass, copper, and manganin were used, and are accurate enough. The temperature coefficient of resistance of the manganin wire at atmospheric pressure was also specially determined. The maximum variation in the constants of the apparatus due to changes of temperature in the range is about 1.5 parts in 1000. The maximum net correction for pure pressure effects on the runs at constant temperature was not much more than 2%, and the maximum net correction on the thermal expansion at constant pressure 5 or 6%. It is thus evident that as far as magnitude of corrections is concerned, the sylphon method is very greatly superior to the previous method of piston displacement.

The method of computation was first to plot the resistance readings as given by the potentiometer on a large scale on millimeter paper, 1000 kg on the scale of abscissas corresponding to 5 cm on the graph. Smooth curves were then drawn through the observed points, and the resistance read off at even intervals of 500 or 1000 kg. The changes of volume corresponding to these even intervals were then calculated with the aid of the various correction formulas. The results so obtained were the actual losses of volume in cm^3 reckoned from the volume at 0° and atmospheric pressure as fiducial. To convert this into fractional changes of volume in terms of the volume at 0° and atmospheric pressure as unity, the density at the fiducial point must be known. This could usually be taken from tables, and usually from International Critical Tables. The values thus used in the final reduction will be given in detail in the following for each individual liquid. The volumes at atmospheric pressure at 50° and 95° were also taken from the tables when possible; this was necessary because, as already explained, the sylphon readings often could not be made at atmospheric pressure at these two temperatures because the thermal expansion was too great, or else because the vapor pressure was above atmospheric.

There are several substances in the list, however, for which the density and thermal expansion at atmospheric pressure has apparently not been determined, notably the isomers of hexane. For these substances the density and thermal expansion at atmospheric pressure was determined by special experiment. For this purpose an alcohol thermometer was emptied, and used as densitometer and dilatometer. The bulb and stem were calibrated by filling with weighed amounts of mercury, and the stem was calibrated for uniformity with a moving length of mercury. Measurements of the volumes of the various liquids were made at three temperatures: 0°, a temperature about 10° below the boiling point, and a temperature half way between the maximum and 0°. Readings were made with both increasing and decreasing temperature as check; this check was always perfect to the limit of reading, which was 0.1° on the scale of the original alcohol thermometer. The results were put in the form of a second degree series in the temperature, and are given in the following. The density was determined by weighing the dilatometer when filled to a known mark at a definite temperature and when empty. The quantities of liquid used were of the order of 1 cm^3, and weighings were to 0.1 milligram, so that the density should be accurate to perhaps 1/30%. I felt that there was no point in trying for greater precision unless a much more exhaustive examination was made than was practical of the purity and reproducibility of the liquids.

If at any time in the future it should be found that better values for the densities or thermal expansions at atmospheric pressure are available, the relative volumes given in the following Tables can be readily corrected in accordance with the improved information. In the Tables the differences of volume from the fiducial volume at 0° and atmospheric pressure are the immediate result of the pressure measurements. If for instance, the future best value for the density at the fiducial point should turn out to be 1% less than the value assumed in the calculations, then the differences of volume reckoned from the fiducial point should all be decreased by 1%, increasing the actual volumes at any definite temperature and pressure. For example, a volume tabulated as 0.900 should be corrected to 0.901 in the hypothetical case.

The liquids measured in the following were obtained from various sources, selected for the greatest probable purity, and will be given in the detailed description. In general, the liquids were used as obtained, with no attempt at further purification; previous experience has shown that compressibility is insensitive to small amounts of impurity.

The detailed presentation of the data for the various liquids now follows.

Detailed Data.

Normal Pentane. The material was from Kahlbaum, their purest, provided in a sealed glass container. The runs at three temperatures and the return check readings were made without incident of any kind on three successive days. The average departure of a single one of the 46 observed points from the smooth curves was 0.11% of the maximum effect.

The final values for the volume are given in Table I. For the

TABLE I.

RELATIVE VOLUME OF N-PENTANE.

Pressure kg./cm.[2]	Temperature 0°	50°	95°
0	1.0000	1.0837	1.1869
1,000	.9021	.9395	.9768
2,000	.8546	.8820	.9078
3,000	.8229	.8454	.8671
4,000	.7997	.8193	.8371
5,000	.7811	.7985	.8125
6,000	.7647	.7807	.7933
7,000	.7506	.7657	.7775
8,000	.7381	.7520	.7641
9,000	.7281	.7409	.7527
10,000	.7192	.7316	.7433

density at atmospheric pressure the following formula was used from International Critical Tables (abbreviated in the following as I. C. T.)

$$\text{Density} = 0.64539 - 0.9398 \times 10^{-3}t - 0.6243 \times 10^{-6}t^2 - 7.53 \times 10^{-9}t^3.$$

The two temperatures, 50° and 95°, listed in the Table are above the boiling point, which is 36°.2. The volumes given in Table I at atmospheric pressure at 50° and 95° were obtained by formal substitution in the formula for density, and must be used merely to get the approximate location of the 50° and 95° volumes at the lower pressure end of the range.

Iso-pentane. This material was from the Eastman Kodak Co., provided in a sealed glass container. Runs were made without incident at 0° and 50°, but at 95° there was an explosion at 10000 kg, so that readings with decreasing pressure were not made at this temperature, nor were the return check readings made. The average departure from a smooth curve of a single one of the 37 readings was 0.17% of the maximum effect.

An experimental point was obtained at 7200 kg at 0° that lay off the curve by three or four times as much as any other point, and was taken as indicating the beginning of freezing. Because of possible damage to the sylphon, the matter was not investigated further. Further consideration makes it very improbable, however, that this irregularity was actually due to freezing, since the melting temperature at atmospheric pressure of iso-pentane is about 28° lower than that of normal pentane, and normal pentane has been carried to 10000 kg at 0° without a trace of freezing. Of course it is not impossible that the melting curves should cross, and there are known examples of it, but certainly 7200 kg should not be accepted as the melting pressure at 0° of iso-pentane without further confirmation.

The values for the relative volumes are given in Table II. The volume of i-pentane as a function of temperature at atmospheric pressure does not seem to be recorded in the literature, and it was therefore specially determined by the method already described. The following results were found:

$$\text{Density} = \frac{.6406}{1 + .001527t + 3.21 \times 10^{-6}t^2}.$$

Because of the low boiling point, the temperature interval was only from 0° to 23°.5, so that the temperature coefficients in this formula have no great accuracy. The relative volumes are given in Table II. The volumes at 50° and 95° at atmospheric pressure were ob-

TABLE II.

RELATIVE VOLUME OF I-PENTANE.

Pressure	Temperature		
kg./cm.^{2}	0°	50°	95°
0	1.0000	1.0843	1.1741
500	.9383	.9883	
1,000	.9028	.9415	.9809
1,500	.8775	.9084	.9405
2,000	.8571	.8845	.9119
3,000	.8264	.8490	.8712
4,000	.8025	.8222	.8403
5,000	.7830	.8014	.8178
6,000	.7670	.7850	.8002
7,000	.7534	.7729	.7863
8,000		.7592	.7706
9,000		.7474	

tained by formal extrapolation of the formula above, and the remarks already made concerning n-pentane apply here also.

Normal Hexane. The material was obtained from the Eastman Kodak Co. Complete runs at three temperatures and the return check runs were made without incident on three successive days. Discarding two points, in which there were probably errors of recording, the average departure from a smooth curve of a single one of the remaining 42 readings was 0.22% of the maximum effect.

The maximum pressure at which readings were attempted at 0° was 6000 kg. From the atmospheric freezing point and a previous determination of the freezing point at 30° as 10600 kg. this was roughly estimated to be the highest safe pressure.

The relative volumes are given in Table III. For the density at

TABLE III.

RELATIVE VOLUME OF N-HEXANE

Pressure kg./cm.²	Temperature 0°	50°	95°
0	1.0000	1.0712	1.1535
1,000	.9191	.9567	
2,000	.8763	.9048	.9304
3,000	.8472	.8720	.8914
4,000	.8259	.8472	.8629
5,000	.8068	.8262	.8404
6,000	.7905	.8091	.8225
7,000		.7943	.8071
8,000		.7817	.7942
9,000		.7707	.7824
10,000		.7615	.7723
11,000		.7529	.7632

atmospheric pressure the following formula from I. C. T. was assumed:

$$\text{Density} = 0.6769 - 0.8486 \times 10^{-3}t - 1.084 \times 10^{-6}t^2 + 0.164 \times 10^{-9}t^3$$

The boiling point is 69°, so that the legitimate use of the formula is restricted to temperatures lower than this; the volume at 95° at atmospheric pressure in the table was obtained by formal substitution in the formula.

2-methyl Pentane. For this isomer of hexane, as well as for the following three others, I am exceedingly indebted to Mr. T. A. Boyd of the Fuel Research Section of the General Motors Corporation. I am also indebted to Professor J. B. Conant for putting me in touch with Mr. Boyd.

The first attempt with this substance at 0° had to be discontinued at 4500 kg because of minor trouble with the insulation, which was speedily repaired. After reassembling, the three regular runs at three temperatures were made on three successive days. The return check readings were inadvertently omitted. The average departure from a smooth curve of a single one of the 42 readings was 0.20% of the maximum effect.

The maximum pressure attempted at 0° was 6000 kg. This was merely estimated as the safe maximum from relative boiling and melting points at atmospheric pressure as compared with n-hexane; no evidence of freezing under pressure was found during the experiment.

The relative volumes are given in Table IV. The density at

TABLE IV.

RELATIVE VOLUME OF 2-METHYL PENTANE.

Pressure	Temperature		
kg./cm.²	0°	50°	95°
0	1.0000	1.0736	1.1549
500	.9460	.9930	
1,000	.9131	.9496	.9851
1,500	.8896	.9203	.9496
2,000	.8711	.8976	.9227
3,000	.8419	.8638	.8826
4,000	.8191	.8382	.8552
5,000	.8009	.8184	.8336
6,000	.7844	.8023	.8179
7,000		.7878	.8022
8,000		.7754	.7886
9,000		.7644	.7763
10,000		.7538	.7658
11,000		.7450	.7564
12,000			.7484

atmospheric pressure, which had to be determined by special experiment as already mentioned, was:

$$\text{Density} = \frac{0.6720}{1 + 0.001297t + 3.52 \times 10^{-6}t^2}.$$

The temperature range of the density measurements was 52°; the value given in the table at atmospheric pressure at 95° is merely a formal extrapolation.

3-methyl Pentane. This also was obtained from the General Motors Corporation. The only incident during the run was a leak in the connecting pipe, which developed at 50° at about 9000 kg. because the washers had been used too long without changing. Pressure was released, the apparatus reassembled, and the run continued. The part of the run below 9000 at 50° was repeated, in perfect agreement. The usual return check readings were made after the three runs at constant temperatures; all measurements were made on three successive days. The average departure from a smooth curve of a single one of the 55 readings was 0.19% of the maximum effect.

The maximum pressure attempted at 0° was much higher than for the two preceding isomers, 11000 against about 6000. That this was safe was suggested by the much lower melting point at atmospheric pressure; no sign of freezing was detected.

The relative volumes are given in Table V. The density at atmospheric pressure, as found by special experiment, was:

TABLE V.

Relative Volume of 3-methyl pentane.

Pressure	Temperature		
kg./cm.²	0°	50°	95°
0	1.0000	1.0727	1.1530
500	.9434	.9884	
1,000	.9120	.9471	.9815
1,500	.8879	.9189	.9462
2,000	.8689	.8971	.9205
3,000	.8408	.8630	.8824
4,000	.8200	.8379	.8553
5,000	.8022	.8189	.8342
6,000	.7866	.8026	.8158
7,000	.7723	.7888	.8006
8,000	.7592	.7761	.7876
9,000	.7476	.7647	.7753
10,000	.7372	.7544	.7649
11,000	.7276	.7455	.7550
12,000		.7373	.7452

$$\text{Density} = \frac{0.6835}{1 + 0.001285t + 3.43 \times 10^{-6}t^2}.$$

The volume given in the table at 95° at atmospheric pressure is a formal extrapolation.

One striking feature of the results with this substance is the way in which the isotherm for 50° is displaced away from the 0° isotherm and toward the 95° isotherm at the upper end of the pressure range. There can be little question but that this effect is genuine, the effect being much beyond the error of the measurements, and the irregularity in the points in the neighborhood being on the whole less than the average.

2-2-dimethyl Butane. This, like the two preceeding isomers, was obtained from General Motors. On the first attempt at 0° there was a leak in the manganin gauge plug at 3000, but this was successfully repaired, and the run continued, the points after reassembly checking those before. Three regular runs were made on three successive days. The return check readings were not made because after the run at 95° the pressure was inadvertently allowed to get so low that the sylphon blew up by boiler pressure. This was particularly unfortunate, because the quantity of this isomer available was smaller than that of any of the others, and there was not enough left for the dilatometer and density measurements. The average departure from a smooth curve of a single one of the 40 readings was 0.15%.

The freezing pressure of this isomer was so unexpectedly low that I was not successful in guessing its location, but freezing was encountered at each of the three temperatures. At 0° freezing was observed at 5400 kg., at 50° at 8000, and at 95° at 10800. It is not pretended that these give accurately the coördinates of the melting curve; it is merely certain that at each temperature the equilibrium pressure is lower than that given. Fortunately, no permanent damage was done to the sylphon by the amount of freezing that took place, as was proved by the perfect check between the points obtained with increasing and decreasing pressure.

The relative volumes are given in Table VI. These values are only approximate, because, as already explained, no measurements were made on the density or dilatation of this isomer. In calculating the values in the table, the density at 0° was assumed to be 0.6693. This value was estimated, on the assumption that the change of density on passing from 2–3 dimethyl butane to 2–2-dimethyl butane

TABLE VI.

RELATIVE VOLUME OF 2–2-DIMETHYL BUTANE.

Pressure	Temperature		
kg./cm.[2]	0°	50°	95°
0	1.0000		
500	.9496	.9955	
1,000	.9154	.9517	.9824
1,500	.8919	.9232	.9478
2,000	.8737	.9016	.9218
3,000	.8473	.8703	.8855
4,000	.8260	.8464	.8594
5,000	.8083	.8269	.8392
6,000		.8106	.8223
7,000			.8079
8,000			.7958
9,000			.7850
10,000			.7756

would be the same as on passing from 3-methyl pentane to 2-methyl pentane. No attempt was made to estimate the dilation, and these places are left vacant in the table.

2–3-dimethyl Butane. This, the last of the isomers of hexane, was also obtained from General Motors. Several vicissitudes were encountered during the runs with this substance. At the termination of the 0° run short circuits developed in the manganin gauge and the sylphon circuits. These were repaired, the apparatus set up again, and perfect check readings obtained at 0°. The 50° run was now completed without incident. The readings from the beginning of the 95° run were unsteady. The matter finally grew so bad that the apparatus was dismantled, and the sylphon set up again in another holder, thus of course changing by a slight amount the zero reading. The 95° run was now completed without incident. The usual check readings at 50° and 0° were finally taken; the difference between these agreed with the value found in the previous holder, as it should. The difference between the 50° readings with the old and new holders gives the difference of zero readings between the two holders. The average departure from a smooth curve of a single one of the 45 readings available for locating the isotherms was 0.23%.

At 0° freezing was probably encountered at 6200 kg.; no permanent distortion in the sylphon seemed to be the result of this freezing. At the other temperatures there was no trace of freezing.

The relative volumes are given in Table VII. The density at

TABLE VII.

RELATIVE VOLUME OF 2–3-DIMETHYL BUTANE.

Pressure	Temperature		
kg./cm.²	0°	50°	95°
0	1.0000	1.0722	1.1496
500	.9485	.9930	
1,000	.9147	.9503	.9841
1,500	.8909	.9202	.9478
2,000	.8695	.8960	.9198
3,000	.8395	.8633	.8836
4,000	.8180	.8383	.8562
5,000	.8004	.8186	.8340
6,000	.7855	.8028	.8162
7,000		.7881	.8008
8,000		.7747	.7874
9,000		.7624	.7757
10,000		.7509	.7645
11,000			.7546

atmospheric pressure, found by special experiment, was:

$$\text{Density} = \frac{0.6808}{1 + .001311t + 2.67 \times 10^{-6}t^2}.$$

The volume at atmospheric pressure at 95° listed in the table is merely a formal extrapolation.

Normal Heptane. The material was obtained from the Eastman Kodak Co. The runs at three temperatures and the return check readings were made on three successive days without incident. This substance was one of the very few for which the return check readings were not satisfactory. The interval between the 0° and the 50° isotherms was satisfactorily checked, but there is some doubt about the relative separation of the 50° and the 95° isotherms. The values given in the table are those which the internal evidence suggests as most probable, but it is barely possible that all the volumes of the 95° isotherm should be decreased by 0.0019. The average departure from a smooth curve of a single one of the 41 readings was 0.12% of the maximum effect.

No trace of freezing was found. The maximum safe pressure was estimated from the atmospheric melting point and the known behavior of n-hexane.

The relative volumes are given in Table VIII. The value of

TABLE VIII

RELATIVE VOLUMES OF N-HEPTANE.

Pressure kg./cm.²	Temperature 0°	50°	95°
0	1.0000	1.0633	1.1350
500	.9535	.9972	
1,000	.9223	.9584	.9919
2,000	.8813	.9083	.9337
3,000	.8531	.8753	.8954
4,000	.8315	.8508	.8676
5,000	.8145	.8312	.8457
6,000		.8146	.8276
7,000		.8001	.8125
8,000		.7875	.7992
9,000		.7763	.7870
10,000		.7659	.7762
11,000			.7660

density used in the calculations was that given in I. C. T. and is:

$$\text{Density} = 0.70048 - 0.8476 \times 10^{-3}t + 0.1880 \times 10^{-6}t^2 - 5.23 \times 10^{-9}t^3.$$

The normal boiling point of heptane is 98°.4, so that the 95° value given in the table at atmospheric pressure is in the range of actual values.

Normal Octane. This material was obtained from the Eastman Kodak Co. The measurements on this were made early in my experience, with a new sylphon, and before the necessity for seasoning a new sylphon to the highest pressure was recognized. There were considerable irregularities on the first applications, so that it was desirable to repeat some of the runs. Two runs were made at 0°, two at 50°, and one at 95°. The first run at 0° was the most irregular, and was not used in the calculations. However, the mean of the points of this discarded run lay on the curve obtained later, and for this reason the return check readings were thought to be unnecessary. The first run at 50° showed less irregularity than the first at 0°, the sylphon by now having been fairly well seasoned, although the maximum pressure had not been reached, and all except two of the points of the first 50° run as well as all of the second were used in the final calculation. The highest point at 95° was also irregular and probably for the same reason. Except for the discarded points just mentioned in detail, the average departure from a smooth curve of a single one of the remaining 50 points was 0.24% of the maximum effect.

No irregularity was found at the highest pressures of the second runs at 0° and 50°. The freezing pressures at 30° and 75° have been previously determined, so that there were data at hand to make it possible to stay within the maximum safe pressure.

The relative volumes are given in Table IX. The density assumed

TABLE IX.

RELATIVE VOLUMES OF N-OCTANE.

Pressure	Temperature		
kg./cm.²	0°	50°	95°
0	1.0000	1.0595	1.1230
500	.9572	1.0005	
1,000	.9311	.9654	.9943
2,000	.8924	.9200	.9422
3,000	.8640	.8882	.9068
4,000		.8639	.8802
5,000		.8428	.8592
6,000		.8251	.8416
7,000		.8103	.8267
8,000			.8134
9,000			.8014
10,000			.7915

in the calculation is that given in I. C. T.:

$$\text{Density} = 0.71848 - 0.8239 \times 10^{-3}t + 0.4459 \times 10^{-6}t^2 - 5.293 \times 10^{-9}t^3.$$

Normal Decane. This was obtained from the Eastman Kodak Co. The initial attempt at 50° was not good because of trouble with the insulation of the manganin gauge. The check readings were made several days after the completion of the regular runs, with the same filling of the sylphon, but with a different holder. The average departure from a smooth curve of a single one of the 34 readings was 0.29% of the maximum effect. This irregularity is apparently a little greater than usual; it is really not so, but the appearance is due to the smaller range possible with this material.

Freezing was not encountered on any of the runs. The safe limits were estimated from the known freezing point at atmospheric pressure and the behavior under pressure of the other hydro-carbons.

The relative volumes are given in Table X. In calculating the

TABLE X.

RELATIVE VOLUMES OF N-DECANE

Pressure kg./cm.^{2}	Temperature 0°	50°	95°
0	1.0000	1.0530	1.1083
500	.9646	1.0029	
1,000	.9383	.9683	.9952
1,500		.9454	.9664
2,000		.9263	.9466
3,000		.8952	.9146
4,000		.8714	.8880
5,000			.8675
6,000			.8481
7,000			.8341
8,000			.8215

results the following formula for the density given in I. C. T. was used:

$$\text{Density} = 0.7455 - 0.7293 \times 10^{-3}t - 0.371 \times 10^{-6}t^2.$$

Benzene. This was Kahlbaum's purest grade, prepared for molecular weight determinations. The pressure range open to investigation with this liquid is very small because of the high freezing point; however the importance of the liquid in various theoretical speculations justified its measurement. The melting curve had been previously exactly determined, so that the maximum safe pressure was known without any guess work. Runs were made at 50° and 95°, and a rather larger number than usual of return check readings were made at 50° without incident. The average departure of a single one of the 25 readings from smooth curves was 0.25% of the maximum effect.

The relative volumes are shown in Table XI. For the density at atmospheric pressure the formula of I. C. T. was used:

$$\text{Density} = 0.90005 - 1.0636 \times 10^{-3}t - 0.0376 \times 10^{-6}t^2 - 2.213 \times 10^{-9}t^3.$$

In Table XI unit volume is taken as the volume at 0°, obtained by extrapolating with the density formula from the melting point down to 0°. In the pressure measurements the volume was not actually measured at atmospheric pressure at 50°, but was obtained by graphical extrapolation from the lowest pressure reading, which was at 370

TABLE XI.

RELATIVE VOLUMES OF BENZENE.

Pressure kg./cm.²	Temperature 50°	95°
0	1.0630	1.1295
500	1.0160	
1,000	.9841	1.0201
1,500	.9591	.9916
2,000		.9684
2,500		.9494
3,000		.9325
3,500		.9177

kg. Such an extrapolation is not very certain, so that it is possible that there may be a slight error in the table in the volume difference between 0 and 500 kg at 50°. The relative volumes in the table at 50° from 500 kg on are based on direct measurement without extrapolation, and these should be relatively exact in terms of the volume at 500 kg and 50° (1.0160), independent of any error in the extrapolated interval.

The volume of benzene has been previously measured by Essex at a number of temperatures between 0° and 100° up to 3000 kg. At the higher temperature the agreement between his results and mine is usually good to one in the third decimal place. At 50° the disagreement is greater, being 3 or 4 in the third place at 500, 1000, and 1500 kg, and 1 in the second place at atmospheric pressure. His volumes are consistently lower than mine.

C_6H_5Cl. This was Kahlbaum's purest. Regular runs at the three temperatures and the return check readings were made on two successive days with no incident. The average departure from smooth curves of a single one of the 39 readings was 0.09%, the measurements on this material being among the most regular of those obtained.

Freezing was not encountered at any temperature.

The relative volumes are given in Table XII. In the calculation

TABLE XII.

RELATIVE VOLUME OF CHLORO-BENZENE.

Pressure	Temperature		
kg./cm.[2]	0°	50°	95°
0	1.0000	1.0502	1.1013
500	.9737	1.0170	
1,000	.9541	.9882	1.0215
1,500	.9380	.9673	.9955
2,000	.9244	.9511	.9755
2,500	.9127	.9384	.9595
3,000		.9268	.9463
4,000		.9068	.9247
5,000		.8903	.9072
6,000		.8762	.8924
7,000		.8652	.8794
8,000			.8675
9,000			.8580
10,000			.8487
11,000			.8406

the density given in I. C. T. was used:

$$\text{Density} = 1.12782 - 1.0664 \times 10^{-3}t - 0.2463 \times 10^{-6}t^2 - 0.53 \times 10^{-6}t^3.$$

C_6H_5Br. This was Kahlbaum's purest. The 0° and the 50° measurements were made without incident. The 95° run, however, as only partially completed, and then measurements were interrupted for nearly three weeks by the illness of my assistant. After this time the 95° run was resumed, repeating the points formerly obtained, and also making the return check readings at 50° and 0°, all in perfect consistency with the first measurements. This constitutes a rather severe test of the reproducibility and single valuedness of the measurements.

Freezing was not encountered at any temperature.

The relative volumes are given in Table XIII. The density used in the calculation was that given in I. C. T.:

$$\text{Density} = 1.5223 - 1.345 \times 10^{-3}t - 0.24 \times 10^{-6}t^2 + 0.76 \times 10^{-9}t^3.$$

The range of application of this formula given in I. C. T. is 0° to 80°. The value at 95° at atmospheric pressure must therefore be regarded

TABLE XIII.

Relative Volumes of Bromo-benzene.

Pressure kg/cm.²	Temperature 0°	50°	95°
0	1.0000	1.0467	1.0940
500	.9763	1.0125	
1,000	.9570	.9891	1.0169
1,500	.9409	.9697	.9953
2,000		.9527	.9765
3,000		.9261	.9460
4,000		.9047	.9229
5,000		.8860	.9033
6,000			.8868
7,000			.8726
8,000			.8604
9,000			.8501

in the nature of an extrapolation, although not nearly so hazardous an extrapolation as one beyond the boiling point.

CCl_4. This was Kahlbaum's purest. The region open to measurement is very small because of the unusually steep melting curve. No run was attempted at 0°, because, although the normal melting temperature is − 23°, the freezing pressure at 0° is only about 500 kg. Regular readings at 50° and 95° and the return check readings were made without incident in one day. The average departure from smooth curves of a single one of the 21 readings was 0.15%. If allowance is made for the smallness of the range, the regularity is rather better than usual.

The relative volumes are given in Table XIV. The density at atmospheric pressure is given by I. C. T., as:

TABLE XIV

Relative Volumes of Carbon Tetrachloride.

Pressure kg./cm.²	Temperature 50°	95°
0	1.0000	
500	.9519	.9928
1,000	.9192	.9540
1,500	.8962	.9262
2,000		.9049
2,500		.8872
3,000		.8726
3,500		.8603

$$\text{Density} = 1.63255 - 1.9110 \times 10^{-3}t - 0.690 \times 10^{-6}t^2,$$

restricted to the range 0° to 40°. However, I used the formula by extrapolation to 50°, and the values in the table are based on the density so obtained. The actual volume was not measured at atmospheric pressure at 50°, but was obtained by extrapolation from the lowest pressure, 130 kg. The volume interval between 0 and 500 kg at 50° in the table is therefore subject to slight uncertainty, and the remarks made with regard to benzene apply.

Bromoform. This was obtained from Kahlbaum. The normal melting point is at +7°.7, so that the possible pressure range was very narrow. A regular run was made at 50°, and on the same day at 95° good points with increasing pressure were obtained, but the decreasing pressure points lay further and further from the curve, and on taking the apparatus apart it was found that the sylphon had developed a leak. On allowing it to stand for several days in the room, the sylphon fell completely to pieces, the bromoform exerting some corrosive action. I am informed by my colleagues among the chemists that this is to be expected. No attempt was made to repeat the run. The average departure from a smooth curve of a single one of the 24 readings up to the maximum pressure at 95° was 0.74% of the maximum effect. This large irregularity is mostly due to the smallness of the range; if the range had been normal, the irregularity would have been only 0.25%, not much larger than usual. There is no reason to think that the results obtained before the leak developed are not essentially correct.

No trace of freezing was encountered.

The relative volumes are given in Table XV. The specific volume

TABLE XV.

RELATIVE VOLUMES OF BROMOFORM

Pressure	Temperature	
kg./cm.²	50°	95°
0	1.0000	1.0427
500	.9628	.9993
1,000	.9369	.9662
1,500	.9158	.9418
2,000		.9225
2,500		.9063
3,000		.8915
3,500		.8782

as a function of temperature at atmospheric pressure is given in a paper by A. and J. Shuman[6] as:

$$V = 0.34204[1 + 9.0411 \times 10^{-4}(t - 7.7) + 6.766 \times 10^{-7}(t - 7.7)^2].$$

Their range was 7°.7 to 50°. I used the formula by extrapolation to 95°. At 50° the volume was not measured at pressure lower than 300 kg, and the volume interval between 0 and 500 kg at 50° given in the table is therefore based on an extrapolation and may be somewhat uncertain. The remarks already made concerning benzene and CCl_4 apply.

Iso-propyl Alcohol. This was originally obtained a number of years ago from Kahlbaum. It had been standing in a cork stoppered bottle and may have absorbed moisture in the intervening time. It was therefore redistilled twice in a fractionating column immediately before the compressibility measurements. As already stated, the reason for including this substance was that measurements had already been made on n-propyl alcohol in the preceding series. The first run at 0° gave highly irregular readings, doubtless because this was a fresh sylphon, not previously seasoned. The runs at 50° and 95° were of normal regularity. The run at 0° was then repeated, with regular readings except at the maximum, 8000 kg, where there was a slight irregularity that had the appearance of freezing. Pressure was not pushed higher for fear of loosing the points already obtained; the points with decreasing pressure lay on a smooth curve with the increasing pressure points. It is somewhat puzzling to understand why on the first application of pressure 10000 was reached at 0° with no appearance of freezing; there is, of course, the possibility of a sub-cooling effect, which may be very capricious. The succession of the melting points of the alcohols at atmospheric pressure would not make it surprising to find freezing at 0° below 12000 kg.

Discarding the first run at 0°, the average departure from smooth curves of a single one of the remaining 42 readings was 0.17%.

The relative volumes are given in Table XVI. The calculations are based on the formula for density given in I. C. T.:

$$\text{Density} = 0.8014 - 0.809 \times 10^{-3}t - 0.27 \times 10^{-6}t^2.$$

Normal-butyl Alcohol. The material was recently obtained from Kahlbaum, and was used without further purification. Considerable trouble was found with this material because of unusual freezing phenomena, and five runs in all were made at 0°, before satisfactory

TABLE XVI.

RELATIVE VOLUMES OF I-PROPYL ALCOHOL.

Pressure	Temperature		
kg./cm.[2]	0°	50°	95°
0	1.0000	1.0540	1.1097
500	.9630	1.0056	
1,000	.9387	.9725	1.0112
1,500	.9200	.9485	.9807
2,000	.9040	.9296	.9562
3,000	.8769	.8995	.9218
4,000	.8571	.8765	.8961
5,000	.8398	.8579	.8750
6,000	.8244	.8415	.8572
7,000	.8116	.8279	.8419
8,000	.8002	.8159	.8289
9,000	.7901	.8052	.8168
10,000	.7813	.7948	.8062
11,000		.7861	.7971
12,000		.7784	.7894

results were obtained, the first two with a different filling of the apparatus from the others. I had not anticipated any trouble from the freezing of this alcohol, so that pressure was pushed to 12000 on the first run at 0°. But the reading at 12000 was the same as at 10200, showing that freezing had taken place somewhere in the interval. On the next application the pressure was kept below 10000, but an irregularity at the maximum, 9700, showed that freezing had again taken place. On the next two applications of pressure freezing phenomena were successively found at 7900 and probably at 7000 kg, and it was only on the fifth attempt, when pressure was not pushed higher than 5000, that no freezing was encountered. The successive lowering of the freezing pressure may have been due to a subcooling effect, there being nuclei of the solid phase left in the liquid, making the new formation of the solid easier at every successive attempt, as might be suggested by my previous experience with ice VI, or it may be that with each successive freezing there was increasing purification of part of the alcohol, complete remixing and diffusion of the segregated impurity being prevented by the high viscosity under pressure and the complex geometrical shape of the sylphon.

The two other runs at 50° and 95° and the return check readings were made without further incident. The average departure from a smooth curve of a single one of the 44 readings of the good runs was 0.27%. This is somewhat higher than usual, and is probably an effect of the freezing, although not the slightest damage to the sylphon was visible to the eye.

The relative volumes are given in Table XVII. The calculations

TABLE XVII.

RELATIVE VOLUMES OF N-BUTYL ALCOHOL.

Pressure kg./cm.^{2}	Temperature 0°	50°	95°
0	1.0000	1.0455	1.0907
500	.9697		
1,000	.9459	.9779	
1,500	.9264	.9557	.9801
2,000	.9111	.9372	.9595
3,000	.8874	.9087	.9276
4,000	.8681	.8867	.9033
5,000	.8512	.8684	.8834
6,000		.8530	.8667
7,000		.8390	.8519
8,000		.8269	.8388
9,000		.8153	.8277
10,000		.8076	.8174
11,000			.8082
12,000			.7999

are based on the formula for density given in I. C. T.:

$$\text{Density} = 0.82390 - 0.699 \times 10^{-3}t - 0.32 \times 10^{-6}t^2.$$

Normal-hexyl Alcohol. This was obtained from Kahlbaum; it was evidently not entirely pure, because it was of a slight yellow color. On the first set-up, runs were attempted at 0° and 50°. On the first run at 0° freezing was encountered, but the amount was so slight that I hoped the sylphon had not been damaged. The results with decreasing pressure at 50°, however, strongly suggested a leak, and on taking the apparatus apart a crease was found in one of the folds of the sylphon which was doubtless the result of the freezing at 0°. A second set-up was therefore made with a new sylphon, and three successful runs at 0°, 50°, and 95°, and the return check readings were obtained without incident. The average departure from smooth curves of a single one of the 39 readings was 0.50% of the maximum effect, considerably larger than usual, but in large part accounted for by the smaller range.

The relative volumes are given in Table XVIII. In making the

TABLE XVIII.

RELATIVE VOLUMES OF N-HEXYL ALCOHOL

Pressure	Temperature		
kg./cm.²	0°	50°	95°
0	1.0000		
500	.9719	1.0078	
1,000	.9488	.9791	1.0052
1,500	.9316	.9583	.9813
2,000	.9183	.9420	.9619
2,500	.9075	.9278	.9467
3,000		.9159	.9337
4,000		.8960	.9118
5,000		.8787	.8931
6,000			.8773
7,000			.8637

calculations the density at atmospheric pressure at 0° was assumed to be 0.8204, this being the figure given in Landolt and Börnstein. The thermal expansion has apparently not been recorded in the literature, and no attempt is made to give values for it in the table.

In addition to these results on new liquids, certain measurements were made on other liquids during the preliminary testing of the apparatus which are worth recording.

Ether. This material was from Mallinckrodt, reagent quality, but it had been standing in a cork stoppered bottle for some time and may have absorbed moisture. The runs with this material were the first successful ones in which the sylphon was carried through a complete cycle without leak. Ether was chosen as the liquid because previous experience had shown that among easily obtainable liquids this, or CS_2, most readily disclosed a leak. In order to check further the reliability of the method, the compressibility of ether was determined numerically, and compared with the results of the previous method. The temperatures for this comparison were chosen as 30° and 75°, which are within the range of the previous work, whereas 0° and 95° both lie outside the previous range. Since the purpose was only comparative, pressure at 75° was not lowered below 2400 kg, the manipulations in this way being somewhat easier. In this preliminary exploration special attention was also given to the method proposed for getting the thermal expansion, the temperature being changed back and forth several times between 30° and 75°. The check here was perfect; it was possible to return exactly to the same isotherm after leaving it at any pressure in the range. The average departure from smooth curves of a single one of the 32 readings was 0.08% of the maximum effect.

In calculating the results the formula given in I. C. T. for the density was used:

$$\text{Density} = 0.73629 - 1.1138 \times 10^{-3}t - 1.237 \times 10^{-6}t^2.$$

The relative volumes are given in Table XIX, and in parallel columns

TABLE XIX.

RELATIVE VOLUMES OF ETHER.

Pressure	Temperature			
	30°		75°	
kg./cm.²	new	previous	new	previous
0	1.0495	*1.0492*		
500	.9761	*.9790*		
1,000	.9364	*.9445*		
1,500	.9085	*.9153*		
2,000	.8858	*.8980*		
2,500	.8671	*.8734*	.8909	*.8970*
3,000	.8511	*.8576*	.8726	*.8788*
4,000	.8255	*.8318*	.8446	*.8496*
5,000	.8055	*.8111*	.8225	*.8268*
6,000	.7888	*.7954*	.8038	*.8098*
7,000	.7742	*.7806*	.7884	*.7940*
8,000	.7616	*.7675*	.7747	*.7800*
9,000	.7504	*.7554*	.7629	*.7672*
10,000	.7399	*.7444*	.7519	*.7560*
11,000	.7305	*.7335*	.7418	*.7457*
12,000	.7225	*.7237*	.7329	*.7353*

in italics the values previously found. The agreement is not by any means as good as could be desired; the maximum discrepancy in the volume is about 0.006 in the neighborhood of 3000 kg. At the maximum pressure the difference becomes much less. It is important to notice that the thermal expansions of the two series differ from the mean by only a few per cent, so that the fundamental question about the functioning of the new apparatus was thereby answered satisfactorily.

These preliminary measurements could be regarded as satisfactory in establishing most of the questionable points with regard to the new method, but in view of the uncertainty introduced by the unknown purity of the ether, did not answer the question as to the consistency of the absolute values obtained by the two methods. Subsequently another attempt was made to check the absolute values by using a liquid easily obtained in sufficient purity. For this purpose water was chosen. From one point of view the test with water was particularly severe, because it is one of the least compressible of liquids.

Water. This was distilled water of ordinary laboratory purity; it would have been useless to have started with conductivity water in view of the fact that it was to be placed in contact immediately with a metal. The three regular runs at 0° 50°, and 95° and the return check readings were made without incident. Except for one discarded point, which was so far from the curve that a blunder in reading is indicated, the average departure from smooth curves of a single one of the remaining 44 points was 0.15%.

The relative volumes are given in Table XX, with the previous values at 0° and 50° for comparison in italics; the previous range

TABLE XX.

RELATIVE VOLUMES OF WATER.

Pressure	Temperature				
	0°		50°		95°
kg./cm.²	new	previous	new	previous	new
0	1.0000				
500	.9771				
1,000	.9567	*.9578*	.9741	*.9743*	.9984
1,500	.9396	*.9424*	.9582	*.9586*	.9812
2,000	.9248	*.9260*	.9439	*.9445*	.9661
3,000	.8996	*.9015*	.9201	*.9205*	.9409
4,000	.8795	*.8807*	.8997	*.8996*	.9194
5,000	.8626	*.8632*	.8824	*.8818*	.9009
6,000			.8668	*.8662*	.8849
7,000			.8530	*.8524*	.8705
8,000			.8407	*.8399*	.8577
9,000			.8296	*.8288*	.8461
10,000			.8192	*.8188*	.8352
11,000					.8256

extended only to 80° so that comparison at 95° is not possible. Because of danger of freezing, no measurements were made at atmospheric pressure at 0°, and in making the calculations the fiducial volume was chosen as that at 500 kg and 0°, using the value 0.9771 obtained in the previous measurements. The agreement is seen to be much closer than for ether, and in particular the agreement at 50° is as good as could be desired. In view of the fact that the new data cover a wider range than the old I think that perhaps the new values should be given the preference, although this is a difficult question.

Finally, during the preliminary examination of the method, an attempt was made to find whether consistent results for compressibility could be obtained irrespective of the amount of liquid in the sylphon. If the cross section is truly independent of the extension of the sylphon, then the compressibility obtained with it should be independent of the amount of liquid in the sylphon, that is, independent of the initial extension of the sylphon. In order to make this test it is obviously necessary to choose a relatively incompressible liquid, because the compression of the more compressible liquids is so great as to use all available stroke of the sylphon. An incompressible liquid has the disadvantage of giving relatively small effects, and so enhancing the error from slight irregularities; there seemed, however, no way of avoiding this source of error.

Glycerine and water. The liquid chosen for this test was a mixture of equal parts by volume of water and glycerine, the glycerine being Kahlbaum's purest. Pure glycerine is much less compressible than water, but its viscosity increases so much under pressure that fear was felt for the safety of the sylphon had it been used pure. The comparison with two fillings of the sylphon was made only at 30°. The compressibilities and dimensions of the apparatus were such that a difference in the two initial amounts of about 25% was possible. Two runs were made with the larger filling, the initial points of the first being irregular, perhaps because the changes of pressure were made too rapidly for this rather viscous liquid. The decreasing points of this run and the second run gave regular points. A single run with the second filling was sufficient, the points lying smoothly. The average departure from a smooth curve of a single one of the 22 readings with the larger filling was 0.44%, and of the 11 readings with the smaller filling 0.25%.

The ratio of the changes of volume obtained with the two fillings should be a constant, independent of pressure. The following values were found for the ratio at successive intervals of 1000 kg, from 1000 to 11000 kg respectively: 1.236, 1.223, 1.230, 1.238, 1.250, 1.259, 1.265, 1.265, 1.265, 1.268, 1.268, 1.266. The ratio is therefore effectively constant beyond 6000 kg. Below 6000, greater deviation would be expected because the effects are absolutely smaller and the chance for relative error greater. In fact, on making a detailed examination of the experimental points, it will be seen that the deviation at the lower pressures is not more than can be accounted for by the irregularity of the experimental points, so that these measurements may be considered to give satisfactory proof of the fact that even under pressure the effective cross section of the sylphon is independent of its extension.

The actual values of the compressibility of the glycerine-water mixture are of some interest, and are given in Table XXI. The

TABLE XXI.

COMPRESSION OF 50 PER CENT GLYCERINE-WATER MIXTURE AT 30° C.

Pressure kg./cm.^{2}	Loss of volume in cm.^{3} per gm.
1,000	.0291
2,000	.0521
3,000	.0715
4,000	.0886
5,000	.1033
6,000	.1159
7,000	.1279
8,000	.1385
9,000	.1479
10,000	.1569
11,000	.1651

density of the mixture was not specially measured, so that the results are best given as loss of volume in cm.^{3} per gm. of mixture. The average of the two fillings is given.

DISCUSSION.

This discussion will touch only some of the main qualitative features of the situation; no attempt will be made to extend the elaborate discussion of the former paper on twelve liquids to these new liquids. All the information to be obtained from these new measurements is implicitly contained in the tables of volumes, from which any one may extricate for himself any information important for his particular purposes.

The new data of this paper allow a very considerable extension in one respect of the information of the previous paper, that is, with regard to the behavior of isomers at high pressures. In the previous list of 12 liquids there were only two isomeric compounds, that is, compounds with the same chemical formula, namely ether and butyl alcohol ($C_4H_{10}O$). These are, however, so different in structure that the chemist is not usually inclined to think of them as isomers. The fact previously found was that the volume of ether and butyl alcohol tends to become more nearly the same at high pressures, the ratio of the volumes of the same number of grams, that is, of the same number of molecules, being 1.106 at 500 kg and 1.038 at 12000 kg at 50. This seems the natural behavior, and it was my expectation that the new data would act in the same way. The number of isomers between which comparison may be made is now considerably increased, there being normal- and iso- propyl alcohols, normal- and iso- butyl alcohols, normal- and iso- pentane, and the 5 isomeric hexanes. These substances are more nearly what the chemist usually thinks of as isomers, the structural differences not being so great as between ether and butyl alcohol.

It turns out that the rule found for ether and butyl alcohol is not followed by these new isomers, but the volume differences between equal weights of the various pairs or sequences of isomers follows no rule. The relations are brought out in the following tables, in which the volumes of equal weights of different isomers are compared at a temperature of 50°. In Table XXII the relative volumes of the

TABLE XXII.

Volumes of Equal Weights of Isomeric Alcohols at 50° C.

Pressure	Propyl Alcohols			Butyl Alcohols		
	Volume of .8179 gm.			Volume of .8239 gm.		
kg./cm.²	normal-	iso-	Ratio	normal	iso-	Ratio
0	1.0493	1.0756	1.0250	1.0455	1.0508	1.0051
2,000	.9293	.9487	1.0209	.9372	.9336	.9962
4,000	.8813	.8945	1.0149	.8867	.8791	.9914
6,000	.8494	.8588	1.0111	.8530	.8445	.9900
8,000	.8250	.8327	1.0093	.8269	.8197	.9913
10,000	.8060	.8114	1.0067	.8076	.7996	.9901
12,000	.7905	.7944	1.0049		.7821	

propyl and butyl alcohols are shown. The ratio of the volumes of the two forms of propyl alcohol approaches more closely to unity at high pressure than at atmospheric pressure, according to expectation, but the reverse is true for the butyl alcohols. The iso- form of propyl alcohol has a larger volume than the normal form over the entire pressure range, whereas the iso- form of butyl alcohol has a larger volume than the normal form at atmospheric pressure, but at high pressures has the smaller volume. The normal form of both alcohols is seen to have a smaller compressibility than the iso- form. In Table XXIII the volumes of equal weights of normal and iso- pentane

TABLE XXIII.

VOLUMES OF EQUAL WEIGHTS OF PENTANES AT 50°

Pressure	Volume of .6454 gm.		
kg./cm.²	normal-	iso-	Ratio
0	1.0837	1.0924	1.0080
2,000	.8820	.8911	1.0103
4,000	.8193	.8283	1.0110
6,000	.7807	.7909	1.0131
8,000	.7520	.7649	1.0172
9,000	.7409	.7530	1.0163

are compared. Iso-pentane has throughout the greater volume, and the difference becomes greater at high pressures than at atmospheric pressure. This means that the normal form of pentane has greater average compressibility than the iso-form, unlike the alcohols. In Table XXIV the volumes of 1 gm. of the isomeric hexanes are given,

TABLE XXIV.

VOLUMES OF 1 GM. OF ISOMERIC HEXANES AT 50°.

Pressure	Volumes					Ratio to n-hexane			
kg/ cm.²	Normal hexane	2-methyl pentane	3-methyl pentane	2.2-dimethyl butane	2.3-dimethyl butane	2-methyl pentane	3-methyl pentane	2.2-dimethyl butane	2.3-dimethyl butane
0	1.5827	1.5973	1.5697		1.5750	1.0092	.9918		.9951
2,000	1.3363	1.3357	1.3125	1.3470	1.3160	.9996	.9897	1.0080	.9851
4,000	1.2517	1.2470	1.2260	1.2647	1.2313	.9962	.9794	1.0104	.9837
6,000	1.1953	1.1940	1.1743	1.2113	1.1793	.9989	.9824	1.0134	.9866
8,000	1.1550	1.1537	1.1357		1.1380	.9989	.9833		.9853
10,000	1.1250	1.1217	1.1037		1.1030	.9971	.9810		.9804
12,000	1.1008*		1.0790				.9802		

* Extrapolated.

89 — 2469

and the ratios to the volume of n-hexane. With the exception of 2-methyl pentane, the volumes of these isomers are more different from the normal form at high pressures than at atmospheric pressure, again contrary to expectation. In general, the compressibility of those compounds is greatest in which the CH_2 radical occupies a 3-position.

In contrast to the previous example of ether and butyl alcohol, the phenomena shown by these isomers are on a much smaller scale, and the volumes are always much more nearly equal. In view of this, I do not believe that the fact that the volumes of these new isomers do not in general tend toward greater equality at high pressures need alter the conclusion previously drawn, namely that because of the compressibility of the molecules there is a tendency to equality of volumes at high pressures. It is obvious however that the comparatively small structural differences of these new isomers persist, and may even become more prominent up to pressures of 12000 kg. Other lines of evidence have already suggested that there are conditions under which liquids may approach a greater semblance of order at high pressures, a fact which is not inconsistent with the discovery by X rays of structure in liquids under ordinary conditions.

Another instructive comparison is afforded by the volumes of the series of normal hydro-carbons, comparing the volumes per gm. molecule. In Table XXV are shown the volumes at 50° of 1 gm.

TABLE XXV.

Volume of 1 gm. Molecule of Normal Hydro-carbons at 50°.

Pressure kg./cm.^{2}	Pentane		Hexane		Heptane		Octane		Decane
0	121.03	*15.24*	136.27	*15.63*	151.90	*16.37*	168.27	*32.40*	200.67
2,000	98.51	*16.54*	115.05	*14.72*	129.77	*16.33*	146.10	*30.50*	176.60
4,000	91.52	*16.25*	107.77	*13.80*	121.57	*15.63*	137.20	*28.93*	166.13
6,000	87.20	*15.73*	102.93	*13.44*	116.37	*14.70*	131.07		
8,000	83.99	*15.46*	99.45	*13.08*	112.53				
10,000	81.72	*15.15*	96.87	*12.58*	109.45				

molecule of the normal hydro-carbons as a function of pressure, together with the successive differences, in italics, on passing from one member of the series to another. In the first place, there are very distinct differences between the odd and even members of the series, the increase of volume when a CH_2 group is added on passing from an odd to an even member being greater than the increase on passing from an even to an odd member. The only exception to this is at atmospheric pressure between pentane and hexane; here there is a special effect. 50° is above the boiling point of pentane, and the volume listed was obtained by extrapolation of the volumes at considerably lower temperatures; the volume so obtained accentuates too much the larger volume of the approaching vapor phase in comparison with the other hydro-carbons, so that the difference between pentane and hexane is too small. The difference between odd and even steps in the series becomes less in going to higher members of the series, but becomes greater on going to higher pressures. This again is not what might at first be expected, and reminds one of the behavior of the isomers. It means that the first effect of high pressure is to accentuate the effect of structural differences in the molecules, in virtue of which they are perhaps squeezed into some semblance to an orderly arrangement. It is already known from X-ray investigations that there are differences between the volumes occupied in the solid crystal form by the odd and even members of the series, and Müller (7) has been able to give a detailed account of this in terms of the shapes of the molecules.

Pass now from a consideration of the volume to the compressibility. The most striking fact found for the previous twelve liquids was the very great decrease of compressibility with increasing pressure, the compressibility at 12000 kg being on the average about 1/15th of the compressibility at atmospheric pressure. This continues to be the fact for these new liquids, and what is more, the details of behavior are the same, in that the compressibility has dropped to about half its initial value in the first thousand kilograms, the rate of decrease becoming slower, until at 6000 pressure must be raised another 6000, that is, from 6000 to 12000, to reduce the compressibility again by a factor of 2.

The compressibilities do not vary through a very wide range, as may be seen by plotting the relative volumes of the various liquids in a single diagram. Of the new liquids, n-pentane is the most compressible. The compressibility in general decreases in a series of compounds on passing in the direction of increasing molecular weight. Thus in the series of normal hydro-carbons, the compressibility of decane is of the order of two-thirds that of pentane; in the series C_6H_6, C_6H_5Cl, C_6H_5Br, the compressibility decreases in the order given. Of the organic liquids previously examined, glycerine stood in a class by itself, with the smallest compressibility. Then came water, also in a class by itself, with a compressibility of the order of twice that of glycerine, and then the twelve liquids, with a compressibility on the average perhaps 50% greater than that of water, although the initial compressibilities were much more than 50% greater. Most of the new liquids fall in the same group as the preceding twelve, but there are two, C_6H_5Cl and C_6H_5Br, which fall in the same class as water; in fact the curves of relative volume of these three liquids very nearly coincide. The compressibility of C_6H_6, on the other hand, is about 60% greater than that of its two substitution products. This very large difference seems surprising. The difference is without doubt due in part to the effect so often discussed by Professor T. W. Richards:[8] the cohesive forces in C_6H_5Cl and C_6H_5Br must be fairly high, as shown by a consideration of the molecular volumes, for the increase of volume on passing from C_6H_6 to C_6H_5Cl and C_6H_5Br is of the order of half the molecular volume of free (solid) Cl or Br, and this means a high internal pressure and small compressibility. But the matter is not as simple as this, because the compressibilities of C_6H_5Cl and C_6H_5Br are very nearly the same, whereas the atomic compression on passing from C_6H_6 to C_6H_5Cl is materially less than on passing to C_6H_5Br.

It is evident that there is an interesting field for future investigation in seeking out organic compounds with abnormally small compressibility; the unique position of glycerine has as yet received no explanation.

The thermal expansion next concerns us. In the previous work on twelve liquids two important facts stood out with regard to thermal expansion; the thermal expansion at 12000 kg is almost always 4 or 5 times smaller than at atmospheric pressure, and at a pressure of a few thousand kilograms there is a reversal of sign of $\left(\frac{\partial^2 v}{\partial \tau^2}\right)_p$, that is, at low pressures the thermal expansion increases with increasing temperature, but at high pressures it decreases with increasing temperature. The new data of this paper are not sufficient to give as much detail about thermal expansion as the preceding data because of the smaller number of temperature intervals, but they are adequate on these two important points. The decrease of thermal expansion with increasing pressure is larger on the average in the series of hydrocarbons than for the preceding twelve liquids; thus for pentane the decrease is by a factor of 7, and for n-hexane by 8. Many of the new data do not extend over as wide a pressure range as the preceding because of freezing, so that a complete comparison cannot be made; the two new isomeric alcohols fall in line with the previous results. The ratio for C_6H_5Cl and C_6H_5Br is smaller than the previous average.

The second important previous feature, the reversal of sign of $\left(\frac{\partial^2 v}{\partial \tau^2}\right)_p$, is also shown by all of these new liquids in which the pressure range attainable at 0° was extensive enough to provide the requisite data. The pressure at which the reversal takes place is also roughly the same as before, that is, in the neighborhood of 2000 or 3000 kg. In the series of hydro-carbons the pressure of reversal tends to become less as the molecular weight increases. Among the isomers of hexane there are some striking examples of strong reversal. 3-methyl pentane is perhaps the most striking: at atmospheric pressure the mean expansion 50°–100° (obtained by increasing the tabulated difference 50°–95° by 11%) is 22.7% greater than the mean expansion between 0° and 50°, whereas at 11000 kg it is 40.7% less. The fact that this reversal of sign is now found again with this fundamentally different apparatus affords welcome presumptive evidence of the essential correctness of the results obtained by both methods.

The suggestion has been made in a previous publication[9] that the probable reason for this reversal is that at high pressures and low temperatures the molecules are crowded so close together that the departures from linearity of the forces between molecules must be greater than at higher temperatures at the same pressure (that is, at larger volumes), since it is known that a large thermal expansion goes with a large departure from linearity, the thermal expansion of a substance with a linear law of force being zero.

Considerable importance attaches to the "pressure coefficient" $\left(\frac{\partial p}{\partial \tau}\right)_v$; the suggestion has been made a number of times that this should be a function of volume only, or an equivalent statement is that at constant volume, pressure is a linear function of temperature. In the previous paper on 12 liquids I showed that this does not agree with the experimental facts, and detailed calculations were presented for CS_2 from which it appeared that $\left(\frac{\partial p}{\partial \tau}\right)_v$ was a complicated function of temperature at constant volume. The theoretical reasons for thinking $\left(\frac{\partial p}{\partial \tau}\right)_v$ to be a volume function only have, however, appealed to some writers as strong enough to justify them in ascribing the complicated variations there found to experimental irregularities, and in continuing to assume that the theoretical relation is correct.

The new data of this paper allow a more decisive inquiry as to the validity of this relation, because the temperature intervals are fewer and wider, thus smoothing out small scale irregularities. The question may at once be answered from the graphs showing volume as

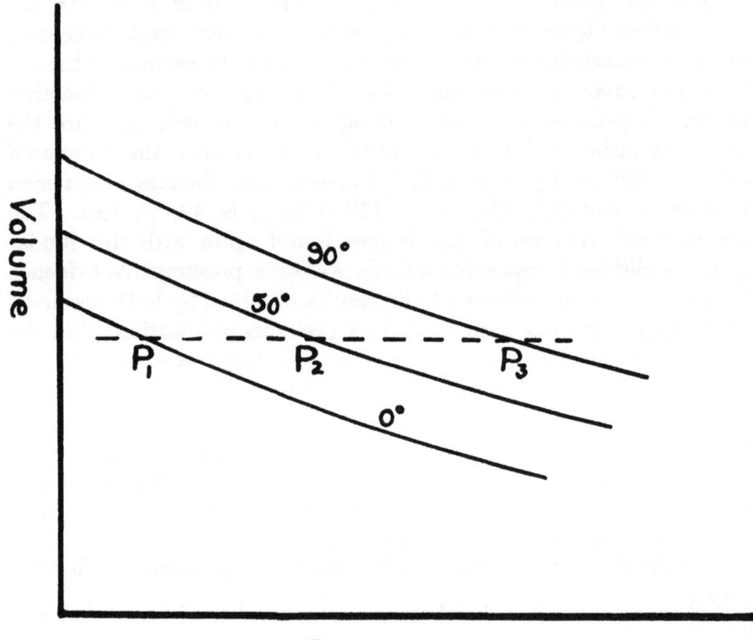

FIGURE 4. Graphical construction for finding whether $\left(\frac{\partial p}{\partial \tau}\right)_v$ is a function of volume only.

a function of pressure at 0°, 50°, and 95°. If $\left(\frac{\partial p}{\partial \tau}\right)_v$ is constant on a line of constant volume then $p_3 - p_2$ should be to $p_2 - p_1$ as 45 to 50, that is $\frac{p_3 - p_2}{p_2 - p_1}$ should equal 0.90 (Fig. 4). This ratio has been calculated at a number of points for many of the liquids of this paper, and the results are shown in Table XXVI. In making the calcula-

TABLE XXVI.

TEST OF THE INDEPENDENCE OF $\left(\frac{\partial p}{\partial \tau}\right)_v$ OF TEMPERATURE AT CONSTANT VOLUME.

Value of ratio $\left(\frac{p_3 - p_2}{p_2 - p_1}\right)_v$

Substance	Rel. vol.	ratio	Rel. vol.	ratio	Rel. vol.	ratio
n-pentane	.92	.84 ±	.75	.98		
i-pentane	.94	.905	.80	.972		
n-hexane	.92	.95 ±	.79	.774		
2-methyl pentane	.94	.897	.80	.910		
3-methyl pentane	.94	.865	.80	.817	.75	.597
2–2-dimethyl butane	.92	.743	.82	.68		
2–3-dimethyl butane	.95	.895	.80	.747		
n-heptane	.99	.914	.82	.832		
n-octane	.99	.806	.87	.767		
n-decane	.94	.837				
C_6H_5Cl	.95	.893				
C_6H_5Br	.95	.870				
i-propyl alcohol	.96	.947	.86	.925	.79	.900
n-butyl alcohol	.96	.837	.86	.850		
n-hexyl alcohol			.80	.845		

tions, the figures given in the tables of relative volume were used by linear interpolation; this accounts for the irregular volumes used, which were so chosen as to be as near as possible to the integral pressures of the tables, thus minimizing the small errors from linear interpolation. Table XXVI shows many deviations of the ratio from 0.90 which are far beyond any experimental error. The ratio usually starts with a value fairly near 0.90 at volumes near 1, that is, at low pressures, but as volume decreases, the ratio nearly always decreases, which means that at high pressures $\left(\frac{\partial p}{\partial \tau}\right)_v$ decreases with increasing temperature at constant volume. This was also the general nature of the results previously found for CS_2. There are, however, a few exceptions to this rule, n- and i- pentane being especially conspicuous for the large increase of the ratio on passing from larger to smaller volumes.

It is worth while to examine a little further the implications in assuming $\left(\frac{\partial p}{\partial \tau}\right)_v$ is a volume function only. If this is the case, integration gives at once

$$p = \tau f(v) + \varphi(v),$$

where f and φ are two arbitrary volume functions. The physical meaning of this equation is easy to see; it states that the pressure exerted by a fluid can be regarded as arising from two mechanisms acting independently of each other. One part, $\varphi(v)$, is a pure volume effect, in virtue of which a fluid exerts as much pressure at any temperature as it would at 0° Abs. at the same volume. This part of the pressure may be thought of as arising from the attractive or repulsive forces between the molecules. The second part is a kinetic part, proportional to absolute temperature, and arising from the change of momentum in unit time of the molecules as they collide with the walls. In a perfect gas, elementary considerations show that the momentum change in unit time is proportional to the kinetic energy, that is, to the temperature, and the volume function, f, becomes very simple. If the molecules have a finite size, f becomes more complicated. The nature of the two terms is clearly shown in van der Waals' equation, which satisfies the demand that $\left(\frac{\partial p}{\partial \tau}\right)_v$ be a function of volume only, and which may be written in the form $p = \tau\frac{R}{v-b} - \frac{a}{v^2}$, where the volume function $\frac{R}{v-b}$ arises from the finite size of the molecules, and $\frac{a}{v^2}$ arises from the mutual attractions. But van der Waals' equation certainly is not correct at high pressures, in that it omits, among other things, the very important factor of the compressibility of the molecules. In fact, one can expect in general that the hypothesis that there are two mechanisms, each acting independently of the other with no interaction effects, can be valid only over a narrow range in which the forces are approximately linear in the changes of volume, and in particular cannot be expected to hold at high pressures where there are certainly departures from linearity. That such is the case may be shown in detail in one very simple case. In the preceding paper on 12 liquids is worked out in detail the characteristic equation of the simplest conceivable substance, consisting of a single molecule. This molecule was, however, deformable, so that during collision with the walls it exerts pressure through an elastic mechanism which could be completely specified. Any ordinary liquid must have features in its mechanism closely corresponding to this. Within a certain range, which alone need concern us here, the characteristic equation of this substance consisting of a single molecule had the form

$$p\{v - b + a\tau^{1/3}\} = 2\tau,$$

where b gives the undeformed size of the molecule, and a depends on its elasticity. This equation is not of the general form given above, and gives at once:

$$\left(\frac{\partial p}{\partial \tau}\right)_v = \frac{2(v-b) + a\tau^{1/3}}{[(v-b) + a\tau^{1/3}]^2}.$$

This is evidently not a function of volume only, and in fact becomes smaller at higher temperature at constant volume. This corresponds to the fact brought out in Table XXVI that in general the ratio $\frac{p_3 - p_2}{p_2 - p_1}$ is less than 0.9. I believe that this feature of the mechanism must be present to a greater or less extent in any actual liquid, and that we are not to expect $\left(\frac{\partial p}{\partial \tau}\right)_v$ to be a volume function only, particularly at high pressures.

Another closely connected topic discussed in the former paper was concerned with the connection between internal energy and volume. Normally, as volume decreases with increasing pressure at constant temperature the internal energy decreases, but eventually passes through a minimum at a pressure equal to $\tau\left(\frac{\partial p}{\partial \tau}\right)_v$, and at higher pressures increases. The interpretation is that at ordinary volumes and temperatures the intermolecular forces are on the average attractive, and at small volumes (high pressures) repulsive; the volume at which $\left(\frac{\partial E}{\partial p}\right)_\tau = 0$ is the volume at which the attractive and the repulsive forces balance. Now at 0° Abs at atmospheric pressure the two sets of forces are in equilibrium, so that if the parts played by the various mechanisms are additive, the volume at which $\left(\frac{\partial E}{\partial p}\right)_\tau$ becomes 0 at room temperature would be expected to be equal to the volume at 0° Abs at atmospheric pressure. I have already examined this question for the solid metals[10] and found, as might be expected, that the critical volume at room temperature is not equal to that at 0° Abs at atmospheric pressure, but in general is less, so that the mechanisms cannot be additive. Recently Hildebrand[11] has applied the same idea to liquids, using in his discussion the data which I had obtained for ether. Most of the others of the twelve liquids were highly polar, and might, therefore, be expected to have complications.

Hildebrand found that the volume at which $\left(\frac{\partial E}{\partial p}\right)_\tau$ becomes 0 for ether at room temperature is very nearly, as well as can be judged from other data, its volume at 0° Abs at atmospheric pressure. Hildebrand also found the same state of affairs approximately for mercury. The liquids of this paper are normal enough to justify an examination of this point for them. In the case of n-pentane, $\left(\frac{\partial E}{\partial p}\right)_\tau = 0$ when relative volume is equal 0.77. The volume at 0° Abs at atmospheric pressure, calculated from the density of the solid[12] by the same argument that Hildebrand used for ether, is 0.80.

It should be noted that these considerations can be good only to the extent that $\left(\frac{\partial p}{\partial \tau}\right)_v$ is a pure volume function, and that therefore they cannot be expected to apply exactly to the liquids of this paper.

Summary.

A new method of measuring compressibility and thermal expansions of liquids has been developed, in which the liquid is enclosed in a sylphon, which is then exposed to external hydrostatic pressure, and the volume change determined from the change of length of the sylphon. This method has been applied to 18 liquids at 0°, 50°, and 95° up to a pressure of 12000 kg, or to the freezing pressure, and the results are collected into extensive tables giving the volume as a function of pressure aand temperature over this range. In the discussion it is shown that small scale differences in the volumes of various isomers persist to high pressures, and there is no simple connection between the relative densities at atmospheric pressure and at high pressure. The compressibility falls off rapidly with rising pressure, as was found in a preceding investigation. Two liquids are found to have the abnormally low compressibility of water. Thermal expansion also drops off by a large factor with increasing pressure, but not as much as the compressibility, as was also found before. The "pressure coefficient" $\left(\frac{\partial p}{\partial \tau}\right)_v$, is not a function of volume only, as has often been supposed, and suggestions are made as to the theoretical significance of this.

I am much indebted to my assistant Mr. W. A. Zisman for making the readings, and to my mechanic Mr. Charles Chase for the construction of the sylphon. I am also indebted to the Milton Fund of Harvard University and to the Rumford Fund of the American Academy of Arts and Sciences for generous financial assistance.

THE JEFFERSON PHYSICAL LABORATORY,
HARVARD UNIVERSITY, CAMBRIDGE, MASS.

REFERENCES 1.

[1] P. W. Bridgman, Proc. Amer. Acad. 44, 1, 1915.
[2] P. W. Bridgman, ZS. f. Krist. 67, 363, 1928.
[3] P. W. Bridgman, Proc. Amer. Acad. 56, 63, 1921.
[4] P. W. Bridgman, Proc. Amer. Acad. 58, 191, 1923.
[5] Harry Essex, Dissertation, Göttingen, 1914.
[6] A. and J. Shuman, Jour. Amer. Chem. Soc. 50, 1118, 1928.
[7] A. Müller, Proc. Roy. Soc. 120, 437, 1928; 124, 317, 1929.
[8] T. W. Richards, Jour. Chim. Phys. 25, 83, 1928.
[9] P. W. Bridgman, Handbuch der Exp. Phys. vol. 8, part II, p. 308.
[10] Reference 9, page 284.
[11] Joel H. Hildebrand, Phys. Rev. 34, 984, 1929.
[12] J. C. McLennan and W. G. Plummer, Trans. Roy. Soc. Can. 21, 99, 1927.

COMPRESSIBILITY AND PRESSURE COEFFICIENT OF RESISTANCE, INCLUDING SINGLE CRYSTAL MAGNESIUM.

By P. W. Bridgman.

Presented Dec. 10, 1930. Received Dec. 12, 1930.

CONTENTS.

Introduction.

During the last year measurements of compressibility have been extended to ten new materials; the effect of pressure on the electrical resistance of two of these has also been measured. The methods are the same as those already described in sufficient detail.[1] All of the materials described in this paper, except metallic magnesium, crystallize in the cubic system, so that measurements on crystalline aggregates give as much information as measurements on single crystals. Magnesium on the other hand, crystallizes in the hexagonal system; single crystals of this were prepared, the compressibility was measured parallel and perpendicular to the axis, and the specific resistance in the same directions was determined.

Detailed Data.

NaF. The source of this material was powdered NaF from Kahlbaum. In order to get it into shape for the compressibility measurements it was compressed into a massive slug at a temperature of 580° C in an arbor press to a pressure of perhaps 10000 or 15000 kg/cm². The specimen was only 0.28 cm. long, so that the relative accuracy of the results was less than usual. Regular measurements were made at 30° and 75°; the average arithmetical deviation of a single one of the 24 observed points from a smooth curve was 1.14% and 1.55% at 30° and 75° respectively. The following are the results calculated for the volume change from the measured change of length:

$$\left.\begin{array}{ll}\text{At } 30^\circ, & -\Delta V/V_0 = 20.7 \times 10^{-7}p - 17.7 \times 10^{-12}p^2 \\ \text{At } 75^\circ, & -\Delta V/V_0 = 20.8 \times 10^{-7}p - 18.1 \times 10^{-12}p^2\end{array}\right\} p \text{ in kg/cm}^2.$$

BaF_2. This was prepared in the same way as NaF, by compressing Kahlbaum's powdered BaF_2 at a temperature of 510° C in an arbor press. Attempts to melt it resulted in decomposition. The length of the slug so produced was about 2.1 cm. After the two compressibility runs the slug was found to be darkly discolored throughout the entire interior, and there was a permanent decrease of length of about 0.0025 cm. Whatever the nature of these effects, they were probably too small to materially affect the accuracy of the compressibility, which is known not to be sensitive to impurity.

At 30° the average departure from a smooth curve of a single one of the 14 readings was 0.64% of the maximum effect, and at 75° the average departure of a single one of the 13 readings, making one discard, was 0.49%. The following results for volume compressibility were calculated from the change of length:

$$\begin{array}{ll}\text{At } 30^\circ, & -\Delta V/V_0 = 19.3 \times 10^{-7}p - 14.8 \times 10^{-12}p^2, \\ \text{At } 75^\circ, & -\Delta V/V_0 = 19.6 \times 10^{-7}p - 14.6 \times 10^{-12}p^2.\end{array}$$

SrF_2. This was prepared from Kahlbaum's powdered material by compressing at 580° in the arbor press. Two cylindrical pieces were made, 0.62 and 0.86 cm. long, which were piled together for measurement. There was no detectible change, either of length or appearance, produced in these specimens by the application of hydrostatic pressure. The average deviations from smooth curves of single readings at 30° and 75° were 0.70% and 0.83% on 14 and 12 readings respectively. The results for change of volume are:

At 30°, $-\Delta V/V_0 = 15.8 \times 10^{-7}p - 10.3 \times 10^{-12}p^2$,

At 75°, $-\Delta V/V_0 = 16.1 \times 10^{-7}p - 10.8 \times 10^{-12}p^2$.

CdF_2. The material was powdered CdF_2 from Kahlbaum, compressed at 560° in the arbor press. It was, however, much more difficult to compact this into a uniform coherent mass, and a number of attempts had to be made. In order to get some idea of the possible error from imperfectly massive specimens, measurements were made on two different specimens. One of these was quite friable, and could easily be scratched to a powder with the point of a knife. The other was harder, and suggested no difficulties of a mechanical character. The length of the specimens was about 0.4 cm.; neither showed any appreciable change of dimensions after exposure to pressure. However, during the measurements on the softer specimen, there was an explosion due to the stem of the insulating plug blowing out, and after this there were large irregularities, so that apart from the softness of the specimen, the measurements with it could not be expected to be as good as those with the harder specimen. The compressibility calculated from the data obtained with the softer specimen was about 2.5% less than that with the harder specimen. The average deviations from smooth curves of single readings on the harder specimen were 1.2% and 0.9% at 30° and 75° respectively, on totals of 14 and 13 readings. The change of volume of the harder specimen was:

At 30°, $-\Delta V/V_0 = 11.02 \times 10^{-7}p - 8.5 \times 10^{-12}p^2$,

At 75°, $-\Delta V/V_0 = 10.96 \times 10^{-7}p - 8.4 \times 10^{-12}p^2$.

The temperature coefficient shown by these results is very uncertain. The results for either temperature by itself would not justify the four significant figures retained, which are given only to permit an estimate of the order of magnitude of the temperature effect.

AlSb. This intermetallic compound has been shown to be cubic by the measurements of Owen and Preston and of Goldschmidt.[2] The specimen was prepared for the compressibility measurements by fusing together in vacuum in a closed iron container weighed amounts of the pure metals, the antimony being Kahlbaum's purest, and the aluminum some specially pure material obtained from the Aluminum Co. of America which I had used in previous work,[3] In order to make the product as homogeneous as possible, it was heated and cooled several times. The specimen was about 1.5 cm. long. The grinding operation by which the specimen was fitted to the compressibility apparatus was not as successful as usual, there probably being some rocking on the end pieces. In consequence the specimen could assume either one of two positions, so that the resulting points lay on two curves instead of one. This sort of behavior has been found before. There need be no error in the final results if enough measurements are taken to give a considerable number of points on both of the curves. At 30° 8 such points were obtained, and at 75° 12 pertaining to the same curve; the average departures of a single one of these points from a smooth curve was 0.64% and 0.48% respectively. The results for volume compressibility were as follows:

$$\text{At } 30°, \quad -\Delta V/V_0 = 18.02 \times 10^{-7}p - 19.0 \times 10^{-12}p^2,$$

$$\text{At } 75°, \quad -\Delta V/V_0 = 20.53 \times 10^{-7}p - 26.9 \times 10^{-12}p^2.$$

CdTe. This has also been shown by Zachariasen[4] to crystallize in the cubic system. It was prepared by melting together in a closed quartz tube weighed amounts of Kahlbaum's Cd and Te, the latter further purified by my method of making single crystals. The specimen, which was about 0.8 cm. long, showed, when mounted in the compressibility apparatus, to a slight extent the same positional instability as the AlSb, it being necessary to discard four of the observed points at 30°; at 75° none were discarded. There was a permanent decrease of length of 0.006 cm. after exposure to hydrostatic pressure; however, a seasoning application was made before the regular runs, so that no error should have been introduced by this effect. At 30° the average departure from a smooth curve of a single one of the 7 readings that were retained was 0.43% of the maximum effect, and at 75° the corresponding departure was 1.2% on 10 readings. The change of volume calculated from the results was:

$$\text{At } 30°, \quad -\Delta V/V_0 = 23.3 \times 10^{-7}p - 12.2 \times 10^{-12}p^2,$$

$$\text{At } 75°, \quad -\Delta V/V_0 = 23.5 \times 10^{-7}p - 11.6 \times 10^{-12}p^2.$$

HgTe. The measurements of Goldschmidt[5] have established that this also is cubic. It was prepared by heating together in a closed quartz tube weighed amounts of Kahlbaum's Te (the same as that used in preparing CdTe), and pure redistilled Hg from the laboratory stock. The specimen was about 1.1 cm. long; there was no perceptible change of length after exposure to pressure. The measurements on this material were more irregular than for any other substance, and in fact were so irregular that it was useless to try to get from them the temperature coefficient of compressibility, but the observations at both temperatures were treated together to get from them the best single smooth curve. The average deviation from a smooth curve of a single one of the 26 observations was 2.1%. The following results were found for the volume compressibility:

$$\text{Mean, } 30° \text{ and } 75°, \quad -\Delta V/V_0 = 19.8 \times 10^{-7}p - 25 \times 10^{-12}p^2.$$

TiN. This material I owe to the kindness of Dr. van Arkel of the Philips Lamp Works, Eindhoven, Holland. It was prepared in the form of a slender rod about 1.0 mm. in diameter by the method of vacuum deposition which he has described.[6] The specimen was not a single grain, but the material is cubic, so that a single grain was not necessary.

Compressibility. The compressibility sample was 1.6 cm. long. Measurements were made as usual at 30° and 75°. This material is less compressible than iron, so that the slope of the curve was abnormal, and furthermore the effect is so small that no departure of the observed points from linearity could be established. At each temperature 13 observations were made; at 30° the average departure of a single point from a smooth curve was 1.2% and at 75° 1.1%. The final results are:

$$\text{At } 30°, \quad -\Delta V/V_0 = 3.32 \times 10^{-7}p - 2.13 \times 10^{-12}p^2,$$

$$\text{At } 75°, \quad -\Delta V/V_0 = 3.51 \times 10^{-7}p - 2.13 \times 10^{-12}p^2.$$

The second degree term in these expressions is contributed by the absolute compressibility of the iron, there being no appreciable departure of the differential compressibility from linearity.

Pressure Coefficient of Resistance. The same specimen was used for the resistance and the compressibility measurements. The potentiometer method was used, connections being made by spring clips, since it is not possible to solder this material. In addition to measurements under pressure, the specific resistance and temperature coefficient of resistance were measured at atmospheric pressure. At 30° at atmospheric pressure the specific resistance was 28.1×10^{-6} ohms per cm. cube. The temperature coefficient at 0° C at atmospheric pressure, obtained by linear extrapolation of measurements at 30° and 75°, was 0.00221.

Pressure measurements were made as usual at 30° and 75°. The effect is unusually small; this, combined with the smallness of the absolute resistance of the specimen, made the percentage irregularity in the results larger than usual. At 30° the average deviation from a straight line of a single one of the 11 readings was 1.2%, and at 75° the corresponding figure was 2.2% on 14 readings. The total displacement of the potentiometer slider for 12000 kg was 1 cm., so that an irregularity of 1% corresponds to 0.1 mm. setting of the slider, indicating that the readings were about as regular as the construction of the apparatus would allow. There was no appreciable departure from linearity at either temperature. The results follow:

Average coefficient to 12000 kg at 30°, -9.35×10^{-7}.

Average coefficient to 12000 kg at 75°, -9.97×10^{-7}.

The most important feature of the results is that the sign is negative, as it is for the majority of the pure metals.

TiC. This also I owe to the kindness of Dr. van Arkel. It was prepared by the same method as TiN; it also is cubic so that a single crystal was not necessary.

Compressibility. The sample was 2.5 cm. long, and 1.5 mm. in diameter; no perceptible alteration of dimensions as produced by pressure. This, like TiN, is less compressible than iron, and again it was not possible to detect any deviation from linearity of the observed differential compressibility. The average deviation from a straight line of a single one of the 13 observed points at 30° was 0.32%, and at 75° 0.41% for 11 points. The results are:

$$\text{At } 30°, \quad -\Delta V/V_0 = 4.72 \times 10^{-7}p - 2.16 \times 10^{-12}p^2,$$
$$\text{At } 75°, \quad -\Delta V/V_0 = 4.78 \times 10^{-7}p - 2.19 \times 10^{-12}p^2.$$

Pressure Coefficient of Resistance. The various details in measuring the resistance were the same as with TiN. The specific resistance at atmospheric pressure at 30° was found to be 158.4×10^{-6}. The temperature coefficient at atmospheric pressure at 0° C, by linear extrapolation of measurements at 30° and 75°, was + 0.000235.

The pressure measurements showed no appreciable departure from linearity. At 30° the average departure from a smooth curve of a single one of the 15 readings was 1.3%, and at 75° the corresponding deviation was 0.7% on 14 readings. The results are:

Average coefficient to 12000 at 30°, -1.375×10^{-6},

Average coefficient to 12000 at 75°, -1.35×10^{-6}.

Again the coefficient is negative and about 50% larger numerically than for TiC. This is to be contrasted with a specific resistance 5.6 fold larger.

Mg. This material crystallizes in the hexagonal system, so that measurements, to be significant, must be made on single crystals. As far as I can find in the literature, single crystals of magnesium have not been hitherto studied. It proved, however, extremely easy to make this in single crystal form by my method of slow lowering of the molten metal from a furnace.[7] For this purpose a special vacuum furnace was constructed, in which a vacuum of the order of 0.01 mm. was maintained during operation. After many trials, iron crucibles were found to be by far the best.

Compressibility. For making the compressibility samples the crucible was about 2 cm. inside diameter and 7.5 cm. long, and was provided with a lid tightly clamped into position. The center of the lid was pierced with a fine steel tube, long enough to reach from the furnace to the colder parts of the enclosure; in this way the interior of the crucible could be exhausted, and at the same time the magnesium prevented from distilling out of the crucible, an effect which may be very annoying. The crucible was lowered at a rate of about 2 cm. per hour from a maximum temperature about 50° above the melting point. The magnesium with which the crucible was loaded was cut from a rod 4 cm. in diameter of commercially pure magnesium, which is usually very pure. The dimensions of the magnesium after casting into single crystal form were 2 cm. diameter and 5 cm. long. The crystal axis was located by optical examination of the reflection pattern This was developed with unusual perfection, all 20 faces being present. These composed a band of 18 faces about the equator, 6 on the equator itself and 6 in a band on either side like the tropics of Cancer and Capricorn, and in addition 2 faces, one at each pole, perpendicular to the hexagonal axis. The crystal axis proved to make an angle of 51° with the length of the casting. Specimens for the compressibility measurements were cut from the casting with a jewellers saw, parallel and perpendicular to the axis. The following results were obtained.

Parallel to the axis. The length of the specimen was 2.0 cm.; there was no appreciable permanent deformation as a result of exposure to hydrostatic pressure. At 30° the average departure from a smooth curve of a single one of the 14 readings was 0.18% and at 75° 0.04% on the same number of readings. The greater regularity of these readings compared with that of the other materials of this paper is what might be expected in view both of the greater geometrical perfection and more perfect homogeneity of the specimens. The numerical results were as follows:

$$\text{At } 30°, \quad -\Delta l/l_0 = 9.842 \times 10^{-7}p - 6.51 \times 10^{-12}p^2,$$

$$\text{At } 75°, \quad -\Delta l/l_0 = 10.154 \times 10^{-7}p - 7.78 \times 10^{-12}p^2.$$

Perpendicular to the axis. The length of this specimen also was 2.0 cm. and it showed no permanent distortion after exposure. At 30° the average departure from a smooth curve of a single one of the 14 readings was 0.08%, and at 75° 0.15% on the same number. The numerical results are:

$$\text{At } 30°, \quad -\Delta l/l_0 = 9.845 \times 10^{-7}p - 9.19 \times 10^{-12}p^2,$$

$$\text{At } 75°, \quad -\Delta l/l_0 = 9.659 \times 10^{-7}p - 6.95 \times 10^{-12}p^2.$$

By combining these results for the linear changes, the following may be found for the volume compressibility:

$$\text{At } 30°, \quad -\Delta V/V_0 = 29.53 \times 10^{-7}p - 28.3 \times 10^{-12}p^2,$$

$$\text{At } 75°, \quad -\Delta V/V_0 = 29.47 \times 10^{-7}p - 24.7 \times 10^{-12}p^2.$$

I have previously measured the volume compressibility of a multicrystalline extruded sample.[8] The results for this were: at 30°, $29.60 \times 10^{-7}p - 20.3 \times 10^{-12}p^2$, and at 75°, $29.97 \times 10^{-7}p - 18.0 \times 10^{-12}p^2$. The agreement is as good as could be expected in view of the indefinite character of the former sample.

Electrical Resistance. It was necessary to make special castings for the resistance measurements, since it would not have been possible without undue distortion to machine sufficiently slender rods out of the massive casting from which the compressibility samples were cut. The material was specially purified magnesium, which I owe to the kindness of Mr. R. S. Archer of the Research Laboratory of the Aluminum Co. of America. The new material was cast in the form of slender rods about 3 mm. in diameter and 5 cm. long in cylindrical iron molds; these molds were tightly closed at both ends. The rate of lowering from the furnace was about the same as for the larger casting. The mold was removed after casting by slitting with a fine saw in the milling machine. It was usually not possible to remove the mold without some slight bending of the specimen. However, the error so introduced was presumably small, as the resistance of this material does not seem sensitive to slight distortions. The orientation of the rods was determined optically; the reflection patterns were not as fully developed as in the larger casting, but enough faces were always present to allow certain identification. Magnesium proved unexpectedly easy to get into single crystal form; none of the rods were multi-crystalline.

The specific resistance at 22.5° was obtained from potentiometer measurements of the resistance, combined with measurements of the

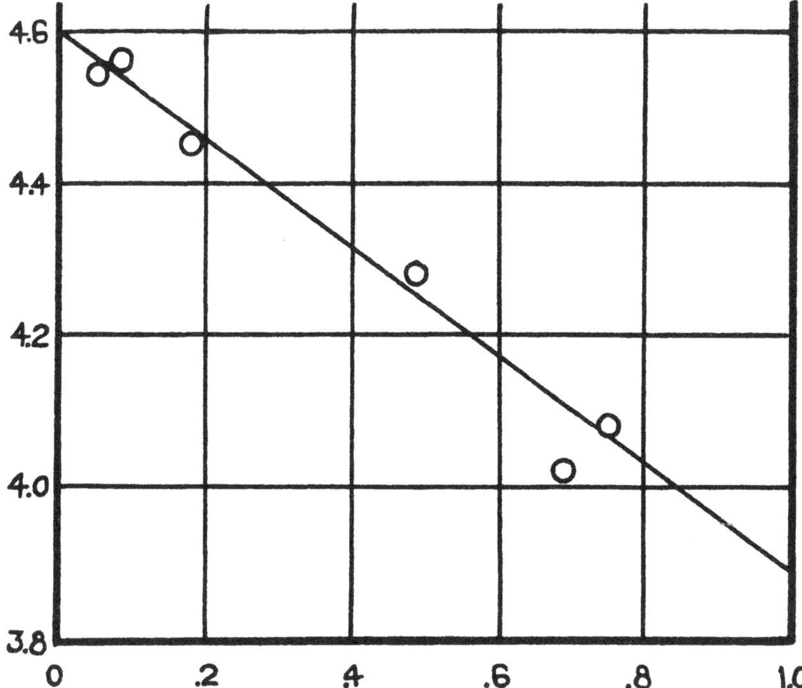

FIGURE 1. The specific resistance at 22.5° C of single crystal magnesium plotted against the square of the cosine of the angle between the hexagonal axis and the direction of current flow.

dimensions. The results are shown with sufficient accuracy in Figure 1, in which specific resistance is plotted against $\cos^2\theta$, where θ is the angle between the length of the rod and the hexagonal axis. Since there is no reason to suspect the correctness of the theoretical result that resistance is a linear function of $\cos^2\theta$, the straight line drawn in the figure may be taken as giving the most probable value of resistance. This means that the specific resistance perpendicular to the hexagonal axis is 4.60×10^{-6} and 3.89×10^{-6} parallel to it. The average specific resistance to be expected in a multi-crystalline casting is therefore 4.36×10^{-6}, or 4.32×10^{-6} corrected to 20°. This may be compared with the value 4.46×10^{-6} at 20° given in International Critical Tables for the polycrystalline metal. The new value is thus somewhat lower; part of this difference is probably to be explained by greater purity of the material, and part by the fact that it is in single crystal form.

The effect of pressure on the resistance up to 12000 kg/cm² at 30° and 75°, and the temperature coefficient of resistance at atmospheric pressure, was determined on three samples, with the axes inclined at 73°, 30°, and 29° to the length. The measurements on the 30° sample were irregular and could be used only to confirm in a general way the results on the 29° sample. There was considerable difficulty in obtaining satisfactory readings, both because of the smallness of the resistance and of the difficulty of making perfect electrical contact. The total resistance of the specimens was about 2×10^{-4}, and the maximum change produced by pressure about 5% of this. Consistency in the pressure readings to 1% demanded therefore resistance measurements correct to 10^{-7} ohms. Electrical connections were made with springs of 0.010 inch piano wire snapped into fine grooves girdling the rods. Since the potentiometer method was used, no error was introduced by resistance at the contacts, but in order to allow high enough currents for measurement, the goodness of the contacts had to be restored at intervals with a high tension current from a small bell ringing magneto.

The following results were found for the effect of pressure:

Hexagonal axis 73° to the length,

$$\text{At } 30^\circ \text{ C,} \quad -\frac{\Delta R}{R(0, 30^\circ)} = 5.61 \times 10^{-6}p - 8.4 \times 10^{-11}p^2.$$

Average coefficient to 12000 = -4.60×10^{-6}.

$$\text{At } 75^\circ \text{ C,} \quad -\frac{\Delta R}{R(0, 75^\circ)} = 5.67 \times 10^{-6}p - 8.35 \times 10^{-11}p^2.$$

Average coefficient to 12000 = -4.65×10^{-6}.

Hexagonal axis 29° to the length,

At 30° C, $-\frac{\Delta R}{R(0, 30°)} = 5.48 \times 10^{-6} - 7.8 \times 10^{-11}p^2.$

Average coefficient to 12000 = -4.55×10^{-6}.

At 75° C, $-\frac{\Delta R}{R(0, 75°)} = 5.99 \times 10^{-6} - 11.8 \times 10^{-11}p^2.$

Average coefficient to 12000 = -4.58×10^{-6}.

The average departures of single readings from smooth curves were 1.2 and 1.5% for the 73° specimen, and 1.4 and 1.6% for the 29° specimen at 30° C and 75° C respectively. In comparing results for different specimens more emphasis should be placed on the average coefficients to 12000 than on the coefficients in the power series expansion.

The average pressure coefficient to 12000 of the polycrystalline metal to be calculated from the above, assuming the pressure coefficient is linear in $\cos^2\theta$, is 4.58×10^{-6}. This may be compared with my best previous value of 4.07×10^{-6} for magnesium of doubtless inferior purity. The direction of variation of pressure coefficient with increasing purity is the same as that always found hitherto, without exception.

The temperature coefficient at atmospheric pressure was obtained by linear extrapolation of the resistance at 30° and 75°. The following are the results:

Hexagonal axis 73° to the length,

Mean coefficient 0° to 100° C, 0.00439.

Hexagonal axis 29° to the length,

Mean coefficient 0° to 100° C, 0.00420.

The mean coefficient for polycrystalline metal, assuming the temperature coefficient to be linear in $\cos^2\theta$, demanded by these values, is 0.00432. The best previous observed value is apparently 0.00412.

Discussion.

NaF was investigated chiefly because it was one of the substances which Slater[10] did not measure in his investigation of the compressibility of the alkali halides, all in single crystal form. The specimen measured here was not a single crystal, but for a cubic crystal this is not important. The value for the compressibility of NaF found here fits smoothly into the series of values found by Slater. This is shown

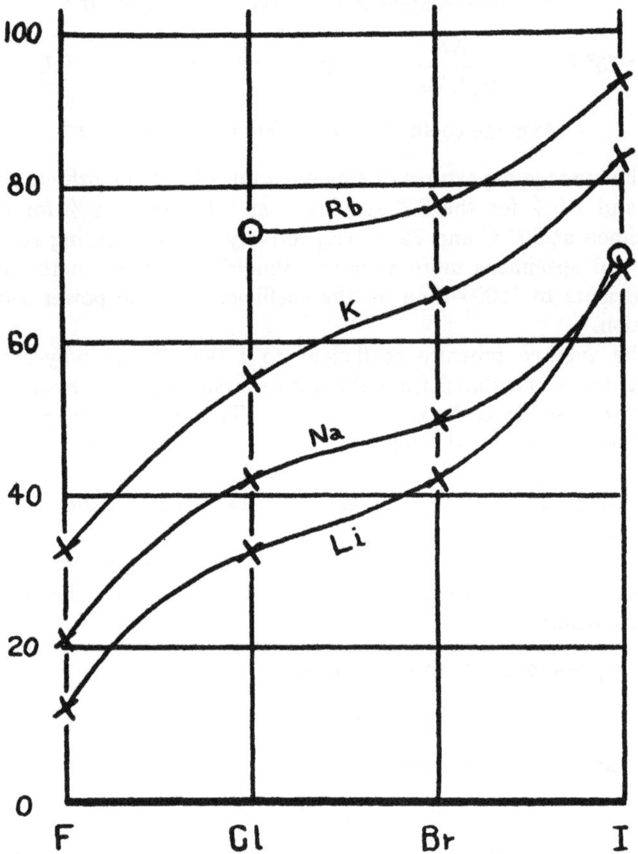

FIGURE 2. The compressibility at room temperature and atmospheric pressure of a number of the alkali halides. The curves connect the points for the compounds with the same alkali metal.

in Figure 2, in which the compressibilities at atmospheric pressure of various alkali halides are plotted with the successive halogens equispaced along the axis of abscissas. The two points for RbCl and LiI, which lie distinctly off the curve, are taken from measurements of Richards and Saerens,[11] which were made by a different method. It would probably be worth while to repeat these measurements.

One might perhaps expect the same sequence of conpressibilities on passing from one halogen compound to another in the series of the alkali earths as in the series of the alkali metals. That this is not the case is shown by Figure 3, in which are plotted the compressibilities found above for BaF_2 and SrF_2 together with the values found by Madelung and Fuchs[12] for $BaCl_2$, $BaBr_2$, $SrCl_2$, and $SrBr_2$. The

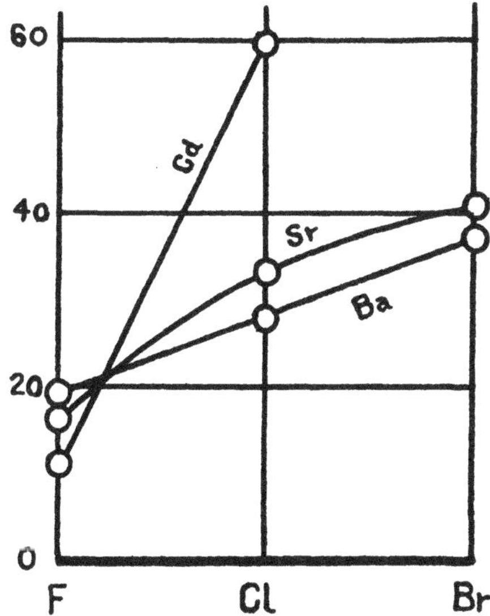

FIGURE 3. The compressibilities at atmospheric pressure and room temperature of the fluorides, chlorides and bromides of Sr and Ba, and the fluoride and chloride of Cd.

sequence is not maintained, but there is a reversal of order at fluorine as compared with chlorine and bromine. The data for CdF_2 and $CdCl_2$ are also plotted in the same diagram; here there is an enormous variation on passing from fluorine to chlorine, which almost might suggest that the crystal structures of these two compounds is not the same.

The compressibility of AlSb might perhaps be expected to be some sort of a mean between the compressibilities of its components. Perhaps the most natural sort of a mean is one in which the compressibilities of the pure metals are weighted according to their atomic volumes. Taking the initial compressibility of Sb as 27.0×10^{-7} and its atomic volume as 18.2, with 13.4×10^{-7} and 10.0 as the corresponding figures for Al, the compressibility to be expected is

$$\left(\frac{18.2}{28.2} \times 27.0 + \frac{10.0}{28.2} \times 13.4\right) \times 10^{-7} = 22.1 \times 10^{-7}.$$

The actual compressibility is 18.0×10^{-7}, 19% less. It is significant that the actual compressibility is less than that calculated; it is highly probable that at least part of this difference is to be explained by the effect of the internal cohesive pressure that the late Professor T. W. Richards was so fond of discussing. This cannot be the whole story, however, as is obvious from the non-cubical character of pure Sb and its greatly differing compressibility in different directions.

The compressibility of the two tellurides, CdTe and HgTe, differs even more from what may be calculated by the rule of mixtures. Taking the compressibilities of Te, Cd, and Hg (solid) as 50.8, 22.8, and 33.1×10^{-7}, and the corresponding atomic volumes as 21.2, 13.0, and 14.3, the calculated compressibility of CdTe is 40.1 and that of HgTe 43.7×10^{-7}, against the experimental values 23.3 and 19.8×10^{-7} respectively, which are thus about one-half the calculated values.

The low compressibilities of TiN and TiC is what would be expected from the high melting points and great mechanical hardness. Again the compressibility is very much less than that of the metallic constituent, the compressibility of pure Ti being about 8×10^{-7} against 3.3 for TiN and 4.7×10^{-7} for TiC. It is interesting that the compound with C is more compressible than that with N, although in the pure state solid N is much more compressible than C either as diamond or as graphite.

The electrical resistance of these two compounds puts them in the metallic class, the specific resistance of TiN being about 0.23, and that of TiC about 1.32 fold that of metallic Bi. The temperature coefficient of resistance of TiN, 0.00221, puts it not far from the pure metals; that of TiC, 0.00024, is in a different class. As far as the effect of pressure on resistance goes, the behavior is again in the class of the metals; the coefficient of TiN, -9.4×10^{-7}, is not far from that of metallic Co, and the coefficient of TiC, -1.37×10^{-6}, is not far from that of Mo. It would appear in general that the temperature coefficient of resistance is a more reliable criterion of the metallic character of a substance than either the specific resistance or the pressure coefficient of resistance.

Magnesium is of interest because it is the first hexagonal metal measured in which the axial ratio corresponds to a piling of *spheres* in close packed hexagonal array. The piling in Zn and Cd, for example, is that of close packed ellipsoids. It is, of course, well known that the close packed hexagonal arrangement of spheres differs little from the close packed face centered cubic arrangement. In fact, if the two arrangements are built up by piling over each other layers perpendicular to the hexagonal axis, the only difference between the two arrangements is a slight relative displacement of the third layers, the first and second layers in the two arrangements being the same. Hence the forces on the atoms in the two arrangements can differ only by the contributions of the more distant atoms, which are unimportant because of the rapid falling off of atomic forces with increasing distance. It follows that a close packed hexagonal arrangement of spheres would not be expected to differ greatly in physical properties from a close packed cubical arrangement, and in particular, the compressibility of the hexagonal arrangement would be expected to be nearly the same parallel and perpendicular to the axis, because the compressibility of a cubic crystal as the same in all directions.

This expectation is strikingly borne out by the behavior of the compressibilities of magnesium; at 30° the initial compressibility in the two directions is practically the same. There are, however, small differences; the second degree terms differ, so that at high pressures magnesium is more compressible parallel to the axis than perpendicular. The temperature behavior in the two directions is also different; perpendicular to the axis the temperature coefficient of compressibility is negative, a rather unusual effect. Furthermore, the negative coefficient perpendicular to the axis more than compensates for the positive coefficient parallel to the axis, so that the initial temperature coefficient of volume compressibility is also negative. The second degree terms act differently, however, so that at high pressures the volume compressibility is greater at the higher temperature, as is normal.

The specific resistance of magnesium differs more in different directions than would be expected from the considerations above and from the fact that the compressibility is so nearly the same in all directions. Furthermore, the resistance is abnormal in that it is least parallel to the hexagonal axis, that is, the resistance is least for flow across the basal plane. The only other example of this sort of behavior is antimony, and here the abnormality disappears at high pressures. The usually greater resistance across the basal plane is associated with the functioning of this plane as the plane of easiest slip or cleavage. The mechanical properties of single magnesium crystals have not yet been sufficiently studied to show whether in it the basal plane is also a slip plane or not.

After the manuscript of this article went to press, a paper has appeared by O. Sckell, Ann. d. Phys. 6, 932, 1930, in which it is shown that mercury (tetragonal) also has a smaller resistance parallel to the crystallographic axis.

The pressure coefficient of resistance does not vary as much with direction as does the specific resistance, but there is nevertheless a noticeable variation, and the variation is the same in character as that of the specific resistance, the pressure coefficient being least numerically in the direction in which the resistance is also least. It will be found on making the numerical calculation that the relative pressure coefficients are such in different directions that the ratio of resistance in the two principal directions is independent of pressure.

The temperature coefficient also varies with direction in the same way, the temperature coefficient being greatest in the direction in which the specific resistance is the greatest. The numerical values are such that the abnormal ratio of resistance in the two principal directions becomes accentuated at high temperatures.

I am much indebted for financial assistance to the Rumford Fund of the American Academy of Arts and Sciences and to the Milton Fund of Harvard University. I am also much indebted to my assistants: Mr. W. A. Zisman for making most of the readings, Mr. George Langreth for preparing the samples of NaF, BaF_2, SrF_2, CdF_2, AlSb, CdTe, and HgTe, and Mr. Sanford Goldman for preparing the single crystals of magnesium.

The Jefferson Physical Laboratory,
Harvard University, Cambridge, Mass.

REFERENCES.

[1] P. W. Bridgman, Proc. Amer. Acad. 52, 573, 1917; 56, 61, 1921; 58, 151, 166, 1923; 59, 109, 1923; 60, 385, 1925; 63, 207, 347, 1928; 64, 50, 1929. Amer. Jour. Sci. 10, 483, 1925; 15, 287, 1928.

[2] E. A. Owen and G. D. Preston, Nat. 113, 914, 1924. V. M. Goldschmidt, Geochemische Verteilungsgesetze der Elemente, VIII, page 32. Oslo, 1927.

[3] P. W. Bridgman, Proc. Amer. Acad. 58, 153, 195, 1923.

[4] W. Zachariasen, ZS. f. phys. Chem. 124, 277, 1926.

[5] Second reference under (2), page 31.

[6] A. E. van Arkel, Chemisch Weekblad, part 24, No. 8, 1927.

[7] P. W. Bridgman, Proc. Amer. Acad. 60, 307, 1924.

[8] Fourth reference under (1), page 209.

[9] P. W. Bridgman, Proc. Amer. Acad. 56, 88, 1921.

[10] J. C. Slater, Phys. Rev. 23, 488, 1924; Proc. Amer. Acad. 61, 135, 1926.

[11] T. W. Richards and E. P. R. Saerens, Jour. Amer. Chem. Soc. 46, 934, 1924.

[12] E. Madelung und R. Fuchs, Ann. Phys. 65, 289, 1921.

THE P-V-T RELATIONS OF NH_4Cl AND NH_4Br, AND IN PARTICULAR THE EFFECT OF PRESSURE ON THE VOLUME ANOMALIES

By P. W. Bridgman

Jefferson Physical Laboratory, Harvard University

(Received May 21, 1931)

Abstract

Measurements have been made of the volume as a function of pressure of NH_4Cl at 75°, 30°, and 0°C, and of NH_4Br at 75°, 0°, and −72°C. The pressure range was 12000 kg/cm^2 except at −72°, where it was 7500. There are discontinuities in the slope of the volume isotherm of NH_4Cl at 3370 kg at 0°, and at 9390 at 30°, and in the isotherm of NH_4Br at 1620 at −72°. The character of the discontinuity of NH_4Br is different from that of NH_4Cl, and is more like that of a polymorphic transition. These discontinuities correspond exactly to the discontinuities in the thermal expansion found by Simon and Ruhemann at atmospheric pressure. The displacement by pressure of the discontinuities to respectively higher and lower temperatures is that demanded by thermodynamics because of the opposite sign of the volume anomalies in the two cases. The thermodynamics of the effect of pressure and temperature on a discontinuity in a volume isotherm is discussed, and a generalized Clapeyron's equation shown to hold. By means of this equation the thermal effect accompanying the change can be calculated, and is found to be about 100 and 160 gm cal per mol for NH_4Cl and NH_4Br respectively. The thermal effect in NH_4Cl is approximately independent of pressure; the volume effect decreases rapidly with increasing pressure, being 0.0030 at 0° and 0.0015 at 30°. The volume effect in NH_4Br is approximately −0.017 at −72°. The explanation proposed by Pauling for the anomalies, namely passing from oscillational to rotational motion by the NH_4 radical, evidently must be supplemented by other considerations to account for the several marked differences between the behavior of the two substances. There are without doubt many other instances of such anomalies spread over a more or less localized range of pressure.

Introduction

In 1922 Simon[1] showed that NH_4Cl has anomalies in its specific heat, and in 1927 Simon, Simson, and Ruhemann[2] showed that NH_4Br has a similar anomaly, the maximum of both anomalies being at temperatures between −30° and −40° C. The anomaly consists of a very abnormally large specific heat distributed over a range of only a few degrees; both above and below this range the heat is of more normal magnitude. The phenomenon is not unlike that of a polymorphic transition spread over a temperature range by dissolved impurity, but x-ray analysis shows that the crystal structure is the same above and below the region of anomaly. In 1930 Simon and Bergmann[3] examined the volume-temperature relations of these substances, and found that the specific heat anomaly is accompanied by a volume anomaly, but this

[1] F. Simon, Ann. d. Physik **68,** 241 (1922).

[2] F. Simon, Cl. v. Simson, und M. Ruhemann, Zeits. f. physik. Chem. **129,** 339 (1927).

[3] F. Simon und R. Bergmann, Zeits. f. physik. Chem. **8,** 255 (1930).

has opposite signs in the two cases, the volume of NH_4Cl increasing and that of NH_4Br decreasing in the region of the specific heat anomaly. Thermodynamic principles at once suggest that it should be possible to displace the temperature of the anomalous region by the application of hydrostatic pressure, the temperature rising with pressure in the case of NH_4Cl because of the increase of volume, but falling with pressure for NH_4Br. These effects have in fact been found, and this paper contains a report of them. The primary purpose of the measurements was to establish the existence of the volume

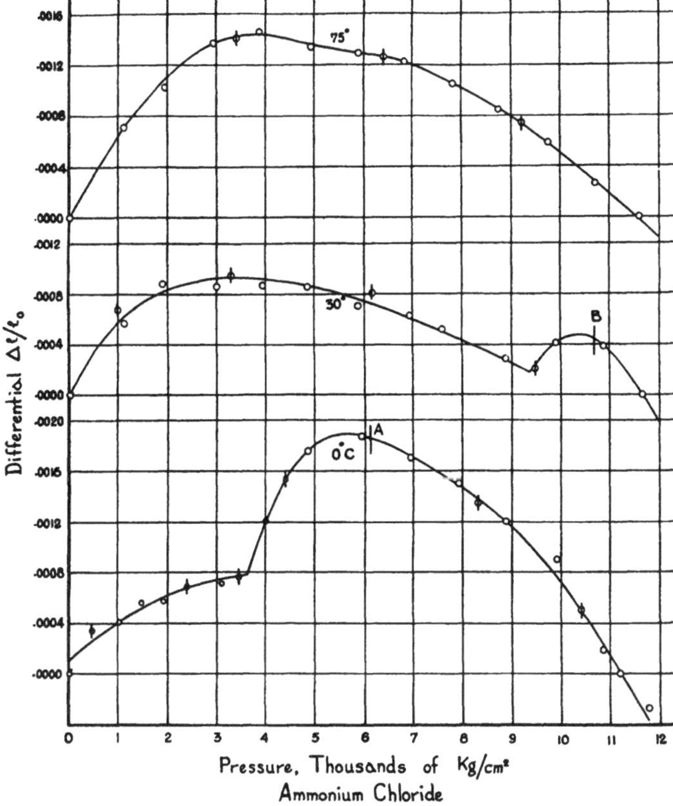

Fig. 1. Difference between the actual and the linear $\Delta l/l_0$ of NH_4Cl as a function of pressure at several temperatures. The open circles are data obtained with increasing pressure, the crossed circles with decreasing pressure.

anomalies at high pressure, but the method of measurement was such that it was possible to find the complete relation between temperature, pressure, and volume on both sides of the anomalous region.

The length, rather than the volume, of these substances was measured as a function of pressure, and from the changes of length the changes of volume were calculated. The materials were originally in the form of finely granulated powder; this was compressed dry at room temperature to a pressure between 20000 and 25000 kg/cm², forming a perfectly coherent mass. This was then turned in the lathe to the requisite dimensions, about 2 cm long and 6 mm in diameter. The manner of formation should ensure that the volume changes take place equally in all directions, and thus justify the calculation of the change of volume from the change of length. There is particularly little danger in this procedure for these two substances, because they both crystallize in the cubic system and are equally compressible in all directions.

The source of the material was as follows: NH_4Cl, Kahlbaum's "K" grade, "for analysis"; NH_4Br, Kahlbaum's best listed grade, Ph. G. VI.

The length was measured at several different temperatures as a function of pressure in the "lever piezometer," with which the compressibility of many other materials has been measured,[4] so that no new description of the apparatus is necessary here.

The Data

NH_4Cl. Measurements were made on this at 30°, 75°, and 0°, in this order. At 30° and 75° nothing very striking displayed itself, but at 0° a very pronounced effect of the sort expected was found, and then reexamination of the 30° measurements showed the same sort of behavior near the end of the pressure range and of much smaller magnitude then at 0°, so that it might have been explained as an experimental irregularity if it had not been expected.

The calculations were made by the regular method. The potentiometer readings were converted into fractional changes of length, which were then plotted against pressure. A straight line was passed between the initial and final points, and the differences between the observed readings and this straight line plotted on a large scale, thus magnifying the effects. These differences, with reversed signs, are plotted in Fig. 1. It is particularly to be noticed that whatever the internal change is which is responsible for the breaks in the curves, it runs reversibly, the points obtained with increasing and decreasing pressure both lying on the curve within the limits of error. The large discontinuities at 0° and 30° are the ones which were looked for, and correspond to that found by Simon at atmospheric pressure, displaced by pressure to higher temperatures. In addition, there seems to be a new anomaly at 75° between 4000 and 6500 kg/cm², much less marked, but still almost certainly beyond experimental error.

The actual volumes of NH_4Cl at 0°, 30°, and 75° at intervals of 2000 kg, calculated from the changes of length, are shown in Table I. The volume at

Table I. *Volume of NH_4Cl as a function of pressure and temperature.*

Pressure kg/cm²	Relative volumes 0°	30°	75°
0	1.00000	1.00496	1.01398
2000	*.98870*	.99317	1.00159
4000	.97730	.98344	.99103
6000	.96586	.97420	.98209
8000	.95714	*.96562*	.97333
10000	.94914	.95599	.96535
12000		.94764	.95783

[4] P. W. Bridgman, Proc. Amer. Acad. 58, 166 (1923).

atmospheric pressure as a function of temperature was taken from International Critical Tables. The horizontal lines in the 0° and the 30° columns indicate the range in which the discontinuity occurs. More accurate coordinates for the points of discontinuity, that is, the point at which the anomaly begins to appear on passing from the high temperature to the low temperature form, are shown in Table II.

TABLE II. *Coordinates of discontinuity of NH_4Cl.*

Pressure kg/cm²	Temperature
0	−30° C (Simon)
3370	0
9390	30

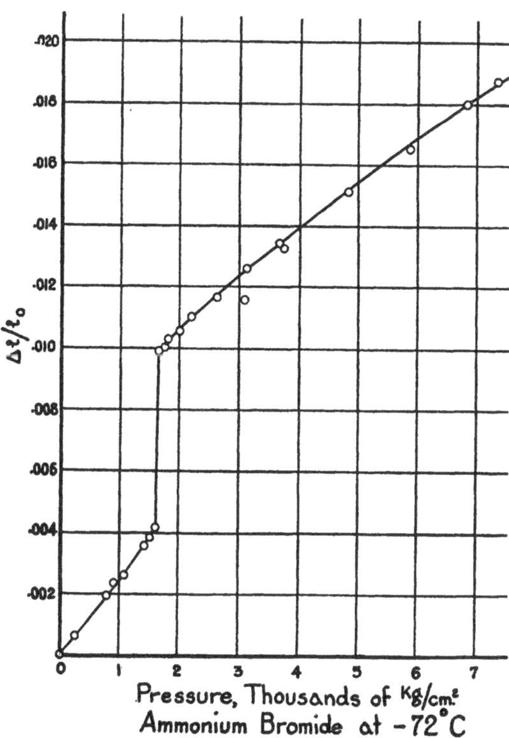

Fig. 2. $\Delta l/l_0$ of NH_4Br as a function of pressure at −72°C.

NH_4Br. Runs were first made at 0° and 75°. The curves of length against pressure at both these temperatures were perfectly smooth. A run was then made at −72°, the temperature of mixed solid CO_2 and ethyl alcohol. Fortunately it proved possible to make the run at this temperature with no essential modification in the regular pressure set-up. The regular temperature bath was used; about 4 gallons of alcohol and 50 pounds of solid CO_2 were sufficient to cool the pressure cylinder and fill the bath; the bath was maintained at temperature for something more than 24 hours, and for this a total of about 100 pounds of solid CO_2 was sufficient. This was purchased in the form of "dry ice" from a commercial company. A more serious consideration than the refrigerant was the pressure transmitting liquid, most liquids freezing at the pressure expected at this temperature. It turned out that iso-pentane, the freezing point of which at atmospheric pressure is 28° lower than that of normal pentane, was capable of transmitting pressures up to more than 7000 kg/cm² without freezing, and this proved more than sufficient.

The effect with NH_4Br is much larger than with NH_4Cl, so that the phenomena can be sufficiently well displayed by plotting against pressure the total change of relative length, instead of the difference between the total and the linear change, as was done for NH_4Cl. The results are shown in Fig. 2. The experimental points include points obtained from five separate excursions, three with increasing and two with decreasing pressure; three of these excursions spanned the break in the curve. It is obvious from the points that the effect is essentially repeatable and reversible. The discontinuity is seen to be very abrupt, quite unlike that of NH_4Cl, and practically indistinguishable from an ordinary polymorphic transition.

In Table III are collected the numerical results for the relative changes of volume as fractional parts of the volume at atmospheric pressure at room temperature. The volume as a function of temperature at atmospheric pressure is not given in International Critical Tables, so that it was not possible

TABLE III. *Fractional volume changes of NH_4Br as a function of pressure on three isotherms.*

Pressure kg/cm²	$\Delta V/V_0$ −72°	0°	75°
0	0.00000	0.00000	0.00000
400	.00264		
800	.00558		
1200	.00876		
1600	*.01233*		
1800	.03021		
2000	.03144		
3000	.03666	.01803	.01932
4000	.04131		
5000	.04548		
6000	.04968	.03297	.03549
7000	.05358		
7500	.05547		
9000		.04617	.04998
12000		.05817	.06324

to give the actual volumes, as in the case of NH_4Cl. The horizontal line in the column at −72° indicates the pressure interval of the discontinuity. More accurately, the major part of the transition takes place between 1613 and 1646 kg/cm², the points at these two pressures lying on the lower and the upper branches of the curve, respectively.

DISCUSSION

It will be convenient first to examine the simple thermodynamics of a discontinuity in the volume-pressure isotherm. In Fig. 3 are represented two isotherms at τ and $\tau+\Delta\tau$, with a discontinuity, the discontinuity on the

$\tau+\Delta\tau$ isotherm occurring at a pressure dp higher than on the τ isotherm. Let (1) and (2) refer to the two parts of the isotherm on the two sides of the discontinuity. An examination of the purely geometrical relations, without the use of any thermodynamics, will yield the relation

$$\frac{d\tau}{dp} = -\left\{\left(\frac{\partial v_2}{\partial p}\right)_\tau - \left(\frac{\partial v_1}{\partial p}\right)_\tau\right\}\Big/\left\{\left(\frac{\partial v_2}{\partial \tau}\right)_p - \left(\frac{\partial v_1}{\partial_\tau}\right)_p\right\}.$$

This may be written in the equivalent form, by means of the thermodynamic relation $(\partial Q/\partial p)_\tau = -\tau(\partial v/\partial \tau)_p$, where Q is heat absorbed

$$\frac{d\tau}{dp} = \tau\left\{\left(\frac{\partial v_2}{\partial p}\right)_\tau - \left(\frac{\partial v}{\partial p}\right)_\tau\right\}\Big/\left\{\left(\frac{\partial Q_2}{\partial p}\right)_\tau - \left(\frac{\partial Q_1}{\partial p}\right)_\tau\right\}.$$

$(\partial v_2/\partial p)_\tau - (\partial v_1/\partial p)_\tau$, the difference of the rate of change of volume with pressure after and before the discontinuity, may be described as the rate of change

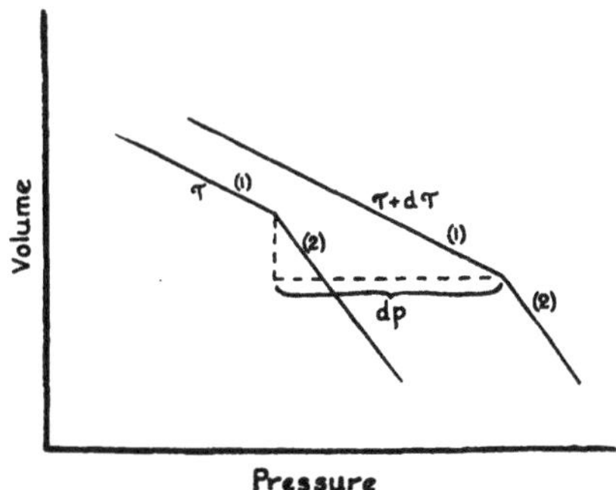

Fig. 3. Figure for computation of the displacement with pressure and temperature of a discontinuity in the volume isotherms.

with pressure of that part of the total volume which arises from whatever agency it is that is responsible for the discontinuity. Similarly, $(\partial Q_2/\partial p)_\tau - (\partial Q_1/\partial p)_\tau$ is the rate of change with pressure of that part of the total heat absorption due to the agency responsible for the discontinuity. The ratio, which may be written as dv/dQ, is merely the ratio of volume change to heat absorbed due to the internal change. The result may then be written

$$\frac{d\tau}{dp} = \tau\frac{dv}{dQ}\cdot$$

This equation is the exact analogue of the ordinary Clapeyron's equation, and also of the more general equation controlling the relation between temperature and pressure in a system in which internal chemical reactions are taking place. In this latter case, $d\tau/dp$ is to be taken at constant composition, and dv/dQ, ordinarily written as $\Delta v/\Delta Q$, is the ratio of volume change to heat for one gram molecule's worth (or any other amount) of the reaction. The requirement that $d\tau/dp$ be taken at constant composition in the conventional analysis for reacting substances is replaced by the requirement, in our case, that $d\tau/dp$ be taken at the initial point of discontinuity, obviously an entirely equivalent requirement. The analysis given above has the advantage for our present purpose that it demands somewhat less specialized assumptions about the nature of the internal changes in the system than the conventional analysis applicable to chemical reactions.

If whatever causes the discontinuity runs its course and is exhausted, so that there is a second discontinuity to mark the completion of the process, then obviously a similar formula applies for the $d\tau/dp$ of the relation between τ and p at the completion of the process. Or a similar formula holds, for example, for the half way completion of the process, dv/dQ now being the ratio of volume change to heat at the half way stage.

This formula may now be applied to NH_4Cl. We have in the first place three points on the P-T curve for the initial point of the discontinuity, as given in Table II, from which a curve may be drawn, and values found for $d\tau/dp$ with some accuracy. The values so found are 0.0067 at 0° and 0.0036 at 30°. The diagrams make it very probable that the internal change is entirely completed in a finite pressure range. The point of completion of the change is of course difficult to locate and indefinite, but it would seem that the points marked A and B at 0° and 30° give fair estimates of such points. This means that the internal change is spread through a pressure range of about 2500 kg at 0° and 1300 kg at 30°. This roughly justifies us in writing the formula as $d\tau/dp = \tau\Delta v/\Delta Q$, where Δv and ΔQ are the total changes of volume and heat accompanying the completed internal change, and $d\tau/dp$ refers to the initial discontinuity. The total Δv at 0° and 30° may also be found roughly by making a small extrapolation of the curves below the discontinuity, and I find in this way $\Delta v/v_0 = 0.0030$ and 0.0015 respectively. Substituted into the generalized Clapeyron's equation these give for the total heat of the change at 0° and 30° respectively 122 and 126 kg cm/cm^3, or 99 and 102 gm cal per mol, or 100 gm cal per mol at both temperatures, within experimental error. The total heat of the change therefore appears to be approximately unaffected by increasing pressure, although the change of volume falls off rapidly.

The values for ΔQ just obtained do not check at all well with those found by Simon at atmospheric pressure. Simon gives for the total ΔQ 356 gm cal per mol, against 100 found above. Part of the discrepancy may be due to my assumption that the transition is completed in a pressure range corresponding to 17° at 0° (0.0067×2500) and 4.7° at 30° (0.0036×1300), whereas at atmospheric pressure Simon found the transition to begin perceptibly at 140° and to end abruptly at 243°K, a range of 103°. However, Simon found that 53 percent of the total thermal effect was concentrated in the 3° region between 240° and 243°, so that the major part of the difference cannot be explained by considerations of this sort. It seems to me that a very appreciable part of Simon's large heat may be due to the contribution made by the internal degrees of vibrational energy within the NH_4 radical, which is not supposed to be the effect concerned at the discontinuity. In any event, the very marked narrowing with increasing pressure and temperature of the temperature range within which the transition is perceptible is worthy of comment.

The drop in Δv with increasing pressure is entirely in line with the difference of compressibility. It is evident at once from Fig. 1 that NH_4Cl is very materially less compressible after the change has been completed than before. At 0° the linear compressibility at 6100 kg, where the transition has been completed, is 1.9×10^{-7} less than at 3600, just before the transition starts. Perhaps a third of this difference is to be ascribed to the natural decrease of compressibility with increasing pressure, leaving outstanding something of the order of 1.2×10^{-7} as the intrinsic difference of compressibility of the two states of NH_4Cl. At 30° the intrinsic difference of compressibility is not far from this same figure.

The discontinuity in $(\partial v/\partial p)_\tau$ at the beginning of the change is doubtless characteristic, and should be recorded; it seems to be very nearly the same at 0° and 30°. At 0°, where the more accurate estimate can be made, the discontinuity is about 18×10^{-7}. This abrupt discontinuity at the low pressure end of the range corresponds precisely to the abrupt discontinuity found by Simon at the *upper* end of the temperature range.

The anomaly of NH_4Br is evidently very different in character from that of NH_4Cl. There is in the first place the fact that the volume change is of the opposite sign. Furthermore, the shape of the discontinuity is different; the discontinuity of NH_4Br appears to end abruptly, when pressure has passed a critical value, but on the low pressure side there is an anomaly in that the curve of length against pressure is concave upward, as may be seen on inspection of Fig. 2. This means that in the low pressure range the compressibility increases with increasing pressure instead of decreasing, as is normal. This region of abnormally changing compressibility may well be a region of some sort of preparation for the approaching internal change. As already mentioned, by far the largest part of the change occurs in a very narrow range, with the abruptness characteristic of a polymorphic transition. The total abrupt fractional volume change is 0.017, almost six times as much arithmetically as the total change for NH_4Cl at 0°. It is a matter of general experience that the volume change usually becomes less as temperature increases, so that if the volume changes were compared at equal temperatures the arithmetical discrepancy between NH_4Cl and NH_4Br would not be quite so extreme.

It is not possible to obtain as accurate values for the heat of the change of NH_4Br as for NH_4Cl, because the change of NH_4Br has been studied at only two temperatures. We may, however, get a rough value by assuming $d\tau/dp$ constant between atmospheric pressure and $-72°$. Taking the temperature of the discontinuity as $-38°$ at atmospheric pressure and $-72°$ at 1600 kg/cm², $d\tau/dp$ is found to be -0.0210. This gives for the heat, $\Delta Q = 203\times.017\div.0210 = 160$ kg.cm/cm³, or 140 gm cal per mol, against 100 for NH_4Cl. The normal correction for the change of $d\tau/dp$ with temperature (or pressure) would raise the figure 140 somewhat, but in any event the heating effects in NH_4Cl and NH_4Br are not widely different, in spite of the difference of sign of the volume effects.

The generally accepted explanation of the anomalies at atmospheric pressure is that of Pauling,[5] namely that the NH_4 radical passes from oscillational to complete rotational freedom as temperature rises, a process which is furthered by the weakening of the restoring forces due to the expansion of the crystal structure with rising temperature. The natural thermal expansion is in its turn increased as rotational freedom is acquired, so that the process is to a certain extent autocatalytic, thus accounting for the comparatively narrow temperature range. I think it very probable, however, in view of the facts detailed above, that this must be recognized to be an over-simplified picture. The small anomaly in NH_4Cl at 75°, not emphasized in the discussion above, suggests the same sort of thing. There can be little doubt that the large features of the phenomenon are the same in NH_4Cl and NH_4Br, as suggested by their chemical similarity and the approximate equality of the temperatures and heats of the transitions. But there must also be important differences, as shown by the difference of sign of the volume effects, and the marked differences in the abruptness and general character of the transitions.

It seems probable that small range anomalies may be of comparatively frequent occurrence in other substances. In many cases these need not involve as radical an internal change as passing from oscillational to rotational motion, but may be due to an intermediate type of change. For example, as the structure opens with thermal expansion, an oscillational motion, which at low temperatures was confined to a definite small amplitude range, may discontinuously acquire the possibility of oscillation through a greater amplitude, with accompanying change of natural frequency and thermal and volume effects. It seems to me highly probable that many of the small range anomalies which I found so characteristic of the *P-V-T* relations in liquids at high pressures[6] are of this character. There are doubtless many similar effects in solids. A possible example is the pressure transition of Ce[7]; this is like the change of NH_4Br in that below the transition pressure the compressibility increases abnormally with pressure. The sign of the volume change is different in the two cases, however. There are many interesting questions here for future experimental examination. For example, the persistently high compressibility of potassium[8] over a wide pressure range may be something of this nature.

In addition to the measurements described above, an exploration was made of $N(CH_3)_4Cl$, a compound very similar chemically to NH_4Cl, but not the slightest irregularity was found at either 0° or 75°. This is consistent with Pauling's picture, for the much greater moment of inertia of $N(CH_3)_4$ as compared with NH_4, should raise the temperature at which the transition might occur, if indeed the transition is not entirely suppressed. The details of the *P-V-T* relations for $N(CH_3)_4Cl$ will be published elsewhere.

Finally, a word of comment may be made on the absolute values of the compressibilities in the regions unaffected by the internal change. The initial

[5] L. Pauling, Phys. Rev. **36**, 430 (1930).

[6] P. W. Bridgman, Proc. Amer. Acad. **49**, 1 (1913).

[7] P. W. Bridgman, Proc. Amer. Acad. **62**, 207 (1927).

[8] P. W. Bridgman, Proc. Amer. Acad. **58**, 204 (1923).

compressibility of NH_4Cl at 30° is about 5.9×10^{-6} and that of NH_4Br 6.1×10^{-6}. The corresponding compressibilities of NaCl and NaBr are 4.2 and 5.0×10^{-6}, and of KCl and KBr 5.5 and 6.6. The general order of the compressibilities of the NH_4 salts is therefore what might be expected. The difference between the compressibility of the chloride and the bromide is unusually small, however, and this again emphasizes the probability that complications obscure any simple picture.

This work is much indebted financially to the Rumford Fund of the American Academy of Arts and Sciences and to the Milton Fund of Harvard University. I am indebted to Mr. L. H. Abbot for the readings.

91 — 2507

Recently Discovered Complexities in the Properties of Simple Substances*

By P. W. Bridgman,† Cambridge, Mass.

(Boston Meeting, September, 1931)

It is a commonplace that experimental physics in the last few decades has discovered manifold complexities in the atomic and subatomic levels, where it was thought for hundreds of years that no structure existed, as is witnessed by the derivation of the word "atom" itself. The field of exploration opened by the discovery of the electrical structure of the atom was so fundamental, rich and stimulating, that it has been almost exclusively cultivated by physicists since they obtained their first intimations as to the nature of the basic facts. Only within the last few years has it begun to dawn on us that we have overshot an enormous domain in which are situated many phenomena of fundamental importance for all the practical uses to which we put matter in our daily lives; the domain, that is, of most of the phenomena of interest to the biologist and the metallurgist. Furthermore, we are beginning to find that this intermediate domain is not entirely simple, but that it contains unsuspected complexities, some of them derived from the complexities of structure on the atomic and subatomic levels, and some of them emergent as matter collects itself into large aggregates, which often offer the key to the explanation of hitherto baffling large-scale properties of matter. Today I want to describe and consider some of these newly discovered complexities in the behavior of matter in bulk. The application to metallurgical problems is not always direct, but I hope to be able to suggest that such general notions as to structure as those here discussed must have important reactions on our attitude toward the problems of metallurgy.

Three different sorts of phenomena of this kind to which I shall direct your attention are (1) the so-called structure sensitive phenomena, (2) complexities of such a nature that they are masked by the presence of extraordinarily small amounts of impurities, and (3) complexities depending on internal molecular rearrangements which can be understood only from the point of view of the quantum theory.

---

* Science Lecture delivered before the Institute of Metals Division and the Iron and Steel Division of the A. I. M. E.

† Hollis Professor of Mathematics and Natural Philosophy, Harvard University.

Structure Sensitive Phenomena

First let us consider the structure sensitive phenomena. It has been known for a long time that some properties of matter are much more definite and clean-cut than others, in the sense that they are easy to measure, and measurements made on different specimens by different observers under different conditions and by different methods yield consistent results, so that such results can justly be regarded as characteristic of the material under investigation. Perhaps the most striking of such properties is the atomic weight of some of the elements; ordinary mechanical density, lattice structure, heat of solution, and optical dispersion are other examples. There are other properties which are less definite, and numerical agreement by different observers is much more difficult to obtain. Sometimes there are technical difficulties in devising the apparatus or in making the measurements; as, for example, thermal conductivity measurements, which are notoriously difficult because of the impossibility of obtaining thermal insulators by which the flow of heat can be compelled to take only the desired channels. Other such effects are the various transverse effects in a magnetic field, such as the Ettingshausen effect, which are exceedingly difficult to measure because of their extreme smallness, and because of the difficulty of eliminating disturbing secondary effects. Sometimes, with the improvement of technique, properties which were at first difficult to measure and which gave conflicting results have become easy and consistent. An example of this sort of thing is cubic compressibility, for which the most discordant results are recorded in the early literature. The difficulty was both in the theory, which at first was not at all well understood, and also in the measurements themselves, because the effects are very small. But the technique has been so improved that now different observers with different apparatus and material from different sources can obtain consistent results, and we can now be assured that recent experimental values for cubic compressibility are truly characteristic of the material.

But apart from difficulties arising from difficulties of technique, there are outstanding certain physical properties for which it is very difficult to get consistent results on material from different sources. Thus although different observers do not find it difficult to agree on the proper value for Young's modulus of iron, various results are to be found for the elastic limit and the ultimate strength. There are many phenomena in the same category, for example, electrical conductivity, diffusion and mixing or unmixing phenomena in the solid state, such optical properties as photoelectric absorption and the electrical conductivity induced in crystals like rock salt by photoelectric action, magnetic susceptibility, and the effect of tension on thermoelectric quality. These properties may be greatly affected by variations in the heat treatment, conditions of crystallization, mechanical working, or by slight impurities.

The first clear recognition that physical phenomena fall into two such groups of sensitive and insensitive properties, and that the sensitive properties demand for their explanation new sorts of consideration not entertained in the ordinary physical theories, was doubtless due to Smekal. The effects are particularly striking with regard to the ultimate strength of single crystals, for example, of a material like NaCl, to take an example not greatly complicated by plastic flow. Certain of the properties of NaCl, as, for example, its heat of solution, or its compressibility, can be calculated theoretically from the lattice structure of the crystal as revealed by X-rays. The calculation involves the assumption of centers of force at the nuclear points indicated by X-rays, and the fact that the results of the calculations agree fairly well with the experiments is pretty good evidence that we have approximately a correct understanding of the situation so far as these phenomena are concerned. But if we know the forces between the ions with sufficient accuracy to calculate the lattice energy and the compressibility, we should also be able to calculate the breaking strength, because this involves only a complete knowledge of the interionic forces when the ions are separated by a great distance. These calculations were made and a result found of the order of 1000 times greater than the experimental values. It was for a time thought that the low experimental values could be explained by the effect of surface imperfections, but after a great deal of discussion it now seems to be accepted by most investigators that the way out is not to be found here, but that the strength of NaCl is enormously lower than it ought to be from considerations which take account merely of its lattice structure.

Smekal recognized that some new physical consideration is demanded to account for the enormous discrepancy between this theory and experiment, and he proposed the new theory that in an actual crystal the lattice structure does not extend indefinitely in every direction, but that it is broken by faults or imperfections into blocks of anywhere from 10^4 to 10^6 atoms each, and that in the places where the blocks fail to join perfectly there are cavities of atomic dimensions left in the crystal structure. The faults between the blocks and the cavities are thought to be responsible for the great discrepancy between some sorts of calculated and experimental phenomena. It is easy to see why certain properties should not be affected by these faults and why the experimental values of these properties should agree with the simple theory. The density, for example, is determined by the mean distance of separation of the atomic units, and since the faults affect only a few of the atoms, the mean distance, and so the density, can be only little affected.

Similarly it is evident that the energy of the lattice, as given, for example, experimentally by the heat of solution, will be little affected by the faults, since the relative position of most of the atoms is unaffected, which means that the simple lattice theory will continue to give the essential explanation of the situation, even when the faults are considered. It is, on the other hand, evident that a phenomenon like the tensile strength of a brittle crystal, to take an example in which complications are not introduced by plastic flow, should be profoundly affected by the internal faults. The mathematical theory of elasticity shows that in the neighborhood of a reentrant angle, or in an interior cavity with sharp angles in a solid material which is the seat of elastic stresses, the stress may build up to infinite values, merely because of geometrical considerations, although the average stress through the body of the solid is finite. In practice this means that a reentrant angle is a source of great weakness, and that rupture takes place in such solids very much more easily than in a geometrically sound piece of the same material. Rupture, in such a material with internal faults, is propagated by a crack, the advancing head of the crack always remaining sharp, and thus providing for the continual existence of the necessary reentrant angle. Theoretically, a perfectly homogeneous piece of a material carrying an imperfection of this sort would rupture with a vanishingly low stress. Actually, substances are not perfectly homogeneous and never have mathematically sharp angles because of the limitations imposed by atomic structure, if for no other reason, but in any event it is easy to see that the existence of imperfections of this general nature will greatly lower the stress needed to produce rupture, and thus account for the discrepancy between the theoretical and the experimental values of the breaking strength.

Relation of Faults to Sensitive Properties of Matter

Smekal has considered a large number of the "sensitive" properties of matter, and has shown in detail how the existence of the faults explains their behavior. His considerations have been mostly confined to nonmetallic crystals, of which rock salt is typical. For example, Gudden and Pohl found that in rock salt there are two kinds of optical absorption present simultaneously; one kind of absorption, by far the most important, is a long-wave absorption, and is to be accounted for by the excitation of the atoms in the conventional way. This part of the absorption is to be thought of as contributed by the unfaulted blocks. Superimposed on this there is a much weaker absorption toward the shortwave side of the spectrum, corresponding to the ordinary photoelectric effect, and which is due to the incident light pulling electrons out of the chlorine ions which abut on the sides of the cavities. These electrons, which are detached in this way inside the cavities, immediately recombine with the positively charged sodium ions which also abut on the cavity walls, and in this way produce neutral sodium atoms. The presence of such neutral atoms imparts to the crystal of NaCl the characteristic purple "radiation" color, so that one has here a most beautiful way of detecting the existence of these cavities and studying the way in which they are affected by various external factors. One would expect, for example, the number of the cavities to be greatly affected when the crystal is strained beyond its elastic limit. If one bends a bar of NaCl enough to give it plastic deformation, it follows from elasticity theory that there will be a relatively unstrained region in the center, and that the upper and lower faces will be permanently affected, the one being plastically deformed in extension, and the other in compression. Such a deformed bar, when illuminated by X-rays to bring out the color, does indeed show a marked coloring of the two deformed faces, while the relatively unstrained center receives only a relatively light coloring. Or if the bar is first colored by being radiated and then deformed, it will be found that the color disappears in the plastically deformed faces, but remains in the elastically unharmed central portion. The explanation is that the deformation produces relative motion in the walls of the cavities, so that the neutral atoms on the walls of the cavities have an opportunity to take their places again as ions in the regular lattice structure, thereby losing their color.

One would expect that the production of electrons in the cavities would have an effect on the electrical conductivity, and this turns out to be the case. Smekal has shown how to analyze the electrical conductivity of NaCl (which is always, of course, exceedingly small) into two parts, one of which is constant irrespective of the degree of purity of the crystal and its thermal treatment, and is to be considered characteristic of the pure lattice structure, while the second part varies greatly from specimen to specimen and is to be ascribed to the free electrons associated with the cavities.

The cavities or imperfections should change in number and importance as the material changes. In the first place, they should be intimately connected with impurities, because every atom of impurity in a crystal usually constitutes a place where the perfect regularity of the crystal lattice is disturbed, and therefore is a potential location for a cavity. It does indeed seem to be true in general that the sensitive properties change in the way that would be expected when the amount of impurity is decreased. However, there are difficulties here because of the difficulty in any practical case of reducing the amount of impurity sufficiently to separate the effect due to it from other effects. It was suggested above that the blocks in the secondary structure might contain possibly 10^6 atoms. Now it is practically impossible in most cases to reduce the impurity to less than one part in 10^6, so that perhaps we have not yet been able to really separate the effect of the other factors from the effect of impurities. There is room for much important future work here, and we are just beginning to recognize that the effect of impurities, which by the old standards would have been considered absolutely negligible in amount, may be very important. There seems to be no doubt that the impurities have a tendency to separate into the faults between the blocks, and in this way to exert a disproportionately great effect on the sensitive properties.

There is also a close connection between the size of the blocks and the heat treatment. It is at first sight somewhat paradoxical that the secondary structure has a more important effect in an NaCl crystal grown by slow solidification from the melt than in one grown from aqueous solution. One would expect microscopic inclusions of the mother liquid to remain in the crystal grown from solution, and so the effect of imperfections to be larger. The fact seems to be that the size of the secondary structure is in some way connected with the thermal fluctuations of density to which the crystal was subject during the process of growth, which are satisfactorily explained by ordinary statistical theory. These statistical fluctuations are greater at higher temperatures, and therefore will give rise to greater effects in the secondary structure in crystals produced from the melt at high temperatures than in those produced from cold aqueous solutions. This point of view may be checked by annealing at high temperature a crystal grown from aqueous solution, when it will be found that the crystal acquires the secondary properties of a crystal deposited from the melt.

Secondary Structure

The assumption of secondary structure makes understandable many hitherto puzzling phenomena. Consider, for example, the complicated phenomena of elastic after-effects and elastic hysteresis. Artificial theories of these have been given in which the material has been supposed to have a complex structure, being composed of different sorts of elementary materials with different properties combined in complicated ways. Thus theories have been made of the elastic after-effects in which mechanisms have been imagined equivalent to perfectly elastic springs tethered to material with the viscosity of molasses. Evidently a much more satisfactory basis is provided for such phenomena by finding the complexities in the small-scale geometrical structure rather than in the properties of the material itself. Thus there would seem to be plenty of opportunity in the neighborhood of the faults of the crystal for those internal strains and local plastic effects necessary to hysteresis and elastic after-effects.

Other sorts of phenomena are also consistent with this point of view. For instance, the fact that thermal conductivity is a sensitive property falls in at once with the modern picture of thermal conductivity as due to elastic waves of microscopic dimensions. We would expect these waves to be scattered by the faults, and so the thermal conductivity to be affected. The picture is also consistent with the fact that electrical conductivity is a sensitive property. The picture that we now have of electrical conductivity involves the assumption that the electrons have what would have been described from the old classical point of view as long free paths. *Long* free paths can be affected by comparatively infrequent faults in the crystal, whereas if the paths are short such faults will be of comparatively little importance; in this way we can see how it is that electrical conductivity can be sensitive.

As already stated, the work of Smekal on this subject was confined mostly to nonmetallic crystals, like NaCl; here the experimental verification is comparatively easy because of the existence of such phenomena as coloring by X-rays. The extension to metallic substances is much more difficult. Very important work on this aspect of the subject is now being done at Pasadena, at the California Institute of Technology, where there is activity on both the experimental and theoretical sides.

On the experimental side, Goetz has made painstaking and elaborate studies of the properties of crystals of bismuth, a material in which the secondary effects are particularly large, and which is therefore a favorable material for investigation. It has been known for a long time that many of the properties of bismuth are exceedingly variable from specimen to specimen, and therefore are presumably structure sensitive. The electrical conductivity is such a property; it is sensitive to small amounts of impurity and is also sensitive to variations in the direction of current flow in the crystal. The conductivity perpendicular to the crystal axis, that is, for directions of current flow in the basal plane, which is the plane of principal cleavage, proves not to be sensitive, and different observers agree rather well on this value; whereas if the current flow is parallel to the crystal axis, which means across the cleavage planes, the disagreement between competent observers may rise to as much as 30 per cent, the conductivity in this direction being less the greater the impurity or the greater the chance for incipient fractures on the cleavage planes. The temperature coefficient of conductivity is also extremely sensitive, a few hundredths of a per cent of lead being sufficient to reduce it to one-half its normal value. Another similar phenomenon is the extraordinary sensitiveness of bismuth crystals to minute stresses during the process of crystallization already studied by Kapitza.

Goetz has used mostly two methods for controlling the secondary structure. In the first place he has altered the conditions during growth by crystallizing from the melt with or without the presence of a magnetic field. The primary properties of the crystal, such for example, as its lattice structure as shown by X-rays, are entirely unaffected by the presence of the magnetic field, and are characteristic only of the material. But other properties, of which the thermoelectric behavior has been specially studied, depend greatly on the field which was applied during crystallization, and therefore are to be classed as structure sensitive. In the second place, the purity has been altered, and some of the properties, as already known, found to vary tremendously with the presence of slight amounts of impurity. The effect of combining an impurity with a magnetic field during crystallization is most complicated. Doubtless the most interesting and important contribution of Goetz is to prove the actual physical existence of the secondary structure on which the sensitive properties are supposed to depend. The structure consists of networks of planes so cutting the basal planes as to give a pattern of equilateral triangles. This substructure can be brought out by suitable etching, and can be shown by photomicrographs, or it can be studied in a most convincing way by watching under a microscope the process of solution in an electrolyte, when it will be found that periodic slight changes are necessary in the solution voltage, corresponding to the arrival of the surface of solution at each of the secondary planes. The most curious feature of this secondary structure is that the spacing of the planes appears to be definite, independent of the conditions of growth or the degree of impurity. There seems to be no doubt that the impurities segregate themselves on the secondary planes; if the amount of impurity is increased, the number of atoms of impurity which separate on any plane increases, accordingly, but the number of planes does not change.

Diamagnetic Susceptibility

Another interesting fact found by Goetz is that the diamagnetic susceptibility is also structure sensitive in a way similar to the electrical conductivity, only here the susceptibility for the magnetic field parallel to the axis is not sensitive, while for the perpendicular direction of the field there is sensitiveness. This again is consistent with recent views as to the nature of diamagnetic action, which is thought not to be an atomic affair, but to involve the cooperation of many atoms, the magnetic electrons effectively describing orbits many atoms in diameter. If these orbits cross the planes of the secondary structure, it is easy to see that the susceptibility for the corresponding direction of the magnetic field will be structure sensitive, whereas for the other direction of the field the orbits will not cross the secondary structure, and so the susceptibility will not be structure sensitive.

Two Theoretical Problems

In addition to the experimental work of Goetz at Caltech, Zwicky has been active in dealing with the theoretical aspects of the problem.

There are two main problems requiring theoretical discussion: (1) to account for the existence of the secondary structure, and (2) given the secondary structure, to show why it should affect the properties of the material in the way it does. Zwicky has been concerned mostly with the first problem, to account for the origin of the structure. His arguments are all dedicated to the thesis that a perfectly regular lattice, either of similar atoms as in a metal, or of molecules as in NaCl, is not the configuration of lowest potential energy, but that a combination of the ordinary lattice shown by X-rays with a superposed secondary structure of much larger scale has smaller potential energy than the unmodified simple lattice. The crystal is supposed, in accordance with the principles of mechanics, to take up automatically the configuration of minimum energy. The calculations are difficult; the argument has taken several different forms, and I do not believe that it can be said even yet to be in its final form. The latest idea of Zwicky is that the relative instability of the simple lattice is connected with the free surface, and that a secondary mosaic structure works its way into the interior from the surface. Considerations are given showing that the order of magnitude of the secondary structure to be expected is the same as that observed. It seems to me that one of the weak points of the theory so far developed is that it is concerned only with the absolutely pure metal, whereas in practice it is known that the secondary structure is intimately connected with slight impurities, and it cannot even be claimed that the existence of a secondary structure has been established experimentally in the absence of slight impurities. At first it might be thought that the regularity of the secondary structure, as shown by Goetz's photographs, can hardly be consistent with an effect of impurity. But there are other phenomena which make this not impossible. For example, the Liesegang rings, which are deposited in gelatin when a solution of a silver salt diffuses into gelatin which has been impregnated with $K_2Cr_2O_7$, have a perfectly definite spacing. It is not difficult to imagine that the deposition of impurities from the metal as it solidifies may be similar to this process.

Complexities Masked by Impurities

We now consider the second group of phenomena, showing that great complexities are possible in the behavior of perfectly pure metals. Here the complexities are of such a character that they are smoothed over and obliterated by the presence of exceedingly slight amounts of impurity. This is somewhat paradoxical and the reverse of the usual state of affairs, for the presence of impurity often makes behavior more complex; as, for example, the sharp freezing of a perfectly pure substance is changed into freezing over a range with a continual change in the composition of the solid separating out, when impurity is added. The experiments to which I refer were made by Schubnikow and DeHaas in the Cryogenic Laboratory at Leiden on the behavior of single crystals of bismuth at low temperatures, and in particular on the effect of a magnetic field on the resistance. The necessity for great care in the purification of bismuth was first shown by the great irregularity and nonreproducibility of the resistance at liquid hydrogen temperatures of bismuth from different sources. For example, the ratio of the resistance at 20° abs. to that at 273° abs. of the purest spectroscopic bismuth available with a total impurity of about 0.006 per cent was 0.11; the corresponding figure for the purest bismuth obtainable from commercial sources was 0.53, whereas another grade of bismuth frequently used in the construction of instruments gave the ratio 1.21. The spectroscopic bismuth was now purified still further by three different procedures. The first was simply repeated recrystallization into single crystal form from the melt, it being well known that many impurities are segregated during solidification. The second and third methods were chemical methods of resolution and reprecipitation. The result of these further purifications was to reduce the ratio 0.11 still further to approximately 0.05, and this ratio was now the same and reproducible for material prepared by all three methods, indicating that the purification had been carried as far as necessary, at least from the point of view of this phenomenon. It is interesting that the residual resistance of the highly purified bismuth at helium temperatures proved to be of the same order of magnitude as that of other pure metals, such, for example, as gold, indicating no essential difference in this phenomenon between other metals and bismuth, although bismuth with an exceedingly slight amount of impurity does act like an essentially different substance.

The effect produced by a magnetic field on the electrical resistance of this ultra pure bismuth was now measured at liquid helium temperature. It should be mentioned that in these experiments the crystal axis was parallel to the direction of current flow, and the magnetic field was perpendicular to it. Definite and reproducible values were obtained, which are doubtless characteristic of pure bismuth, with no extraneous effects, although similar measurements on ordinary bismuth are very far from reproducible. There are many surprising features about the results. Perhaps the most spectacular result of all was the enormous magnitude of the effect, a field of 30,000 Gauss increasing the resistance by a factor of almost 150,000 at liquid helium temperature. Another surprising result was that the effect varies greatly, sometimes by a factor of nearly 3, depending on the orientation of the *secondary* crystal axis to the magnetic field. (The secondary axis has no connection with the secondary structure discussed above.) This means that if the rod is rotated about its length in a magnetic field perpendicular to the length the resistance may vary by as much as a factor of 3, depending on the orientation. But to me the most surprising result of all was that the effect does not increase smoothly with the magnetic field, unlike all other known magnetic phenomena, but the dependence is far from simple, with at least half a dozen points of inflection in the curve between 0 and 30,000 Gauss. To express the complete relation between magnetic field and orientation with respect to the secondary axis, seven terms in the expansion in Fourier's series are necessary, and each of the coefficients is itself a highly complicated function of the field.

So far as I know, no attempt has yet been made to find a theoretical explanation of these magnetic phenomena, and I shall certainly not attempt it. The point which I want to emphasize is that this is a very striking example of great complexity in the behavior of a pure substance, which may be entirely altered in appearance and obscured by the presence of amounts of impurity so small as to be beyond treatment by ordinary methods.

Complexities Connected with Quantum Effects

The third group of phenomena revealing unsuspected complexities in the behavior of ordinary substances are definitely known to be connected with quantum effects. It is now beginning to appear that there are a great many phenomena in this class; most of these are important only at low temperatures, where we are perhaps not surprised to find them, but it is also becoming evident that important phenomena of this character may be expected under more ordinary conditions. Perhaps the most definite and striking suggestion of what may be expected here goes back to the discovery by Simon in 1922 of certain anomalies in the specific heat of NH_4Cl. Since that time, Simon, with various collaborators, has shown that in the first place there is a similar anomaly in the specific heat of NH_4Br, and secondly that connected with the specific heat anomaly there is also an anomaly in the thermal expansion. The specific heat anomaly consists in a gradual rise of the specific heat above that to be normally expected. This rise first becomes just perceptible at about 150° abs., but then grows rapidly until at 243° abs. (about −30° C.) the specific heat is about twice the value to be expected. At this temperature it drops in a range of a few degrees back to its normal value. Fig. 1 shows the nature of the effect. The anomaly in the specific heat of NH_4Br is very similar, and the maximum departure from normal occurs at very nearly the same temperature. NH_4I also has

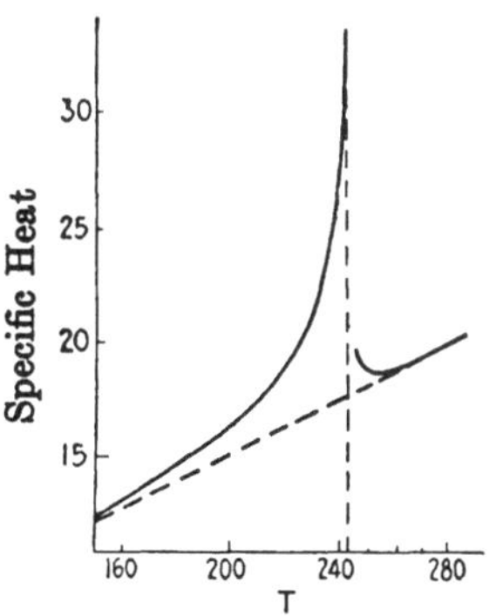

Fig. 1.—Molecular specific heat of NH_4Cl plotted against absolute temperature. [*From paper by F. Simon: Ann. Phys.* (1922) **68**, 241.]

a similar effect, but at a slightly lower temperature. The volume anomaly of NH_4Cl is similar in character. This may best be studied by measuring the thermal expansion, which in the same way as the specific heat increases from its normal value, at first slowly, and then rapidly, until at the same temperature as the maximum anomaly in the specific heat it is of the order of 15 times the normal value; from this point it drops back in a few degrees to normal. The net result of the anomaly is that above the temperature of the anomaly the volume is somewhat greater than would normally be expected. The changes of

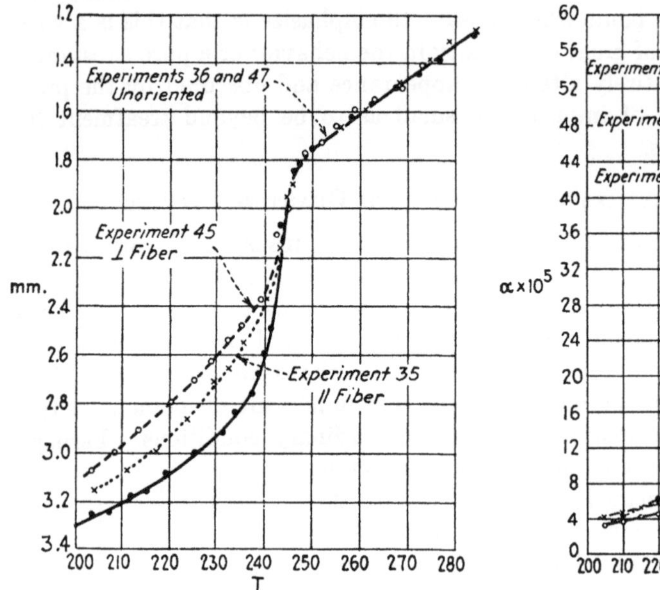

Fig. 2.—Change of length of NH_4Cl plotted against absolute temperature [*From paper by F. Simon and R. Bergmann: Ztsch phys. Chem.* (1930) **8**, 255.]

Fig. 3.—Thermal expansion of NH_4Cl plotted against absolute temperature. Ordinates of this curve are slopes of curve of Fig. 2. (*From the same paper as Fig.* 2.)

length and the thermal expansion are shown in Figs. 2 and 3. The volume anomaly of NH_4Br is similar in some respects and different in others. There is similarity in that the abnormal behavior is spread over a range of temperature, rising to a maximum at the same temperature as the maximum of the specific heat anomaly, and then dropping back rapidly to normal, but the behavior is different in that the anomaly is of the opposite sign, the thermal expansion being negative in the region of the anomaly, with the net result that the volume is less above the temperature of the anomaly than would be normally expected.

A combination of a volume effect with a thermal effect is of course exactly what occurs during a polymorphic transition, and it is at first natural to seek the explanation in this way. The outstanding difference between this and an ordinary polymorphic transition is that instead of being sharply discontinuous the anomaly is spread over a range of temperature. Such effects occur, however, in polymorphic transitions in the presence of impurities unequally dissolved in the two phases, and the explanation might be sought here. However, an X-ray determination of the lattice structure made by Simon above and below the temperature of anomaly shows no change whatever in the crystal lattice, so that this possibility must be discarded, and we are evidently confronted with something quite different from the ordinary polymorphic transition. It may be mentioned that both NH_4Cl and NH_4Br do have ordinary polymorphic transitions, which are sharp, as is to be expected; these occur at temperatures between 100° and 200° C., 200° or so above the range of the anomaly so that there can be no confusion between the two phenomena. NH_4I, however, does have the corresponding polymorphic transition at a temperature of $-18°$ C., which is so near the temperature of the new anomaly that they could easily be confused, so that Simon did not make as exhaustive an examination of NH_4I as of the two others.

It is well known that the temperature of an ordinary polymorphic transition is displaced by the application of hydrostatic pressure, just as the temperature of melting or vaporization is displaced, and furthermore, the magnitude of the displacement involves the volume difference and the latent heat between the two phases by the well-known formula of Clapeyron. Now this new anomaly involves also volume differences and differences of total heat, the distinction being that the effects are now spread over a range of temperature instead of being sharply discontinuous. Analogy leads one to expect that the temperature range in which the anomalies occur might be similarly displaced by the action of hydrostatic pressure. Thermodynamic analysis does show that there is a precise parallelism between the two cases and that an equation the exact analogue of Clapeyron's equation governs the displacement by pressure of the temperature range in which the anomaly is situated. In the case of the ordinary transition the temperature is raised by pressure if the phase stable at the higher temperature has a larger volume than the phase stable at the lower temperature, which is normal for most substances, but the transition temperature is depressed by pressure if the low-temperature phase has the larger volume, as in the case of abnormal substances like water and ice. Applied to our examples, this means that the temperature of the anomalous region of NH_4Cl should be raised by pressure, because NH_4Cl expands on passing from the low-temperature to the high-temperature form through the region of anomaly, but that on the other hand the temperature of the anomaly of NH_4Br should be displaced by pressure to lower temperatures, because NH_4Br contracts on passing through the anomaly from lower to higher temperatures.

Effect of Pressure on Temperature

Recently, I carried out experiments to test these predictions, and have obtained results exactly as expected. The effect was first found with NH_4Cl, where one expects the temperature of the anomaly to be raised by pressure. At 0° C., 30° above the temperature of the maximum of the specific heat anomaly, the length was measured as a function of pressure, and an abrupt break was found in the curve at a pressure of 3370 kg. per sq. cm. Passing through the break the compressibility becomes abnormally high, and then after an interval of 2500 kg. per sq. cm. or so, assumes again its normal value. At the still higher temperature of

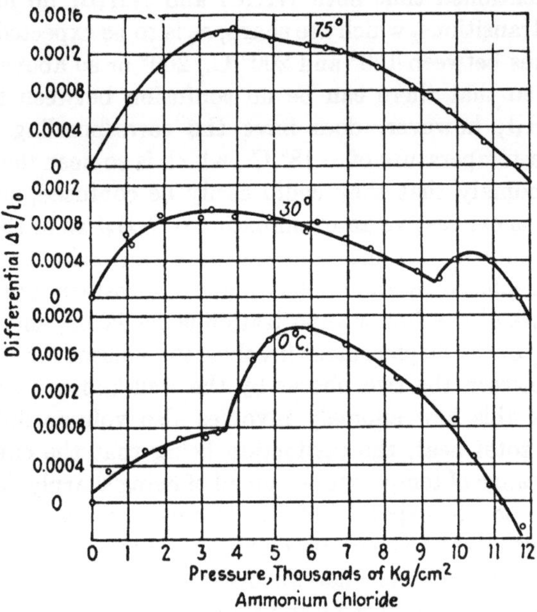

FIG. 4.—EFFECT OF PRESSURE IN DISPLACING TO HIGHER TEMPERATURES THE ANOMALY OF NH_4Cl.

In order to magnify the effect the ordinates are not the actual changes of length, but the difference between the actual change and the change that would have taken place if the relation with pressure had been linear.

30° C. a break was also found, but now at 9390 kg. per sq. cm. a considerably higher pressure, as is to be expected, because the temperature and pressure rise together. The behavior at 30° is much like that at 0°, except that the discontinuity is less in amount, and is not spread over so wide a pressure range. A third set of measurements at 75° disclosed no break of this kind up to a pressure of 12,000 kg. per sq. cm., the range of the apparatus, which is what would be expected, because extrapolation of the curve drawn through the points at −30°, 0°, and 30° indicates that at 75° a considerably higher pressure than 12,000 would be necessary to produce the effect. The results are indicated in Fig. 4, in which are plotted the differences between the actually observed changes of length and the changes which would have been observed if the relation had been linear. It is to be particularly noticed that the relation between length and pressure is completely reversible, and that there are no hysteresis or lag effects, the experimental points lying on the same smooth curve whether they are obtained with increasing or decreasing pressure. Although the curve at 75° does not show the same anomaly as do those at 0° and 30°, it does nevertheless show a minor anomaly in the region between 4000 and 7000 kg. per sq. cm., which seems to be definitely beyond experimental error, and is doubtless another sort of complication than that now under discussion.

Turning now to NH_4Br, we expect pressure to depress the temperature of the anomaly to below −30°. This was first checked by measurements

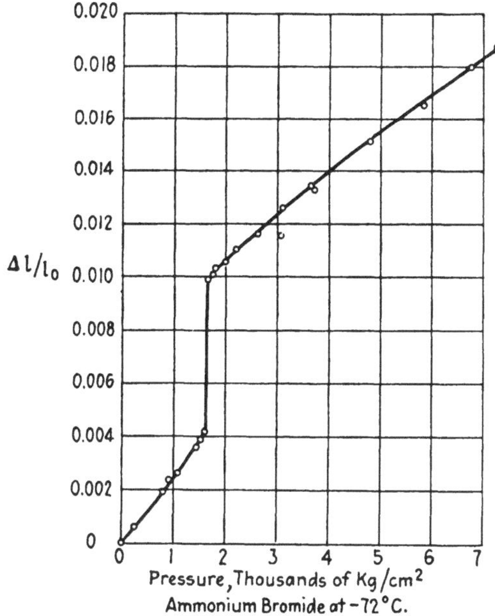

FIG. 5.—EFFECT OF PRESSURE IN DISPLACING TO LOWER TEMPERATURES THE ANOMALY OF NH_4Br.

The effect is much larger than for NH_4Cl and is sufficiently well shown by plotting the total fractional change of length against pressure.

of the volume as a function of pressure at 0° and 75° C. At both temperatures perfectly smooth curves were obtained, with no discontinuity of any kind, as was to be expected. Measurements were then made at −72°, the temperature of mixed solid CO_2 and alcohol. It is fortunate that the commercial demand for solid CO_2 which has sprung up in the last years makes it possible to obtain this substance in sufficient quantity and economically enough to make it feasible to cool the rather heavy pressure cylinder to as low a temperature as −72° C. One hundred pounds of CO_2

and 4 gal. of alcohol proved to be sufficient. Another technical difficulty was in obtaining a liquid suitable to transmit pressure, practically all ordinary liquids being frozen solid by the pressure at so low a temperature. Compressed gases, which would not freeze, have the disadvantage that large volumes are necessary, and furthermore are dangerous. Fortunately it proved that iso-pentane is capable at this temperature of supporting a pressure of more than 7000 kg. per sq. cm. without freezing, and this proved to be ample. In Fig. 5 the results are shown which were obtained for the change of length at $-72°$ as a function of pressure. There is, as was to be expected, a discontinuity in the length, and therefore also in the volume; this occurs at about 1630 kg. per centimeter. The discontinuity is now so sharp as to be indistinguishable from the ordinary polymorphic transition. The only effect at this temperature corresponding to the range of temperature over which the anomaly is spread at atmospheric pressure is the anomalous upward curvature of the graph below the pressure of discontinuity.

It is possible to go further with the information that may be obtained from these curves and to calculate, for example, the total heat absorbed in the anomalous range. For NH_4Cl it turns out that this is approximately independent of pressure and temperature, and is about 100 cal. per mol. The total heat of NH_4Br is also not greatly different from this. However, although these questions are interesting for themselves, they are not immediately concerned with our main thesis, and I do not discuss them further.

Explanation of Effect of Absorption of Heat

The most interesting question here is the explanation of the existence of the effect. Evidently some sort of internal change must be taking place in the material. What sort of change is there that can be responsible for the absorption of an amount of heat of the same order of magnitude as that of an ordinary polymorphic transition and at the same time be consistent with the experimental fact that there is no change in the space lattice? Doubtless the most important feature in the situation is correctly reproduced in the explanation offered by Pauling. The fundamental idea of this explanation is that at low temperatures the NH_4 radical is more or less tightly restrained by its neighbors, so that in addition to the ordinary vibrational to and fro motion of its center of mass, it can only oscillate back and forth through a restricted range of angle. A crude picture of the situation would be that the hydrogen atoms project like prongs from the heavy nitrogen atom in their center, and because of their entanglement with their neighbors prevent angular oscillations of more than a certain amplitude. As the substance expands with rising temperature, however, the prongs eventually get free enough so that the radical may rotate continuously instead of only oscillating. The temperature range during which the radical is acquiring freedom to make complete revolutions is the range of the anomaly. How does this picture explain the specific heat phenomena? According to the classical pre-quantum mechanics, this would have been no explanation at all, because the energy of a degree of freedom is a fixed quantity, so that in passing from oscillational to rotational motion, since there was no change in the number of degrees of freedom, there would be no change in the energy of the motion, and so no thermal effects. According to the quantum mechanics, however, the energy of a degree of freedom has a unique value only at high temperatures, and at a low temperature a degree of freedom may have less than this energy. The most familiar manifestation of this is the falling off of the specific heat curves of ordinary substances from their constant value at ordinary temperatures to zero at 0° abs., as shown in Fig. 6. The temperature at which the falling off begins depends on the tightness with which the vibrating system is bound in position; the smaller this binding force, the lower the temperature. Thus in Fig. 6, curve 1 corresponds to a degree of freedom with relatively tight binding, and curve 2 to one with relatively loose binding. At the temperature τ in the diagram the degree of freedom corresponding to the curve 2 has absorbed in warming from 0° abs. a total amount of heat equal to the area under the curve 2, while the degree of freedom corresponding to 1 has absorbed a considerably smaller amount of heat; namely, the area under the curve 1. In spite, however, of the difference in the total amounts of heat absorbed, the specific heats at the temperature corresponding to the curves 1 and 2 are the same (same ordinate of the two curves).

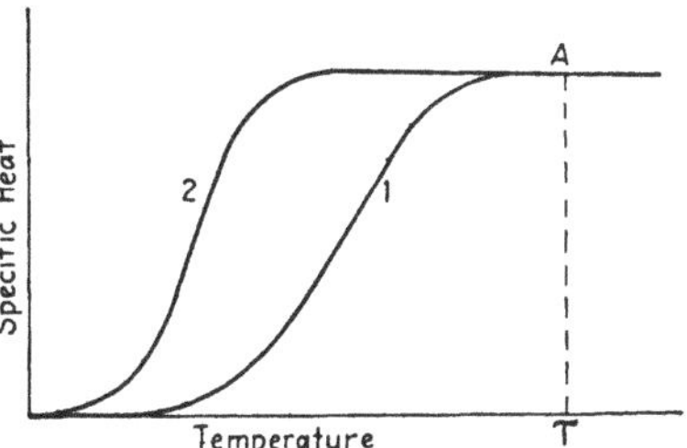

FIG. 6.—DIFFERENCE IN SPECIFIC HEAT CURVES FOR A SUBSTANCE WITH TIGHTLY BOUND ATOMS (1) COMPARED WITH ONE WITH MORE LOOSELY BOUND ATOMS (2).

The application to our problem is now obvious. In the position in which only oscillational motions are possible, the binding forces are relatively tight, and the heat curve corresponds to curve 1, whereas when rotational motion becomes possible, the binding forces become less and curve 2 describes the relations. In the temperature domain in which we are passing from oscillational to rotational motion we are passing from curve 1 to curve 2 and must absorb a total amount of heat equal to the difference of the areas under the two curves. It is this extra amount of heat absorbed during the transition, which is entirely absent from the classical picture, which accounts for the abnormally high specific heat. When the transition from oscillational to rotational has been completed we return again to the normal specific heat corresponding to the common ordinate A on the two curves 1 and 2.

I believe that you all will feel that this explanation is very beautiful, and I think we must consider that it is essentially correct. The same sort of ideas are also competent to explain other anomalies of some of the metals and of hydrogen gas at low temperatures, and so they acquire enhanced probability. I think it is clear, however, that this simple explanation will have to be supplemented in some respects, for in its present form it offers no explanation of the striking difference in the volume changes of NH_4Cl and NH_4Br which accompany the anomaly, one an expansion and the other a contraction.

Anomalies in Thermal and Volume Effects

There is no reason to think that there may not be other examples of anomalies in thermal and volume effects spread over a range instead of being sharply located, for mechanisms similar to that suggested by Pauling must be of fairly frequent occurrence, particularly in complicated compounds. There are, in fact, a large number of effects already known which doubtless come under this or some similar category. Thus, it has been very recently discovered at the Geophysical Laboratory in Washington, by Kracek, that $NaNO_3$ has thermal and volume anomalies very much like those of NH_4Cl, except that the region of the anomaly is in the neighborhood of 250° C. instead of at −30°. The explanation may well be similar to that of Pauling for NH_4Cl; namely, that the NO_3 radical, which is similar in its structure to the NH_4 radical, passes from oscillatory to rotational motion. The moment of inertia of the NO_3 group is so much greater than that of the NH_4 group that it is to be expected that the corresponding temperature would be much higher, and this agrees with experiment. Another example is NH_4F, studied also by Simon, who found again an anomaly, although very much less pronounced than that for the chloride. An interesting negative example is that of $N(CH_3)_4Cl$. This is similar in structure to NH_4Cl, the H being replaced by the CH_3 group, and the two compounds are very closely related chemically. The moment of inertia of the $N(CH_3)_4$ radical is, however, so much higher than that of the NH_4 radical that it is to be expected that very much higher temperatures would be required to produce corresponding effects; in fact, it may easily be that the necessary temperature would be above the temperature of decomposition. I made a search for the existence of the anomaly at room temperature up to a pressure of 12,000 kg. per sq. cm., but with negative results as would be expected.

The possibility of transitions of this kind evidently opens the whole problem of polymorphic transitions. It is not at all inconceivable that the major part of the region throughout which the anomaly is situated should be so narrow as to simulate, under ordinary conditions of temperature and pressure, the sharpness of an ordinary polymorphic transition.

A very strong suggestion in this direction is contained in the observation above, that the transition of NH_4Br when running under pressure at $-72°$ is very much sharper than when taking place at atmospheric pressure and $-30°$. With so many possibilities as are now being recognized in the way of transitions, it is evident that the completest possible study should be made of the properties of the substance above and below the transition range, and, in particular, a study should be made of the lattice structure by means of X-rays. It is not at all impossible that some of the new transitions which I have found at high pressures are of this new quantum type. For example, the transition of metallic cerium is like that of NH_4Br in that the graph of volume against pressure has the same abnormal upward curvature below the transition pressure, which

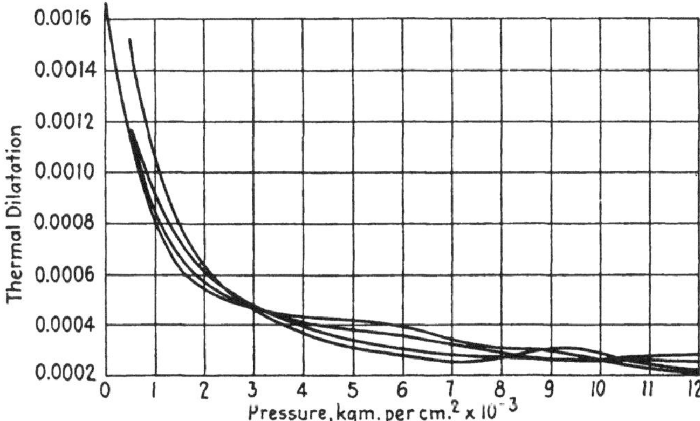

FIG. 7.—ANOMALIES IN THERMAL EXPANSION OF ETHER AT VARIOUS TEMPERATURES PRODUCED BY PRESSURE.

At pressures below 2000 kg. per sq. cm. the order of the curves reading from the bottom up is 20°, 40°, 60°, 80°. Crossing of curves or a curve rising with increasing pressure indicates an anomaly.

means that the compressibility increases with increasing pressure, instead of decreasing, as for the vast majority of substances. Another transition, hitherto only realized under pressure, which may be of this type is that of metallic cadmium, the chief argument being merely that it is so different in some of its properties from an ordinary polymorphic transition. There is no reason why anomalies of this sort should be confined to solids, but one might expect very complicated possibilities in organic liquids of complicated compounds, especially now that it is recognized on X-ray evidence that structures do exist in the liquid not greatly unlike crystal structures. There do not seem to be many such anomalies that can be definitely identified under ordinary conditions of temperature increase at atmospheric pressure, but if pressure is added to the temperature variable, giving two degrees of freedom, and increasing the range in the same ratio that a plane area is greater than a line, it is evident that the possibilities might be enormously increased. This does in fact correspond to what I have found experimentally. Practically all the organic liquids examined with sufficient precision show ranges in which their behavior is anomalous; the anomaly may consist in a compressibility increasing with increasing pressure or falling with increasing temperature, or a thermal expansion increasing with increasing pressure. Fig. 7 shows one example of such behavior. Corresponding to these volume anomalies, there are also anomalies in the specific heats; these, however, at present can be detected only indirectly by thermodynamic methods from the volume anomalies, and not by direct measurement. I have no doubt that the explanation of effects of this sort, when finally arrived at, will be found to be in many respects like that of Pauling for the anomalies of NH_4Cl.

The clear appreciation that the simple materials of ordinary experience may exhibit under ordinary conditions the various kinds of complexity in their behavior which have been discussed will evidently have a far-reaching effect on our theoretical ideas of the structure of ordinary matter. The complete working out of such theoretical implications, however, is for the future, and I shall content myself with the suggestions already made. But there is another direction in which the recognition of the widespread existence of such complexities will affect our actions, and this is in the recording and reduction of our observational data. It is usually accepted as axiomatic that scattering observations which do not fall on a smooth curve are probably incorrect, and that only smooth curves are to be employed in reporting experimental data. In fact, apparently there has been growing up among physicists a sort of sixth intuitive sense as to what sort of smoothness a physical curve is most likely to have, and various people interested in the discussion of errors of observation have been concerned to get some mathematical formulation of what the experimental physicist does in drawing free hand the curve which seems to him best. The mathematical suggestions have been most varied in character, ranging from simple suggestions demanding the continuity of perhaps the third derivative to more complicated suggestions demanding the maximizing or minimizing of perhaps the entire area under the curve of the second derivative. The value of all such intuitive and instinctive demands is obviously much lessened with the recognition that anomalies may be expected which are truly characteristic of the material, and which become more accentuated as the purity of the material becomes greater. These possibilities evidently make the task of the experimental physicist much more difficult. He can no longer blithely discard a discordant observation as probably due to experimental error and so as of no significance; neither must he go to the opposite extreme of keeping in his final result all discordant individual measurements, because the chances of observational and instrumental error have not been at all reduced by the recognition of the possible existence of real anomalies. Every discordant observation, when greater than the normal recognized instrumental or observational error, must be treated with more respect than hitherto; observations must be multiplied in the neighborhood of the discordant point to find whether the effect repeats and is in so far real. If the irregularity does repeat, one must make sure that it cannot be explained by an instrumental imperfection. If there are left unexplained irregularities in the data which one has not opportunity or time to clear up, these must be recorded meticulously in the final report of the experiment, in order that other experimenters may know exactly what the situation is. But this must all be done with a fine discretion, fortified by an understanding of the theoretical possibilities, or else the experimental papers of the future will be cluttered with the announcement of possible but unproved revolutionary new effects.

VOLUME-TEMPERATURE-PRESSURE RELATIONS FOR SEVERAL NON-VOLATILE LIQUIDS.

By P. W. Bridgman.

Presented Oct. 14, 1931. Received Oct. 16, 1931.

TABLE OF CONTENTS.

Introduction.

The material of this paper constitutes an extension to new liquids of measurements of volume as a function of pressure and temperature by the sylphon method recently developed and described.[1] The apparatus, and methods of experiment and calculation were the same as in the first paper and need not be described again in detail.

Most of the liquids of this paper were chosen particularly with a view to finding organic liquids which should be as incompressible as possible. One of the interesting results which had appeared in the previous paper on summarizing all available data for the volume of organic liquids at high pressures was the exceptionally low compressibility of glycerine, which stands in a class by itself with a compressibility about one half as great as that of water, which in its turn is of the order of only half as compressible as the majority of all other liquids hitherto measured. The problem that suggested itself, therefore, was to find if possible other organic liquids as incompressible as glycerine. This was interesting on its own account as throwing some light on the reasons for low compressibility in organic liquids, and also had an incidental interest in that the sylphon method of measuring compressibilities could be extended to solid substances if liquids could be found of high incompressibility with which to transmit pressure to the solids in the sylphon. Pursuant to this program, four liquids were investigated in the same chemical series as glycerine, and in addition, glycerine itself was remeasured, the previous measurements having been made at only a single temperature. These measurements showed nothing less compressible than glycerine; the measurements made evident in what direction in the series incompressibility increases, and also that glycerine is the most incompressible liquid available in the series, the next members in the direction of increasing incompressibility being solid rather than liquid. Another clue was therefore followed in the further search for incompressibility. Low compressibility might reasonably be associated with low thermal expansion, and this in its turn is known to be associated with low vapor pressure. The Eastman Kodak Co., in their search for liquids suitable for diffusion pumps, has recently issued a list of organic liquids with vapor pressures less than that of mercury. Half a dozen of these liquids, of widely varying chemical composition, were selected for measurement. These all proved to be much more incompressible than the ordinary organic liquid, but nothing was found superior to glycerine.

In addition to the eleven liquids just described and chosen with a view to high incompressibility, three other liquids were investigated for various reasons. Eugenol was measured because of the interesting effects found by W. E. Danforth, Jr.,[2] in measurements of the effect of pressure on its dielectric constant, which demanded a knowledge of volume as a function of pressure to bring out their full significance. Iso-octane which has recently become available through the activities of one of the large commercial research laboratories, was measured in order to supplement the measurements of the preceding paper on normal octane. Isoprene was measured in order to supplement the studies previously made by Professor J. B. Conant, C. O. Tongberg, and W. R. Peterson[3] on the effect of pressure in causing it to polymerize to rubber.

Details of Experiment or Calculation.

The measurements given in the following differ in only two important respects from those in the preceding paper. In the first place, a knowledge of the density of the liquid at 0° C at atmospheric pressure is necessary in order to reduce the sylphon readings to fractional changes of volume. The necessary data have been previously determined in only a few cases, so that special measurements had to be made. These usually consisted of a determination of the density at room temperature at atmospheric pressure by weighing a mass of copper immersed in the liquid. To reduce this density to 0° the thermal expansion is also necessary. This was not directly measured at atmospheric pressure, but was found by an extrapolation of the measurements under pressure of the volume at 0° and 50°. At 0° the volume was measured in the sylphon as a function of pressure down to atmospheric pressure, but at 50° the lowest pressure was usually about 200 kg/cm^2. The thermal expansions between 0° and 50° at 200, 400, 600, 800, and 1000 kg were obtained from the sylphon readings, and extrapolated back to atmospheric pressure. This could be done with very little uncertainty because the curvature was slight, and even sometimes the thermal expansion was sensibly a linear function of pressure over this range. The density at 0° was then calculated from the observed density at 25° at atmospheric pressure with the value just obtained for the thermal expansion, assuming that at atmospheric pressure the thermal expansion is linear in temperature between 0° and 50°. This again is not quite exact, but the error so introduced cannot be large.

In the following the necessary data are given so that in the future correction can be made for the various effects just discussed if it should prove necessary. The tables of volume given in the following ostensibly contain the relative volumes in terms of the volume at 0° C and atmospheric pressure as unity. The tables were actually constructed, however, in the first place by measuring the volume decrements of a certain weight of liquid, reducing these volume decrements by division to values corresponding to the weight of liquid which by calculation would occupy 1 cm^3 at 0°C at atmospheric pressure, and then finding the actual volumes by subtracting these volume decrements from 1.0000.

The way in which corrections should be applied in case of future improvements in the thermal expansion at atmospheric pressure may be shown by a detailed example. Consider methyl oleate. The decrements for this are stated to be for 0.8950 gm, which means that the density at 0° C at atmospheric pressure, D_0, was taken to be 0.8950. The measured density at atmospheric pressure at 26.0° is stated to be 0.8768. The extrapolated thermal expansion must therefore have been such as to increase the density from 0.8768 at 26° to 0.8950 at 0°. The mean thermal expansion between 0° and 26° was therefore $\left(\frac{1}{.8768} - \frac{1}{.8950}\right)\frac{1}{26} = \frac{1.14051 - 1.11732}{26} = \frac{.02319}{26}$ $= .000893$. Suppose now that future experiment shows that a better value for the mean thermal expansion in this range is 0.00095. Then $\frac{1}{D_0}[1 + 26 \times .00095] = \frac{1}{.8768}$. (The density at 26° must not be adjusted even if future measurements on methyl oleate give a better value, because the error of measurement of the density was here low enough to give valid values, and future disagreement must be ascribed to differences in the purity of the liquid.) Solving the last equation for D_0 gives $D_0 = .8768 \times 1.02470 = .8984$. The Table should therefore have been constructed for 0.8984 gm rather than 0.8950 gm. The volume decrements in the table must therefore be multiplied by $\frac{.8984}{.8950}$. This would give, for example, a volume at 0° and 1500 kg of $\left[1.0000 - \frac{.8984}{.8950} \times .0638\right] = 1.0000 - .0640 = .9360$. Again, to get the correct volume at 95° and 1500 kg the volume difference in the table between 95° and 0° at 1500 kg (i. e. $0.9886 - 0.9362 = 0.0524$) is to be multiplied by $\frac{.8984}{.8950}$ $(= 0.0526)$, and this added to the corrected volume at 0° and 1500 kg, giving for the final volume $0.9360 + 0.0526 = 0.9886$, accidentally the same as the original volume. The volume differences in the tables were constructed as far as possible on the isotherms at 0°; at pressures higher than the maximum pressure allowed at 0°, they were constructed from the maximum pressure of the 0° isotherm, which in the case of methyl oleate is 1500 kg, and the appropriate temperature as a fiducial point. Thus in the example above, the corrected volume at 95° and 10000 kg would be $.9886 - [.9886 - .8373]\frac{.8984}{.8950} = .8367$, instead of .8373.

In a second respect the experimental procedure was different from that in the preceding paper. The maximum pressures attainable are in many cases limited by the freezing of the liquid. In the previous paper there was often sufficient data already available to allow some guess to be made at what the freezing pressures might be, but so few of the physical properties of the liquids of this paper have been tabulated that a direct determination of the freezing pressures was undertaken for some of the liquids. This usually took the form of a freezing point determination at atmospheric pressure by the method of temperature arrest, the substance being first solidified at liquid air temperature, and then the rise of temperature, as given by a thermocouple imbedded in the substance, observed as a function of time, melting being indicated in the conventional way by a flat place in the curve. The freezing under pressure was usually determined at room temperature. This was done as simply as possible. A cylinder with a hole 5/16 of an inch, 0.79 cm, in diameter was filled with the liquid, the manganin pressure gauge being directly immersed in the liquid. The liquids were in all cases good enough insulators so that this could be done with no appreciable amount of short circuiting. Pressure was then increased by pushing in the piston, and the piston position was plotted against the pressure as given by the manganin gauge. Freezing was indicated by a flat place in the curve. After the substance had been frozen by increasing pressure, readings were usually repeated with decreasing pressure, in this way getting a better value for the freezing pressure without error from subcooling. The two points, at atmospheric pressure and at room temperature, were usually good enough to give a sufficient indication of the range of pressure at any temperature that might safely be applied without runnuing into the freezing curve.

Detailed Presentation of Data.

```
                    H   H
                    |   |
Ethylene Glycol. H — C — C — H
                    |   |
                    OH  OH
```

Ethylene Glycol. This material was obtained from the Eastman Kodak Co., and was used without further purification. In order to avoid the possibility of freezing, runs were made first at 50° and 95° to 12000 kg, starting the run at 50° at atmospheric pressure. These runs were accomplished without incident, the points with increasing and decreasing pressure lying on the same smooth curve. After the run at 95° it was necessary to take the apparatus apart to renew some of the insulation, and on reassembling the apparatus and starting the run at 0° it was evident that some permanent distortion had been introduced into the sylphon, probably by releasing the pressure to atmospheric before the sylphon had sufficiently cooled. Furthermore, the run at 0° encountered stationary readings, indicating freezing, at about 6000 kg. This, however, was not the true freezing pressure, but was too high because of subcooling; in fact pressure was released as far as 3300 kg without signs of melting. So great a subcooling was not anticipated, and the apparatus was again dismantled for inspection, with the expectation of finding the sylphon ruined. No difficulty was obvious, however, and weighing showed that there had been no leak. The apparatus was then again reassembled, and the 0° readings checked. The run was then concluded with several readings at 50° and 95° and then back to 0°, in order to establish the temperature expansion, which could not be obtained between 50° and 95° because of the distortion on the first disassembly. This experiment provides a rather extreme example of the amount of abuse a sylphon will sometimes support without being put out of commission. The average deviation from a smooth curve of a single one of the 70 readings was 0.26% of the maximum effect, treating the readings after the distortion as belonging to a different series from those before it.

TABLE I.

RELATIVE VOLUMES OF ETHYLENE GLYCOL.

Pressure	Temperature		
kg/cm²	0°	50°	95°
0	1.0000	1.0300	
500	.9841	1.0133	
1,000	.9703	.9979	1.0278
1,500	.9589	.9846	1.0091
2,000	.9486	.9724	.9938
3,000	.9304	.9519	.9708
4,000	.9154	.9343	.9515
5,000	.9032	.9197	.9352
6,000		.9073	.9215
7,000		.8959	.9089
8,000		.8851	.8982
9,000		.8753	.8883
10,000		.8664	.8789
11,000		.8585	.8706
12,000		.8503	.8627

The relative volumes are given in Table I. The volume decrements in the table are for 1.1263 gm. The density, as directly determined from the loss of weight of an immersed copper weight was 1.1137 at 20.25°. In reducing this to the density at 0°, 1.1263, the total thermal expansion of 1 gm of the liquid between 0° and 50° at atmospheric pressure was taken as 0.0267 cm^3, this being the value given directly by the sylphon measurements.

It is not at all improbable that the melting pressures at 50° and 95° may be below the maximum of 12000 kg reached in these measurements, the substance having remained liquid by subcooling. A similar effect is known to exist in glycerine, in which the subcooling is very marked, and it is not unreasonable to expect the same effect here.

Tri-methylene Glycol.

```
    H   H   H
    |   |   |
H — C — C — C — H
    |   |   |
    OH  H   OH
```

This was obtained from the Eastman Kodak Co., and was most kindly further purified by fractional distillation by Professor Conant, who did not find, however, any marked change in its boiling point introduced by the redistillation.

Runs were made at 0° to a maximum of 5000 kg, and then at 50° and 95° to 12000 kg, with no sign of freezing at any temperature. There was an unexplained break between 0° and 50°, not due to the sylphon, which was taken down and checked between the two runs; the break was probably due to an inadvertent and unrecorded change in the electrical constants of the potentiometer. The return check readings were made at 50° and 0° as usual, and the thermal expansion between 0° and 50° taken from the check readings. The average departure from smooth curves of a single one of the 61 readings was 0.28% of the maximum effect.

The relative volumes are shown in Table II. The volume decrements listed in the table are for 1.063 gm. The density at 20.2° at atmospheric pressure determined by buoyancy measurements was 1.0528. This substance is listed in International Critical Tables, for which the density at 20° is given as 1.053, very good agreement.

Propylene Glycol.

```
    H   H   H
    |   |   |
H — C — C — C — H
    |   |   |
    OH  OH  H
```

This material was obtained from the Eastman Kodak Co., and was used without further purification. Readings were made first at 0° to a maximum of 4200 kg, then at 50° to 12000, at 95° to 12000, a return check reading at 50° at 900 kg, and then the run at 0° was repeated to a maximum of 12000 and back, with perfect agreement with the first run in the common interval, and no indication of freezing at any temperature. The average departure from smooth curves of a single one of the 73 readings was 0.13% of the maximum effect.

TABLE II.

RELATIVE VOLUMES OF TRI-METHYLENE GLYCOL.

Pressure kg/cm²	Temperature 0°	50°	95°
0	1.0000	1.0274	
500	.9830	1.0082	
1,000	.9671	.9929	1.0165
1,500	.9547	.9799	1.0016
2,000	.9432	.9678	.9879
3,000	.9237	.9472	.9640
4,000	.9072	.9300	.9450
5,000	.8922	.9150	.9292
6,000		.9020	.9154
7,000		.8898	.9030
8,000		.8792	.8919
9,000		.8707	.8817
10,000		.8630	.8729
11,000		.8553	.8655
12,000		.8474	.8582

The relative volumes are given in Table III; the volume decrements are for 1.0486 gm. The measured density at 20° was 1.0340. International Critical Tables gives 1.038 at 23°, not especially good agreement. The assumed value for the density at 0°, 1.0486, was obtained by taking the total expansion at atmospheric pressure between 0° and 50° to be 0.0325 cm³, a value obtained by slight extrapolation of the sylphon readings.

Diethylene Glycol.

```
     H   H       H   H
     |   |       |   |
H — C — C — O — C — C — H
     |   |       |   |
    OH   H       H  OH
```

This was obtained from the Eastman Kodak Co., and was used without further purification. The first run was at 0° to a maximum of 5800 kg. At this pressure a stationary reading was obtained, indicating freezing. Temperature was raised at once to 50° in order to avoid possible damage to the sylphon, but in spite of this precaution, some permanent set was produced. At 50°, pressure was lowered to 300 kg, increased back to 12000, and then lowered back to 400; all the points at 50° lay on the same smooth curve, indicating no further damage to the sylphon. The 95° run was next made, then the check reading at 50°, with perfect agreement, and finally a new curve was obtained at 0°, up to 4500 and back. The average deviation from smooth curves of a single one of the 62 readings was 0.25% of the maximum effect.

TABLE III.

RELATIVE VOLUMES OF PROPYLENE GLYCOL.

Pressure	Temperature		
kg/cm²	0°	50°	95°
0	1.0000		
500	.9819	1.0129	
1,000	.9664	.9947	1.0201
1,500	.9540	.9794	1.0027
2,000	.9432	.9659	.9877
3,000	.9237	.9454	.9637
4,000	.9070	.9266	.9430
5,000	.8935	.9120	.9257
6,000	.8809	.8989	.9110
7,000	.8679	.8864	.8985
8,000	.8569	.8750	.8867
9,000	.8472	.8652	.8763
10,000	.8380	.8563	.8672
11,000	.8294	.8477	.8589
12,000	.8219	.8400	.8514

The relative volumes are given in Table IV. The density at 0° at atmospheric pressure was taken as 1.1298, which is therefore the number of grams to which the volume decrements of the table refer. The directly measured density at 25° was 1.1122. This does not agree particularly well with the value of International Critical Tables, which is 1.132 at 20°. In calculating the density at 0°, the total expansion of one gram between 0° and 25° was taken to be 0.0142 cm³, a value calculated, as explained, by a slight extrapolation of the sylphon readings at 50°.

TABLE IV.

RELATIVE VOLUMES OF DIETHYLENE GLYCOL.

Pressure kg/cm²	Temperature 0°	50°	95°
0	1.0000	1.0320	
500	.9845	1.0135	
1,000	.9710	.9987	1.0223
1,500	.9604	.9861	1.0077
2,000	.9492	.9729	.9833
3,000	.9320	.9523	.9705
4,000	.9174	.9352	.9522
5,000	.9047	.9207	.9368
6,000		.9075	.9235
7,000		.8963	.9113
8,000		.8859	.9005
9,000		.8762	.8902
10,000		.8677	.8810
11,000		.8608	.8728
12,000		.8542	.8648

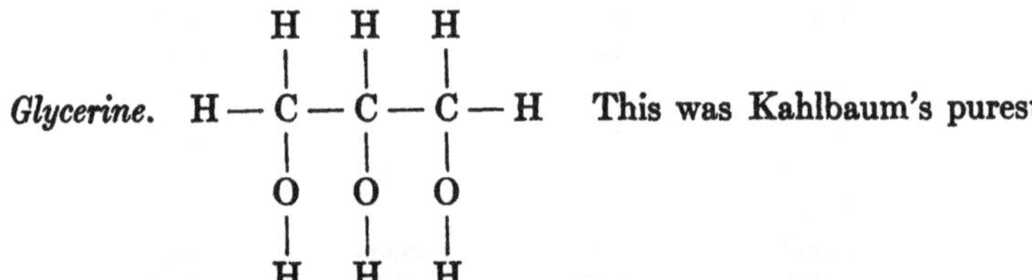

Glycerine. This was Kahlbaum's purest grade, "Zur Analyse," used without further purification. It was, however, doubtless, somewhat impure with water, as shown by the low value of the density, discussed in the next paragraph. This material had been previously measured, but only at a single temperature, and with the free piston piezometer, which is not capable of such great accuracy as the sylphon. Readings were first made at 0° to a maximum of 5000, then at 50° to 12000 and back, at 95° to 12000 and back, and then the run at 0° was repeated, this time to 12000 and back, with a perfect check on the common portions of the two 0° curves. Particular care was taken to change the pressure slowly, because the viscosity of gylcerine is known to increase greatly under pressure. The average deviation from smooth curves of a single one of the 65 readings was 0.31% of the maximum effect. Two thirds of the total irregularity was contributed by the 13 points of the 0° run above 6000 kg, and is without doubt the expected effect of viscosity.

The relative volumes are given in Table V. The density at atmospheric pressure at 0° was taken as 1.2385, and the volume decrements of the table are for this number of grams. The measured density at atmospheric pressure at 24.6° was 1.2226. International Critical Tables gives at 20° 1.260. The low value for this glycerine is to be explained by the presence of water. That this was the case was checked later by Mr. W. E. Danforth, Jr. who tried to measure the effect of pressure on the dielectric constant of this same material without success, because of its high conductivity, which means water. He was able to remove the water sufficiently for his purposes by chemical means; it might pay at some future time to redetermine the compressibility of entirely water-free glycerine. The effect of the water cannot be large, however, as may be seen by a comparison of the values of Table V with those given in the preceding paper for a 50% water-glycerine mixture.

TABLE V.

RELATIVE VOLUMES OF GLYCERINE.

Pressure kg/cm²	Temperature 0°	50°	95°
0	1.0000	1.0266	
500	.9900	1.0136	
1,000	.9806	1.0025	1.0240
1,500	.9721	.9930	1.0125
2,000	.9641	.9843	1.0024
3,000	.9501	.9688	.9853
4,000	.9373	.9548	.9700
5,000	.9264	.9423	.9565
6,000	.9157	.9310	.9447
7,000	.9057	.9211	.9342
8,000	.8958	.9121	.9244
9,000	.8867	.9036	.9152
10,000	.8783	.8955	.9070
11,000	.8712	.8879	.8994
12,000	.8648	.8800	.8925

The density at 0° at atmospheric pressure was calculated from the measured density at 24.6° by taking 0.0215 cm³ for the total expansion between 0° and 50° of one gram, a value obtained by extrapolation of the sylphon readings at 50°.

Comparison with the previous values of the compressibility of glycerine from the same source gives rather good agreement within the limits of error of the previous method. The former values for the relative volumes at 30° at 0, 2000, 4000, 6000, 8000, 10000, and 12000 kg respectively are: 1.000, 0.958, 0.932, 0.911, 0.893, 0.879, and 0.866. The values obtained by linear interpolation between 0° and 50° of the present measurements are: 1.000, 0.960, 0.933, 0.910, 0.891, 0.875, and 0.860. Compared with the sylphon, the former free piston seems to give volume changes which are too large at low pressures and too small at high pressures.

Tri-o-cresyl-Phosphate.

```
     ⎡   H\       /CH3    ⎤
     ⎢     C — C          ⎥
     ⎢   /       \        ⎥
     ⎢ H — C       C — O  ⎥ ≡ P = O
     ⎢   \       /        ⎥
     ⎢     C — C          ⎥
     ⎣   H/       \H      ⎦3
```

This was obtained from the Eastman Kodak Co. and was used without further purification. The first run at 0° indicated freezing somewhere between 2200 and 3000 kg. At the latter pressure temperature was raised immediately to 50°, and the run at 50° made to a maximum of 4900 kg, and then the run at 95° to a maximum of 6200. The return check reading at 0° now showed that a permanent distortion had taken place at the first freezing; the run at 0° was repeated to a maximum of 1800, giving a curve of the same shape as the first curve, but displaced by a constant amount. Points on the 50° and the 95° curve were then repeated, giving agreement with the first determinations. The average deviation from smooth curves of a single one of the 36 readings was 0.54% of the maximum effect. The greater irregularity than usual is to be explained in part by the much narrower range than usual.

TABLE VI.

RELATIVE VOLUMES OF TRI-O-CRESYL PHOSPHATE.

Pressure	Temperature		
kg/cm²	0°	50°	95°
0	1.0000	1.0302	
500	.9831	1.0109	
1,000	.9687	.9938	1.0163
1,500	.9573	.9790	.9994
2,000	.9478	.9669	.9858
3,000		.9468	.9630
4,000		.9308	.9448
5,000		.9180	.9300
6,000			.9166

The relative volumes are given in Table VI. The density at 0° at atmospheric pressure was taken to be 1.1987, which means that the volume decrements listed in the table are for this number of grams. The directly measured density was 1.1803 at 25°. The volume expansion of one gram at atmospheric pressure between 0° and 50° was taken to be 0.02528, obtained by extrapolation of the sylphon readings.

$$\begin{array}{l} CH_3COOCH_2 \\ \quad\quad\quad\;\; | \\ CH_3COOCH \\ \quad\quad\quad\;\; | \\ CH_3COOCH_2 \end{array}$$

Tri-acetin. The material was obtained from the Eastman Kodak Co. and used without further purification. Runs were made at 0° to a maximum of 4900 kg, at 50° to 10000, and at 95° to 12000, with a return check reading at 0° with good agreement. The maxima at 0° and 50° were as high as it was thought safe to go in view of a preliminary freezing point examination in the special small apparatus; the indications of freezing were not very sharp, however, and it should not be assumed without further examination that the limits mentioned above are actually very close to the freezing points. The average departure from smooth curves of a single one of the 44 readings was 0.13%.

TABLE VII.

RELATIVE VOLUMES OF TRI-ACETIN.

Pressure kg/cm²	Temperature 0°	50°	95°
0	1.0000	1.0441	
500	.9793	1.0198	
1,000	.9627	.9981	1.0289
1,500	.9492	.9805	1.0071
2,000	.9374	.9653	.9893
3,000	.9177	.9412	.9628
4,000	.9012	.9225	.9432
5,000	.8861	.9068	.9258
6,000		.8932	.9098
7,000		.8807	.8956
8,000		.8694	.8831
9,000		.8591	.8721
10,000		.8498	.8624
11,000			.8533
12,000			.8445

The relative volumes are given in Table VII. The density at 0° at atmospheric pressure was taken as 1.2157, so that the volume decrements in the table are for this number of grams. The directly measured density at 26.9° was 1.1613; in reducing this to the density at 0° the expansion of one gram between 0° and 26.9° was taken to be 0.0194 cm^3, obtained by extrapolation of the sylphon readings. This substance is unusual in that the thermal expansion between 0° and 50° is linear in the pressure up to 1100 kg/cm^2 within the limits of experimental error.

Ethyl Dibenzyl Malonate. $(C_6H_5CH_2)_2C(COOC_2H_5)_2$ This was obtained from the Eastman Kodak Co. and used without further purification. A preliminary freezing exploration indicated that the freezing pressure was very low, and the limits of the measurements were set by this exploration. The first run was made at 50° to a maximum of 2100 kg, then at 95° to 5000, at 50° again to a maximum of 3000, at 0° to 1000, with a final check reading at 50° in perfect agreement. The average departure from smooth curves of a single one of the 31 readings was 0.48% of the maximum effect. The irregularity, which is somewhat greater than usual, is to be entirely explained by the small range, the average absolute irregularity in the readings being no greater than usual.

TABLE VIII.

Relative Volumes of Ethyl Dibenzyl Malonate.

Pressure	Temperature		
kg/cm^2	0°	50°	95°
0	1.0000	1.0361	
500	.9775	1.0111	
1,000	.9620	.9912	1.0172
1,500		.9748	.9979
2,000		.9616	.9823
3,000		.9412	.9581
4,000			.9381
5,000			.9179

The relative volumes are shown in Table VIII. The density at 0° at atmospheric pressure was taken to be 1.1143, so that the volume decrements in the table refer to this number of grams. The directly measured density at 25.9° was 1.0943; the density at 0° was calculated from this, using for the total expansion of one gram between 0° and

25.9° 0.0167 cm³, a value obtained by extrapolation of the sylphon readings. The thermal expansion between 0° and 50° is again linear in the pressure up to 1000 kg.

Methyl Oleate. $CH_3(CH_2)_7CH = CH(CH_2)_7COOCH_3$. This was obtained from the Eastman Kodak Co. and used without further purification. A preliminary freezing point determination gave an unusually sharp freezing point between 2900 and 3300 kg at 24°; this may be accepted with some confidence as the proper freezing coordinate. In the compressibility measurements the range was fixed by this information, together with the freezing point at atmospheric pressure listed in tables of physical constants. Two sets of runs were made; the first was terminated by damage to the sylphon before the return check readings could be made. The second set-up with a new sylphon gave first a run at 95° to 10000 kg, one at 50° to 5900, one at 0° to 1800, and finally a return check reading at 95° with perfect agreement. The average deviation from smooth curves of a single one of the 30 readings, discarding a single one where there must have been a blunder in recording, was 0.20%.

TABLE IX.

RELATIVE VOLUMES OF METHYL OLEATE.

Pressure kg/cm²	Temperature 0°	50°	95°
0	1.0000	1.0401	
500	.9740	1.0103	
1,000	.9532	.9859	1.0110
1,500	.9362	.9669	.9886
2,000		.9509	.9711
3,000		.9251	.9421
4,000		.9044	.9159
5,000		.8874	.9004
6,000		.8716	.8845
7,000			.8703
8,000			.8583
9,000			.8475
10,000			.8373

The relative volumes are given in Table IX. The density at 0° at atmospheric pressure was taken as 0.8950, so that the volume decrements of the table are for this number of grams. The directly measured density at atmospheric pressure at 26.0° was 0.8768. In calculating the density at 0°, the total expansion of one gram between 0° and 26° was taken to be 0.0234 cm³, as given by extrapolation of the sylphon readings.

$CH_3(CH_2)_4COOCH_2$
|
Tri-caproin. $CH_3(CH_2)_4COOCH$ This was obtained from the
|
$CH_3(CH_2)_4COOCH_2$

Eastman Kodak Co. and used without further purification. The first run at 0° was to a maximum of 7500 kg, then at 50° to 11000, at 95° to 12000, and finally a check reading at 0° in perfect agreement. The average departure from a smooth curve of a single one of the 49 readings was 0.16%.

TABLE X.

RELATIVE VOLUMES OF TRI-CAPROIN.

Pressure kg/cm²	Temperature 0°	50°	95°
0	1.0000	1.0412	
500	.9741	1.0095	
1,000	.9547	.9850	1.0120
1,500	.9388	.9656	.9905
2,000	.9251	.9490	.9723
3,000	.9025	.9229	.9431
4,000	.8846	.9031	.9208
5,000	.8698	.8862	.9024
6,000	.8565	.8714	.8867
7,000	.8442	.8584	.8726
8,000	.8333	.8471	.8601
9,000		.8373	.8493
10,000		.8285	.8395
11,000		.8207	.8309
12,000			.8230

The relative volumes are given in Table X. The density at 0° at atmospheric pressure was taken to be 0.9970 so that the volume decrements of the table are for this number of grams. The measured density at atmospheric pressure at 27.0° was 0.9754. In computing the density at 0° the thermal expansion at atmospheric pressure of one gram between 0° and 50° was taken to be 0.0413 cm³, as given by extrapolation of the sylphon readings.

Normal-butyl Phthalate. $C_6H_4\begin{matrix}-COO(CH_2)_3CH_3\\-COO(CH_2)_3CH_3\end{matrix}$, $C_{16}H_{22}O_4$ This was obtained from the Eastman Kodak Co.; it had been specially freed from all volatile impurities so as to be ready for immediate use in diffusion pumps, and it was used directly in the sylphon with no attempt at further purification. A first attempt with this failed because of the development of a leak in the sylphon. The first set-up established, however, the safe pressure limits. The second set-up gave first a run at 0° to 8200 kg, then at 50° to 12000, at 95° to 12000, and then return check readings at both 0° and 95° again with good agreement. The average departure from smooth curves of a single one of the 48 readings was 0.16%.

TABLE XI.

RELATIVE VOLUMES OF NORMAL-BUTYL PHTHALATE.

Pressure		Temperature	
kg/cm²	0°	50°	95°
0	1.0000	1.0369	
500	.9768	1.0095	
1,000	.9575	.9872	1.0172
1,500	.9419	.9690	.9928
2,000	.9297	.9540	.9759
3,000	.9096	.9301	.9489
4,000	.8931	.9125	.9277
5,000	.8779	.8964	.9099
6,000	.8649	.8811	.8945
7,000	.8534	.8682	.8809
8,000	.8447	.8574	.8690
9,000		.8479	.8586
10,000		.8392	.8494
11,000		.8313	.8409
12,000		.8241	.8340

The relative volumes are given in Table XI. The density at 0° at atmospheric pressure was taken to be 1.0643, so that the volume decrements of the table are for this number of grams. The measured density at atmospheric pressure at 25.0° was 1.0453. In reducing this to the density at 0° the thermal expansion at atmospheric pressure of one gram between 0° and 25.0° was taken to be 0.0173 cm³, as given by extrapolation of the sylphon readings.

Eugenol. $-CH_2-CH=CH_2$, $C_{10}O_2H_{12}$ This mate-

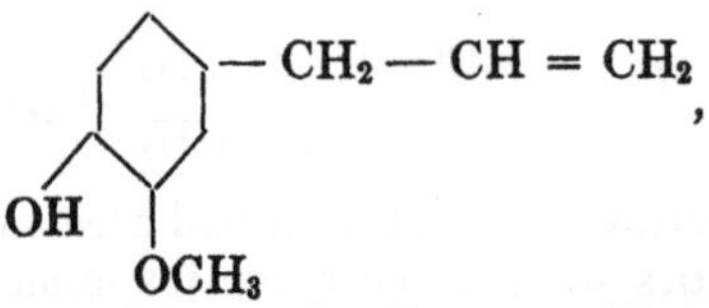

OH

OCH_3

rial was obtained from the Eastman Kodak Co. and used without further purification. It was colorless in appearance, contrasted with the yellow color of the material which had been previously used in measurements of the viscosity under pressure.[4] As already mentioned, the compressibility of this was measured to give greater point to the measurements by W. E. Danforth, Jr., of the effect of pressure on dielectric constant, who in turn was interested in it because of the previously discovered enormously large increase of viscosity produced by pressure. For the purpose in mind, it was sufficient to determine the compressibility at a single temperature, 0°, because this was the temperature at which the interesting dielectric effects had been found. The maximum pressure at this temperature was only 4600 kg, this being as high as it was thought safe to go because of the greatly increased viscosity. As it was, the sylphon was damaged, the points with decreasing pressure lying off the curve, and the zero was greatly displaced. On opening the apparatus there was a distinct smell of eugenol, showing that the sylphon had leaked. This one run gave sufficient data for the purpose in hand, however, and no attempt was made to repeat the measurements. This run gave 8 points with increasing pressure, which lay remarkably well on a smooth curve, the average departure of a single reading being only 0.03% of the maximum effect.

TABLE XII.

RELATIVE VOLUMES OF EUGENOL.

Pressure kg/cm²	Temperature 0°
0	1.0000
500	.9801
1,000	.9639
1,500	.9501
2,000	.9382
3,000	.9187
4,000	.9028
5,000	.8896

The relative volumes are given in Table XII. The density at

atmospheric pressure at 25.9° was measured as 1.0515. No sylphon measurements were made from which the thermal expansion could be calculated, and there seem to be no values recorded in the literature. For the construction of the table the density at atmospheric pressure at 0° was assumed to be 1.065, using for the thermal expansion a mean from other organic compounds.

Iso-octane (2, 2, 4 tri-methyl pentane). This material I owe to the courtesy of Mr. A. E. Becker of the Standard Oil Development Co. The material had been refined in the research laboratory of this company in connection with experiments on motor fuels. They give for it the following physical constants. Boiling point: 99.3° within a range of 0.3°; specific gravity 0.6917 at 20° referred to water at 4°; freezing point — 107° C; aniline point 84°. It was used in the sylphon without further purification. A preliminary freezing point exploration indicated sharp freezing, evidence of high purity. Runs were made with a single filling of the apparatus, to 5600 kg at 0°, 8700 at 50°, 10200 at 95°, with a return check reading at 0° with very good agreement. The pressure limits of these runs were set by the preliminary freezing point examination. The average departure from smooth curves of a single one of the 40 readings was 0.075% of the maximum effect.

TABLE XIII.

RELATIVE VOLUMES OF ISO-OCTANE.

Pressure	Temperature		
kg/cm²	0°	50°	95°
0	1.0000	1.0607	
500	.9529	.9949	
1,000	.9221	.9560	.9851
1,500	.9001	.9286	.9531
2,000	.8825	.9077	.9292
3,000	.8558	.8764	.8946
4,000	.8356	.8529	.8684
5,000	.8161	.8318	.8453
6,000	.8037	.8176	.8307
7,000		.8038	.8161
8,000		.7911	.8024
9,000		.7797	.7901
10,000			.7795

The relative volumes are given in Table XIII. The density at 0° at atmospheric pressure was taken to be 0.7086, and the volume decrements given in the table are for this number of grams. The directly measured density at atmospheric pressure was 0.6897 at 22.5°. In calculating the density at 0° the thermal expansion at atmospheric pressure between 0° and 22.5° was taken as 0.0385 cm^3 per gm, obtained by extrapolation of the sylphon readings.

$$CH_2 = \underset{\displaystyle CH_3}{\underset{|}{C}} - CH = CH_2$$

Isoprene. This material was synthesized at the Chemical Laboratory of Harvard University under the direction of Professor J. B. Conant. The effect of pressure in polymerizing this substance to a rubber-like material had already been extensively studied by Professor Conant[3], and it was of interest to determine the compressibility in connection with these studies. The material was used only a short time after it had been synthesized, and during this time as well as during the process of filling the sylphon it was maintained continuously at 0° in order to avoid danger of polymerization. The compressibility was attempted only at 0°, because at higher temperatures polymerization is known to proceed much more rapidly. No appreciable amount of polymerization was produced by the compressibility run, as shown by the agreement of the points obtained with increasing and decreasing pressure. The pressure range at 0° was safely taken to be the full 12000 kg, isoprene being a very volatile substance of low freezing point and low viscosity. The average departure from a smooth curve of a single one of the 18 readings was 0.16% of the maximum effect.

The relative volumes are given in Table XIV. The density at 0° at atmospheric pressure was taken to be 0.691. For this figure I am indebted to Dr. W. R. Peterson, who made a direct determination in the laboratory of Professor Conant.

Discussion.

This discussion will be mainly occupied with the attempt to correlate differences in the compressibility with differences in the structural formulas. It is worth mentioning, however, that certain general features of the p-v-t- relations of liquids already emphasized in two preceding papers[5] continue to hold for these liquids also. There is in the first place the reversal in the sign of $\left(\frac{\partial^2 v}{\partial \tau^2}\right)_p$ at high pressures. At atmospheric and low pressures the sign of this is positive, but at pressures of the order of 3000 and over it had been found to become

TABLE XIV.

RELATIVE VOLUMES OF ISOPRENE.

Pressure kg/cm²	Temperature 0°
0	1.0000
500	.9524
1,000	.9208
1,500	.8973
2,000	.8781
3,000	.8478
4,000	.8252
5,000	.8070
6,000	.7911
7,000	.7776
8,000	.7658
9,000	.7551
10,000	.7450
11,000	.7357
12,000	.7272

negative, which means that at high pressures the thermal expansion of a liquid is less at high temperatures than at low temperatures at the same pressure. Inspection of the tables of volume of the new liquids of this paper will show that this continues to be true. A second general feature previously found was that $\left(\frac{\partial p}{\partial \tau}\right)_v$ is not a function of volume only, as had been assumed in various theoretical discussions by several authors. An application of the same method of analysis as that used in the preceding paper will show that for these new liquids also there are failures of the relation just as pronounced as found before.

Turning now to the relative behavior of the various compounds, the series terminating with glycerine is perhaps the most interesting. In Figure 1 are plotted the relative volume decrements at 50° of the five members of this series as a function of pressure, in terms of the respective volumes at 50° and atmospheric pressure as unity. 50° was chosen in order to allow comparison over a wider pressure range than would have been possible at 0°, where the range is often restricted by freezing. Included in the same diagram are the volumes of two of the alcohols and n-pentane for comparison. The most striking feature of the diagram is that the volume decrements of the four members of the series containing two OH groups in the molecule are very nearly alike.

The diagram also brings out in a very striking way that when an OH group is substituted into a hydrocarbon by far the most important factor in determining the compressibility is the number of OH's in

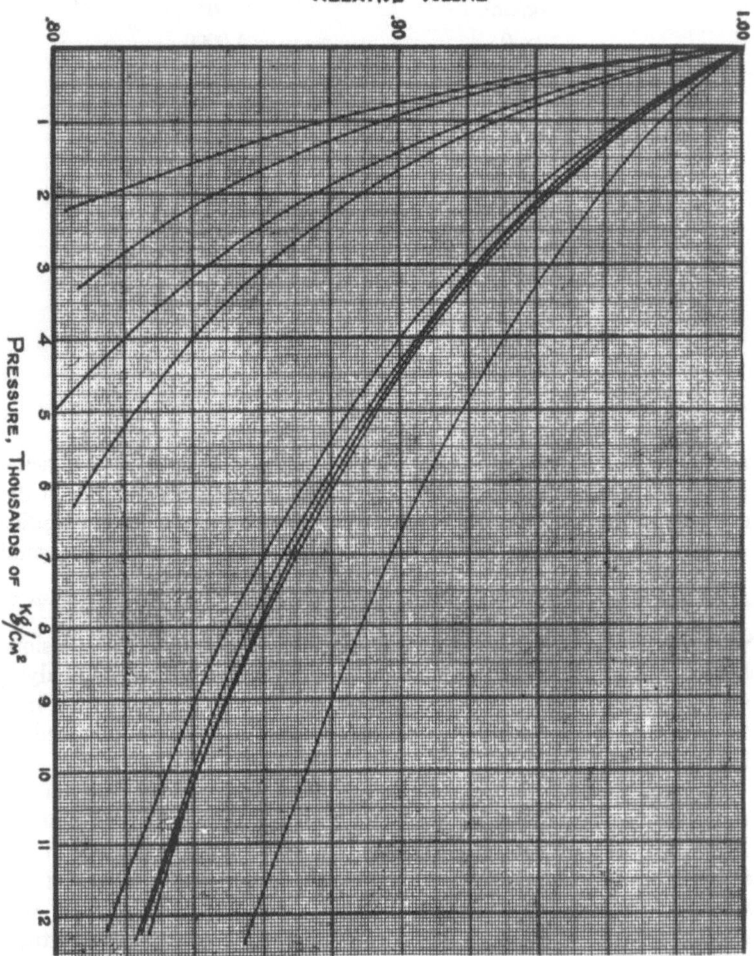

FIGURE 1. The relative volumes at 50° C of a number of liquids as a function of pressure. At 2000 kg/cm² the order of the curves, reading from the top down, is: glycerine, ethylene glycol, diethylene glycol, tri-methylene glycol, propylene glycol, amyl alcohol, methyl alcohol, n-hexane, and n-pentane.

93 — 2552

the molecule, the compressibility decreasing as the number of OH's increases. This is shown in the diagram by the very marked tendency of the curves to fall into families, the unaltered hydrocarbons with no OH groups, typified by n-pentane, constituting the first family with the highest compressibility, the alcohols, which are hydrocarbon chains with one OH group, constituting the next family with lower compressibility, the four glycols of this paper with two OH groups, constituting the next family with still lower compressibility which is nearly the same for all members of the family, and finally glycerine, with three OH groups, and with the lowest compressibility of any organic liquid yet measured. It obviously suggests itself that in the search for organic compounds with low compressibility the natural course is to try those with a large number of OH groups in the molecule. Unfortunately, glycerine appears to be the last compound which is liquid under ordinary conditions, and even with it the liquidity is due only to its remarkable capacity for subcooling. It would be of interest, however, to study for its own sake the compressibility of solid compounds whose molecules contain more than three OH groups, for the same regularities which attach to the compressibility of the liquid phases will doubtless carry over to the solid phase. The study of solids is complicated, however, by the fact that for an exhaustive characterization of their behavior it will be necessary to measure the compressibility of single crystals in different directions.

Any detailed numerical study of the data presented in Figure 1 will encounter complications from temperature effects, the curves being in most cases drawn for the same temperature (50°) whereas doubtless a better method would have been to draw them for the same reduced temperature. This accounts for part of the effect of adding OH groups, for the addition of an OH group increases the critical temperature, so that 50° C becomes an increasingly smaller fraction of the critical temperature, and compressibility normally decreases on decreasing the reduced temperature. The same can be said for the behavior in a single family. On increasing the size of the molecule, keeping the number of OH groups constant, as, for example, in passing from methyl to amyl alcohol, the effect of increasing the length of the hydrocarbon chain is to raise the critical temperature. This is not by any means the entire effect, however, and the addition of the OH group does of itself markedly decrease the compressibility irrespective of the temperature effect, as shown, fór example, by methyl alcohol and hexane, the critical temperature of the former being only 1% higher than that of the latter, while its compressibility is 16% less.

In seeking for an explanation of the effect of the OH group it is important to notice that the effect cannot be regarded as a specific contribution of the OH group only, without interaction with the rest of the molecule. If such were the case the compressibility of ethylene glycol [$C_2H_4(OH)_2$] should be practically the same as that of glycerine, the OH groups having .55 of the total mass of the molecule in the former compound, and .56 in the latter. Another example of the same sort is afforded by trimethylene and diethylene glycols. The OH groups constitute .46 of the weight of the molecule of the first and .32 of the second, yet the compressibilities are nearly the same. It is the number of OH groups in the molecule primarily rather than the fraction of the molecule occupied by OH groups which is determinative, although doubtless in very complicated compounds a single OH group may be swamped by the rest of the molecule. These facts suggest that in the molecule the OH groups interact with all the other parts of the molecule, consolidating them and decreasing the compressibility. Such effects are strongly suggested by wave mechanics.

A detailed comparison of the curves for the four glycols shows certain irregularities, of which the most prominent are the crossing of the ethylene and the diethylene glycol curves at 10000 kg, and the relatively large downward bulge in the trimethylene glycol curve between 6000 and 9000. Such irregularities are not at all unexpected in view of previous experience. A detailed explanation of them would be very complicated and doubtless beyond present day possibilities; I have suggested in a recent paper[6] that effects may be involved similar to the anomalies in the behavior of NH_4Cl and NH_4Br.

An interesting minor structural difference appears between propylene and trimethylene glycol, which differ only in the position of the OH groups; in the latter the two OH groups occupy the two ends of the chain, whereas in the former one OH occupies an end position and the other a mid position. The diagram shows that the propylene compound is distinctly more compressible.

Turning now to a comparison of the other substances, there are few structural similarities to suggest analogies of behavior. Of these, triacetin and tricaproin are the only ones with strictly similar structures. Both of these are in turn similar to glycerine, the difference being that in triacetin each OH group of glycerine is replaced by a CH_3COO group, and in tricaproin each OH group is replaced by a $CH_3(CH_2)_4COO$ group. One would expect similarities in the behavior of these three substances. This, however, is not the case. In the first place both triacetin and tricaproin are much more compressible than glycerine, and in fact their compressibility is in excess of the compressibility of the glycols by a quantity of the same order of magnitude as that by which the compressibility of the glycols is in excess of that of glycerine. This again emphasizes the abnormally low compressibility of the OH group. Furthermore, triacetin is noticeably less compressible than tricaproin, the difference being of the order of 15%. This is not what might at first be expected, the usual rule being for the compressibility to decrease as one proceeds in a family of chemical compounds from those with small to those with larger and more complicated molecules. The evident significance of this abnormal behavior is that the compressibility of the $(CH_2)_4$ group is greater than the average compressibility of the CH_3COO group, an effect for which we are already prepared from our discussion of the glycerine family, where it was suggested that the compressibility of the hydrocarbon chain is rather high. In general, the presence of an O atom in a molecule seems to decrease the compressibility.

N-butyl phthalate and eugenol are similar in that they are substitution products of the benzene ring. In the former two H's of benzene have been replaced by two $COO(CH_2)_3 CH_3$ groups, $C_{10}H_{18}O_4$ altogether, and in eugenol three H's have been replaced by CH_3O, OH, and $— CH_2 — CH = CH_2$, or $C_4H_8O_2$ altogether. Because of the somewhat larger percentage of O in the eugenol replacement group, and particularly because of the presence of the OH group, one would expect the compressibility of eugenol to be less than that of the phthalate. This indeed proves to be the case, the compressibility of the latter being about 11% greater than that of the former. That the OH group does not impart a still more marked incompressibility is evidently explained by the fact that it is swamped by the rest of the molecule, which is much more complex in the ring compounds than in the members of the glycerine family.

Two other benzene substitution compounds have been previously investigated,[7] C_6H_5Cl and C_6H_5Br. These two compounds can be measured only over a small range because of freezing by pressure, but in the short range open to comparison the compressibilities of both are very nearly equal to that of the phthalate.

The structure of ethyl dibenzyl malonate is complicated and involves two benzene rings in the same molecule. It is liquid over only a small range of pressure; in this range its compressibility is very nearly the same as that of eugenol.

Tri-ortho-cresyl-phosphate contains three benzene rings, but there is a new element, phosphorus, so that we do not know exactly what to expect. As a matter of fact it is one of the more incompressible compounds, and is very close indeed to ethylene glycol. It would be interesting to examine phosphates containing OH groups if such exist.

Methyl oleate is essentially a long hydro-carbon chain with a $COOCH_3$ termination. The proportion of O atoms is small, so that one expects a compressibility of the same order as that of the higher straight hydro-carbons. The compressibility of the hydro-carbons has been measured as far out as n-decane. If the volume decrements at 0° produced by 1000 kg are plotted for these hydro-carbons as a function of the number of C's in the chain, and if due account is taken of the fact that the odd members of the family follow a different progression from the even members, the difference between odd and even vanishing asymptotically for long chains, it will be found that a natural and unforced extrapolation of the curve can be made from n = 10 (decane) to the volume decrement of methyl oleate at n = 19, 19 being the total number of C atoms in methyl oleate.

It is a curious fact that the volumes of methyl oleate and tricaproin are almost identical functions of pressure and temperature. This must be to a certain extent coincidence, the structure of the two being unlike.

Isoprene, C_5H_8, is essentially an iso-chain compound, with the number of H atoms cut down by the presence of two double bonds. The effect of a double bond seems to be in general to cut down the compressibility. One therefore expects a compressibility not greatly different from that of i-pentane, but somewhat less. This turns out to be the case, the volume decrement of isoprene at 0° at 7000 kg, for example, being about 10% less than that of i-pentane, and almost exactly the same as that of n-hexane.

The reason for measuring such an unusual compound as isoprene was, as already mentioned, that the data was of interest in connection with the work of Professor Conant on polymerization by pressure. The polymerization product has under normal conditions a volume of the order of 20% less than isoprene, so that it is at first natural to think that the polymerization is a pure pressure effect connected with the change of volume. The rate of polymerization increases rapidly with increasing pressure, however, and at the same time the volume of isoprene approaches that of the polymerization product, because of the very much greater compressibility of isoprene, now established by these measurements. In fact, special experiments in which the polymerization was conducted at 20000 kg showed that at this pressure the rate is more rapid than at any lower pressure, but that the change of volume is not perceptible. It is therefore evident that some other effect than that of pure volume change must be responsible for the polymerization. The situation has been studied in full detail, and a solution proposed, in papers by Professor Conant.[3]

Finally, the measurements on i-octane must be mentioned. One would expect a compressibility very nearly the same as that of n-octane, but somewhat less, judging from the effect of the iso-structure as shown by the behavior of n- and i- pentane, for example. This turns out to be the case, the volume decrement at 0° and 5000 kg, for example, of i-octane, being 0.9% less than that of n-octane.

It is a pleasure to acknowledge generous assistance from the Rumford Fund of the American Academy of Arts and Sciences and from the Milton Fund of Harvard University. For the readings I am indebted to my assistant Mr. L. H. Abbot.

THE JEFFERSON PHYSICAL LABORATORY,
HARVARD UNIVERSITY, CAMBRIDGE, MASS.

REFERENCES.

[1] P. W. Bridgman, Proc. Amer. Acad. 66, 185, 1931.

[2] W. E. Danforth, Jr., Phys. Rev. 38, 1224, 1931.

[3] P. W. Bridgman and J. B. Conant, Proc. Nat. Acad. 15, 680, 1929.
J. B. Conant and C. O. Tongberg, Jour. Amer. Chem. Soc. 52, 1659, 1930.
J. B. Conant and W. R. Peterson, forthcoming paper in Jour. Amer. Chem. Soc.

[4] P. W. Bridgman, Proc. Amer. Acad. 61, 57, 1926.

[5] Reference 1, and
P. W. Bridgman, Proc. Amer. Acad. 49, 1, 1913.

[6] P. W. Bridgman, Phys. Rev. 38, 182, 1931.

[7] Reference 1, pages 210 and 211.

Bei Fragen zur Produktsicherheit wenden Sie sich bitte an:
If you have any questions regarding product safety,
please contact:

Walter de Gruyter GmbH
Genthiner Straße 13
10785 Berlin
productsafety@degruyterbrill.com